D1619437

Nanomaterials for the Life Sciences Volume 6
Semiconductor Nanomaterials

Edited by Challa S. S. R. Kumar

Related Titles

Kumar, C. S. S. R. (ed.)
Nanotechnologies for the Life Sciences
10 Volume Set
ISBN: 978-3-527-31301-3

Kumar, C. S. S. R. (Ed.)
Nanomaterials for the Life Sciences (NmLS)
Book Series, 10 Volumes

Vol. 6
Semiconductor Nanomaterials
2010
ISBN: 978-3-527-32166-7

Vol. 7
Biomimetic and Bioinspired Nanomaterials
2010
ISBN: 978-3-527-32167-4

Vol. 8
Nanocomposites
2010
ISBN: 978-3-527-32168-1

Vol. 9
Carbon Nanomaterials
2011
ISBN: 978-3-527-32169-8

Vol. 10
Polymeric Nanomaterials
2011
ISBN: 978-3-527-32170-4

Nanomaterials for the Life Sciences
Volume 6

Semiconductor Nanomaterials

Edited by
Challa S. S. R. Kumar

WILEY-VCH

WILEY-VCH Verlag GmbH & Co. KGaA

The Editor

Dr. Challa S. S. R. Kumar
CAMD
Louisiana State University
6980 Jefferson Highway
Baton Rouge, LA 70806
USA

■ All books published by Wiley-VCH are carefully produced. Nevertheless, authors, editors, and publisher do not warrant the information contained in these books, including this book, to be free of errors. Readers are advised to keep in mind that statements, data, illustrations, procedural details or other items may inadvertently be inaccurate.

Library of Congress Card No.:
applied for ‥

British Library Cataloguing-in-Publication Data
A catalogue record for this book is available from the British Library.

Bibliographic information published by the Deutsche Nationalbibliothek
The Deutsche Nationalbibliothek lists this publication in the Deutsche Nationalbibliografie; detailed bibliographic data are available on the Internet at <http://dnb.d-nb.de>.

© 2010 WILEY-VCH Verlag GmbH & Co. KGaA, Weinheim

All rights reserved (including those of translation into other languages). No part of this book may be reproduced in any form – by photoprinting, microfilm, or any other means – nor transmitted or translated into a machine language without written permission from the publishers. Registered names, trademarks, etc. used in this book, even when not specifically marked as such, are not to be considered unprotected by law.

Composition   Toppan Best-set Premedia Limited
Printing and Bookbinding   Strauss GmbH, Mörlenbach
Cover Design   Schulz Grafik Design, Fußgönheim

Printed in the Federal Republic of Germany
Printed on acid-free paper

ISBN: 978-3-527-32166-7

Contents

List of Contributors XIII

Part I Imaging and Diagnostics 1

1 **Quantum Dots for Cancer Imaging** 3
Yan Xiao and Xiugong Gao
1.1 Introduction 3
1.2 Cancer 5
1.2.1 A Primer on Cancer Biology 5
1.2.2 The Importance of Early Cancer Detection and Diagnosis 6
1.2.3 The Role of Biomarkers in Cancer Early Detection and Diagnosis 7
1.3 Quantum Dots: Physics and Chemistry 8
1.3.1 Photophysical Properties of QDs 8
1.3.2 Quantum Dot Chemistry 11
1.3.2.1 Synthesis 11
1.3.2.2 Surface Passivation 12
1.3.2.3 Water Solubilization 12
1.3.2.4 Bioconjugation 14
1.4 Cancer Imaging with QDs 15
1.4.1 In Vitro Screening and Detection of Cancer Biomarkers Using Microarrays 15
1.4.2 In Vitro Cellular Labeling of Cancer Biomarkers 19
1.4.2.1 Labeling of Fixed Cells and Tissues 19
1.4.2.2 Live Cell Imaging in Cancer Cells 22
1.4.3 In Vivo Cancer Imaging 28
1.4.3.1 In Vivo Tracking of Cancer Cells 29
1.4.3.2 Tumor Vasculature Imaging 31
1.4.3.3 Sentinel Lymph Node Mapping and Fluorescence Lymphangiography 33
1.4.3.4 In Vivo Whole-Body Tumor Imaging in Animals 35
1.4.4 Multimodality Tumor Imaging 39
1.4.5 Dual-Functionality QDs for Cancer Imaging and Therapy 41
1.5 Quantum Dot Cytotoxicity and Potential Safety Concerns 45

Nanomaterials for the Life Sciences Vol.6: Semiconductor Nanomaterials.
Edited by Challa S. S. R. Kumar
Copyright © 2010 WILEY-VCH Verlag GmbH & Co. KGaA, Weinheim
ISBN: 978-3-527-32166-7

1.6	Concluding Remarks and Future Perspectives *47*
	Abbreviations *48*
	Acknowledgments *50*
	References *50*

2 Quantum Dots for Targeted Tumor Imaging *63*
Eue-Soon Jang and Xiaoyuan Chen

2.1	Introduction *63*
2.2	Types of Quantum Dots (QDs) *64*
2.2.1	Type I Core–Shell QDs *73*
2.2.1.1	Group II–VI Semiconductor Core–Shell QDs *74*
2.2.1.2	Group III–V Semiconductor Core–Shell QDs *76*
2.2.1.3	Group IV–VI Semiconductor Core–Shell QDs *78*
2.2.2	Reverse Type I and Type II Core–Shell QDs *78*
2.2.3	Core–Shell–Shell (CSS) QDs *81*
2.3	Surface Modifications of QDs for Bioapplications *83*
2.3.1	Preparation of Water-Soluble QDs *83*
2.3.2	Bioconjugation of QDs *88*
2.4	Methods of Delivering QDs into Tumors *91*
2.4.1	Passive Targeting *91*
2.4.2	Active Targeting *95*
2.4.2.1	Peptide-Mediated Tumor Uptake *95*
2.4.2.2	Antibody-Mediated Uptake *97*
2.5	Recent Advances in QDs Technology *99*
2.5.1	Bioluminescence Resonance Energy Transfer (BRET) *99*
2.5.2	Non-Cd-Based QDs *100*
2.5.3	Moving Towards Smaller QDs *102*
2.5.4	Multifunctional Probes *102*
2.6	Conclusion and Perspectives *104*
	Acknowledgments *105*
	References *105*

3 Multiplexed Bioimaging Using Quantum Dots *115*
Richard Byers and Eleni Tholouli

3.1	Introduction *115*
3.2	The Need for Novel Multiplex Imaging Systems *115*
3.3	Quantum Dots *117*
3.3.1	Optical Properties *117*
3.3.2	Manufacture *119*
3.4	Bioimaging Applications of Quantum Dots *120*
3.4.1	Quantum Dot Use for Immunohistochemistry *120*
3.4.2	Quantum Dot Use for *In Situ* Hybridization *125*
3.4.3	Quantum Dot Use in Solid- and Liquid-State Multiplex Detection Platforms *129*
3.5	Translation to Clinical Biomarker Measurement *131*
3.5.1	Quantitation *131*

3.5.2	Imaging Analysis	*132*
3.5.3	Combinational Microscopy Methods	*134*
3.5.4	Clinical and Mechanistic Biological Applications	*134*
3.5.5	Cancer Molecular Profiling	*137*
3.6	Summary and Future Perspectives	*139*
	References	*142*

4 Multiplexed Detection Using Quantum Dots *147*
Young-Pil Kim, Zuyong Xia and Jianghong Rao

4.1	Introduction	*147*
4.2	*In Vitro* Multiplexed Analysis Using QDs	*150*
4.2.1	DNA Hydridization	*150*
4.2.2	Immunoassay	*153*
4.2.3	Assaying the Activity of Enzymes	*157*
4.2.3.1	FRET-Based Protease Detection with QDs as the Donor	*158*
4.2.3.2	BRET-Based Protease Detection with QDs as Acceptor	*161*
4.2.4	Other *In Vitro* Multiplexed Detection Systems	*163*
4.3	Multiplexed Imaging Using QDs	*165*
4.3.1	Multiplexed Cellular Imaging	*165*
4.3.2	Multiplexed Imaging in Small Animals	*168*
4.4	Summary	*171*
	References	*171*

Part II Therapy *177*

5 Medical Diagnostics of Quantum Dot-Based Protein Micro- and Nanoarrays *179*
Anisha Gokarna and Yong-Hoon Cho

5.1	Introduction	*179*
5.1.1	Invention of Protein Microarrays	*180*
5.1.2	Optical Readout of Protein Arrays: Traditional versus New Labels	*182*
5.2	Protein Arrays	*185*
5.2.1	The Various Types of Protein Array	*185*
5.2.2	Methods of Optical Detection in Protein Arrays Using Labeled Probes	*186*
5.3	From Microarrays to Nanoarrays: Why Nanoarrays?	*188*
5.4	Functionalization of QDs for Attachment to Proteins or Other Biomolecules	*189*
5.5	Fabrication of Protein Biochips	*193*
5.5.1	Printing of QDs-Conjugated Protein Microarray Biochips	*193*
5.5.2	Nanoarray Protein Chips Using QDs	*194*
5.6	QDs Probes in Clinical Applications	*197*
5.6.1	Detection of Disease (Cancer) Biomarkers Using QD Labels	*197*
5.6.2	Certain Limitations in QDs Labeled Microarrays: Nonspecific Binding Effects	*201*
5.6.3	QDs versus Organic Dyes in Protein Microarrays	*203*

5.6.4	Multiplexing in Protein Microarrays Using QDs 206
5.7	QDs-Labeled Protein Nanoarrays 207
5.8	Summary and Future Perspectives 212
	Acknowledgments 212
	References 212
6	**Imaging and Tracking of Viruses Using Quantum Dots** 219
	Kye-Il Joo, April Tai and Pin Wang 219
6.1	Introduction 219
6.2	Quantum Dot–Biomolecule Hybrids 220
6.2.1	Optical Properties of QDs 220
6.2.2	Functionalized QDs Conjugated with Biomolecules 220
6.2.3	Formation of Virus–QD Networks 222
6.2.4	Decoration of Discretely Immobilized Virus with QDs 222
6.2.5	QD Encapsulation in Viral Capsids 224
6.3	Quantum Dots for Single Virus Tracking in Living Cells 225
6.3.1	Conventional Labeling for Single-Virus Tracking in Live Cells 228
6.3.2	Problems Encountered in Single-Virus Tracking 229
6.3.3	Use of QDs for Labeling Enveloped Viruses 230
6.3.4	Use of QDs for Labeling Nonenveloped Viruses 231
6.4	Quantum Dots for the Sensitive Detection of Virus and Infection 239
6.4.1	Monitoring the Progression of Viral Infection with Fluorescent QD Probes 239
6.4.2	The Use of Two-Color QDs for the Real-Time Detection of Viral Particles and Viral Protein Expression 241
6.4.3	Viral Detection with pH-Sensitive QDs in a Biological Motor 242
6.5	Conclusions 245
	Acknowledgments 245
	References 246
7	**Nanomaterials for Radiation Therapy** 251
	Ke Sheng and Wensha Yang
7.1	Introduction 251
7.2	A Brief Introduction to Radiation Therapy, and its Limitations 252
7.3	Physical Radiosensitizers 255
7.3.1	Enhanced Radiation Therapy Using High Z but Non-Nanoscaled Materials 255
7.3.2	Enhanced Radiation Therapy by Gold Nanoparticles 257
7.3.3	Dose Enhancement of Physical Radiosensitizers 259
7.3.4	Radiation Therapy Enhancement Using Nonionizing Radiation 260
7.4	Radiation Therapy in Combination with Photodynamic Therapy Using Semiconductor Nanoparticles as the Energy Mediator 261
7.4.1	Photodynamic Therapy 261
7.4.2	Semiconductor Nanoparticles as the Energy Mediator for Photodynamic Therapy 262

7.4.3	Quantum Dots 263
7.4.4	Photoluminescent Nanoparticles in Radiation Therapy 264
7.5	Nanobrachytherapy 271
7.5.1	Liposomes 273
7.5.2	Nanoparticles 274
7.5.3	Dendrimers 275
7.6	Nanoparticles as Radioprotectors 277
7.7	Radiation Dosimeters Using Semiconductor Nanomaterials 280
7.8	Conclusions 282
	References 283

8	**Prospects of Semiconductor Quantum Dots for Imaging and Photodynamic Therapy of Cancer** 291
	Vasudevanpillai Biju, Sathish Mundayoor, Abdulaziz Anas and Mitsuru Ishikawa
8.1	Introduction 291
8.2	Basic Principles and Challenges in PDT 292
8.3	Advantages of Quantum Dots for PDT 294
8.4	Synthesis of Quantum Dots 295
8.4.1	Synthesis of Visible QDs 295
8.4.1.1	Synthesis of Cadmium-Based QDs 295
8.4.1.2	Synthesis of InP QDs 296
8.4.2	Synthesis of NIR QDs 296
8.4.2.1	Synthesis of Core-Only NIR QDs 297
8.4.2.2	Synthesis of Core–Shell QDs 297
8.4.2.3	Synthesis of CdTe/CdSe QDs 298
8.5	Optical Properties of Quantum Dots 299
8.5.1	Absorption and Fluorescence Properties 299
8.5.2	Photostability of QDs 301
8.5.3	Two-Photon Absorption by QDs 301
8.6	Preparation of Biocompatible Quantum Dots 303
8.7	Nontargeted Intracellular Delivery of QDs 304
8.7.1	Physical Methods of Intracellular Delivery 305
8.7.2	Biochemical Techniques 307
8.7.2.1	Nonspecific Intracellular Delivery of QDs 307
8.7.2.2	Cell-Penetrating Peptide-Mediated Delivery 307
8.7.2.3	Chitosan- and Liposome-Mediated Delivery 309
8.8	Targeting Cancer Cells with QDs 309
8.8.1	*In Vitro* Targeting of Cancer Cells with QDs 310
8.8.1.1	Targeting Cancer Cells with QD–Antibody Conjugates 310
8.8.1.2	Targeting Cancer Cells with QD–Ligand Conjugates 312
8.8.2	*In Vivo* Targeting and Imaging Cancer with QDs 313
8.8.3	*In Vivo* Targeting of Tumor Vasculature and Lymph Nodes with QDs 314
8.9	Quantum Dots for Photodynamic Therapy of Cancer 316
8.9.1	Quantum Dot Alone for PDT 316

8.9.2	Potentials of QD–Photosensitizer Conjugates for PDT	318
8.10	Toxicity of QDs	320
8.11	Conclusions	321
	Abbreviations	322
	References	323

Part III Synthesis, Characterization, and Toxicology 329

9 Type-I and Type-II Core–Shell Quantum Dots: Synthesis and Characterization 331
Dirk Dorfs, Stephen Hickey and Alexander Eychmüller

9.1	Introduction	331
9.2	Core–Shell UV-Vis Nanoparticulate Materials	332
9.2.1	Type-I Core–Shell Structures	332
9.2.2	Type-II Core–Shell Structures	334
9.2.3	Multiple Shell Structures	337
9.2.4	Nonspherical Nanoheterostructures	349
9.3	Characterization of Nanoheterostructures	350
9.4	Core-Shell Infrared Nanoparticulate Materials	352
9.5	Type I Core–Shell Infrared Structures	353
9.6	Type II Core–Shell Infrared Structures	355
9.7	Summary and Conclusions	361
	References	362

10 Nanowire Quantum Dots 367
Thomas Aichele, Adrien Tribu, Gregory Sallen, Catherine Bougerol, Régis André, Jean-Philippe Poizat, Kuntheak Kheng and Serge Tatarenko 367

10.1	Introduction	367
10.2	Quantum Dots	368
10.3	Growth of Quantum Dots, Nanowires and Nanowire Heterostructures	371
10.4	Applications for Nanowires and Nanowire Quantum Dots	378
10.4.1	Nanowires and Quantum Dots in the Life Sciences	378
10.4.2	Nanowire Electronic Devices	379
10.4.3	Single-Photon Sources	380
10.4.3.1	Single-Photon Generation	380
10.4.3.2	High-Temperature Single-Photon Emission from Nanowire QDs	381
10.5	Conclusions	386
	References	387

11 Quantum Dot–Core Silica Glass–Shell Nanomaterials: Synthesis, Characterization, and Potential Biomedical Applications 393
Norio Murase

11.1	Introduction	393
11.2	Historical Overview	394

11.2.1	Quantum Dots and their Incorporation into Glass Matrices	394
11.2.2	Single-Particle Spectroscopy and its Application to Biological Systems	396
11.3	Incorporation of Quantum Dots in Glass Beads for Bioapplications	397
11.3.1	The Stöber Method	398
11.3.2	Reverse Micelle Method	398
11.3.3	Bifunctional Glass Beads Derived from the Reverse Micelle Method	404
11.3.4	Complex Structures Created in Small Glass Beads with Novel Photoluminescent Properties	409
11.3.5	Emitting Glass Fibers Created by Self-Assembly of the Complex Structured Beads	414
11.4	Preparation of Cadmium-Free Quantum Dots With Water Dispersibility	414
11.5	Summary and Future Perspective	422
	Acknowledgments	423
	References	423
12	**Toxicology and Biosafety Evaluations of Quantum Dots**	**429**
	Pinpin Lin, Raymond H.S. Yang, Chung-Shi Yang, Chia-Hua Lin and Louis W. Chang	
12.1	Introduction	429
12.2	The Physico-Chemical Properties of Quantum Dots	432
12.2.1	Surface Chemistry and Coatings	433
12.2.2	Charge and Size of QDs in Relationship to Potential Toxicity	435
12.2.3	Core Metals and Biodegradation of QDs	437
12.3	Toxicity and Role of Oxidative Stress	441
12.4	ADME and Biosafety Evaluation	445
12.5	The Mitochondrion as a Prime Target for QD Toxicity	450
12.6	Environmental and Ecological Concerns	456
12.7	Concluding Comments and Future Perspectives	456
	References	458

Index 465

List of Contributors

Thomas Aichele
Humboldt Universität zu Berlin
Institut für Physik
Hausvogteiplatz 5-7
10117 Berlin
Germany

Abdulaziz Anas
National Institute of Advanced
Industrial Science and
Technology (AIST)
Nanobioanalysis Team
Health Technology Research
Center
2217-14 Hayashi-Cho
Takamatsu
Kagawa 761-0395
Japan

Régis André
CEA/CNRS/Université Joseph
Fourier
Nanophysics and Semiconductor
Group, Institut Néel
25 rue des Martyrs
58042 Grenoble cedex 9
France

Vasudevanpillai Biju
National Institute of Advanced
Industrial Science and
Technology (AIST)
Nanobioanalysis Team
Health Technology Research
Center
2217-14 Hayashi-Cho
Takamatsu
Kagawa 761-0395
Japan

Catherine Bougerol
CEA/CNRS/Université Joseph
Fourier
Nanophysics and Semiconductor
Group, Institut Néel
25 rue des Martyrs
58042 Grenoble cedex 9
France

Richard Byers
University of Manchester
School of Cancer and
Imaging Sciences
Faculty of Medical and
Human Sciences
Stopford Building
Oxford Road
Manchester M13 9PT
UK

List of Contributors

Louis W. Chang
National Health Research
Institutes
Division of Environmental Health
& Occupational Medicine
Zhunan, Miaoli County
Taiwan

Xiaoyuan Chen
Stanford University
School of Medicine
Molecular Imaging Program
at Stanford (MIPS) and
Bio-X Program
Department of Radiology
1201 Welch Road
Stanford, CA 94305
USA
National Institute of Biomedical
Imaging and Bioengineering
(NIBIB)
Laboratory of Molecular Imaging
and Nanomedicine
National Institutes of Health
(NIH)
31 Center Drive, 31/1C22
Bethesda, MD 20892
USA

Yong-Hoon Cho
Korea Advanced Institute of
Science and Technology (KAIST)
National Research Laboratory for
Nano-Bio-Photonics, Department
of Physics and Graduate School
of Nanoscience & Technology
(WCU)
Daejeon 305-701
Korea

Dirk Dorfs
Istituto Italiano Di Tecnologia
Via Morgeo 30
16163 Genova
Italy

Alexander Eychmüller
Physikalische Chemie
TU Dresden
Bergstrasse 66b
01062 Dresden
Germany

Xiugong Gao
Translabion
Clarksburg, MD 20871
USA

Anisha Gokarna
Korea Advanced Institute of
Science and Technology (KAIST)
National Research Laboratory for
Nano-Bio-Photonics, Department
of Physics and Graduate School
of Nanoscience & Technology
(WCU)
Daejeon 305-701
Korea

Stephen Hickey
Physikalische Chemie
TU Dresden
Bergstrasse 66b
01062 Dresden
Germany

Mitsuru Ishikawa
National Institute of Advanced
Industrial Science and
Technology (AIST)
Nanobioanalysis Team
Health Technology Research
Center
2217-14 Hayashi-Cho
Takamatsu
Kagawa 761-0395
Japan

Eue-Soon Jang
Stanford University
School of Medicine
Molecular Imaging Program
at Stanford (MIPS) and
Bio-X Program
Department of Radiology
1201 Welch Road
Stanford, CA 94305
USA

Kye-Il Joo
University of Southern California
Department of Chemical
Engineering and Materials
Science
925 Bloom Walk, HED 216
Los Angeles, CA 90089-1211
USA

Kuntheak Kheng
CEA/CNRS/Université Joseph
Fourier
Nanophysics and Semiconductor
Group, Institut Néel
25 rue des Martyrs
58042 Grenoble cedex 9
France

Young-Pil Kim
Stanford University
Molecular Imaging Program
at Stanford
Department of Radiology
1201 Welch Road
Stanford, CA 94305-5484
USA

Chia-Hua Lin
National Health Research
Institutes
Division of Environmental Health
& Occupational Medicine
Zhunan
Miaoli County
Taiwan

Pinpin Lin
National Health Research
Institutes
Division of Environmental Health
& Occupational Medicine
Zhunan
Miaoli County, Taiwan

Sathish Mundayoor
National Institute of Advanced
Industrial Science and
Technology (AIST)
Nanobioanalysis Team
Health Technology Research
Center
2217-14 Hayashi-Cho
Takamatsu
Kagawa 761-0395
Japan

Norio Murase
National Institute of Advanced
Industrial Science and
Technology (AIST)
Photonics Research Institute
Ikeda
Osaka 563-8577
Japan

Jean-Philippe Poizat
CEA/CNRS/Université Joseph
Fourier
Nanophysics and Semiconductor
Group, Institut Néel
25 rue des Martyrs
58042 Grenoble cedex 9
France

Jianghong Rao
Stanford University
Molecular Imaging Program
at Stanford
Department of Radiology
1201 Welch Road
Stanford, CA 94305-5484
USA

Gregory Sallen
CEA/CNRS/Université Joseph
Fourier
Nanophysics and Semiconductor
Group, Institut Néel
25 rue des Martyrs
58042 Grenoble cedex 9
France

Ke Sheng
University of Virginia
Department of Radiation
Oncology
P. O. Box 800383
1215 Lee Street
Charlottesville, VA 22908
USA

April Tai
University of Southern California
Department of Chemical
Engineering and Materials
Science
925 Bloom Walk, HED 216
Los Angeles, CA 90089-1211
USA

Serge Tatarenko
CEA/CNRS/Université Joseph
Fourier
Nanophysics and Semiconductor
Group, Institut Néel
25 rue des Martyrs
58042 Grenoble cedex 9
France

Eleni Tholouli
Manchester Royal Infirmary
Department of Haematology
Oxford Road
Manchester M13 9WL
UK

Adrien Tribu
CEA/CNRS/Université Joseph
Fourier
Nanophysics and Semiconductor
Group, Institut Néel
25 rue des Martyrs
58042 Grenoble cedex 9
France

Pin Wang
University of Southern California
Department of Chemical
Engineering and Materials
Science
925 Bloom Walk, HED 216
Los Angeles, CA 90089-1211
USA

Zuyong Xia
Stanford University
Molecular Imaging Program
at Stanford
Department of Radiology
1201 Welch Road
Stanford, CA 94305-5484
USA

Yan Xiao
National Institute of Standards
and Technology
Gaithersburg, MD 20899
USA

Chung-Shi Yang
National Health Research
Institutes
NanoMedicine Research
Zhunan
Miaoli County
Taiwan

Raymond H.S. Yang
Colorado State University
Department of Environmental &
Radiological Health Sciences
Fort Collins, CO 80523-1680
USA

Wensha Yang
University of Virginia
Department of Radiation
Oncology
P. O. Box 800383
1215 Lee Street
Charlottesville, VA 22908
USA

Part I
Imaging and Diagnostics

1
Quantum Dots for Cancer Imaging
Yan Xiao and Xiugong Gao

1.1
Introduction

Although significant progress has been made in both the understanding and treatment of cancer during the past 30 years, it remains the second leading cause of death in the United States (US). The cancer community has been called to eliminate cancer-related suffering and death by 2015, but to achieve this goal, groundbreaking innovations in both the diagnosis and treatment of cancer are needed. The noninvasive detection of cancer in its early stages is of great interest since early cancer diagnosis, in combination with precise cancer therapies, could significantly increase the survival rate of patients and eventually save millions of lives. Nanomedicine, an emerging research area that integrates nanomaterials and biomedicine, has come into the spotlight because of its potential to provide novel diagnostic tools for the detection of primary cancers at their earliest stages, and to provide improved therapeutic protocols for the effective and highly selective extermination of malignant cells. Although it could be simply described as the application of nanotechnology to medicine, nanomedicine is actually a multidisciplinary science that involves materials science, nanotechnology, physics, chemistry, biology, and medicine [1]. Research into nanomedicine will lead to the understanding of the intricate interplay of nanomaterials with components of biological systems.

Interest in nanomedicine burgeoned after the US National Institutes of Health (NIH) announced a four-year program for nanoscience and nanotechnology in medicine in December 2002 [2]. Nanoscience and nanotechnology, with their potential applications in biomedicine, have been highlighted as "Priority Areas" for funding opportunities by the National Science Foundation (NSF) since 2001 [3]. The NIH Roadmap's new Nanomedicine Initiatives, first released in late 2003, started funding a five-year plan to set up eight Nanomedicine Development Centers from 2005 [4]. To harness the power of nanotechnology to radically change the way in which cancer is diagnosed, treated and prevented, the National Cancer Institute (NCI) announced in July 2004 the Cancer Nanotechnology Plan, which

Nanomaterials for the Life Sciences Vol.6: Semiconductor Nanomaterials.
Edited by Challa S. S. R. Kumar
Copyright © 2010 WILEY-VCH Verlag GmbH & Co. KGaA, Weinheim
ISBN: 978-3-527-32166-7

is outworked by the NCI Alliance for Nanotechnology in Cancer [5]. With the grand level of attention and fervent study in the related areas, it is no longer a distant dream that nanotechnology will radically transform clinical oncology by offering high sensitivity and localized treatment for cancers with increasing variety of nanomaterials.

"Nanomaterials" is a general term used to describe structures with morphological features smaller than a one-tenth of a micrometer in at least one dimension [6]. In recent years the arsenal of nanomaterials has continued to expand, and now include seven major categories based on their composition: carbon-based nanomaterials, nanocomposites, metals and alloys, biological nanomaterials, nanopolymers, nanoglasses, and nanoceramics. On the other hand, nanomaterials can be classified according to their shape as nanoparticles, nanotubes, nanorods, or nanowires, and so on. The heart of nanomedicine lies in the ability to shrink the size of the tools and devices, based on various nanomaterials, to the nanometer size range. Indeed, the elementary functional units of biological systems – proteins/enzymes, nucleic acids (DNAs, RNAs), membranes, and many other cellular components – are all on the nanometer scale. Therefore, there are many points of intersection between nanoscience and nanotechnology and the biological sciences. It is predicted that many nanotechnologies will soon become translational tools for medicine, and move quickly from laboratory discoveries to clinical tests and therapies.

Among various nanomaterials, quantum dots (QDs) distinguish themselves in their far-reaching possibilities in many avenues of biomedicine. QDs are nanometer-sized semiconductor crystals with unique photochemical and photophysical properties that are not available from either isolated molecules or bulk solids. QD research started in 1982 with the realization that the optical and electric properties of small semiconductor particles were heavily dependent on particle size due to quantum confinement of the charge carriers in small spaces [7, 8]. During the next two decades, extensive research was carried out for potential applications in optoelectronic devices, QD lasers and high-density memory. In 1998, two seminal reports simultaneously demonstrated that QDs could be made water-soluble and could be conjugated to biological molecules, providing the first glimpse of the vast potential of QDs as probes for studying biological systems [9, 10]. In comparison with organic dyes and fluorescent proteins, QDs have the advantages of improved brightness, resistance against photobleaching, and multicolor fluorescence emission. These properties could improve the sensitivity of biological detection and imaging by at least one to two orders of magnitude. Significant improvements have been made in the synthesis, surface modification, and biofunctionalization of QDs in the following years and, indeed, the current literature is rife with examples of QDs used in various biomedical applications. It can now be said with confidence that QDs have completed the transition from a once curious demonstration of quantum confinement in semiconductors to ubiquitous fluorophores providing unique insights into biological investigations [11].

In this chapter, an attempt will be made to provide a comprehensive, state-of-the-art overview of QD applications in cancer imaging. Following a brief introduction of the topic within the broad background of biomedicine, cancer and its imaging requirements will be briefly overviewed. The following section describes the photophysics and chemistry of QDs, and a provides a clear understanding of the merits of using QDs in bioimaging, as well as the requirements and challenges in the synthesis, surface modification and bioconjugation of QDs in order to make them amenable to bioimaging applications. Next, some recent advances in the use of QDs in various imaging applications for cancer detection and diagnosis are detailed, both *in vivo* and *in vitro*. The literature cited in this section is confined to reports that are germane to cancer studies. Finally, the issue of QD cytotoxicity and potential safety concerns is briefly covered, followed by a summary of the chapter and some future perspectives on QD applications in cancer imaging.

1.2
Cancer

The word "cancer" derives from the Latin term for crab, the basis if which related to the fourteenth century when a Greek physician first used the term to describe a disease of which the swollen mass of blood vessels around a malignant tumor resembled the shape of a crab. In modern dictionaries, cancer is referred to as a malignant tumor of potentially unlimited growth that expands locally by invasion, and systemically by metastasis.

1.2.1
A Primer on Cancer Biology

Cancer is not a single disease but rather a family of different diseases. In humans, there may be more than 200 different cancers [12], based primarily on the part of the body in which the cancer first develops, referred to as the primary site. The human body, and any other multicellular organism alike, can be viewed as a society or ecosystem, the individual members of which–the cells–reproduce by cell division and organize into collaborative assemblies or tissues. Cancer is the result of mutation, competition, and natural selection that occurs in the population of somatic cells, leading to an abnormal state in which individual mutant cells not only prosper at the expense of their neighbors but also ultimately destroy the whole cellular society and die. Research has indicated that most cancers derive from a single abnormal cell, and are probably initiated by a change in the DNA sequence of that cell [13]. Cancerous cells are, on the one hand, self-sufficient in growth signals and, on the other hand, insensitive to anti-growth signals and can evade apoptosis–a process by which a cell is commanded by the environment to die [14]. As a result, these cells multiply at an exponential growth rate in an uncontrolled fashion, leading to the formation of a tissue lump called a *malignant tumor*. Gradually, the tumor tissue grows bigger and invades the adjacent tissues and organs, obstructing normal physiological functions and usually causing great pains in the patient. In many cases, the invading tumor can break loose and enter the bloodstream or lymphatic vessels and form secondary tumors at other sites of the body – a process termed *metastasis*. The more widely a cancer metastasizes, the more difficult it becomes to eradicate. Over time, the tumor may cause malfunctioning of various organs, and ultimately proves to be fatal.

1.2.2
The Importance of Early Cancer Detection and Diagnosis

Despite rapid advances in global cancer research towards the understanding and treatment of cancer during the past several decades, the disease remains the second leading cause of death in the US, and has become the leading cause of death for patients under the age of 85 years since 1999. It is estimated that, in 2008, there will have been 1 437 180 new cases and 565 650 deaths of all cancers in the US [15]. Cancer primary sites with the highest incidence rate and/or death rate include the breast for women, prostate gland for men, and colon and rectum, lung and bronchus, non-Hodgkin lymphoma, pancreas, urinary bladder for both sexes, based on the data during the period 2000–2004 in the US [15].

The clinical outcome of cancer diagnosis is strongly related to the stage at which the malignancy is detected. The mortality of cancer would decrease substantially if it were to be detected at an early stage, when the disease could be more effectively treated. This is especially true for breast and cervical cancer in women and colorectal and prostate cancer in men. When diagnosed early, the tumor is confined to the organ of origin, and can be either resected surgically, or treated with radiation therapy and/or systemic chemotherapy. However, most solid tumors are currently detectable only after they reach a size of ~1 cm in diameter, by which time they contains millions of cells. Sadly, by this point metastasis may have already taken place so that treatment becomes much more difficult, if not impossible. The five-year relative survival rate for all cancers increased from 50% in 1975–1977 to 66% in 1996–2003 [15], due to progress in diagnosing certain cancers at an earlier stage, as well as improvements in treatment.

Cancer detection and diagnosis means an attempt to accurately identify the anatomical site of origin of the malignancy, and the type of cells involved. The most commonly used techniques in current clinical practice are tissue biopsy and medical imaging, and sometimes a bioanalytic assay of the body fluids. *Biopsy* can provide information about the histological type, classification, grade, potential aggressiveness and other information that may help to determine the best treatment regime, and also has the advantage that any tissue abnormality can be examined at the cellular level. However, biopsy is practical only if the tumor is located near the surface of the body, where a tissue sample can be easily retrieved for microscopic evaluation. When the tumor is inaccessible for a biopsy, then reliance must be placed on *medical imaging* techniques to detect the location of the tumor. Existing noninvasive imaging techniques include computed tomography (CT), magnetic resonance imaging (MRI), positron emission tomography (PET), single photon emission CT (SPECT), ultrasound, X-ray imaging, and optical imaging. These imaging methods are effective for the macroscopic visualization of tumors and, if combined with biopsies, they are able not only to confirm the presence of tumors but also to pinpoint the primary and/or secondary sites of the lesion.

However, none of the existing techniques is sufficiently sensitive and/or specific to detect abnormalities at the microscopic level, and is thus unable to identify tumors at their incipient stage – not to mention precancerous lesions at the molecular level. In recent years, substantial research efforts have been applied to the development of better cancer imaging techniques [16], and significant progress has been seen in the development of nanoparticle-based, highly sensitive contrast agents [17]. Nanoparticle-based contrast agents such as QDs have strong potential for early cancer detection and diagnosis as they are bright and photostable, and can offer the possibility for multiplex biomarker imaging.

1.2.3
The Role of Biomarkers in Cancer Early Detection and Diagnosis

Today, it is well recognized by the cancer community that biomarkers play an increasingly important role in the effective treatment and management of cancer. *Biomarkers* are cellular indicators of physiologic state, and also of changes during a disease process. Carcinogenic transformation often results in the secretion of elevated levels of abnormal molecules or biomarkers into the body fluids, and these markers can provide information on the occurrence and progression of the disease. Many proteins have been identified as specific markers for a variety of cancers [18]. For example, serum levels of prostate-specific antigen (PSA) are elevated in prostate cancer patients, and today this is the best-known cancer biomarker in clinical use [19]. Cancer antigen (CA)-125 is recognized as an ovarian cancer-specific protein, which is present in a subset of ovarian cancer patients [20]. The identification of such biomarkers is of major interest for cancer early detection and diagnosis.

During the past few years, there has been increasing interest and enthusiasm in molecular markers as tools for cancer detection and diagnosis. Targeted cancer imaging using biomarkers is usually more sensitive and specific than nontargeted imaging. In addition, recent advances in high-throughput technologies in genomics, proteomics and metabolomics have facilitated the discovery of cancer biomarkers. These biomarkers may be genes or genetic variations, changes in mRNA and/or protein expression, post-translational modifications of proteins, or differences in metabolite levels. Some newly identified biomarkers and their use in cancer research have been described in recent reviews [21, 22]. Depending on their type, these biomarkers can be potentially useful for predicting several outcomes during the course of the disease, including risk assessment, early detection, prognosis, disease progression, recurrence prediction, and therapeutic responses. For instance, one of the most important uses of biomarkers is the classification of a specific tumor (when one has been detected) into its various subtypes since, for many cancers, each of these subtypes has a drastically different prognosis and a preferred method of treatment. In spite of the critical importance of the diagnosis of cancer subtypes, many cancers do not yet have a reliable test to differentiate between their many different subtypes, for example, highly invasive tumors versus their less-fatal counterparts. The final judgment is usually left to the expert opinion of a pathologist who examines the tumor biopsy, with such judgment being unavoidably subjective. With the advent of high-throughput genomic and proteomic data of cancer specimen, it is becoming apparent that many cancer subtypes can be distinguished by the expression level of a specific biomarker or the expression profile of a panel of biomarkers [21].

In the pursuit of sensitive and quantitative methods for biomarker analysis, nanotechnology has been heralded as a field of great promise. Semiconductor QDs could enable the detection of tens to hundreds of cancer biomarkers in blood assays, and on cancer tissue biopsies. With the emergence of gene and protein profiling and microarray technology, the high-throughput screening of biomarkers has generated databases of genomic and expression data for certain cancer types, and this has led to the identification of new, cancer-specific markers [21]. Today, QDs are being used to expand this *in vitro* analysis and to extend it to cellular, tissue and whole-body multiplexed cancer biomarker imaging. QDs also continue to show promise in personalized medicine, which may be highly desirable to treat a patient uniquely for his or her distinct cancer phenotype [22]. Therefore, QDs bear potential impact for improving the diagnosis and treatment of cancer.

1.3
Quantum Dots: Physics and Chemistry

QDs are semiconductor nanocrystals which have a near-spherical shape and a diameter in the range of 1 to 10 nm [23]. By virtue of this nanoscale size, which is comparable to electron delocalization lengths (tens to hundreds of Angstroms), QDs possess certain unique physical properties that are unavailable in either individual atoms or bulk semiconductor solids. These properties, which are desirable for biological imaging, and when combined with the development of ways to solubilize QDs in solution and to conjugate them with biological molecules, have led to an explosive growth in their biomedical applications.

1.3.1
Photophysical Properties of QDs

In physics, solid-state materials are classified into three categories based on their electrical conductivity, namely insulators, semiconductors, and conductors. The conductivity is often rationalized as the energy difference between the valence band and the conduction band, called bandgap energy (E_g). When an electron in the valence band gains energy (either thermally or by the absorption of a photon) that is equal to or greater than E_g, it enters the conduction band, leaving behind a positively charged hole in the valence band, thus forming electron-hole pairs which are charge carriers; this process is called *excitation*. The bandgaps in semiconductor materials are small enough that some electrons may be excited thermally at room temperature to form charge carriers. An electron in the excited state at the conduction band may relax back to its ground state at the valence band through recombination with its hole, resulting in the emission of a photon, in the form of fluorescence, with the same energy as the bandgap E_g. Thus, the wavelength of the fluorescence emission is determined by the E_g of the semiconductor material. For bulk semiconductor materials, E_g is dictated by its composition only.

When one or more dimensions of a semiconductor are reduced under the Bohr exciton radius, which is typically a few nanometers, the so-called "quantum confinement effect" is observed [24]. The reduction of one dimension leads to a structure called a *quantum well*, where the semiconductor material is in the form of thin sheets only a few atoms thick; a reduction in two dimensions leads to the structure of *quantum wires*. When all the three dimensions are reduced to the nanometer scale, tiny spherical semiconductors–QDs–are formed. Under such situations, the bandgap energy E_g is quantized, with the value being directly related to the QD's size. In other words, the energy is no longer only a function of the semiconductor composition, but also of its size. The quantum confinement effect could be most accurately explained by quantum mechanical models, which may be enigmas for most readers with a background in biology and are certainly beyond the scope of this chapter. However, the interested reader may find in the literature a simplified explanation using the "particle in a box" analogy [25].

Due to the quantum confinement effect, it is possible to systematically control the electronic energy level spacings and, as a result, the wavelength of light emission, by adjusting the size of the semiconductor. This is the fundamental principle behind the unique photophysical properties of QDs. With different sizes and compositions, QDs can emit fluorescent light with wavelengths over a wide range spanning regions from the ultraviolet, through the visible, into the infrared [26–30] (Figure 1.1a). In addition, as QD emission is due to a radiative recombination of an exciton, it is also characterized by a long lifetime (>10 ns), and a narrow and symmetric energy band [32] (Figure 1.1b). In comparison with traditional organic fluorescent molecules, which are usually characterized by red-tailed broad emission band and short fluorescence lifetimes, QDs have several attractive optical features that are desirable for long-term, multitarget and highly sensitive bioimaging applications.

First, QDs have very large molar extinction coefficients [33], typically of the order of $0.5–5 \times 10^6 \, M^{-1} cm^{-1}$; this is approximately 10- to 50-fold higher than those of organic dyes, which means that QDs are capable of absorbing excitation photons much more efficiently. The higher rate of absorption is directly correlated to the QD brightness, and it has been found that QDs are approximately 10- to 20-fold brighter than organic dyes [9, 10, 34]. As a result, QDs are highly sensitive fluorescent agents, allowing for the highly efficient fluorescence labeling of cells and tissues. This is of particular importance for cancer diagnosis, as many cancer biomarkers exist in low copy numbers and/or at low concentrations.

Second, QDs have excellent photostability, being typically several thousandfold more stable against photobleaching than organic dyes [35]. This feature allows the

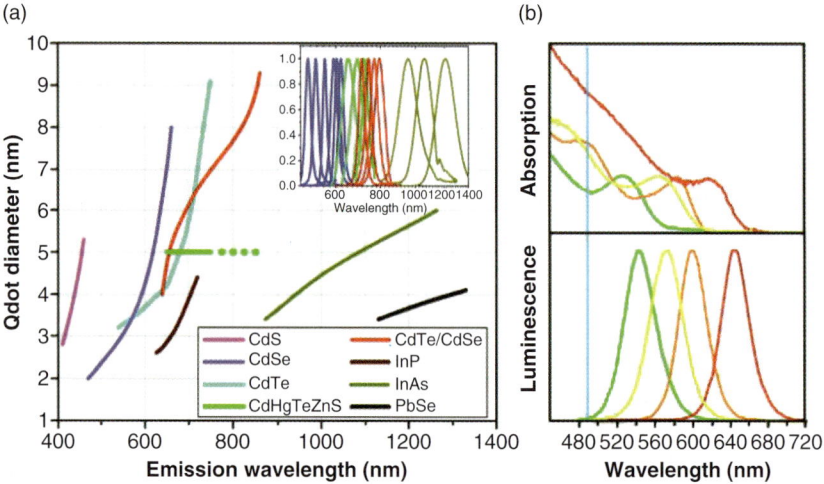

Figure 1.1 (a) Size and composition dependence of QD emission maxima. QDs can be synthesized from various types of semiconductor materials (II–VI: CdS, CdSe, CdTe ... ; III–V: InP, InAs ... ; IV–VI: PbSe ...) characterized by different bulk band gap energies. The curves represent experimental data from the literature on the dependence of peak emission wavelength on QD diameter. The range of emission wavelength is 400 to 1350 nm, with size varying from 2 to 9.5 nm (organic passivation/solubilization layer not included). All spectra are typically around 30 to 50 nm (full width at half-maximum). Inset: Representative emission spectra for some materials; (b) Absorption (upper curves) and emission (lower curves) spectra of four CdSe/ZnS QD samples. The blue vertical line indicates the 488 nm line of an argon-ion laser, which can be used to efficiently excite all four types of QDs simultaneously. Adapted from Ref. [31]; © 2005, American Association for the Advancement of Science.

real-time monitoring of biological processes over long periods of time, and is essential for cancer biomarker assays and *in vivo* imaging where much longer times are needed.

In addition, QDs have a longer excited state lifetime (20–50 ns), which is about one order of magnitude longer than that of organic dyes [32]. This allows the effective separation of QD fluorescence from background fluorescence by using a time-gated or time-delayed data acquisition mode [36]. In this mode, signal acquisition is not started until the background autofluorescence disappears largely, since this way the signal-to-noise ratio (SNR) will be increased dramatically, resulting in a significantly improved image contrast [37, 38]. Therefore, QDs are highly suitable for time-correlated lifetime imaging spectroscopy.

Moreover, QDs have broad absorption bands, narrow emission bands, and a large Stokes shift (Figure 1.1b). Unlike organic dyes, the excitation and emission spectra of QDs are well separated. Depending on the wavelength of the excitation light, the Stokes shift of QD may be as large as 400 nm, and this allows further improvement of detection sensitivity in imaging tissue biopsies and living organisms [39]. Biological specimens contain a variety of intrinsic fluorophores such as proteins and cofactors, yielding a significant background signal that decreases probe detection sensitivity. The peak of the intrinsic fluorescence falls in the blue to green spectral region. Due to the broad absorption band and the large Stokes shift, QDs can be tuned to emit with longer wavelengths in the red or infrared spectra, where autofluorescence is diminished.

Lastly – and most importantly – the wavelength of QD emission is size-tunable, which is a unique feature of QD materials in comparison with organic dyes. The size-dependent emission of QDs allows the imaging and tracking of multiple targets simultaneously, using a single excitation source. This is of particular importance in cancer detection and diagnosis, since it has been realized that a panel of disease-specific molecular biomarkers can provide more accurate and reliable information with regards to the disease status and progression than can any single biomarker [21]. Although theoretically possible, the multiplexing of QD signals would not be practically feasible without the combination of their characteristic broad absorption bands and narrow emission bands. While broad absorption bands allow multiple QDs to be excited with a single light source of short wavelength – thus simplifying instrumental design and increasing detection speed – their narrow emission bands (as small as 20 nm in the visible range) allow for distinct signals to be detected simultaneously, with very little overlapping (Figure 1.1b). In stark contrast, organic dyes and fluorescent proteins have narrow absorption bands but relatively wide emission bands, rendering the detection of separated signals from distinct fluorophores more difficult.

1.3.2
Quantum Dot Chemistry

The development of QD-based fluorescent probes involves a multistep process that includes synthesis, surface passivation, solubilization, and bioconjugation.

1.3.2.1 Synthesis
QD crystals are composed of hundreds to thousands of atoms that typically belong to Group II and VI elements (e.g., CdSe, CdTe, ZnSe) or Group III and V elements (e.g., InP and InAs) in the Periodic Table. Among the various synthetic methods that have been reported, two are most popular. Early studies used aqueous-phase synthesis in reverse micelles; this produces water-soluble QDs but they are usually of low quality [40]. In order to produce high-quality QDs, later studies used a method in which the synthesis was carried out at an elevated temperature in an organic solvent, and this has now becomes the most widely used technique of synthesizing highly homogeneous and highly crystalline QDs. The solvent is nonpolar with a high boiling point, and is most commonly a mixture of trioctylphosphine, trioctylphosphine oxide (TOPO), and hexadecylamine, all of which have long alkyl chains [26, 41, 42]. The purpose of using hydrophobic organic molecules as mixed solvents is two-fold. First, the mixture serves as a robust reaction medium, and second, their basic moieties also coordinate with unsaturated metal atoms on the QD surface to prevent the formation of bulk semiconductors. As a result the ligand alkyl chains are directed away from the QD surface, forming

a highly hydrophobic monolayer that is only soluble in nonpolar solvents such as chloroform and hexane.

1.3.2.2 Surface Passivation

The QD nanocrystal core (e.g., CdSe) is often coated with a shell of a wider bandgap semiconductor material (e.g., ZnS or CdS). This process, which often is called *surface passivation*, is extremely important as it serves multifold purposes [43, 44]. First, it protects the core from oxidation, and prevents the leaching of highly toxic Cd^{2+} ions. More importantly, it drastically improves the quantum yield by reducing surface defects and thus preventing nonradiative decay. Nonradiative recombination events occurring at surface defects decrease fluorescence efficiency. Due to the large surface-to-volume ratio of QD nanoparticles, surface-related recombination resulting from the localized trapping of carriers is considered to be one of the main factors that reduce the emission efficiencies of QDs [45]. The use of a higher bandgap semiconductor as a passivation shell confines the charge carriers to the core QD, thus minimizing surface recombination and leading to a significant improvement (up to >50%) in fluorescence quantum yields [44]. CdSe is normally passivated with ZnS, resulting in a structure referred to as (CdSe)ZnS or CdSe/ZnS, but ZnSe and CdS are also commonly used [44, 46, 47].

1.3.2.3 Water Solubilization

QDs produced using most of the synthesis strategies (especially the hot solvent method) are water-insoluble. For biological applications, these hydrophobic particles must first be rendered hydrophilic so that they are soluble in aqueous buffers. Two general strategies have been developed for the surface modification (Figure 1.2), namely ligand or cap exchange, and native surface modification by amphiphilic polymer coating [48].

In ligand or cap exchange the hydrophobic ligands on the QD surface are replaced with bifunctional ligands such as mercaptoacetic acid or mercaptosilane [9, 10]. The bifunctional ligands have two moieties, one which binds strongly to the QD surface (e.g., thiol), and the other which points out from the QD surface, forming a hydrophilic shell (e.g., hydroxyl, carboxyl). This method, which often is called "cap exchange," has two disadvantages: (i) an increased tendency to form particle aggregation, thus decreasing fluorescent efficiency; and (ii) an increased potential toxicity due to the exposure of toxic QD elements as a result of desorption of labile ligands from the QD surface.

In native surface modification, the native hydrophobic ligands (e.g., TOPO) are retained on the QD surface, but an amphiphilic polymer is introduced to interact favorably with the hydrophobic alkyl chains, while simultaneously leaving the hydrophilic segment directed away from the QD surface, thus rendering the QDs water-soluble. Several such polymers have been reported, including octylamine-modified polyacrylic acid [49], polyethylene glycol (PEG)-derivatized phospholipids [50], block copolymers [39], and amphiphilic polyanhydrides [51]. As the original hydrophobic ligands (TOPO) are retained on the inner surface of QDs, the QD core is shielded from the outside environment by a hydrocarbon bilayer. Thus,

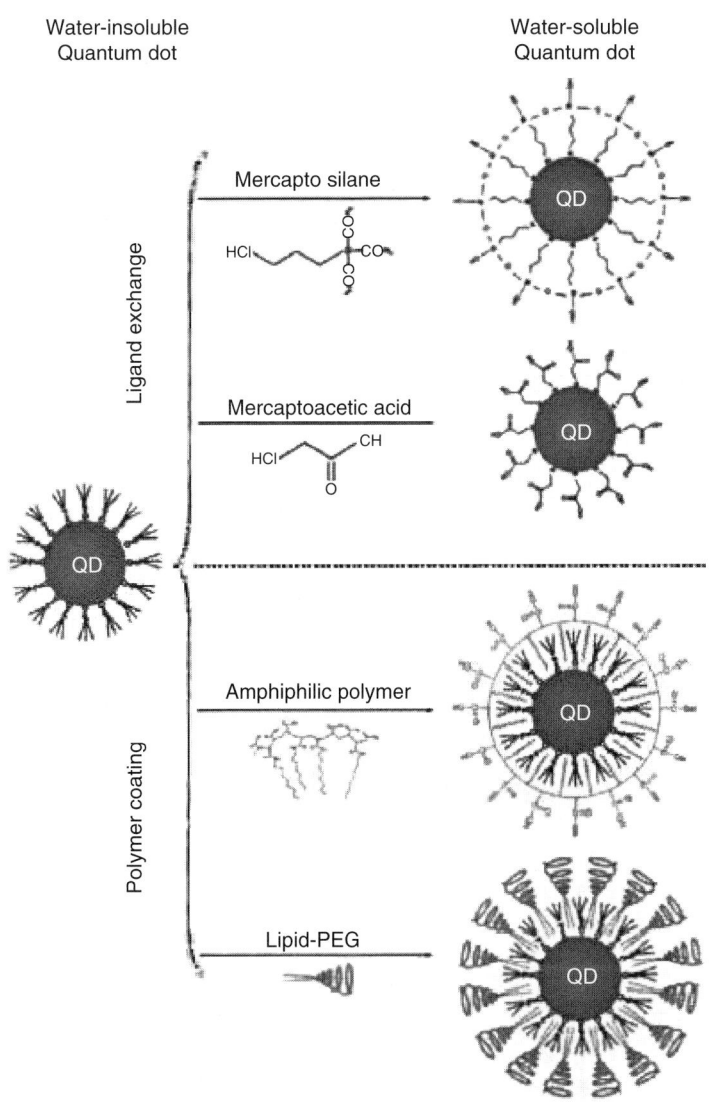

Figure 1.2 Diagram of two general strategies for phase transfer of TOPO-coated QD into aqueous solution. The ligands are drawn disproportionately large for detail, but the ligand-polymer coatings are usually only 1–2 nm in thickness. The top two panels illustrate the ligand exchange approach, where TOPO ligands are replaced by heterobifunctional ligands such as mercapto silanes or mercaptoacetic acid. This scheme can be used to generate hydrophilic QD with carboxylic acids or a shell of silica on the QD surfaces. The bottom two panels illustrate the polymer coating procedure, where the hydrophobic ligands are retained on the QD surface and rendered water soluble through micelle-like interactions with an amphiphilic polymer or lipids. From Ref. [48]; © 2007, Springer-Verlag.

this latter method is more advantageous than the first at maintaining the QD optical properties and storage stability in aqueous buffer over a broad range of pH and salt concentrations [39], as well as having a reduced cytotoxicity [51], but unfortunately increases the overall size of the QD probes. To date, however, this is the preferred procedure for the commercial production of water-soluble QDs.

1.3.2.4 Bioconjugation

For biological applications such as cellular imaging, QDs must be conjugated to biomolecules (e.g., proteins, antibodies, peptides, DNA, etc.), and for this several approaches have been developed (Figure 1.3) [52]. A commonly used strategy is to use QDs that contain streptavidin, which provides a convenient and indirect link to a variety of biotin-tagged biomolecules [49, 53–56] (Figure 1.3a). However, since the streptavidin-biotin complex is quite bulky, in cases where QD particle size is a consideration, for example when intracellular pores have to be crossed or cellular transport proteins are targeted, small covalent linkages are often preferred. As most solubilization methods results in QDs capped with ligands containing

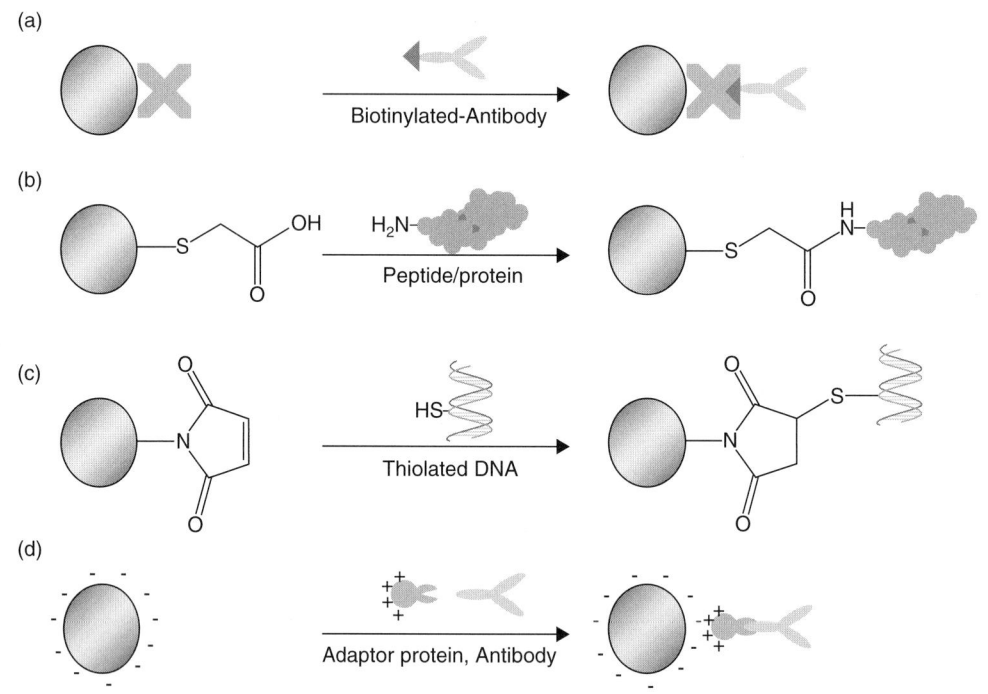

Figure 1.3 Examples of the bioconjugation of differently functionalized QDs to biomolecules by various strategies. (a) Streptavidin and biotin affinity binding; (b) Amide bond formation between carboxylic acid groups on QD surface and amine groups on proteins; (c) Thioether bond formation between maleimide groups on QD and thiol groups on thiolated DNA; (d) Electrostatic interaction between negatively charged QDs and positively charged biomolecules. From Ref. [52]; © 2007, Elsevier Inc.

hydrophilic groups such as carboxylic acids and maleimides, biomolecules containing basic functional groups such as amines and thiols can form stable amide or thioether bonds directly with QDs (Figure 1.3b,c) or through heterobifunctional crosslinkers [10, 49, 50, 57–60]. QDs in aqueous buffer solutions are negatively charged; therefore, positively charged molecules, such as cationic avidin proteins or recombinant maltose-binding proteins fused with positively charged peptides, could be conjugated to QDs through electrostatic interactions [53, 61–63] (Figure 1.3d). More detailed reviews on this topic can be found elsewhere [31, 52, 64].

It should be pointed out that none of these methods can control the number of proteins per QD. There is also a lack of experimental tools to control the orientation of a protein immobilized on the QD surface. For specific targeting, it is highly desirable that the delivery proteins (e.g., antibody) are properly oriented and fully functional. Goldman *et al.* first explored this area by using a fusion protein as an adapter for immunoglobulin G antibody coupling to QDs [63]. The adapter protein has a protein G domain for binding to the antibody Fc region, and a positively charged leucine zipper domain for electrostatic interaction with negatively charged QDs. As a result, the Fc end of the antibody is attached to the QD surface, leaving the target-specific F(ab')$_2$ domain facing outwards. As the bioconjugation step is critical to the success of bioimaging, the surface engineering of QD nanoparticles represents a field that will require much further study in the future.

1.4
Cancer Imaging with QDs

Bioconjugated QD fluorescent probes offer a promising and powerful imaging tool for cancer detection and diagnosis. Their much greater brightness, rock-solid photostability and unique capabilities for multiplexing, combined with their intrinsic symmetric and narrow emission bands, have made them far better substitutes for organic dyes in existing diagnostic assays. Following the two seminal reports in *Science* during 1998, demonstrating the feasibility of using QDs in biological environments [9, 10], many new techniques have been developed during the past decade, utilizing the unique photophysical properties of QDs, for the *in vitro* biomolecular profiling of cancer biomarkers, *in vivo* tumor imaging, and dual-functionality tumor-targeted imaging and drug delivery (Table 1.1). Some of these emerging technologies are currently being improved and integrated into clinical practice in oncology, and may have important implications for the diagnosis, prognosis, and therapeutic management of cancer patients in the near future.

1.4.1
In Vitro Screening and Detection of Cancer Biomarkers Using Microarrays

As the clinical outcome of cancer patients is heavily dependent on the stage at which the malignancy is detected, early screening is extremely important in any type of cancer. Techniques currently available in clinical practice for cancer

Table 1.1 Summary of quantum dots applications in cancer imaging.

Imaging type	Applications		References
In vitro	High-throughput screening and detection of cancer biomarkers	Profiling cancer biomarker expression on tissue microarrays (TMAs)	[65–67]
		Cancer biomarker detection in fluidic biological specimens using microarrays	[68, 69]
	Cellular labeling of cancer biomarkers	Cancer biomarker labeling by immunohistochemistry (IHC) in fixed cells and tissue sections	[49, 66, 70–85]
		Cancer genetic mapping by fluorescence *in situ* hybridization (FISH)	[84, 86–91]
	Live cell imaging	Labeling and imaging of cell-surface receptors in cancer cells	[49, 54, 70, 81, 92–100]
		Labeling and imaging of intracellular targets in cancer cells	[10, 101–129]
In vivo	*In vivo* tracking of cancer cells	Cancer cell tracking for homing, migration, engrafting, and metastasis potential.	[39, 114, 115, 130, 131]
	Tumor vasculature imaging	Nontargeted imaging of tumor vasculature	[130, 132]
		Targeted imaging of tumor vasculature	[133, 134]
	Sentinel lymph node (SLN) mapping	SLN mapping in tumor-bearing animals	[135–137]
		Fluorescence lymphangiography for studying the lymphatic drainage networks	[138, 139]
	Whole-animal tumor imaging	Targeted molecular imaging of tumors in animals	[39, 60, 100, 107, 125, 133, 140–143]
		Nontargeted imaging of tumors in animals	[130, 144]

Table 1.1 Continued

Imaging type	Applications		References
Multimodality	Positron emission tomography (PET)	Dual-modality imaging probe for both NIR fluorescence imaging and PET	[145, 146]
	Magnetic resonance imaging (MRI)	Dual-modality imaging probe for both NIR fluorescence imaging and MRI	[147, 148]
Multifunctionality	Drug delivery	QD-based nanoparticles for imaging and drug delivery	[107, 149]
	Small interfering RNA (siRNA) delivery	QD-based nanoparticles for imaging and siRNA delivery	[128, 150]

diagnosis such as medical imaging and tissue biopsy are neither highly sensitive nor highly specific for the detection of cancers at their incipient stage. In comparison, cancer screening and diagnosis using molecular biomarkers is usually more sensitive and specific, and is at the same time less invasive, less laborious, and less expensive. There is also a growing belief that a panel of markers rather than one marker alone will predict a more accurate outcome. Herein, lies the great potential offered by QDs, which can be used as sensitive probes for the screening and detection of cancer markers in clinical specimens. The intense and stable fluorescent properties of QDs, coupled with their multiplexing capacity, could enable the screening and detection of tens to hundreds of cancer biomarkers in *in vitro* diagnostic assays, promising for rapid and accurate clinical diagnostics.

In theory, QDs are particularly well suited to microarray applications, especially when multiplexing is needed. However, in most array applications developed to date, organic fluorescent dyes such as Cy3, Cy5, fluorescein isothiocyanate (FITC) and the Alexa series were used. One possible reason for this lies in the fact that array imaging is less demanding on fluorescence stability, by virtue of the high-speed laser scanners used. Nonetheless, it has been shown that Alexa Fluor 488 faded quickly during laser scanning cytometry (LCS) rescanning, with only 75% fluorescence intensity remaining after five scans; in contrast, QDs remained almost unchanged (98%) after the same number of scans [86].

During the past few years, many reports have been made where QDs were used as fluorescence probes in array imaging. Among these, several have been related to cancer biomarker identification and detection [65–69]. Ghazani *et al*. first demonstrated the use of QD bioconjugates as fluorescence probes for profiling the protein expression of tumor antigens on tissue microarrays (TMAs) [65]. Multiplexed staining with QDs of different emission spectra allowed the simultaneous

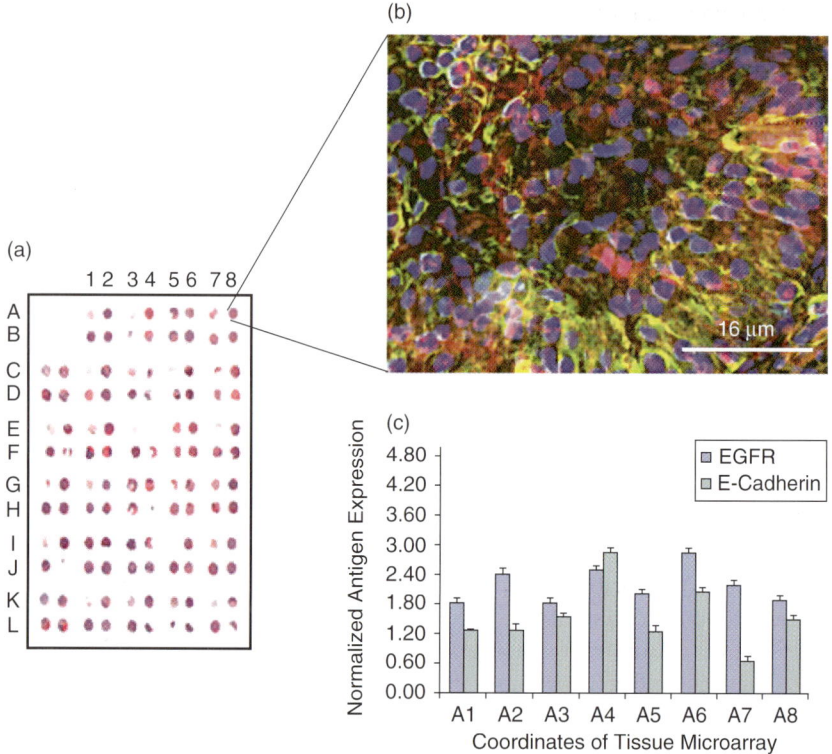

Figure 1.4 Quantification analysis of cancer-derived antigens in tissue microarray. (a) Tissue cores on a tissue microarray (hematoxylin and eosin image) were stained for EGFR, cytokeratin, and E-cadherin; (b) The fluorescent image (40×) is a composite picture of the antigens detected by individual QD immunostaining (of different color emission) for each of the targets. 4′,6-diamidino-2-phenylindole (DAPI) staining is used to locate nuclei in blue; (c) The antigen (EGFR and E-cadherin) expression values (normalized to cytokeratin) in each lung cancer xenograft cores. From Ref. [65]; © 2006, American Chemical Society.

detection of epidermal growth factor receptor (EGFR), E-cadherin, and cytokeratin on formalin-fixed and paraffin-embedded (FFPE) A549 lung carcinoma tissue sections (Figure 1.4). In this study, the use of QDs in conjunction with optical spectroscopy provided a tool to obtain a sensitive, accurate and quantitative measurement on a continuous scale that was not attainable by using traditional methods. Subsequently, QD-based tissue microarrays have been used successfully to distinguish tumor tissues from normal tissues in clinical samples [66, 67]. Zajac et al. developed a QD-based microarray platform that is potentially useful for cancer biomarker detection in fluidic biological specimens such as serum, plasma, and body fluids [68]. For this, the QDs were conjugated to the detector antibody either directly or via streptavidin–biotin interaction. In a series of multiplexing experiments, these authors were able to detect six different cytokines down to picomolar concentrations, thus demonstrating the high sensitivity of the QD-based detection system. In another study, Gokarna et al. reported the fabrication of QD-based protein biochips for the detection of the prostate cancer biomarker, PSA [69]. Here, the QDs were conjugated directly to anti-PSA antibodies in order to minimize any nonspecific binding, while retaining the high affinity of the QD bioconjugates. With an array spot size down to the nanometer scale, the authors were able to demonstrate the potential of nanoarrays for the detection of biomolecular interactions at the molecular level.

Although published reports in this area are sparse, the results of these studies have proved that QDs are highly promising for the high-throughput sensitive and specific detection of cancer biomarkers, and will undoubtedly inspire further investigation in the near future that, ultimately, will lead to efficient the *in vitro* screening and detection of various types of cancers.

1.4.2
In Vitro Cellular Labeling of Cancer Biomarkers

Determination of the expression and spatial distribution of molecular biomarkers in patient tissue specimens has substantially improved the pathologist's ability to classify disease processes. Many disease pathophysiologies, including cancers, are marked by the characteristic increased or decreased expression of certain biomarkers. Diagnostic and prognostic classifications of human tumors are currently based on immunohistochemistry (IHC) or fluorescence *in situ* hybridization (FISH), both of which have been used for almost a century. However, traditional IHC or FISH is mostly qualitative rather than quantitative, and is consequently subjective to an individual pathologist's bias. Moreover, traditional methods using organic dyes have a single-color nature and are unable to perform the multiplexed detection of several biomarkers simultaneously. In recent years, QDs have been used successfully in place of organic dyes in IHC and FISH analysis for labeling cancer biomarkers in fixed cells or tissue specimens. In many cases, the quantitative and simultaneous profiling of multiple biomarkers has been achieved by virtue of their intense and stable fluorescence, coupled with their multiplexing capacity. In addition, QDs have been applied successfully to live cell labeling for basic biological studies, as well as for developing clinically relevant applications for cancer detection and therapy.

1.4.2.1 Labeling of Fixed Cells and Tissues
The application of bioconjugated QDs for specific cancer biomarker labeling by IHC was first reported in 2003, when Wu et al. used QDs to label the breast cancer marker HER2 on the surface of SK-BR-3 cells, and in tissue sections from mouse mammary tumors [49]. These authors showed the labeling signals to be specific for the intended target, and to be brighter and considerably more photostable than comparable organic dyes. These results indicated that QD-based probes could be very effective in cellular imaging, and offered substantial advantages over organic dyes in multiplex target detection. This study was the first step towards using

QD-based diagnostics to help guide therapeutic decisions for cancer patients. Subsequently, many different cellular biomarkers, both general and/or specific to various cancers, have been labeled using QDs in fixed cells and tissues, including HER2 [49, 66, 70–73], estrogen receptor (ER) [71, 73] and EGFR [71] for breast cancer, CA125 for ovarian cancer [74], PSA for prostate cancer [75], (potentially useful) biomarkers of risk for colorectal cancer [76], tumor vasculature cluster of differentiation (CD) markers [77, 78], cytokeratin 18 on ovarian tumors [78], adrenocorticotropic hormone (ACTH) in pituitary adenoma [71], as well as mortalin [79], *p*-glycoprotein [80], carcinoembryonic antigen (CEA) [81], telomerase [66] and tumor-associated carbohydrate antigens [82] that each have differential expression levels and/or expression patterns in various cancer cells in comparison with normal cells. These studies have made it clear that QD are superior to organic dyes for cellular biomarker labeling in fixed cells and in FFPE tissue sections of tumor biopsies.

As no single biomarker can effectively predict all stages of cancer, an accurate diagnosis and prognosis of cancer must rely on a suite of several markers. Therefore, there is a growing need for multiplex staining due to the limited quantity of clinical samples. The optical properties of QDs have engendered considerable interest in their application to multiplex IHC for the simultaneous identification and colocalization of multiple biomarkers. By using QDs with different emission spectra, the practicality of the simultaneous detection of two cellular targets with one excitation wavelength was first demonstrated by Wu *et al.* [49]. In this case, HER2 on the cell surface and nuclear antigens in the nucleus of SK-BR-3 cells were detected simultaneously with QDs of different wavelengths conjugated to different secondary detection reagents. Subsequently, Kaul *et al.* studied the subcellular distribution of two heat-shock proteins, mortalin and heat-shock protein (HSP) 60, by double-labeling cells with red- and green-emission QDs, respectively [83]. These authors were able to visualize minute differences in the subcellular niche of these two proteins in normal and cancer cells. Fountaine *et al.* reported that five different cellular markers could be analyzed simultaneously on the same section of routinely processed FFPE tissues by using QDs with distinct emission spectra [77]. Later, Bostick *et al.* described a QD-based IHC system that could detect, simultaneously, up to six biomarkers that were potentially useful for predicting the risk of colorectal cancer [76]. Itoh *et al.* simultaneously detected two breast cancer biomarkers, HER2 on the cell membrane and ER in the nucleus, in human breast tissues [71]. These authors showed that very weak signals produced by conventional immunofluorescence and/or enzyme-labeled antibody methods could be significantly enhanced using QD labeling. Very recently, Xu *et al.* also reported the simultaneous detection of HER2 and ER in FFPE human breast cancer tissue using QDs emitting at 550 and 610 nm [73]. Sensitive spectra and images obtained in this study made possible the quantitative measurement of subcellular proteins inside the tumor tissues. Whilst all of the above-described reports on multiplex labeling have relied on sequential staining, which is both time-consuming and laborious, Sweeney *et al.* have recently developed a single-step, multiplex staining method in which streptavidin-conjugated QDs are conjugated to biotinylated primary antibodies; this enabled the multiplexed labeling of up to three cellular biomarkers on FFPE samples [78]. Shi *et al.* also reported the concurrent staining of three proteins in fixed cells [75].

Although the high intensity and photostability of QDs make the quantitative analysis of immunofluorescence signals possible, until now QD-based IHC for cancer biomarker detection has been mostly qualitative, with very few studies having attempted the quantitative analysis of biomarker expression [70, 73, 75, 76, 78]. Robust quantitation allows biomarker validation, as well as an accurate diagnosis of cancer progression, and to this end the present authors' group has been investigating the development of a quantitative IHC imaging system that combines the advantages of QD fluorescence imaging with the high sensitivity and specificity of IgY antibodies developed for cancer biomarkers. As a consequence, the relative quantitation of cancer biomarkers HER2 and telomerase has been demonstrated in tissue microarrays of patient and control tissues [66] (Figure 1.5a). Although the patient numbers were small, the study demonstrated the feasibility of the relative quantitation of cancer biomarkers with IgY and QD fluorophores, and showed promise for rigorous clinical validation in large patient cohorts. By using SK-BR-3 and MCF-7 as reference control cell lines, this assay platform allows for the accurate quantitative measurement of HER2 expression in breast cancer specimens [84]. Karwa *et al.* also reported quantitative multiplexed detection of inflammation biomarkers in tissue sections using QD conjugates where the quantitation of biomarker expression by fluorescence intensity correlated well with clinical disease severity in a mouse colitis model [85].

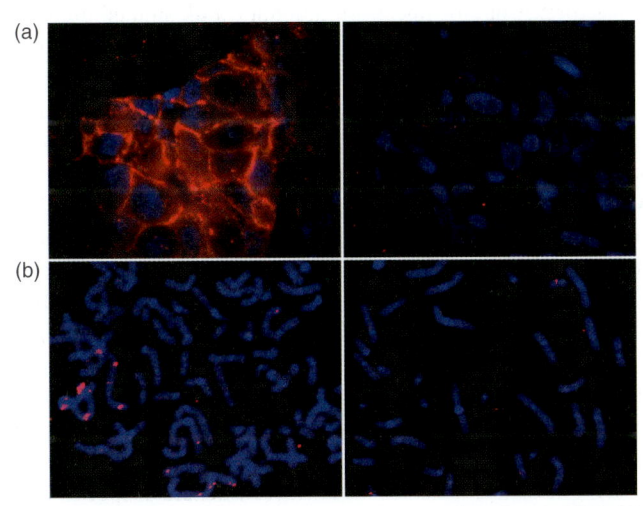

Figure 1.5 (a) Immunohistochemistry detection of breast cancer biomarker HER2 with anti-HER2 IgY antibody and QD fluorophores in breast cancer TMA sections with HER2 overexpression (left) and with normal HER2 expression (right); (b) FISH with BAC CTD-2019C10 probe and QD fluorophores for *HER2* gene amplification detection in fixed cell culture of breast cancer SK-BR-3 cells (left) and of normal PB-6 cells (right).

Today, FISH is used routinely in research and clinical laboratories for applications such as gene mapping, quantitation of gene copy number in tumors with gene amplification, or quantitation of the density of telomere repeats at the ends of human and mammalian chromosomes. Unfortunately, reports on the use of QDs (conjugated to oligonucleotide probes) as fluorescent tags in FISH have been relatively few in number, reflecting the technical difficulties of this endeavor, such as the optimization of QD/oligonucleotide conjugation, signal stability, tissue autofluorescence, and quantitation of signal. Initially, it was shown that QD fluorescent tags were better than standard FISH for detecting *HER2* oncogene amplification in breast cancer cells, by showing that fluorescence from QD fluorophores was significantly brighter and more photostable than the organic fluorophores Texas Red and fluorescein [87]. In this experiment, biotinylated DNA probes were detected using streptavidin-conjugated QDs (Figure 1.5b), and it was also shown that pH of the final incubation buffer had a significant effect on the fluorescence of QD-detected hybridization signals in the FISH experiments [86]. Coupled with the use of control cell lines, this method promises the accurate determination of *HER2* gene copy number in fixed cells and FFPE tissue samples [84].

In order to allow for the detection of multiple genes by exploiting the multiplexing capability of QDs, the labeling of DNA probes directly by QDs is required, and this has since been demonstrated in several studies [88–90]. For this, a faster, single-step FISH protocol was employed which coupled hybridization with detection and eliminated the need for any laborious secondary amplification. Chan *et al.* directly labeled oligonucleotide probes by QDs through streptavidin and biotin interactions, and used the probes for the simultaneous study of the subcellular distribution of multiple mRNA targets in mouse brain sections [88]. Bentolila *et al.* developed direct QD-probes by attaching short DNA oligonucleotides onto QDs and using the probes to target repetitive satellite noncoding DNA sequences [89]. In order to improve probe sensitivity and specificity, Jiang *et al.* used large genomic DNA probes directly labeled with QDs to visualize gene amplification in the cells of lung cancer specimens [90]. The results of these studies have suggested that the QD-FISH probes may offer an effective approach to analyze cancer-related genomic aberrations in basic research and clinical applications. More recently, Byers *et al.* developed a multiplex QD-based FISH that was both semiautomated and quantitative, and could be applied to the high-throughput processing of FFPE tissue samples [91]. The use of spectral imaging for the detection and subsequent deconvolution of multiple signals has also enabled the sensitive colocalization of multiple genes, and facilitated the quantitation of fluorescence signals from each gene.

Interestingly, both Chan *et al.* [88] and Byers *et al.* [91] combined QD FISH with QD IHC in the same experiment, which facilitated the simultaneous study of multiple mRNA and protein markers in tissue culture and histological sections.

1.4.2.2 Live Cell Imaging in Cancer Cells

Live cell imaging with QDs can provide information that is not possible with imaging in fixed cells, such as monitoring several intercellular and intracellular interactions in real time over a period as long as several days. On the other hand, the imaging of live cells is a more challenging task than imaging fixed cells, in that care must be taken to keep the cells alive, and that efficient methods are required to deliver QD probes across the cell membrane for imaging intracellular targets. Moreover, it has been commonly observed that QDs tend to aggregate in the cytoplasm and are often trapped in endocytotic organelles such as endosomes, lysosomes, and other intracellular vesicles. Nevertheless, significant progress has been made during the past few years in the use of QDs as fluorescent probes for labeling and imaging cell-surface proteins as well as intracellular proteins in live cells.

1.4.2.2.1 **Labeling and Imaging of Cell-Surface Receptors** Imaging membrane receptors on the cellular surface is relatively easier than imaging those inside the cells, as these proteins can often be labeled using techniques similar to those used for fixed cells through antibody–antigen or receptor–ligand interactions. Considerable success has been achieved in using QD bioconjugates as imaging agents for the specific targeting and imaging of cellular surface antigens on live cancer cells. Wu et al. first demonstrated the use of QD–streptavidin conjugates as probes together with Herceptin (a humanized anti-HER2 antibody) and biotinylated goat anti-human IgG, for the specific detection of HER2 on the surface of live SK-BR-3 cells [49]. Lidke et al. coupled QDs directly to epidermal growth factor (EGF), a small protein with a specific affinity for the erbB/HER receptors. After adding the EGF-QDs to cultured human cancer cells, it was possible to observe EGF binding to the receptor and subsequent internalization of the receptor conjugate in real time, at the single-molecule level [54]. Minet et al. used a QD–streptavidin conjugate and biotinylated concanavalin A (ConA) to image glycoproteins on the plasma membrane, in order to study the microdosimetry of heat stress in a tumor model of breast cancer cells [92]. QD fluorescence from labeled cells allowed the observation of alterations in plasma membrane organization and integrity as a result of the thermal effects from the heat stress. These applications have inspired the subsequent use of QDs for monitoring various plasma membrane proteins such as EGFR [93–96], G-protein-coupled receptors [97], prostate-specific membrane antigen (PSMA) [98], transferrin receptor [99], CEA [81], and dansyl (DNS) receptors [100].

The live cell labeling of membrane proteins by QDs may have implications for important biological applications. The dynamics of membrane proteins in living cells has become a major issue to understand important biological questions such as chemotaxis, synaptic regulation, or signal transduction. QDs have opened new perspectives for the study of membrane properties, as they enable measurements at the single molecule level with a high SNR. Li-Shishido et al. labeled HER2-overexpressing human breast cancer cell line KPL-4 with QDs conjugated to anti-HER2 antibody Herceptin, and traced the movement of the QDs on the cell membrane over periods up to 50s, during which 2500 images were taken and subsequently superimposed [70]. Traces of centers of fluorescence spots from the superimposed images showed the detailed movement of the QDs (Figure 1.6).

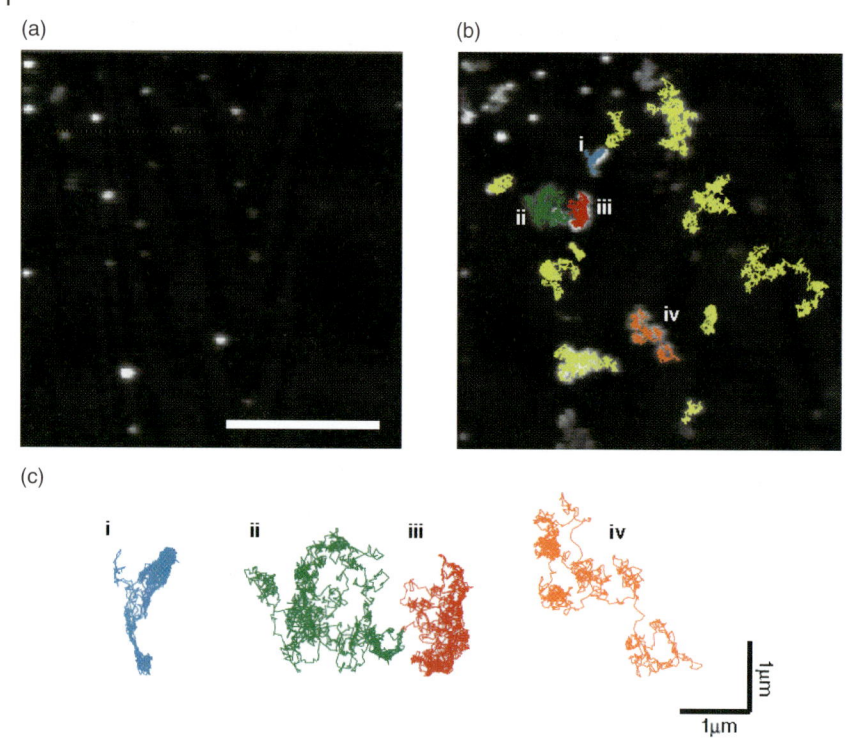

Figure 1.6 Movement of the QDs on the cell membrane. QDs labeled with anti-HER2 antibody (Herceptin) were bound to living KPL-4 cells overexpressing HER2. (a) Fluorescence images of QDs were taken under a total internal reflection microscope at an exposure time of 20 ms and laser power of ~70 W mm^{-1} [2]. Scale bar = 5 μm; (b) The QDs bound to cells were taken at a higher magnification. To trace the movement of QDs, 2500 images were collected over a 50 s period and then superimposed. Colored lines indicate the traces of single QDs; (c) Traces indicate movements of QDs that occurred on the cell membrane for a 50 s period. Adapted from Ref. [70]; ©, 2006 Elsevier Inc.

This study also opens up the application of QDs as tools for nanometer-scale measurements of positions.

1.4.2.2.2 Labeling and Imaging of Intracellular Targets

The first problem to tackle for the intracellular imaging of live cells using QD bioconjugates is how to circumvent the plasma membrane barrier. Several strategies have been employed for the delivery of QD probes across the membrane into cancer cells:

1. **Receptor-mediated endocytosis.** In the seminal studies of Chan and Nie [10], QDs were conjugated to transferrin and spontaneously entocytosed by HeLa cells. This receptor-mediated endocytosis mechanism was used in several subsequent studies [101–107]. By targeting receptors that are overexpressed specifically on the surface of cancer cells, but not on normal cells, QDs can be selectively delivered into tumor cells [108, 140, 151]. However, a recent study

has indicated that QDs may have severe consequences on cell physiology, as it was found that QDs coupled to ligands such as transferrin can be arrested within endosomes and somehow perturb the normal endosomal sorting in cells [106].

2. **Spontaneous endocytosis.** Many cell types are also able to engulf QDs spontaneously through nonspecific uptake [109–111, 152]. It has been noted that the surface chemistry of QDs has profound effects on their cellular uptake [112]. In addition, differences in cancer cell phenotypes can lead to significant differences in the intracellular sorting, trafficking, and localization of QDs [113].

3. **Chemical-mediated transfection.** QDs can be delivered into intracellular spaces by attaching to cationic lipids [114–116] or cationic cell-penetrating peptides such as polyarginine and Tat [117–120], or being encapsulated within a lipid emulsion [153]. The mechanism of Tat-mediated cellular uptake of QD bioconjugates has recently been studied systematically in living cancer cells under different endocytosis-inhibiting conditions [121]. The results suggest that Tat-QDs internalize through lipid-raft-dependent macropinocytosis, which is different from that of Tat conjugates with organic dyes such as FITC.

4. **Microinjection.** QDs can be delivered into the cytoplasm of a single cell through microinjection [122]. This technique is obviously useful for studying single cells, but it is labor-intensive and requires the delicate manipulation of one cell at a time. Therefore, it is difficult to obtain statistically relevant data because of the small number of cells that could be realistically injected.

Several new methods have also been developed in recent years for the intracellular delivery of QD bioconjugates. Kaul *et al.* used internalizing antibodies against mortalin/HSP70, a member of the heat-shock protein family that has a dynamic subcellular distribution, as QD caters for the active delivery of QDs into live cancer cells [123]; the molecular mechanism of this internalization has yet to be resolved, however. QD-based mortalin staining can be used to study senescence in cancer cells [124]. The specific intracellular uptake of QD-antibody conjugates by pancreatic cancer cells [103], liver cancer cells [125] and mesothelioma cells [126] has also been reported. A novel method for the intracellular delivery of QDs to image subcellular structures in live cells has been reported recently by Kim *et al.* [127]. For this, QDs conjugated to antibodies were first incorporated into biodegradable polymeric nanospheres. Then, upon cellular internalization, the pH-dependent reversal of surface charge of the nanospheres enabled their escape from the endolysosomal compartments to the cytoplasmic space, where a controllable release of the functionalized QD probes could be achieved through hydrolysis of the polymeric capsule. By attaching QDs of different emission spectra to antibodies against varying cellular targets, the multiplexed labeling of subcellular structures inside live cells was also demonstrated [127]. This approach allowed the noninvasive, high-throughput cytoplasmic delivery of QDs, with minimal toxicity to the cell. Yezhelyev *et al.* designed a QD-based nanoparticle with proton-absorbing polymeric coatings (proton sponges) on the QD surface for the efficient delivery of QDs bound with small interfering RNA (siRNA) into the cytoplasm [128].

It is interesting to note that Zhang et al. have demonstrated a potentially useful method for the specific delivery of QDs into cancer cells by using the protease-modulated cellular uptake of QD conjugates [129]. Here, the QDs were conjugated to cationic peptides which were in turn linked to blocking peptides through a linker peptide (protease substrate). The negatively charged blocking peptides could then prevent the cationic peptide-mediated cellular uptake of the QD conjugates. As proteases such as matrix metalloproteases (MMPs) are greatly involved with tumor formation and progression in many types of human cancers, the overexpression of these proteases in cancer cells can remove the negatively charged groups by enzymatic hydrolysis of the linker peptide, leading to uptake of the QD conjugates specifically into cancer cells. Modulation of the uptake of nanoparticles into cells with tumor-specific enzymes may lead to a selective accumulation of nanoparticles in tumor cells, a procedure which may in time find wide applications in nanomedicine.

QD-based fluorescent probes for intracellular labeling and imaging has been instrumental in basic biological studies, such as the molecular mechanisms of intracellular transport process involving motor proteins. Nan et al. measured the motions of the motor proteins kinesin and dynein along microtubules by following the movements of endocytic vesicles that contain QDs [117]. By virtue of the exceptional brightness and photostability of the QDs, it was possible to record individual microtubule motor steps with 300 μs time resolution and 1.5 nm spatial precision. These studies demonstrated the ability of QDs to probe the operation of motor proteins at the molecular level in living cells, under physiological conditions. While the calculation used in this study was based on two-dimensional (2-D) trajectories of QD-containing vesicles, a more recent investigation conducted by Watanabe et al. involved monitoring the stepwise movements generated by myosin, dynein, and kinesin in three dimensions [102]. By using QDs conjugated to HER2 and a three-dimensional (3-D) confocal microscope, it was possible to watch the QD-enclosing vesicles (after they had been endocytosed into the cells) be moved along the membrane by transferring actin filaments, along microtubules toward the nucleus, or away from the nucleus back to the cell membrane, and with time resolution and spatial precision similar to those reported in the previous study. This study added further information towards an understanding of the molecular mechanisms underlying traffic to and from cellular membranes. By using dual-focus imaging optics, the time interval between data points of the displacements may be as short as 2 ms [104]. Very recently, the same group used QD bioconjugates to study intracellular protein movement over long periods of time and showed that, in contrast to the smoothly continuous movement of kinesin found in *in vitro* assays, kinesin in live cells displayed a "stop-and-go" behavior, where kinesin came to an almost stop, paused for a few seconds, and then moved once again. The maximum velocity of kinesin observed in live cells was also faster than that in the *in vitro* assays [116].

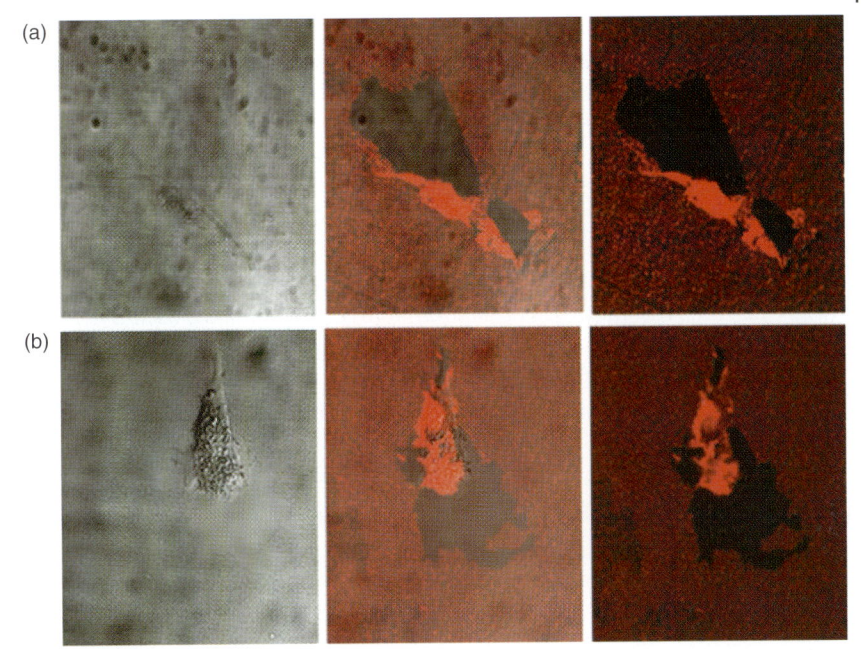

Figure 1.7 Phagokinetic tracks of the highly metastatic human mammary gland adenocarcinoma cell line MDA-MB-231 (a) and the human mammary ductal carcinoma cell line MDA-MB-435S (b) grown on a collagen substrate that had been coated with a layer of silanized, water-soluble QDs Images were collected with a confocal microscope using a fluorescence detector to record the nanocrystal trails (right) and a transmitted light detector to visualize the cells (left); the merged pictures (middle) colocalize the cells and the layer. After 24 h, sizable regions free of nanocrystals, larger than the cells themselves, were detected. From Ref. [109]; © 2003, International Society of Differentiation.

Some of the QD applications for intracellular targeting and imaging are also relevant to clinical cancer diagnosis and drug discovery. One such application is the cell motility assay developed by Alivisatos and colleagues [109–111], in which the migration of cells over a homogeneous layer of QDs was measured in real time. As the cells moved across, they endocytosed the QDs and, as a consequence, left behind a fluorescence-free trail (Figure 1.7). By subsequently determining the ratio of cell area to fluorescence-free track area, it was possible to differentiate between invasive and noninvasive cancer cells. As the motility of cancer cells *in vitro* is strongly correlated with their metastatic potential *in vivo*, this assay method could aid in the clinical classification of cancers for better diagnosis and management. In another application, Chen *et al.* coupled QDs to molecular beacons (MBs) to measure the expression of the endogenous proto-oncogene *c-myc* in MCF-7 breast cancer cells [122]. MBs are retained in the cytoplasmic compartment after being linked to QDs. Consequently, false-positive signals are reduced to marginal levels, as such nonspecific signals only arise in the nucleus of living cells. By quantifying the total fluorescent signal emanating from individual cells, accurate measurements of RNA expression of the oncogene at the single-cell level were made possible. Garon *et al.* used QDs attached to membrane-translocating peptides to label hematologic cells in malignant and nonmalignant patient samples [120], and showed that the QD-peptide conjugates could be taken up by a diverse range of hematologic cells, and followed through many divisions and through differentiation. Taken together, these results establish QDs as extremely useful molecular imaging tools for the study of hematologic cells. Yet another clinically relevant application was developed by Wylie *et al.*, of a multiplexed assay to determine the effects of drugs on different cell lines in high-throughput format [118]. By labeling live cells with QDs of different emission spectra and identifying cell proliferation using a microplate cytometer, it was possible to determine the differential rates of cell proliferation of the individual cell lines in the same well over time. Determining the differential responses of normal cells compared to cancerous cells in response to a chemical stimulus by using this assay method might prove valuable in selecting compounds that have maximal antitumor activity, while incurring minimal toxicity to normal tissues.

1.4.3
In Vivo Cancer Imaging

The visualization of tissue, with its anomalies, can provide information that allows certain pathologies to be eliminated, and the most probable pathology to be diagnosed. In today's oncological investigations, *in vivo* imaging can provide such a large amount of information that diagnoses are subject to much less uncertainty, and clues are also provided for the optimization of treatment. Some of the major imaging techniques used routinely in hospital set-ups for cancer diagnosis include CT, MRI, PET, SPECT, ultrasound and X-ray imaging. These techniques rely on signals that can transmit through thick tissue, and generate image contrast from the differences in signal attenuation through different tissue types. As tumors differ from normal tissues in both their structure and anatomy, many tumors can be identified based on the image contrast generated with or without exogenous contrast agents. However none of these modalities has a sufficiently high spatial resolution that is capable of detecting cancers at early stages, when the size of the tumors is very small. Moreover, most of these techniques are very expensive.

Optical imaging – particularly fluorescence imaging – is a sensitive and relatively inexpensive imaging technique that has higher intrinsic spatial resolution and for which the potential for cancer diagnosis has been demonstrated in living animal models [154, 155]. In contrast to *in vitro* fluorescence microscopic studies on cell cultures or thin tissue sections, *in vivo* fluorescence imaging functions at a macroscopic level on the whole body of animals. Whilst this enables the visualization of biology in its intact and native physiological state, it also presents some technical challenges. For example, thick, opaque animal tissues absorb and scatter photons and generate a strong autofluorescence, all of which obscure signal collection and quantification [156]. In order to overcome these problems, the near-infrared (NIR) optical window (650–950 nm) can be exploited for deep-tissue optical imaging

[157]. This is because the penetration of NIR light into tissue is significantly higher than light of shorter wavelengths, as the Rayleigh scattering decreases with increasing wavelength, and most tissue chromophores have a weak absorption in the NIR range. In recent years, several organic dyes have become available that can be used as NIR probes for *in vivo* imaging [155], but unfortunately these suffer from the same photobleaching problems as their visible counterparts. QD emission wavelengths can be tuned throughout the NIR spectrum by adjusting their composition and size; this results in photostable NIR-emitting QDs that can greatly expand their potential for *in vivo* cancer imaging. Consequently, by using appropriate hardware and software, multispectral imaging or spectral unmixing can be achieved so that the signal degradation caused by autofluorescence can be removed while adding enhanced multiplexing capabilities [39, 73, 77, 114, 158–160].

1.4.3.1 *In Vivo* Tracking of Cancer Cells

Cell tracking has the goal of determining the fate of a particular cell population within a heterogeneous environment. Cancer cell tracking can provide information on the homing, migration and engrafting of tumor cells, and thus further the current understanding of the critical stages of cancer pathology, such as metastasis. The unique photophysical properties of QDs make them desirable for the long-term tracking of cancer cells *in vivo*, and to this end the cancer cells can be labeled with QDs *in vitro*, administered to animals, and then followed using fluorescence microscopy. As an example, Gao *et al.* labeled human cancer cells with QDs and injected the cells subcutaneously into an immunecompromised mouse [39]. Subsequent *in vivo* whole-animal imaging indicated that, whilst a subcutaneous deposit of ~1000 QD-labeled cancer cells could be easily visualized, green fluorescent protein (GFP) stably transfected cells could not be detected under the same conditions. Voura *et al.* used QDs to track metastatic tumor cell extravasation *in vivo*. For this, melanoma cells labeled with QDs were injected intravenously into mice, and followed as they were extravasated into the lung tissues. The QDs and spectral imaging allowed the simultaneous identification of five different populations of cells using multiphoton laser excitation [114]. This approach allowed the study of single cells at the early stages of metastasis, and the process to be examined at the single-cell level in a natural tissue environment. A simultaneous identification and study of the interactions of multiple different populations of tumor cells and tissue cells within the same animal was also possible.

The homing mechanism of bone marrow-derived cells, and their contribution to tumor neovascularization, have been the subjects of intense debate [161]. Indeed, there has been a general paucity of such information obtainable from *in vivo* cell tracking studies. To this end, Stroh *et al.* successfully monitored the recruitment of QD-labeled bone marrow-derived precursor cells to the tumor vasculature [130]. Here, the cells were labeled *ex vivo* with QDs and imaged *in vivo* as they flowed, rolled over, and adhered to tumor blood vessels following intravenous administration (Figure 1.8). *In vivo* cell tracking studies can also aid in the behavioral profiling of cancers. It has been revealed from outcome studies of many types of cancer, that tumors of indistinguishable histologic appearance may differ

Figure 1.8 Tracking of QD-labeled bone marrow-derived precursor cells to the tumor vasculature. Seven images are superimposed in time as a single bone marrow lineage-negative cell labeled with QD590-Tat (orange) navigated the tumor vessels highlighted with QD470 micelles (blue) *in vivo*. The image represents seven repeated scans at a fixed depth (~100 µm) taken at 1 s intervals. Scale bar = 50 µm. Adapted from Ref. [130]; © 2005, Nature Publishing Group.

significantly in their degree of aggression and in their response to therapy. To enable an early identification of patients at high risk for disease progression, and to allow for the screening of multiple therapeutic agents simultaneously for their efficacy, Estrada *et al.* developed an orthotopic organ culture model of bladder cancer to obtain quantitative measurements of tumor cell behavior [131]. For this, human transitional cell carcinoma cells were first labeled with QDs; the cells were then instilled into the rat bladder *in vivo*, after which the bladder was excised and cultured *ex vivo*. By monitoring cell implantation, proliferation, and invasion into the organ wall, it was possible to assign distinct phenotypes to two metastatic bladder cancer cell lines, based on their different patterns of invasiveness into the bladder wall. These findings suggest that this assay system could recapitulate the salient aspects of tumor growth in the host, and be amenable to the behavioral profiling of human cancer.

Recently, Chang *et al.* reported using lipid-enclosed QDs as contrast agents to track tumor cells in a subcutaneous mouse model with epi-detection third harmonic generation (THG) microscopy [115]. Here, the QDs were mixed with a mixture of lipids to form a cationic lipid coating in order to improve their solubility and cell uptake. The epi-THG intensities were 20-fold stronger than the corresponding fluorescence intensities which, when combined with a high-penetration 1230 nm laser, provided a method that would allow for cell tracking in deep tissues.

Surface coatings on QDs may sometimes render them too large to enter into cells, unless they are taken up by naturally occurring internalization mechanisms (e.g., phagocytosis). To facilitate cell labeling, one solution to this problem is to couple the QDs to targeting molecules that facilitate cellular uptake; examples are Tat peptide [39, 130] and the proprietary Q-Tracker from Invitrogen [162]. Negatively charged QDs can also be efficiently transduced into tumor cells using cationic lipids [114, 115]. Recent studies of cell-penetrating surface coatings have offered promise for additional solutions to this problem [163]. The encapsulation of QDs in the internal aqueous phase of lipid bilayer vesicles (liposomes) to form QD–liposome hybrid nanoparticles might offer yet another solution to the problem. More recently, Al-Jamal *et al.* showed that such QD–liposome hybrids, when injected intratumorally into solid tumor models, led to an extensive fluorescence staining of tumor cells compared to injections of QDs alone [164]. These hybrid nanoparticles constitute a versatile tool for the very efficient labeling of cells both *ex vivo* and *in vivo*, and particularly when long-term imaging and tracking of cells is sought. Moreover, such a system offers many opportunities for the development of combinatory imaging and therapeutic modalities by incorporating both drug molecules and QDs within the different compartments of a single vesicle.

1.4.3.2 Tumor Vasculature Imaging

Cancerous cells can induce capillary (blood vessel) growth in the tumor by secreting various growth factors such as vascular endothelial growth factor (VEGF). This process, termed *tumor angiogenesis*, is believed not only to supply the required nutrients that allow for tumor expansion, but also to serve as a waste drainage, removing the biological end-products from rapidly dividing cancer cells. Angiogenesis is also required for the spread of a tumor, or metastasis. Consequently, the noninvasive imaging of tumor angiogenesis has many clinical applications, including lesion detection, patient stratification, new drug development and validation, treatment monitoring, and dose optimization.

Several reports have been made during the past few years using QDs as a contrast agent for the nontargeted imaging of normal vascular systems in small animals. Foe example: Larson *et al.* showed that green-light emitting QDs remained both fluorescent and detectable in the capillaries of adipose tissue and the skin of a live mouse following intravenous injection [165]; Lim *et al.* used NIR QDs to image the coronary vasculature of a rat heart [166]; and Smith *et al.* imaged the blood vessels of chicken embryos with a variety of NIR and visible QDs [167]. Recently, Jayagopal *et al.* demonstrated the potential for QDs to serve as molecular imaging agents for targeted vascular imaging [168]. For this, spectrally distinct QDs were first conjugated to antibodies against three different cell adhesion molecules (CAMs), and then injected intravenously into a rat model of diabetes. Fluorescence angiography of the retinal vasculature revealed CAM-specific increases in fluorescence, and allowed imaging of the inflammation-specific behavior of individual leukocytes as they floated freely in the vessels, rolled along the endothelium, and underwent leukostasis. The unique spectral properties of QDs allowed the simultaneous imaging of up to four spectrally distinct QD tags. It was also found in this study that, by incorporating PEG crosslinkers and Fc-shielding antibody fragments into the QD–antibody conjugates, their *in vivo* circulation times and targeting efficiency could be increased.

In 2005, two groups reported the use of QDs for the nontargeted *in vivo* imaging of tumor vasculature in mice. Morgan *et al.* coated NIR QDs with bovine serum albumin (BSA) and injected them either subcutaneously or intravenously into mice bearing murine squamous cell carcinoma (SCC). This allowed the blood vessels surrounding and traversing a SCC tumor that was growing in the right hind leg of a C3H mouse, and which had a diameter on the order of ~100 μm and was located at a depth of several hundred microns, to be clearly visualized [132]. These data showed that the QDs represent a valuable angiographic contrast agent for blood vessels surrounding and penetrating the tumor tissues. Stroh *et al.* used PEG-coated QDs and two-photon microscopy to image blood vessels within the microenvironment of subcutaneous tumors in mice, and to concurrently image and differentiate tumor vessels from both the perivascular cells and the matrix through green fluorescent protein (GFP) in the perivascular cells and autofluorescence from collagen in the extracellular matrix (ECM). A stark contrast between the cells, ECM, and the erratic, leaky vasculature was evident [130]. Taken together, the results of these studies pointed towards the use of QDs as fluorescence contrast agents for the high-resolution, noninvasive imaging of tumor vasculatures.

The first attempt to use QD for the targeted *in vivo* imaging of blood vessels in tumors was reported by Cai *et al.* [133]. For this, NIR QDs labeled with arginine-glycine-aspartate (RGD) peptide were used for the active targeting and imaging of blood vessels expressing $\alpha_v\beta_3$ integrins in a murine xenograft model of subcutaneous human glioblastoma tumors. The tumor location could be identified by virtue of the specific binding of the RGD-labeled QDs to the tumor vasculature. More recently, Smith *et al.* used the same methodology to target RGD-labeled QDs to newly formed/forming blood vessels expressing $\alpha_v\beta_3$ integrins in a SKOV-3 ear tumor mouse model [134]. By using high-resolution (~0.5 μm) intravital microscopy, it was possible to observe the binding of QD conjugates to the tumor blood vessels, with details at the cellular level (Figure 1.9). These authors showed that the QDs did not extravasate, but bound as aggregates rather than individually; the latter finding was of critical relevance to the regulatory approval of nanoparticles in human clinical applications for disease diagnostics and therapeutics, as concerns have been expressed that the aggregation of QDs might contribute to their cytotoxicity. The RGD–QD conjugates used in these studies were prepared from commercially available PEG-coated QDs and thiolated RGD peptides through a heterobifunctional linker, 4-maleimidobutyric acid *N*-succinimidyl ester [169]. It has been suggested recently that peptides containing isoaspartate-glycine-arginine (*iso*DGR) may also be conjugated to QDs as binding ligands for the targeted imaging of tumor vasculature [170].

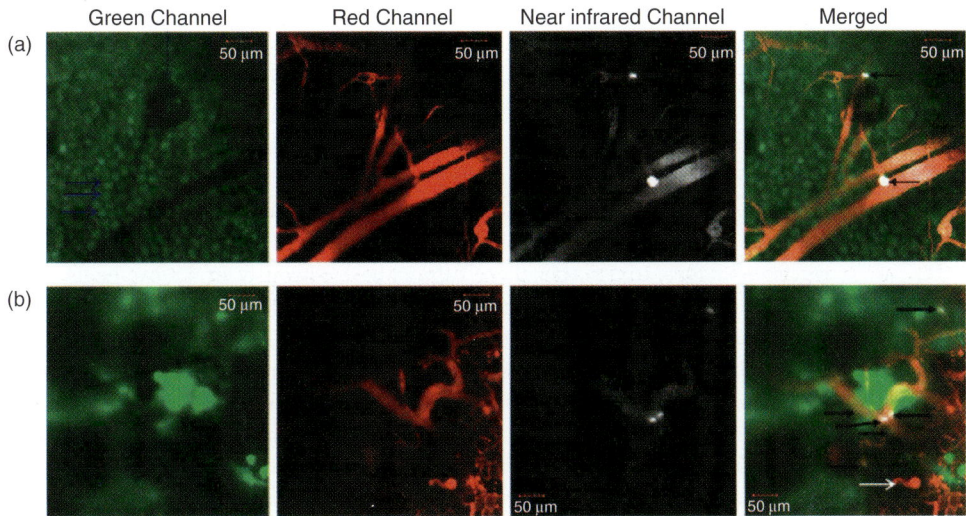

Figure 1.9 Direct visualization of binding of RGD-QDs to tumor vessel endothelium. (a) These panels display different output channels of the identical imaging plane along the row with scale bars. In the green channel, individual EGFP-expressing cancer cells are visible (marked by thick horizontal blue arrows; the vertical blue arrow points to a hair follicle), while the red channel outlines the tumor's vasculature via injection of Angiosense dye. The NIR channel shows intravascularly administered QDs which remain in the vessels (i.e., they do not extravasate). Binding events are visible by reference to bright white signal. These are demarcated by arrows in the rightmost merged image, in which all three channels have been overlaid; (b) These panels display the same as panels (a) in a different mouse, except that a sixfold higher RGD-QD dose has been injected. Individual cells are not generally visible. Six binding events are observed in this field-of-view, as marked by arrows in the merged image at right. White arrows in the bottom merged image designate areas of tissue autofluorescence. Adapted from Ref. [134]; © 2008, American Chemical Society.

1.4.3.3 Sentinel Lymph Node Mapping and Fluorescence Lymphangiography

Sentinel lymph node (SLN) mapping has revolutionized the intraoperative staging of many solid tumors, and is now the standard of care in breast cancer and melanoma [171]. The underlying hypothesis of SLN mapping is that sampling of the first lymph node to receive lymphatic drainage from a solid tumor is sufficient to assess for the presence or absence of lymphatic metastasis [172]. If no malignant cells are found in the SLN, the patient is spared the morbidity of radical lymph node dissection. The current clinical practice of SLN mapping is performed with a combination of preoperative radiocolloid injection and intraoperative injection of a visible blue dye. Unfortunately, this procedure suffers from the drawbacks of potential radiation hazards, along with the extra cost of the radioisotope, allergic reactions after blue dye injection, the overall long duration of the procedure, and technical difficulties in performing the mapping. In particular, the blue dye used for locating the SLN is extremely difficult to see in the presence of blood or anthracosis. Thus, technological innovation is clearly needed.

NIR QDs have innate features that make them an excellent choice as lymphatic tracers for SLN mapping. Most QDs have a hydrodynamic diameter in the range of 10–20 nm and a negatively charged coating, which not only allows them to undergo rapid uptake into lymphatics but also provides them with excellent retention in the lymph nodes. The NIR emission permits fluorescence detection from deep inside the tissue, with a low background autofluorescence. Frangioni and colleagues have pioneered the use of QDs for NIR fluorescence SLN mapping, by using an imaging system that simultaneously displays color video and NIR fluorescence images. In this way, these authors have shown that QDs can be used for real-time intraoperative imaging of SLN in the skin [173], breast [173], lung [174], esophagus [175], pleural space [176], small intestine [177, 178], stomach [178], and colon [178] in small animals (mouse, rat), and also in large animals approaching human size (e.g., pigs). Although these studies did not incorporate a model system with spontaneous cancer metastatic to regional lymph nodes, they have shown that QD-mediated SLN mapping overcomes the limitations of currently available methods, permits patient-specific imaging of lymphatic flow and sentinel nodes, and provides highly sensitive, real-time *in situ* visual guidance for cancer surgery that would enable surgeons to identify and excise nodes draining from primary metastatic tumors, both quickly and accurately. More recently, the value of this technology for SLN mapping in tumor settings has also been demonstrated in pigs with spontaneous melanoma [135] and in mice bearing subcutaneous tumors [136, 137]. Here, the hydrodynamic diameter of QDs was found to have a profound impact on the tracer behavior *in vivo*, with the average time for QDs to be detected in SLN after injection increasing in line with the increasing QD diameter [135, 137].

The multiplexing capabilities of QDs can also be exploited for fluorescence lymphangiographic studies of the lymphatic drainage networks. Due to their small size and poor access, the lymphatic system has been difficult to study *in vivo*, especially when mapping lymphatic drainage simultaneously from multiple basins. However, by injecting QDs of different colors at two different locations, Hama *et al.* were able to observe the QDs draining to a common node, using wavelength-resolved spectral fluorescence lymphangiography [138]. In a subsequent study, the same group demonstrated the simultaneous visualization in real time of five separate lymphatic flows *in vivo* and their trafficking to distinct lymph nodes, using QDs with similar physical sizes but different emission spectra [139] (Figure 1.10). These studies have important implications for predicting the route of cancer metastasis into the lymph nodes.

Robe *et al.* recently studied the selective accumulation of QDs in axillary lymph nodes, and their biodistribution in different tissues in mice [179]. Here, the QDs were detected in the nodes as soon as 5 min and up to 24 h after the injection. Maximum amounts of QDs in the nodes were detected at 60 min after injection, and this corresponded to 2.42% of the injected dose. Most of the injected QDs remained at the injection site, with none being detected in other tissues, nor in the plasma, urine, and feces.

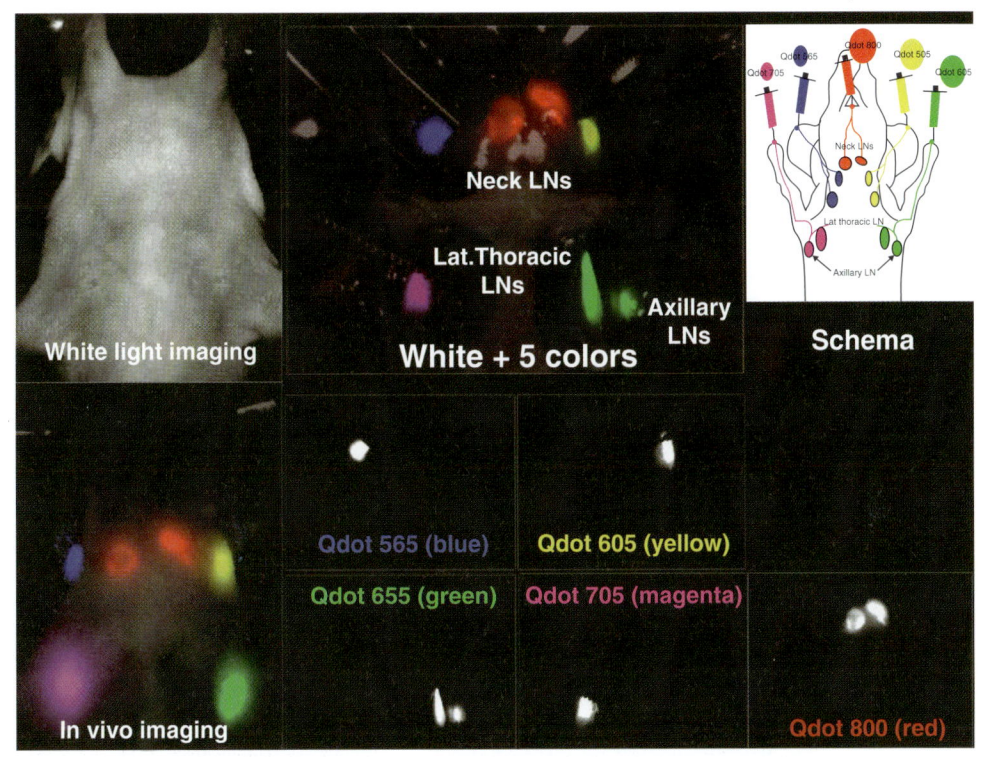

Figure 1.10 *In vivo* five-color lymphatic drainage imaging showing five distinct lymphatic drainages of a mouse injected with five carboxyl QDs (565, blue; 605, green; 655, yellow; 705, magenta; 800, red) intracutaneously into the middle digits of the bilateral upper extremities, the bilateral ears, and at the median chin, as shown in the schema. Five primary draining lymph nodes were simultaneously visualized with different colors through the skin. Adapted from Ref. [139]; © 2007, American Chemical Society.

1.4.3.4 *In Vivo* Whole-Body Tumor Imaging in Animals

Although the ability to visualize and identify tumors in living organisms is invaluable for clinical diagnostic applications, the imaging of tumors presents certain challenges, not only from the need for sensitive and specific imaging agents but also from the fundamental barriers to optical imaging in biological tissues. On the other hand, tumor tissues possess certain unique biological attributes that can facilitate optical imaging using nanoparticles such as QDs. During tumor-induced angiogenesis, blood vessels are formed abnormally with erratic architectures and wide endothelial pores. The highly permeable vasculature, in combination with a lack of effective lymphatic drainage, allow large molecules and particulates (up to ~400 nm in size) to extravasate and accumulate in the tumor microenvironment, a phenomenon called the "enhanced permeability and retention" (EPR) effect [180]. The EPR effect has facilitated the use of a variety of nanoparticles, including QDs, for cancer imaging and targeting.

1.4.3.4.1 QD Conjugates as Active Targeting Probes for Molecular Imaging Due to the high permeability of tumor vasculature, cancer cells are effectively exposed to the constituents of the bloodstream. Surface antigens on cancerous cells may therefore be used as active targets for molecular imaging using bioaffinity fluorescence probes. *Molecular imaging* is the generation of image contrast due to the molecular differences in tissue (e.g., presence or absence of tumor antigens), rather than to differences in tissue-induced radiation attenuation, as are used in other imaging techniques. Molecular imaging has become an area of tremendous interest in oncology because of its potential to detect early-stage cancers and their metastases. In this respect, QDs have demonstrated great superiority over organic fluorescent dyes for these applications, due to their intense fluorescence signals, long-lasting photostabilities, and unique capabilities for multiplexing, and for enabling the long-term, simultaneous imaging of multiple cancer biomarkers with high degrees of sensitivity and selectivity.

In 2202, Akerman *et al.* first reported the use of QD conjugates for the specific targeting of tumors [60], where QDs attached with tissue-specific peptides were injected intravenously into mice bearing human tumor xenografts. Although the probe detections were not performed in living animals, but rather on embedded tissue specimens, the *in vitro* histological results revealed that the QDs homed to tumor vessel guided by the peptides, and were able to escape clearance by the reticuloendothelial system (RES). It was also shown in this study that the addition of PEG to the QD coating prevented the nonselective accumulation of QDs in the RES.

The first demonstration of targeted molecular imaging using QDs in live animals was made by Gao *et al.* in 2004 [39]. For this, the authors used a new class of multifunctional QD probe for the simultaneous targeting and imaging of subcutaneous human prostate tumors in mice after intravenous injection of QDs conjugated to an antibody against PSMA. The QD conjugate contained an amphiphilic triblock copolymer for *in vivo* protection, and also targeting ligands for tumor antigen recognition and multiple PEG molecules for improved biocompatibility and circulation. Subsequent tissue section microscopy and whole-animal spectral imaging revealed that, whilst QD accumulation in the tumor was primarily due to antibody–antigen binding, it was also aided by the EPR effect. These early studies also introduced the concept of multicolor imaging of QD-tagged cancer cells, opening new possibilities for the ultrasensitive and simultaneous imaging of multiple biomarkers involved in cancer metastasis and invasion. The method used to prepare the QD probe has been published [181]. Similarly, Cai *et al.* used NIR QDs labeled with RGD peptide for the active targeting and imaging of subcutaneous human glioblastoma tumors in mice [133]. Subsequently, Yu *et al.* also achieved active targeting and imaging of human hepatoma in mice with QDs conjugated to an antibody against alpha-fetoprotein (AFP) [141], which is an important biomarker for hepatocellular carcinoma. Very recently, Cheng *et al.* conjugated QDs with DNS and injected the QD probes intravenously into mice bearing CT26/DNS tumors [100]. Subsequent whole-body imaging at 2h after injection revealed the DNS-QDs to be retained preferentially in the tumor location, which in turn indicated that the QD conjugates could be used for the targeted imaging of DNS receptor expression *in vivo*.

Tada *et al.* studied the biological processes involved in the active targeting of tumors by QDs [142]. For this, QDs labeled with anti-HER2 monoclonal antibody were used to target human HER2-overexpressing breast cancer in mice. After systemic QD administration, the molecular process was examined in terms of its mechanistic delivery to the tumor by using a high-speed confocal microscope fitted with a high-sensitivity camera. The movement of single QD–antibody complexes could be clearly observed as they circulated in the bloodstream, extravasated into the tumor, diffused into the ECM, bound to their receptors on the tumor cells, and then translocated into the perinuclear region of the cells. The image analysis of the delivery processes of single particles *in vivo* provides valuable information on antibody-conjugated therapeutic nanoparticles, which will undoubtedly be useful for increasing therapeutic efficacies.

Diagaradjane *et al.* reported the first pharmacokinetic and biodistribution study of QD imaging probes for targeted *in vivo* tumor imaging [143]. Here, the NIR QDs were coupled to EGF and injected systemically into mice bearing HCT116 xenograft tumors. *In vivo* imaging showed the EGF-QDs to be mainly distributed in the liver and lymph nodes shortly after injection, and then to accumulate progressively in the tumors, presumably due to a specific binding of the QD–EGF conjugates to EGFR on the tumor cells. The maximum contrast was reached at 4h after systemic administration. Subsequent immunofluorescence images showed the diffuse colocalization of EGF–QD fluorescence within EGFR-expressing tumor parenchyma, compared to a patchy perivascular sequestration of unconjugated QDs. These results implied that the QD–EGF nanoprobe might permit the quantifiable and repetitive imaging of EGFR expression in tumors. In a similar pharmacokinetic study, Weng *et al.* reported that the systemic administration of liposomes tethered with QDs and an anti-HER2 antibody single-chain F_v fragment resulted in tumor localization and *in vivo* fluorescence imaging in a MCF-7/HER2 xenograft mouse model (Figure 1.11) [107]. These authors found that, although the conjugation of QDs to liposomes increased the average diameter of the labeled liposomal nanoparticles, it effectively eliminated the nonspecific binding observed with QDs alone and enabled an extended circulation time *in vivo*. Chen *et al.* used QDs linked to monoclonal antibody for AFP as a probe for the targeted imaging of human hepatocellular carcinoma xenograft growing in nude mice after injection of the QD probes into a tail vein [125]. On examining the hemodynamics and tissue distribution of the QD probes, they were found to be mainly distributed in the liver and spleen, both of which incorporate the RES. Some QDs were also found in the kidneys, although this may be related to their elimination. Similar findings were later reported by Yang *et al.*, who used QDs conjugated with a single-chain anti-EGFR antibody for targeted imaging in nude mice bearing intrapancreatic human xenograft tumors [140].

1.4.3.4.2 QDs as Nontargeting Contrast Agents for Optical Imaging
In the case where no cellular biomarker is available for the targeted molecular imaging of

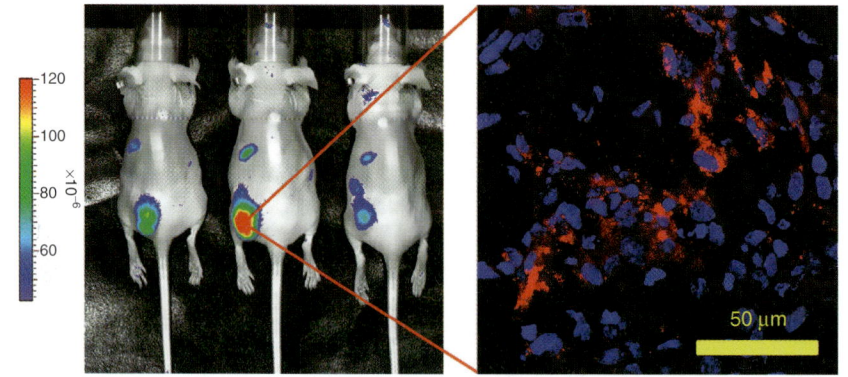

Figure 1.11 *In vivo* targeted fluorescence imaging in a MCF-7/HER2 xenograft mouse model. Left: *In vivo* fluorescence imaging of three nude mice bearing MCF-7/HER2 xenografts implanted in the lower back 30 h after intravenous injection with anti-HER2 QD-ILs. Imaging showed that QD-ILs had localized prominently in tumors, as well as in mononuclear phagocytic system (MPS) organs. Units: efficiency (the fractional ratio of fluorescence emitted per incident photon). Right: A 5 μm section cut from frozen tumor tissues harvested at 48 h post-injection and examined by confocal microscopy using a 63× oil immersion objective (image size, 146 μm × 146 μm). QD-ILs are likely internalized to the cytosol of MCF-7/HER2 cells. From Ref. [107]; © 2008, American Chemical Society.

cancer, nontargeted passive imaging may be attempted. Currently, QDs can be used as powerful imaging contrast agents for studying the complex anatomy and pathophysiology of cancer in animal models.

A solid tumor is composed of cancer and host cells embedded in an ECM and nourished by blood vessels. Whilst a prerequisite to understanding tumor pathophysiology is an ability to distinguish and monitor each component in dynamic studies, standard fluorophores hamper the simultaneous intravital imaging of these components. Stroh *et al.* showed that QDs would greatly enhance current intravital microscopy techniques for the imaging of tumor microenvironment [130], by using QDs as fluorescent contrast agents for blood vessels using two-photon excitation, and concurrently imaging and differentiating tumor vessels from both the perivascular cells and the ECM. The use of QDs allowed a stark contrast to be made between the tumor constituents, due to their intense brightness, tunable wavelengths, and reduced propensity to extravasate into the tumor in comparison with organic dyes. These authors also used QD-tagged beads of varying sizes to model the size-dependent distribution of nanoparticles in tumors.

For nontargeted tumor imaging, it is critical that ways are developed to allow QDs to accumulate at the tumor site. Recent studies on the biodistribution of QDs in living mice have revealed that they are cleared from the circulation primarily by phagocytosis of the nanoparticles by the RES in the liver, spleen, and lymph nodes [182, 183]. The primary phagocytic cells of the RES are monocytes and/or macrophages, including circulating macrophages, perivascular macrophages, and tissue macrophages. Experimental data have suggested that peripheral tissue macrophages are capable of phagocytosing QDs, whilst in the brain a macrophage-derived cell – the microglia – is also capable of the phagocytosis of nanoparticles [184]. QDs with surface modifications that improve their blood circulation half-life might escape from the reticuloendothelial cells of the liver, spleen, and lymph nodes, allowing tissue macrophages to phagocytose the circulating QDs. When Muhammad et al. used QDs to image tumors in a rat glioma model [144] they found that, after intravenous injection, the QDs were uptaken by macrophages that colocalized within the experimental glioma. In this way, the deposition of QD-laden macrophages within the experimental glioma allowed optical detection of the glioma in the animals. The adaptation of these techniques to the surgical management of gliomas has the potential to reduce the operating time, and to improve not only the diagnostic accuracy but also patient outcomes in glioma therapy.

1.4.4
Multimodality Tumor Imaging

In vivo imaging using QD-based fluorescent probes is limited by the depth of tissue penetration, the lack of anatomic resolution and spatial information, and difficulties in quantitation. In order to overcome these obstacles, the QD surface can be modified (through versatile chemistry) to accommodate multiple imaging (radionuclide or paramagnetic) probes so as to allow for multimodality imaging, by coupling QD-based optical imaging with other imaging modalities such as PET, SPECT, and MRI. One of the most promising applications for QDs is the development of multifunctional QD-based probes for multimodality cancer imaging *in vivo*. A multimodality approach would make it possible to image targeted QDs at all scales, from whole-body down to nanometer resolution, and using a single probe. This would in turn permit elucidation of the targeting mechanisms, biodistribution, and dynamics in living animals with higher sensitivity and/or accuracy.

To allow for quantitative targeted imaging in deep tissues, Chen and colleagues developed a dual-modality imaging probe for both NIR fluorescence imaging and PET, by attaching ^{64}Cu to the polymeric coating of QDs through a covalently bound chelating compound [145]. The targeted *in vivo* imaging of a subcutaneous tumor in mouse, by using this probe, was achieved by simultaneously attaching $\alpha_v\beta_3$ integrin-binding RGD peptides onto the QD surface. The quantification ability and ultrahigh sensitivity of PET imaging enabled the quantitative analysis of the biodistribution and targeting efficacy of this dual-modality imaging probe in glioblastoma-bearing mice. However, the full potential of *in vivo* dual-modality imaging was not realized in this study, as fluorescence was used only as an *ex vivo* imaging tool to validate the *in vivo* results of PET imaging, primarily due to the lower sensitivity of optical imaging in comparison with PET. In a more recent study, the same group constructed a similar QD conjugate and achieved *in vivo* dual PET and NIR fluorescence imaging in the same animal model (Figure 1.12) [146]. This time-targeted imaging of tumor vasculature was achieved by attaching the VEGF protein onto the QD surface so as to specifically target the VEGF receptors

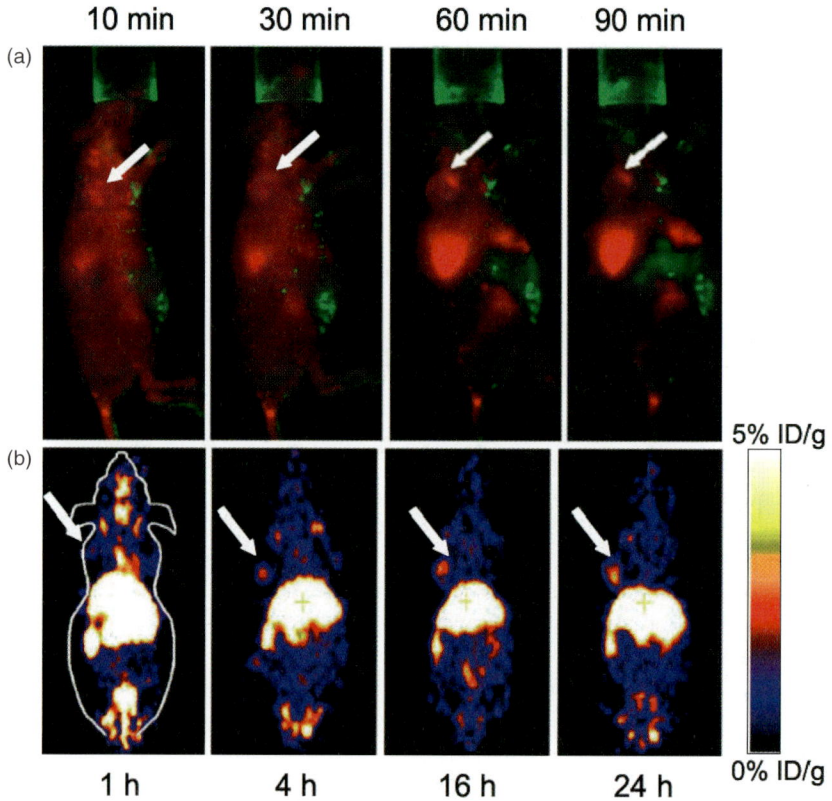

Figure 1.12 Dual modality *in vivo* fluorescence and PET imaging of U87MG tumor-bearing mice. (a) NIR fluorescence imaging at 10, 30, 60, and 90 min post-injection of 200 pmol of DOTA-QD-VEGF; (b) Whole-body coronal PET images at 1, 4, 16, and 24 h post-injection of ca. 300 µCi of ^{64}Cu-DOTA-QD-VEGF. The arrows indicate the tumor. Adapted from Ref. [146]; © 2007, Springer-Verlag.

(VEGFRs) through strong VEGF–VEGFR interaction. VEGFR, which is expressed almost exclusively on the vasculature, serves as a prime target for QD-based imaging, since extravasation is not required to observe the signal. The success of this bifunctional imaging approach may render a higher degree of accuracy for the quantitative targeted NIR fluorescence imaging in deep tissues.

As apoptosis plays an important role in the etiology of a variety of diseases, including cancer, its visualization would allow both an early detection of therapy efficiency and an evaluation of disease progression. To this end, van Tilborg *et al.* developed a dual-modality imaging probe for both optical imaging and MRI by encapsulating QDs in a paramagnetic micelle containing gadolinium [147]. By attaching the nanoparticles with annexin A5, the value of this probe for labeling apoptotic cells *in vitro* could be demonstrated, with a significant T_1 contrast enhancement with a brightening effect in MRI, as well as an easily detectable fluorescence signal from QDs, being observed. The results of the study suggested a high potential for QD-based dual-modality nanoprobes for the *in vivo* detection of apoptosis with both intravital fluorescence microscopy and MRI.

MRI is preferable for molecular imaging due to its excellent spatial resolution and soft tissue contrast. Although molecular MRI potentially allows the direct covisualization of tumor angiogenic activity with anatomy, its inherently low sensitivity may be problematic due to the typically low abundance of upregulated biomolecules. However, this difficulty can be overcome by using large-molecular-weight constructs capable of carrying a high payload of paramagnetic dopants, and multiple targeting ligands to enhance the relaxivity and targeting efficacy, respectively, of the MRI probe. Oostendorp *et al.* developed a new class of paramagnetic QDs (pQDs) by attaching biotin-poly(lysine) dendritic gadolinium to the surface of streptavidin-bound QDs. The probe was labeled with cyclic Asn-Gly-Arg (cNGR) for the noninvasive assessment of tumor angiogenic activity, using quantitative *in vivo* MRI [148]. cNGR colocalizes with an aminopeptidase (CD13) that is highly overexpressed on the angiogenic tumor endothelium. An intravenous injection of cNGR–pQDs in tumor-bearing mice resulted in an increased quantitative contrast, allowing *in vivo* quantification and accurate localization of angiogenic activity (Figure 1.13). Since, similar to the previously mentioned PET study [145], QD fluorescence was used only *ex vivo*, and not *in vivo*, the full potential of *in vivo* dual-modality imaging was not realized in this study.

1.4.5
Dual-Functionality QDs for Cancer Imaging and Therapy

Drug-laden nanoparticles have shown great promise for targeted drug delivery into tumors. A premise of nanomedicine is that it may be feasible to develop multifunctional constructs that combine diagnostic and therapeutic capabilities, thus leading to a better targeting of drugs to diseased cells. The large surface area of QDs, combined with their versatile surface chemistry, makes them convenient scaffolds to accommodate anticancer drugs, either through chemical linkage or by simple physical immobilization, leading in turn to the development of nanostructures with integrated imaging and therapy functionalities. Bagalkot *et al.* reported a multifunctional system which comprised a QD, RNA aptamers, and the anticancer drug doxorubicin (Dox) for targeted cancer imaging, drug delivery, and sensing [149]. The RNA aptamers were attached covalently to the surface of the QDs to serve as targeting molecules for the extracellular domain of PSMA. The intercalation of Dox in the double-stranded stem of the aptamer resulted in a targeted tertiary conjugate with reversible self-quenching properties based on a bi-FRET (fluorescence resonance energy transfer) mechanism, one between QD and Dox and another between Dox and the aptamer. As demonstrated in the *in vitro* experiment, the multifunctional nanoparticle system can first deliver Dox to the targeted prostate cancer cells, and then sense the release of Dox by activating the fluorescence of QD, which concurrently images the cancer cells. By

42 *1 Quantum Dots for Cancer Imaging*

cNGR-pQDs   Unlabeled pQDs

(a)

$\Delta R1\ (s^{-1})$
0 — 0.5

(b)

$\Delta S_0\ (\%)$
0 — -50

(c)

(d)

Figure 1.13 T_2-weighted anatomic MRI images with color overlay of ΔR_1 (a) and ΔS_0 (b) for tumor (T) and muscle (M) tissue of mice injected with cNGR-labeled or unlabeled pQDs ($n = 7$ for both groups). Changes in R_1 were most pronounced at the tumor rim for cNGR-pQDs. Although an R_1 increase in the tumor rim was also observed for unlabeled pQDs, the average response was threefold lower when compared to cNGR-pQDs, indicating a high specificity of cNGR for angiogenic tumor endothelium. This is further supported by the low changes in R_1 found in muscle tissue. Changes in S_0 (b) colocalized almost completely with changes in R_1 (a). Representative TPLSM images of tumor (c) and muscle tissue (d) showing pQD signal (red) and EC-specific αCD31-FITC (green). The cNGR-pQDs accurately colocalized with tumor ECs, indicating binding of the contrast agent to the tumor endothelium (c). The cNGR-pQDs were also detected in muscle tissue with TPLSM (arrows in d), although to a much lesser extent than in tumor tissue. The cNGR-pQDs did not display any colocalization with muscle ECs, and were only found intraluminally. Unlabeled pQDs were not or only sparsely detected in both tumor and muscle tissue. Scale bar = 50 µm. From Ref. [148]; © 2008, American Association for Cancer Research.

---

incorporating multiple CG sequences within the stem of the aptamers, the drug-loading capacity of the system can be further increased, thereby enhancing the therapeutic efficacy of the conjugates. While this study was at a proof-of-concept stage, using only cultured cancer cells, a subsequent study by Weng et al. proved to be more immediately relevant to the *in vivo* imaging and treatment of solid tumors [107]. Here, a multifunctional QD-conjugated immunoliposome system (QD–ILs) was developed for tumor-cell targeting, imaging, and drug delivery, where anti-HER2 single chain Fv fragments were attached to the surface of the nanoparticle for targeted delivery to HER2-expressing breast cancer cells. The anticancer drug Dox was encapsulated in the aqueous interior of the liposome (Figure 1.14). *In vitro* experiments indicated that Dox-loaded QD–ILs were internalized by HER2-expressing cancer cells through receptor-mediated endocytosis, and showed an efficient anticancer activity. In MCF-7/HER2 xenograft mouse models, the localization of QD-ILs at tumor sites was confirmed by *in vivo* fluorescence imaging (see Figure 1.11). It was also shown that QD–ILs could significantly prolong the circulation of QDs in the bloodstream. Although the anticancer activity of the Dox-loaded QD–ILs was not demonstrated *in vivo*, these studies will nonetheless guide the future design and optimization of multifunctional nanoparticle agents for *in vivo*-targeted tumor imaging and therapy.

Recently, RNA interference (RNAi) has become a powerful technology for sequence-specific gene suppression. Although RNAi-effectuated oncogene silencing using siRNA represents an effective means of targeted gene therapy for various cancers [185], the application of RNAi *in vivo* has been hampered by its rapid excretion and nontargeted tissue distribution during systemic delivery. In order to develop and optimize methods for the efficient delivery of siRNA into tumor cells, QDs may also provide a versatile nanoscale scaffold to develop multifunctional nanoparticles for targeted siRNA delivery and imaging of the delivery process. In this regard, Derfus et al. used QDs as a scaffold to conjugate siRNA against enhanced green fluorescent protein (EGFP) and tumor-homing peptide F3

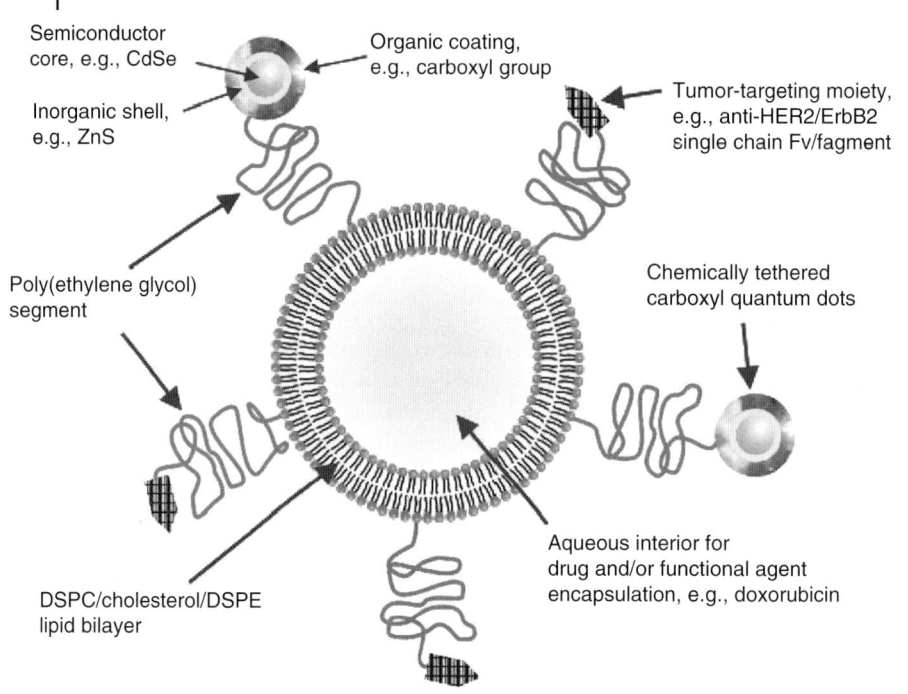

Figure 1.14 Schematic showing the structure of a QD–IL nanoparticle. The liposomal core is about 100 nm in diameter, as visualized by freeze-fracture electron microscopy (ff-EM). Derivatized CdSe/ZnS core–shell QDs are represented as a sphere with a layer of organic coating (gray) covering the outer surface of the inorganic shell (yellow) and the semiconductor core (orange). They were also characterized by ff-EM, indicating an average diameter of ~11 nm. Surface-derivatized QDs were chemically linked to functionalized PEG-DSPE incorporated in extruded liposomes. Anti-HER2 single chain Fv fragments (scFv, arrowheads) are attached to the end of PEG chains. scFv moieties (MW ~ 25 kDa) are not drawn to scale. The QD–ILs retain an aqueous interior for loading and delivery of drugs/probes. From Ref. [107]; © 2008, American Chemical Society.

to functional groups on the QD surface. The F3 peptide, targeting cell-surface nucleolin, was attached to achieve targeted internalization by tumor cells. It was shown that the delivery of these F3/siRNA–QDs to EGFP-transfected HeLa cells, and release from their endosomal entrapment, led to a significant knockdown of the EGFP signal [150]. Although this study was only a proof-of-concept, by designing the siRNA sequences against oncogenes instead of EGFP, this technology might ultimately be adapted for the simultaneous imaging and treatment of cancers. More recently, Yezhelyev *et al.* reported a multifunctional nanoparticle system for siRNA delivery and imaging using QDs with proton-absorbing polymeric coatings ("proton sponges") [128]. By optimizing the proton-absorbing capacity through balancing the composition of the tertiary amine and carboxylic acid groups on the QD surface, an endosomal release of the siRNA was achieved via the proton sponge effect. As a result, a dramatic (10–20-fold) improvement in gene silencing efficiency, and a simultaneous five- to sixfold reduction in cellular toxicity, compared to existing transfection agents, was observed in MDA-MB-231 cells. The QD–siRNA nanoparticles were also dual-modality optical and electron-microscopy probes, allowing real-time tracking and the ultrastructural localization of QDs during delivery and transfection. These new insights and capabilities represent a major step towards nanoparticle engineering for combined imaging and therapeutic applications.

1.5
Quantum Dot Cytotoxicity and Potential Safety Concerns

One major obstacle to fully exploring the *in vivo* applications of QDs in biomedical imaging is the concern regarding their possible cytotoxicity. Compared to gold and iron oxide nanoparticles (which have been used for several decades and have proved to be biocompatible), QDs are relatively new materials and their toxicity has not yet been fully characterized. Prior to any clinical applications on human subjects, the biocompatibility of QDs must be characterized and any potential safety concerns clarified.

A detailed discussion on the topic of QD toxicity may warrant a separate chapter, or even a book. In addition, a number of dedicated reviews have been published that summarize much of what is known in this area. Hoet *et al.* reviewed most of the data available up to 2004 on the health effects of nanoparticles in general [186], and some later reviews have since updated the topic, with data extended to 2007 [6, 187]. In 2006, Hardman published one of the most extensive reviews on the toxicity of QDs in particular, in which the cumulative results from almost all previous *in vivo* studies (both cellular and small animal) were summarized [188]. Therefore, only a brief outline of previous major findings will be provided here, and some of the most recent studies highlighted.

One source of QD cytotoxicity derives from the semiconductor materials that are commonly found in the QD core, such as the heavy metals Cd and Se, the toxicities of which are well known. Under certain circumstances, these elements may leach from the QDs. By using hepatocytes, Derfus *et al.* showed that the oxidation of Cd on the QD surface and subsequent Cd^{2+} release, mediated by oxygen or ultraviolet light, is one possible mechanism responsible for QD cytotoxicity [189]. The protective shell, which in most cases is ZnS, has a much lower toxicity than the core; thus, adding a protective shell may result in a significant reduction in the cytotoxic effects of QDs, as well as improving their optical properties [189, 190]. Most current studies have indicated that, when properly capped by both ZnS and hydrophilic shells, no acute and obvious QD-induced toxicity was detected in studies of cell proliferation and viability or systemic toxicity in mice [191]. For example, *in vivo* studies performed by Ballou *et al.* indicated that stably protected QDs had no apparent toxicity in mice over long periods of time [192]. However, the introduction of capping layers may still be insufficient to solve the problem of cytotoxicity completely, as various other factors, including the aggregation of particles on the cell surface [51] and even the stabilizing QD surface ligands

[193], have been shown to impair cell viability. Hence, the choice of an appropriate surface coating has also been shown to be a critical parameter, since simple coatings (e.g., thiol-containing carboxylic acids) present only a minor diffusion barrier for Cd^{2+}. Thiol-coated QDs are also known for their poor colloidal stability [194, 195]; consequently, the introduction of inert and stable, macromolecular or crosslinked surface coatings, may be an important improvement [196].

QD cytotoxicity should not be attributed solely to the toxic effect of Cd^{2+} released from the particle core [197]. Another source of QD toxicity derives from the reactive oxygen species (ROS) generated during excitation [197–200], since QDs can transfer absorbed optical energy to oxygen molecules. Free radicals can cause damage to DNA [201] and other cellular components and, as a consequence, induce apoptosis and necrosis.

It should be pointed out that the unique structure of QDs presents a complex set of physico-chemical characteristics that compounds studies in this area. The QD cores can be constituted from different combinations of binary semiconductors such as CdSe, CdTe, CdS, and InP. Further, the cores are commonly encapsulated with a protective layer and are then functionalized with a variety of surface-coating ligands that include small thiolated molecules or larger amphiphilic polymers for aqueous compatibility. In fact, surface coatings have been found to be determinants of QD cytotoxicity [112, 202, 203]. Additionally, the QDs may exist in a wide range of sizes, with diameters ranging from a few nanometers to more than 10 nm, which may also affect their cytotoxicity [112]. For biological use, QDs can be further modified with various proteins or nucleic acids. Cumulatively, this combination of materials and physical properties serves to confound any systematic study of toxicity, even before issues such as dosage or exposure time are formally addressed. As a result, most of the studies in the past have primarily been observational in nature, where authors have reported the effects of a given QD material on a particular cell line or animal, at some specified concentration(s) for some exposure time [186–188]. Consequently, the results were mixed with some which reported no visible toxicity [123, 204], while others reported high cytotoxicity [205]. This can be interpreted to reflect what is posited above: the choice of core–shell material and solubilization cap, in conjunction with the dosage/exposure time, will obscure any simple assessments of toxicity. Therefore, more systematic and extensive studies are required in order to fully understand the toxicity of QDs. Recently, Chen *et al.* conducted a systematic study of the effect of dosage on cytotoxicity using a QD–antibody conjugate [125]. The results from this *in vitro* study on the human hepatocellular carcinoma cell line HCCLM6 indicated that the QD cytotoxicity was dose-dependent. For example, with a dose of $1 \times 10^7 \,\mathrm{mol\,l^{-1}}$ or less, there were no discernable adverse effects on cell growth and development, but when the dose exceeded $1 \times 10^7 \,\mathrm{mol\,l^{-1}}$ a significant decrease in cell viability was observed. *In vivo* studies with nude mice at a dose of $1 \times 10^5 \,\mathrm{mol\,l^{-1}}$ showed no evidence of acute toxicity in the test group as compared to the control groups.

Besides cytotoxicity, another concern over QD safety for clinical applications is their degradation, metabolism, and body clearance, which has not been studied until recently. When Fischer *et al.* [206] investigated the distribution, sequestration, and clearance of mercaptoundecanoic acid-functionalized QDs in rats, the

QDs coated with BSA were shown to be cleared more rapidly from serum than those coated with crosslinked lysine. Further, almost all (99%) of the BSA-coated QDs were found in the liver after 90 min, as compared to only 40% of the lysine QDs. These results may reflect one of the primary metabolic roles of endogenous serum albumin. Interestingly, these QDs were not excreted by rats, even after 10 days. Recent studies on the biodistribution of QDs in live mice have revealed that QDs can accumulate in the liver, spleen, and lymph nodes for significant periods of time after systemic injection [107, 125, 140, 143, 182, 183].

Since QDs contains toxic materials, an understanding of the toxicity mechanisms and the process of clearing nanoparticles from animal and human bodies should be given serious attention in the near future. Recent studies have indicated that even though QDs did not affect cell morphology, they might alter the expression of specific genes [207, 208], and therefore the potential risk at the molecular level and the long-term effects of QDs on humans and the environment should be carefully and extensively evaluated. However, there is at present a general paucity of information on molecular mechanistic studies of QD toxicity. When Choi et al. examined the cytotoxicity of QDs in human neuroblastoma cells, they found that CdTe QD-induced toxicity was correlated with Fas upregulation on the surface of treated cells and an increased membrane lipid peroxidation, that may lead to an impaired mitochondrial function (Figure 1.15) [209]. It was also found that QDs modified by *N*-acetylcysteine (NAC), an antioxidant, were internalized to a lesser extent by the cells and were less cytotoxic than unmodified QDs.

Currently, several strategies have been developed to overcome the potential toxicity of QDs. More stable and robust coatings can be developed to protect the QD surfaces from oxidative environments, and QDs may also be encapsulated inside polymeric nanoparticles for further protection. For example, Pan and Feng used nanoparticles composed of a mixture of two vitamin E-containing copolymers to encapsulate QDs in order to reduce their side effects, as well as to improve their imaging effects. Compared to the free QDs, the QDs formulated in polymeric nanoparticles showed a lower *in vitro* cytotoxicity for both MCF-7 cells and NIH-3T3 cells [151]. In addition, more effective targeting systems can be employed to increase the detection efficiency and reduce the required dose of QDs [200]. Another possible strategy might be to develop new, high-quality QD systems that do not contain potentially biologic toxic components. For example, Pradhan et al. have synthesized Cu- or Mn-doped ZnSe QDs with acceptable quantum efficiency and optical properties [190]. Clearly, progress made using these new strategies will ultimately lead to improved cytotoxicity profiles for QDs.

1.6
Concluding Remarks and Future Perspectives

QDs, as a novel class of fluorescent probes, have lived up to many of their promises for the molecular imaging of cancers. With rapid advances in their synthesis, surface modification and bioconjugation, significant progress has been made in applying QDs to potentially useful clinical applications, such as profiling cancer

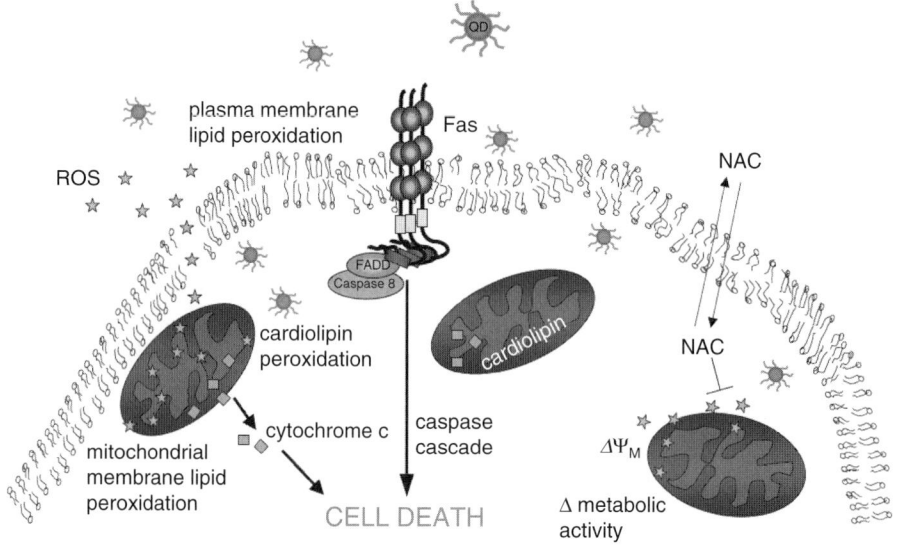

Figure 1.15 Proposed mechanism of QD-induced cell death involving Fas, lipid peroxidation, and mitochondrial impairment. Cells exposed to cadmium telluride QDs (unmodified and NAC-modified) induce ROS which causes Fas upregulation and plasma membrane lipid peroxidation. Apoptotic cell death is induced by activation of Fas and its downstream effectors. Lipid peroxidation also occurs at the mitochondrial membranes, degrading cardiolipin, changing the mitochondrial membrane potential, eventually leading to the release of cytochrome c, and promoting apoptotic cascades. NAC bound to the QD surface, modifies the extent of QD internalization, which is correlated with cell death, upregulation of Fas, and ROS induced lipid peroxidation. NAC treatment (2–5 mM) abolishes oxidative stress, induces antioxidant enzymes, and attenuates mitochondrial impairment. From Ref. [209]; © 2007 Choi et al.).

biomarkers in pathologic specimens, the *in vivo* imaging of cancer and metastasis, and monitoring the clinical responses of tumors to therapy. In addition, the potential for using QDs in multimodality imaging and for combined imaging and therapy has also been demonstrated. Despite all of these successes, several challenges remain for enhancing sensitivity, maximizing specificity, and minimizing toxicity, all of which are areas to which future research should be directed. In particular, the question of intrinsic toxicity and long retention times of the QD probes within the body represent significant challenges to their medical use, and these problems must be solved before clinical applications can proceed. Nonetheless, this exciting new technology holds great promise for improving the diagnosis and treatment of patients with cancers.

Abbreviations

ACTH	adrenocorticotropic hormone
AFP	alpha-fetoprotein

BSA	bovine serum albumin
CA	cancer antigen
CAM	cell adhesion molecule
CD	cluster of differentiation
CEA	carcinoembryonic antigen
cNGR	cyclic asparagine-glycine-arginine
CT	computed tomography
DAPI	4′,6-diamidino-2-phenylindole
DNA	deoxyribonucleic acid
DOTA	1,4,7,10-tetraazacyclododecane-1,4,7,10-tetraacetic acid
Dox	doxorubicin
EGF	epidermal growth factor
EGFP	enhanced green fluorescent protein (GFP)
EGFR	epidermal growth factor receptor
EPR	enhanced permeability and retention
ER	estrogen receptor
ff-EM	freeze-fracture electron microscopy
FFPE	formalin-fixed and paraffin-embedded
FISH	fluorescence *in situ* hybridization
FITC	fluorescein isothiocyanate
FRET	fluorescence resonance energy transfer
GFP	green fluorescent protein
HER	human epidermal growth factor receptor
HER2	human epidermal growth factor receptor 2
HSP	heat-shock protein
IHC	immunohistochemistry
IL	immunoliposome
*iso*DGR	isoaspartate-glycine-arginine
LSC	laser scanning cytometry
MB	molecular beacon
MMP	matrix metalloprotease
MPS	mononuclear phagocytic system
MRI	magnetic resonance imaging
mRNA	messenger ribonucleic acid (RNA)
NAC	*N*-acetylcysteine
NFS	National Science Foundation
NIH	National Institutes of Health
NIR	near-infrared
PEG	polyethylene glycol
PEG-DSPE	*N*-(polyethylene glycol)-1,2-distearoyl-*sn*-glycero-3-phosphoethanolamine
PET	positron emission tomography
pQD	paramagnetic quantum dot (QD)
PSA	prostate-specific antigen
PSMA	prostate-specific membrane antigen

QD	quantum dot
RES	reticuloendothelial system
RGD	arginine-glycine-aspartate
RNA	ribonucleic acid
RNAi	RNA interference
ROS	reactive oxygen species
SCC	squamous cell carcinoma
siRNA	small interfering RNA
SLN	sentinel lymph node
SPECT	single photon emission computed tomography (CT)
THG	third harmonic generation
TMA	tissue microarray
TOPO	trioctylphosphine oxide
TPLSM	two-photon laser scanning microscopy
VEGF	vascular endothelial growth factor
VEGFR	vascular endothelial growth factor (VEGF) receptor

Acknowledgments

Certain commercial equipment or materials have been identified in this chapter in order to specify adequately the experimental procedures. Such identification does not imply recommendation or endorsement by the National Institute of Standards and Technology, nor does it imply that the materials or equipment identified are necessarily the best available for the purpose.

References

1 Freitas, R.A. Jr (2005) What is nanomedicine? *Nanomedicine*, 1, 2–9.
2 Editorial (2003) Nanomedicine: grounds for optimism, and a call for papers. *Lancet*, 362, 673.
3 The National Science Foundation (NSF) (2004) NSF Priority Areas, http://www.nsf.gov/od/lpa/news/publicat/nsf04009/cross/priority.htm (accessed 31 March 2009).
4 The National Institutes of Health (NIH) (2009) NIH roadmap for medical research: Nanomedicine–Overview, http://nihroadmap.nih.gov/nanomedicine/ (accessed 31 March 2009).
5 The National Cancer Institute (NCI) (2004) Cancer nanotechnology plan, http://nano.cancer.gov/about_alliance/cancer_nanotechnology_plan.asp (accessed 31 March 2009).
6 Buzea, C., Pacheco, I.I. and Robbie, K. (2007) Nanomaterials and nanoparticles: sources and toxicity. *Biointerphases*, 2, MR17–71.
7 Efros, Al L. and Efros, A.L. (1982) Interband absorption of light in a semiconductor sphere. *Soviet Physics: Semiconductors*, 16, 772–5.
8 Ekimov, A.I. and Onushchenko, A.A. (1982) Quantum size effect in the optical-spectra of semiconductor micro-crystals. *Soviet Physics: Semiconductors*, 16, 775–8.
9 Bruchez, M. Jr, Moronne, M., Gin, P., Weiss, S. and Alivisatos, A.P. (1998) Semiconductor nanocrystals as fluorescent biological labels. *Science*, 281, 2013–16.

10 Chan, W.C. and Nie, S. (1998) Quantum dot bioconjugates for ultrasensitive nonisotopic detection. *Science*, **281**, 2016–18.

11 Medintz, I.L., Mattoussi, H. and Clapp, A.R. (2008) Potential clinical applications of quantum dots. *International Journal of Nanomedicine*, **3**, 151–67.

12 Tannock, I.F., Hill, R.P., Bristow, R.G. and Harrington, L. (eds) (2005) *The Basic Science of Oncology*, 4th edn, McGraw-Hill Professional, New York.

13 Alberts, B., Bray, D., Lewis, J., Raff, M., Roberts, K. and Watson, J.D. (eds) (1994) *Molecular Biology of the Cell*, 3rd edn, Garland Publishing, New York.

14 Schuiz, W.A. (2005) *Molecular Biology of Human Cancers: An Advanced Student's Textbook*, Springer, New York, pp. 11–17.

15 American Cancer Society (2008) Cancer Facts & Figures 2008, http://www.cancer.org/downloads/STT/2008CAFFfinalsecured.pdf (accessed 31 March 2009).

16 Frangioni, J.V. (2008) New technologies for human cancer imaging. *Journal of Clinical Oncology*, **26**, 4012–21.

17 Debbage, P. and Jaschke, W. (2008) Molecular imaging with nanoparticles: giant roles for dwarf actors. *Histochemistry and Cell Biology*, **130**, 845–75.

18 Chatterjee, S.K. and Zetter, B.R. (2005) Cancer biomarkers: knowing the present and predicting the future. *Future Oncology*, **1**, 35–50.

19 Stamey, T.A., Yang, N., Hay, A.R., McNeal, J.E., Freiha, F.S. and Redwine, E. (1987) Prostate-specific antigen as a serum marker for adenocarcinoma of the prostate. *The New England Journal of Medicine*, **317**, 909–16.

20 Devine, P.L., McGuckin, M.A. and Ward, B.G. (1992) Circulating mucins as tumor markers in ovarian cancer. *Anticancer Research*, **12**, 709–17.

21 Maruvada, P., Wang, W., Wagner, P.D. and Srivastava, S. (2005) Biomarkers in molecular medicine: cancer detection and diagnosis. *Biotechniques*, **38**, S9–15.

22 Duffy, M.J. and Crown, J. (2008) A personalized approach to cancer treatment: how biomarkers can help. *Clinical Chemistry*, **54**, 1770–9.

23 Goldstein, A.N., Echer, C.M. and Alivisatos, A.P. (1992) Melting in semiconductor nanocrystals. *Science*, **256**, 1425–7.

24 Alivisatos, A.P. (1996) Semiconductor clusters, nanocrystals, and quantum dots. *Science*, **271**, 933–7.

25 Sapra, S. and Sarma, D.D. (2004) Evolution of the electronic structure with size in II-IV semiconductor nanocrystals. *Physical Review B*, **69**, 125304.

26 Qu, L. and Peng, X. (2002) Control of photoluminescence properties of CdSe nanocrystals in growth. *Journal of the American Chemical Society*, **124**, 2049–55.

27 Zhong, X., Han, M., Dong, Z., White, T.J. and Knoll, W. (2003) Composition-tunable $Zn_xCd_{1-x}Se$ nanocrystals with high luminescence and stability. *Journal of the American Chemical Society*, **125**, 8589–94.

28 Kim, S., Fisher, B., Eisler, H.-J. and Bawendi, M. (2003) Type II quantum dots: CdTe/CdSe(core/shell) and CdSe/ZnTe(core/shell) heterostructures. *Journal of the American Chemical Society*, **125**, 11466–7.

29 Zhong, X., Feng, Y., Knoll, W. and Han, M. (2003) Alloyed $Zn_xCd_{1-x}S$ nanocrystals with highly narrow luminescence spectral width. *Journal of the American Chemical Society*, **125**, 13559–63.

30 Pietryga, J.M., Schaller, R.D., Werder, D., Stewart, M.H., Klimov, V.I. and Hollingsworth, J.A. (2004) Pushing the band gap envelope: mid-infrared emitting colloidal PbSe quantum dots. *Journal of the American Chemical Society*, **126**, 11752–3.

31 Michalet, X., Pinaud, F.F., Bentolila, L.A., Tsay, J.M., Doose, S., Li, J.J., Sundaresan, G., Wu, A.M., Gambhir, S.S. and Weiss, S. (2005) Quantum dots for live cells, *in vivo* imaging, and diagnostics. *Science*, **307**, 538–44.

32 Efros, A.L. and Rosen, M. (2000) The electronic structure of semiconductor nanocrystals. *Annual Review of Materials Science*, **30**, 475–521.

33 Leatherdale, C.A., Woo, W.-K., Mikulec, F.V. and Bawendi, M.G. (2002) On the absorption cross section of CdSe nanocrystal quantum dots. *The Journal of Physical Chemistry B*, **106**, 7619–22.

34 Dabbousi, B.O., Rodriguez-Viejo, J., Mikulec, F.V., Heine, J.R., Mattoussi, H., Ober, R., Jensen, K.F. and Bawendi, M.G. (1997) (CdSe)ZnS core-shell quantum dots: synthesis and characterization of a size series of highly luminescent nanocrystallites. *The Journal of Physical Chemistry B*, **101**, 9463–75.

35 Gao, X., Yang, L., Petros, J.A., Marshall, F.F., Simons, J.W. and Nie, S. (2005) In vivo molecular and cellular imaging with quantum dots. *Current Opinion in Biotechnology*, **16**, 63–72.

36 Dahan, M., Laurence, T., Pinaud, F., Chemla, D.S., Alivisatos, A.P., Sauer, M. and Weiss, S. (2001) Time-gated biological imaging by use of colloidal quantum dots. *Optics Letters*, **26**, 825–7.

37 Pepperkok, R., Squire, A., Geley, S. and Bastiaens, P.I. (1999) Simultaneous detection of multiple green fluorescent proteins in live cells by fluorescence lifetime imaging microscopy. *Current Biology*, **9**, 269–72.

38 Jakobs, S., Subramaniam, V., Schönle, A., Jovin, T.M. and Hell, S.W. (2000) EFGP and DsRed expressing cultures of *Escherichia coli* imaged by confocal, two-photon and fluorescence lifetime microscopy. *FEBS Letters*, **479**, 131–5.

39 Gao, X., Cui, Y., Levenson, R.M., Chung, L.W. and Nie, S. (2004) In vivo cancer targeting and imaging with semiconductor quantum dots. *Nature Biotechnology*, **22**, 969–76.

40 Kortan, A.R., Hull, R., Opila, R.L., Bawendi, M.G., Steigerwald, M.L., Carroll, P.J. and Brus, L.E. (1990) Nucleation and growth of CdSe on ZnS quantum crystallite seeds, and vice versa, in inverse micelle media. *Journal of the American Chemical Society*, **112**, 1327–32.

41 Murray, C.B., Norris, D.J. and Bawendi, M.G. (1993) Synthesis and characterization of nearly monodisperse CdE (E = S, Se, Te) semiconductor nanocrystallites. *Journal of the American Chemical Society*, **115**, 8706–15.

42 Talapin, D.V., Rogach, A.L., Komowski, A., Haase, M. and Weller, H. (2001) Highly luminescent monodisperse CdSe and CdSe/ZnS nanocrystals synthesized in a hexadecylamine–trioctylphosphine oxide–trioctylphospine mixture. *Nano Letters*, **1**, 207–11.

43 Yang, H. and Holloway, P.H. (2004) Efficient and photostable ZnS-passivated CdS:Mn luminescent nanocrystals. *Advanced Functional Materials*, **14**, 152–6.

44 Hines, M.A. and Guyot-Sionnest, P. (1996) Synthesis and characterization of strongly luminescing ZnS-capped CdSe nanocrystals. *The Journal of Physical Chemistry*, **100**, 468–71.

45 Nirmal, M. and Brus, L. (1999) Luminescence photophysics in semiconductor nanocrystals. *Accounts of Chemical Research*, **32**, 407–14.

46 Peng, X., Schlamp, M.C., Kadavanich, A.V. and Alivisatos, A.P. (1997) Epitaxial growth of highly luminescent CdSe/CdS core/shell nanocrystals with photostability and electronic accessibility. *Journal of the American Chemical Society*, **119**, 7019–29.

47 Reiss, P., Bleuse, J. and Pron, A. (2002) Highly luminescent CdSe/ZnSe core/shell nanocrystals of low size dispersion. *Nano Letters*, **2**, 781–4.

48 Gao, X. and Dave, S.R. (2007) Quantum dots for cancer molecular imaging. *Advances in Experimental Medicine and Biology*, **620**, 57–73.

49 Wu, X., Liu, H., Liu, J., Haley, K.N., Treadway, J.A., Larson, J.P., Ge, N., Peale, F. and Bruchez, M.P. (2003) Immunofluorescent labeling of cancer marker Her2 and other cellular targets with semiconductor quantum dots. *Nature Biotechnology*, **21**, 41–6.

50 Dubertret, B., Skourides, P., Norris, D.J., Noireaux, V., Brivanlou, A.H. and Libchaber, A. (2002) In vivo imaging of quantum dots encapsulated in phospholipid micelles. *Science*, **298**, 1759–62.

51 Kirchner, C., Liedl, T., Kudera, S., Pellegrino, T., Javier, A.M., Gaub, H.E., Stölzle, S., Fertig, N. and Parak, W.J. (2005) Cytotoxicity of colloidal CdSe and CdSe/ZnS nanoparticles. *Nano Letters*, **5**, 331–8.

52 Hild, W.A., Breunig, M. and Goepferich, A. (2008) Quantum dots – nano-sized probes for the exploration of cellular and intracellular targeting. *European Journal*

of Pharmaceutics and Biopharmaceutics, **68**, 153–68.

53 Goldman, E.R., Balighian, E.D., Mattoussi, H., Kuno, M.K., Mauro, J.M., Tran, P.T. and Anderson, G.P. (2002) Avidin: a natural bridge for quantum dot-antibody conjugates. *Journal of the American Chemical Society*, **124**, 6378–82.

54 Lidke, D.S., Nagy, P., Heintzmann, R., Arndt-Jovin, D.J., Post, J.N., Grecco, H.E., Jares-Erijman, E.A. and Jovin, T.M. (2004) Quantum dot ligands provide new insights into erbB/HER receptor-mediated signal transduction. *Nature Biotechnology*, **22**, 198–203.

55 Dahan, M., Lévi, S., Luccardini, C., Rostaing, P., Riveau, B. and Triller, A. (2003) Diffusion dynamics of glycine receptors revealed by single-quantum dot tracking. *Science*, **302**, 442–5.

56 Jaiswal, J.K., Mattoussi, H., Mauro, J.M. and Simon, S.M. (2003) Long-term multiple color imaging of live cells using quantum dot bioconjugates. *Nature Biotechnology*, **21**, 47–51.

57 Zhang, C., Ma, H., Ding, Y., Jin, L., Chen, D. and Nie, S. (2000) Quantum dot labeled trichosanthin. *Analyst*, **125**, 1029–31.

58 Srinivasan, C., Lee, J., Papadimitrakopoulos, F., Silbart, L.K., Zhao, M. and Burgess, D.J. (2006) Labeling and intracellular tracking of functionally active plasmid DNA with semiconductor quantum dots. *Molecular Therapy*, **14**, 192–201.

59 Parak, W.J., Gerion, D., Zanchet, D., Woerz, A.S., Pellegrino, T., Micheel, C., Williams, S.C., Seitz, M., Bruehl, R.E., Bryant, Z., Bustamante, C., Bertozzi, C.R. and Alivisatos, A.P. (2002) Conjugation of DNA to silanized colloidal semiconductor nanocrystalline quantum dots. *Chemistry of Materials*, **14**, 2113–19.

60 Akerman, M.E., Chan, W.C., Laakkonen, P., Bhatia, S.N. and Ruoslahti, E. (2002) Nanocrystal targeting in vivo. *Proceedings of the National Academy of Sciences of the United States of America*, **99**, 12617–21.

61 Mattoussi, H., Mauro, J.M., Goldman, E.R., Anderson, G.P., Sundar, V.C., Mikulec, F.V. and Bawendi, M.G. (2000) Self-assembly of CdSe-ZnS quantum dot bioconjugates using an engineered recombinant protein. *Journal of the American Chemical Society*, **122**, 12142–50.

62 Hanaki, K.I., Momo, A., Oku, T., Komoto, A., Maenosono, S., Yamaguchi, Y. and Yamamoto, K. (2003) Semiconductor quantum dot/albumin complex is a long-life and highly photostable endosome marker. *Biochemical and Biophysical Research Communications*, **302**, 496–501.

63 Goldman, E.R., Anderson, G.P., Tran, P.T., Mattoussi, H., Charles, P.T. and Mauro, J.M. (2002) Conjugation of luminescent quantum dots with antibodies using an engineered adaptor protein to provide new reagents for fluoroimmunoassays. *Analytical Chemistry*, **74**, 841–7.

64 Parak, W.J., Pellegrino, T. and Plank, C. (2005) Labelling of cells with quantum dots. *Nanotechnology*, **16**, R9–25.

65 Ghazani, A.A., Lee, J.A., Klostranec, J., Xiang, Q., Dacosta, R.S., Wilson, B.C., Tsao, M.S. and Chan, W.C. (2006) High throughput quantification of protein expression of cancer antigens in tissue microarray using quantum dot nanocrystals. *Nano Letters*, **6**, 2881–6.

66 Xiao, Y., Gao, X., Gannot, G., Emmert-Buck, M.R., Srivastava, S., Wagner, P.D., Amos, M.D. and Barker, P.E. (2008) Quantitation of HER2 and telomerase biomarkers in solid tumors with IgY antibodies and nanocrystal detection. *International Journal of Cancer*, **122**, 2178–86.

67 Caldwell, M.L., Moffitt, R.A., Liu, J., Parry, R., Sharma, Y. and Wang, M.D. (2008) Simple quantification of multiplexed Quantum Dot staining in clinical tissue samples. *Conference Proceedings IEEE Engineering in Medicine and Biology Society*, **2008**, 1907–10.

68 Zajac, A., Song, D., Qian, W. and Zhukov, T. (2007) Protein microarrays and quantum dot probes for early cancer detection. *Colloids and Surfaces B, Biointerfaces*, **58**, 309–14.

69 Gokarna, A., Jin, L.H., Hwang, J.S., Cho, Y.H., Lim, Y.T., Chung, B.H., Youn, S.H., Choi, D.S. and Lim, J.H. (2008)

Quantum dot-based protein micro- and nanoarrays for detection of prostate cancer biomarkers. *Proteomics*, **8**, 1809–18.
70 Li-Shishido, S., Watanabe, T.M., Tada, H., Higuchi, H. and Ohuchi, N. (2006) Reduction in nonfluorescence state of quantum dots on an immunofluorescence staining. *Biochemical and Biophysical Research Communications*, **351**, 7–13.
71 Itoh, J. and Osamura, R.Y. (2007) Quantum dots for multicolor tumor pathology and multispectral imaging. *Methods in Molecular Biology*, **374**, 29–42.
72 Sun, J., Zhu, M.Q., Fu, K., Lewinski, N. and Drezek, R.A. (2007) Lead sulfide near-infrared quantum dot bioconjugates for targeted molecular imaging. *International Journal of Nanomedicine*, **2**, 235–40.
73 Xu, H., Peng, J., Tang, H.W., Li, Y., Wu, Q.S., Zhang, Z.L., Zhou, G., Chen, C. and Li, Y. (2009) Hadamard transform spectral microscopy for single cell imaging using organic and quantum dot fluorescent probes. *Analyst*, **134**, 504–11.
74 Wang, H.Z., Wang, H.Y., Liang, R.Q. and Ruan, K.C. (2004) Detection of tumor marker CA125 in ovarian carcinoma using quantum dots. *Acta Biochimica et Biophysica Sinica (Shanghai)*, **36**, 681–6.
75 Shi, C., Zhou, G., Zhu, Y., Su, Y., Cheng, T., Zhau, H.E. and Chung, L.W. (2008) Quantum dots-based multiplexed immunohistochemistry of protein expression in human prostate cancer cells. *European Journal of Histochemistry*, **52**, 127–34.
76 Bostick, R.M., Kong, K.Y., Ahearn, T.U., Chaudry, Q., Cohen, V. and Wang, M.D. (2006) Detecting and quantifying biomarkers of risk for colorectal cancer using quantum dots and novel image analysis algorithms. *Conference Proceedings IEEE Engineering in Medicine and Biology Society*, **1**, 3313–16.
77 Fountaine, T.J., Wincovitch, S.M., Geho, D.H., Garfield, S.H. and Pittaluga, S. (2006) Multispectral imaging of clinically relevant cellular targets in tonsil and lymphoid tissue using semiconductor quantum dots. *Modern Pathology*, **19**, 1181–191.
78 Sweeney, E., Ward, T.H., Gray, N., Womack, C., Jayson, G., Hughes, A., Dive, C. and Byers, R. (2008) Quantitative multiplexed quantum dot immunohistochemistry. *Biochemical and Biophysical Research Communications*, **374**, 181–6.
79 Kaul, Z., Yaguchi, T., Kaul, S.C., Hirano, T., Wadhwa, R. and Taira, K. (2003) Mortalin imaging in normal and cancer cells with quantum dot immuno-conjugates. *Cell Research*, **13**, 503–7.
80 Li, Z., Wang, K., Tan, W., Li, J., Fu, Z., Ma, C., Li, H., He, X. and Liu, J. (2006) Immunofluorescent labeling of cancer cells with quantum dots synthesized in aqueous solution. *Analytical Biochemistry*, **354**, 169–74.
81 Yang, D., Chen, Q., Wang, W. and Xu, S. (2008) Direct and indirect immunolabelling of HeLa cells with quantum dots. *Luminescence*, **23**, 169–74.
82 Chatterjee, U., Bose, P.P., Dey, S., Singh, T.P. and Chatterjee, B.P. (2008) Antiproliferative effect of T/Tn specific Artocarpus lakoocha agglutinin (ALA) on human leukemic cells (Jurkat, U937, K562) and their imaging by QD-ALA nanoconjugate. *Glycoconjugate Journal*, **25**, 741–52.
83 Kaul, Z., Yaguchi, T., Kaul, S.C. and Wadhwa, R. (2006) Quantum dot-based protein imaging and functional significance of two mitochondrial chaperones in cellular senescence and carcinogenesis. *Annals of the New York Academy of Sciences*, **1067**, 469–73.
84 Xiao, Y., Gao, X., Maragh, S., Telford, W.G. and Tona, A. (2009) Cell lines as candidate reference materials for quality control of HER2 amplification and expression assays in breast cancer. *Clinical Chemistry*, **55**, 1307–15.
85 Karwa, A., Papazoglou, E., Pourrezaei, K., Tyagi, S. and Murthy, S. (2007) Imaging biomarkers of inflammation in situ with functionalized quantum dots in the dextran sodium sulfate (DSS) model of mouse colitis. *Inflammation Research*, **56**, 502–10.
86 Xiao, Y., Telford, W.G., Ball, J.C., Locascio, L.E. and Barker, P.E. (2005)

Semiconductor nanocrystal conjugates, FISH and pH. *Nature Methods*, **2**, 723.

87 Xiao, Y. and Barker, P.E. (2004) Semiconductor nanocrystal probes for human metaphase chromosomes. *Nucleic Acids Research*, **32**, e28.

88 Chan, P., Yuen, T., Ruf, F., Gonzalez-Maeso, J. and Sealfon, S.C. (2005) Method for multiplex cellular detection of mRNAs using quantum dot fluorescent in situ hybridization. *Nucleic Acids Research*, **33**, e161.

89 Bentolila, L.A. and Weiss, S. (2006) Single-step multicolor fluorescence in situ hybridization using semiconductor quantum dot-DNA conjugates. *Cell Biochemistry and Biophysics*, **45**, 59–70.

90 Jiang, Z., Li, R., Todd, N.W., Stass, S.A. and Jiang, F. (2007) Detecting genomic aberrations by fluorescence in situ hybridization with quantum dots-labeled probes. *Journal of Nanoscience and Nanotechnology*, **7**, 4254–9.

91 Byers, R.J., Di Vizio, D., O'Connell, F., Tholouli, E., Levenson, R.M., Gossage, K., Twomey, D., Yang, Y., Benedettini, E., Rose, J., Ligon, K.L., Finn, S.P., Golub, T.R. and Loda, M. (2007) Semiautomated multiplexed quantum dot-based in situ hybridization and spectral deconvolution. *The Journal of Molecular Diagnostics*, **9**, 20–9.

92 Minet, O., Dressler, C. and Beuthan, J. (2004) Heat stress induced redistribution of fluorescent quantum dots in breast tumor cells. *Journal of Fluorescence*, **14**, 241–7.

93 Nida, D.L., Rahman, M.S., Carlson, K.D., Richards-Kortum, R. and Follen, M. (2005) Fluorescent nanocrystals for use in early cervical cancer detection. *Gynecologic Oncology*, **99**, S89–94.

94 Rahman, M., Abd-El-Barr, M., Mack, V., Tkaczyk, T., Sokolov, K., Richards-Kortum, R. and Descour, M. (2005) Optical imaging of cervical pre-cancers with structured illumination: an integrated approach. *Gynecologic Oncology*, **99**, S112–15.

95 Lidke, D.S., Lidke, K.A., Rieger, B., Jovin, T.M. and Arndt-Jovin, D.J. (2005) Reaching out for signals: filopodia sense EGF and respond by directed retrograde transport of activated receptors. *The Journal of Cell Biology*, **170**, 619–26.

96 Lidke, D.S., Nagy, P., Jovin, T.M. and Arndt-Jovin, D.J. (2007) Biotin-ligand complexes with streptavidin quantum dots for *in vivo* cell labeling of membrane receptors. *Methods in Molecular Biology*, **374**, 69–79.

97 Young, S.H. and Rozengurt, E. (2006) Qdot nanocrystal conjugates conjugated to bombesin or ANG II label the cognate G protein-coupled receptor in living cells. *American Journal of Physiology Cell Physiology*, **290**, C728–32.

98 Chu, T.C., Shieh, F., Lavery, L.A., Levy, M., Richards-Kortum, R., Korgel, B.A. and Ellington, A.D. (2006) Labeling tumor cells with fluorescent nanocrystal-aptamer bioconjugates. *Biosensors & Bioelectronics*, **21**, 1859–66.

99 Liu, T.C., Wang, J.H., Wang, H.Q., Zhang, H.L., Zhang, Z.H., Hua, X.F., Cao, Y.C., Zhao, Y.D. and Luo, Q.M. (2007) Bioconjugate recognition molecules to quantum dots as tumor probes. *Journal of Biomedical Materials Research Part A*, **83**, 1209–16.

100 Cheng, C.M., Chu, P.Y., Chuang, K.H., Roffler, S.R., Kao, C.H., Tseng, W.L., Shiea, J., Chang, W.D., Su, Y.C., Chen, B.M., Wang, Y.M. and Cheng, T.L. (2009) Hapten-derivatized nanoparticle targeting and imaging of gene expression by multimodality imaging systems. *Cancer Gene Therapy*, **16**, 83–90.

101 Grecco, H.E., Lidke, K.A., Heintzmann, R., Lidke, D.S., Spagnuolo, C., Martinez, O.E., Jares-Erijman, E.A. and Jovin, T.M. (2004) Ensemble and single particle photophysical properties (two-photon excitation, anisotropy, FRET, lifetime, spectral conversion) of commercial quantum dots in solution and in live cells. *Microscopy Research and Technique*, **65**, 169–79.

102 Watanabe, T.M. and Higuchi, H. (2007) Stepwise movements in vesicle transport of HER2 by motor proteins in living cells. *Biophysical Journal*, **92**, 4109–20.

103 Qian, J., Yong, K.T., Roy, I., Ohulchanskyy, T.Y., Bergey, E.J., Lee, H.H., Tramposch, K.M., He, S., Maitra, A. and Prasad, P.N. (2007) Imaging

pancreatic cancer using surface-functionalized quantum dots. *The Journal of Physical Chemistry B*, **111**, 6969–72.
104 Watanabe, T.M., Sato, T., Gonda, K. and Higuchi, H. (2007) Three-dimensional nanometry of vesicle transport in living cells using dual-focus imaging optics. *Biochemical and Biophysical Research Communications*, **359**, 1–7.
105 Erogbogbo, F., Yong, K.T., Roy, I., Xu, G., Prasad, P.N. and Swihart, M.T. (2008) Biocompatible luminescent silicon quantum dots for imaging of cancer cells. *ACS Nano*, **2**, 873–8.
106 Tekle, C., Deurs, B., Sandvig, K. and Iversen, T.G. (2008) Cellular trafficking of quantum dot-ligand bioconjugates and their induction of changes in normal routing of unconjugated ligands. *Nano Letters*, **8**, 1858–65.
107 Weng, K.C., Noble, C.O., Papahadjopoulos-Sternberg, B., Chen, F.F., Drummond, D.C., Kirpotin, D.B., Wang, D., Hom, Y.K., Hann, B. and Park, J.W. (2008) Targeted tumor cell internalization and imaging of multifunctional quantum dot-conjugated immunoliposomes *in vitro* and *in vivo*. *Nano Letters*, **8**, 2851–7.
108 Schroeder, J.E., Shweky, I., Shmeeda, H., Banin, U. and Gabizon, A. (2007) Folate-mediated tumor cell uptake of quantum dots entrapped in lipid nanoparticles. *Journal of Controlled Release*, **124**, 28–34.
109 Pellegrino, T., Parak, W.J., Boudreau, R., Le Gros, M.A., Gerion, D., Alivisatos, A.P. and Larabell, C.A. (2003) Quantum dot-based cell motility assay. *Differentiation*, **71**, 542–8.
110 Gu, W., Pellegrino, T., Parak, W.J., Boudreau, R., Le Gros, M.A., Gerion, D., Alivisatos, A.P. and Larabell, C.A. (2005) Quantum-dot-based cell motility assay. *Science's STKE*, **2005**, l5.
111 Gu, W., Pellegrino, T., Parak, W.J., Boudreau, R., Le Gros, M.A., Alivisatos, A.P. and Larabell, C.A. (2007) Measuring cell motility using quantum dot probes. *Methods in Molecular Biology*, **374**, 125–31.
112 Clift, M.J., Rothen-Rutishauser, B., Brown, D.M., Duffin, R., Donaldson, K., Proudfoot, L., Guy, K. and Stone, V. (2008) The impact of different nanoparticle surface chemistry and size on uptake and toxicity in a murine macrophage cell line. *Toxicology and Applied Pharmacology*, **232**, 418–27.
113 Barua, S. and Rege, K. (2009) Cancer-cell-phenotype-dependent differential intracellular trafficking of unconjugated quantum dots. *Small*, **5**, 370–6.
114 Voura, E.B., Jaiswal, J.K., Mattoussi, H. and Simon, S.M. (2004) Tracking metastatic tumor cell extravasation with quantum dot nanocrystals and fluorescence emission-scanning microscopy, *Nature Medicine*, **10**, 993–8.
115 Chang, C.F., Chen, C.Y., Chang, F.H., Tai, S.P., Chen, C.Y., Yu, C.H., Tseng, Y.B., Tsai, T.H., Liu, I.S., Su, W.F. and Sun, C.K. (2008) Cell tracking and detection of molecular expression in live cells using lipid-enclosed CdSe quantum dots as contrast agents for epi-third harmonic generation microscopy. *Optics Express*, **16**, 9534–48.
116 Yoo, J., Kambara, T., Gonda, K. and Higuchi, H. (2008) Intracellular imaging of targeted proteins labeled with quantum dots. *Experimental Cell Research*, **314**, 3563–9.
117 Nan, X., Sims, P.A., Chen, P. and Xie, X.S. (2005) Observation of individual microtubule motor steps in living cells with endocytosed quantum dots. *The Journal of Physical Chemistry B*, **109**, 24220–4.
118 Wylie, P.G. (2007) Multiple cell lines using quantum dots. *Methods in Molecular Biology*, **374**, 113–23.
119 Xue, F.L., Chen, J.Y., Guo, J., Wang, C.C., Yang, W.L., Wang, P.N. and Lu, D.R. (2007) Enhancement of intracellular delivery of CdTe quantum dots (QDs) to living cells by Tat conjugation. *Journal of Fluorescence*, **17**, 149–54.
120 Garon, E.B., Marcu, L., Luong, Q., Tcherniantchouk, O., Crooks, G.M. and Koeffler, H.P. (2007) Quantum dot labeling and tracking of human leukemic, bone marrow and cord blood cells. *Leukemia Research*, **31**, 643–51.
121 Chen, B., Liu, Q., Zhang, Y., Xu, L. and Fang, X. (2008) Transmembrane delivery of the cell-penetrating peptide

conjugated semiconductor quantum dots. *Langmuir*, **24**, 11866–71.

122 Chen, A.K., Behlke, M.A. and Tsourkas, A. (2007) Avoiding false-positive signals with nuclease-vulnerable molecular beacons in single living cells. *Nucleic Acids Research*, **35**, e105.

123 Kaul, Z., Yaguchi, T., Harada, J.I., Ikeda, Y., Hirano, T., Chiura, H.X., Kaul, S.C. and Wadhwa, R. (2007) An antibody-conjugated internalizing quantum dot suitable for long-term live imaging of cells. *Biochemistry and Cell Biology*, **85**, 133–40.

124 Kaul, Z., Yaguchi, T., Chiura, H.X., Kaul, S.C. and Wadhwa, R. (2007) Quantum dot-based mortalin staining as a visual assay for detection of induced senescence in cancer cells. *Annals of the New York Academy of Sciences*, **1100**, 368–72.

125 Chen, L.D., Liu, J., Yu, X.F., He, M., Pei, X.F., Tang, Z.Y., Wang, Q.Q., Pang, D.W. and Li, Y. (2008) The biocompatibility of quantum dot probes used for the targeted imaging of hepatocellular carcinoma metastasis. *Biomaterials*, **29**, 4170–6.

126 Bidlingmaier, S., He, J., Wang, Y., An, F., Feng, J., Barbone, D., Gao, D., Franc, B., Broaddus, V.C. and Liu, B. (2009) Identification of MCAM/CD146 as the target antigen of a human monoclonal antibody that recognizes both epithelioid and sarcomatoid types of mesothelioma. *Cancer Research*, **69**, 1570–7.

127 Kim, B.Y., Jiang, W., Oreopoulos, J., Yip, C.M., Rutka, J.T. and Chan, W.C. (2008) Biodegradable quantum dot nanocomposites enable live cell labeling and imaging of cytoplasmic targets. *Nano Letters*, **8**, 3887–92.

128 Yezhelyev, M.V., Qi, L., O'Regan, R.M., Nie, S. and Gao, X. (2008) Proton-sponge coated quantum dots for siRNA delivery and intracellular imaging. *Journal of the American Chemical Society*, **130**, 9006–12.

129 Zhang, Y., So, M.K. and Rao, J. (2006) Protease-modulated cellular uptake of quantum dots. *Nano Letters*, **6**, 1988–92.

130 Stroh, M., Zimmer, J.P., Duda, D.G., Levchenko, T.S., Cohen, K.S., Brown, E.B., Scadden, D.T., Torchilin, V.P., Bawendi, M.G., Fukumura, D. and Jain, R.K. (2005) Quantum dots spectrally distinguish multiple species within the tumor milieu *in vivo*. *Nature Medicine*, **11**, 678–82.

131 Estrada, C.R., Salanga, M., Bielenberg, D.R., Harrell, W.B., Zurakowski, D., Zhu, X., Palmer, M.R., Freeman, M.R. and Adam, R.M. (2006) Behavioral profiling of human transitional cell carcinoma ex vivo. *Cancer Research*, **66**, 3078–86.

132 Morgan, N.Y., English, S., Chen, W., Chernomordik, V., Russo, A., Smith, P.D. and Gandjbakhche, A. (2005) Real time *in vivo* non-invasive optical imaging using near-infrared fluorescent quantum dots. *Academic Radiology*, **12**, 313–23.

133 Cai, W., Shin, D.W., Chen, K., Gheysens, O., Cao, Q., Wang, S.X., Gambhir, S.S. and Chen, X. (2006) Peptide-labeled near-infrared quantum dots for imaging tumor vasculature in living subjects. *Nano Letters*, **6**, 669–76.

134 Smith, B.R., Cheng, Z., De, A., Koh, A.L., Sinclair, R. and Gambhir, S.S. (2008) Real-time intravital imaging of RGD-quantum dot binding to luminal endothelium in mouse tumor neovasculature. *Nano Letters*, **8**, 2599–606.

135 Tanaka, E., Choi, H.S., Fujii, H., Bawendi, M.G. and Frangioni, J.V. (2006) Image-guided oncologic surgery using invisible light: completed pre-clinical development for sentinel lymph node mapping. *Annals of Surgical Oncology*, **13**, 1671–81.

136 Ballou, B., Ernst, L.A., Andreko, S., Harper, T., Fitzpatrick, J.A., Waggoner, A.S. and Bruchez, M.P. (2007) Sentinel lymph node imaging using quantum dots in mouse tumor models. *Bioconjugate Chemistry*, **18**, 389–96.

137 Takeda, M., Tada, H., Higuchi, H., Kobayashi, Y., Kobayashi, M., Sakurai, Y., Ishida, T. and Ohuchi, N. (2008) *In vivo* single molecular imaging and sentinel node navigation by nanotechnology for molecular targeting drug-delivery systems and tailor-made medicine. *Breast Cancer*, **15**, 145–52.

138 Hama, Y., Koyama, Y., Urano, Y., Choyke, P.L. and Kobayashi, H. (2007) Simultaneous two-color spectral fluorescence lymphangiography with near infrared quantum dots to map two lymphatic flows from the breast and the upper extremity. *Breast Cancer Research Treatment*, **103**, 23–8.

139 Kobayashi, H., Hama, Y., Koyama, Y., Barrett, T., Regino, C.A., Urano, Y. and Choyke, P.L. (2007) Simultaneous multicolor imaging of five different lymphatic basins using quantum dots. *Nano Letters*, **7**, 1711–16.

140 Yang, L., Mao, H., Wang, Y.A., Cao, Z., Peng, X., Wang, X., Duan, H., Ni, C., Yuan, Q., Adams, G., Smith, M.Q., Wood, W.C., Gao, X. and Nie, S. (2009) Single chain epidermal growth factor receptor antibody conjugated nanoparticles for *in vivo* tumor targeting and imaging. *Small*, **5**, 235–43.

141 Yu, X., Chen, L., Li, K., Li, Y., Xiao, S., Luo, X., Liu, J., Zhou, L., Deng, Y., Pang, D. and Wang, Q. (2007) Immunofluorescence detection with quantum dot bioconjugates for hepatoma *in vivo*. *Journal of Biomedical Optics*, **12**, 014008.

142 Tada, H., Higuchi, H., Wanatabe, T.M. and Ohuchi, N. (2007) In vivo real-time tracking of single quantum dots conjugated with monoclonal anti-HER2 antibody in tumors of mice. *Cancer Research*, **67**, 1138–44.

143 Diagaradjane, P., Orenstein-Cardona, J.M., Colón-Casasnovas, N.E., Deorukhkar, A., Shentu, S., Kuno, N., Schwartz, D.L., Gelovani, J.G. and Krishnan, S. (2008) Imaging epidermal growth factor receptor expression *in vivo*: pharmacokinetic and biodistribution characterization of a bioconjugated quantum dot nanoprobe. *Clinical Cancer Research*, **14**, 731–41.

144 Muhammad, O., Popescu, A. and Toms, S.A. (2007) Macrophage-mediated colocalization of quantum dots in experimental glioma. *Methods in Molecular Biology*, **374**, 161–71.

145 Cai, W., Chen, K., Li, Z.B., Gambhir, S.S. and Chen, X. (2007) Dual-function probe for PET and near-infrared fluorescence imaging of tumor vasculature. *Journal of Nuclear Medicine*, **48**, 1862–70.

146 Chen, K., Li, Z.B., Wang, H., Cai, W. and Chen, X. (2008) Dual-modality optical and positron emission tomography imaging of vascular endothelial growth factor receptor on tumor vasculature using quantum dots. *European Journal of Nuclear Medicine and Molecular Imaging*, **35**, 2235–44.

147 van Tilborg, G.A., Mulder, W.J., Chin, P.T., Storm, G., Reutelingsperger, C.P., Nicolay, K. and Strijkers, G.J. (2006) Annexin A5-conjugated quantum dots with a paramagnetic lipidic coating for the multimodal detection of apoptotic cells. *Bioconjugate Chemistry*, **17**, 865–8.

148 Oostendorp, M., Douma, K., Hackeng, T.M., Dirksen, A., Post, M.J., van Zandvoort, M.A. and Backes, W.H. (2008) Quantitative molecular magnetic resonance imaging of tumor angiogenesis using cNGR-labeled paramagnetic quantum dots. *Cancer Research*, **68**, 7676–83.

149 Bagalkot, V., Zhang, L., Levy-Nissenbaum, E., Jon, S., Kantoff, P.W., Langer, R. and Farokhzad, O.C. (2007) Quantum dot-aptamer conjugates for synchronous cancer imaging, therapy, and sensing of drug delivery based on bi-fluorescence resonance energy transfer. *Nano Letters*, **7**, 3065–70.

150 Derfus, A.M., Chen, A.A., Min, D.H., Ruoslahti, E. and Bhatia, S.N. (2007) Targeted quantum dot conjugates for siRNA delivery. *Bioconjugate Chemistry*, **18**, 1391–6.

151 Pan, J. and Feng, S.S. (2009) Targeting and imaging cancer cells by folate-decorated, quantum dots (QDs)-loaded nanoparticles of biodegradable polymers. *Biomaterials*, **30**, 1176–83.

152 Hyun, B.R., Chen, H., Rey, D.A., Wise, F.W. and Batt, C.A. (2007) Near-infrared fluorescence imaging with water-soluble lead salt quantum dots. *The Journal of Physical Chemistry B*, **111**, 5726–30.

153 Liu, S., Lee, C.M., Wang, S. and Lu, D.R. (2006) A new bioimaging carrier for fluorescent quantum dots: phospholipid nanoemulsion mimicking natural lipoprotein core. *Drug Delivery*, **13**, 159–64.

154 Ballou, B., Ernst, L.A. and Waggoner, A.S. (2005) Fluorescence imaging of tumors *in vivo*. *Current Medicinal Chemistry*, **12**, 795–805.

155 Rao, J., Dragulescu-Andrasi, A. and Yao, H. (2007) Fluorescence imaging *in vivo*: recent advances. *Current Opinion in Biotechnology*, **18**, 17–25.

156 Cheong, W.F., Prahl, S.A. and Welch, A.J. (1990) A review of the optical properties of biological tissues. *IEEE Journal of Quantum Electronics*, **26**, 2166–95.

157 Weissleder, R. (2001) A clearer vision for *in vivo* imaging. *Nature Biotechnology*, **19**, 316–17.

158 Mansfield, J.R., Gossage, K.W., Hoyt, C.C. and Levenson, R.M. (2005) Autofluorescence removal, multiplexing, and automated analysis methods for in-vivo fluorescence imaging. *Journal of Biomedical Optics*, **10**, 41207.

159 Gao, X. (2007) Multifunctional quantum dots for cellular and molecular imaging. *Conference Proceedings IEEE Engineering in Medicine and Biology Society*, **2007**, 524–5.

160 True, L.D. and Gao, X. (2007) Quantum dots for molecular pathology: their time has arrived. *The Journal of Molecular Diagnostics*, **9**, 7–11.

161 Jain, R.K. and Duda, D.G. (2003) Role of bone marrow-derived cells in tumor angiogenesis and treatment. *Cancer Cell*, **3**, 515–16.

162 Das, B., Tsuchida, R., Malkin, D., Koren, G., Baruchel, S. and Yeger, H. (2008) Hypoxia enhances tumor stemness by increasing the invasive and tumorigenic side population fraction. *Stem Cells*, **26**, 1818–30.

163 Duan, H. and Nie, S. (2007) Cell-penetrating quantum dots based on multivalent and endosome-disrupting surface coatings. *Journal of the American Chemical Society*, **129**, 3333–8.

164 Al-Jamal, W.T., Al-Jamal, K.T., Bomans, P.H., Frederik, P.M. and Kostarelos, K. (2008) Functionalized-quantum-dot-liposome hybrids as multimodal nanoparticles for cancer. *Small*, **4**, 1406–15.

165 Larson, D.R., Zipfel, W.R., Williams, R.M., Clark, S.W., Bruchez, M.P., Wise, F.W. and Webb, W.W. (2003) Water-soluble quantum dots for multiphoton fluorescence imaging *in vivo*. *Science*, **300**, 1434–6.

166 Lim, Y.T., Kim, S., Nakayama, A., Stott, N.E., Bawendi, M.G. and Frangioni, J.V. (2003) Selection of quantum dot wavelengths for biomedical assays and imaging. *Molecular Imaging*, **2**, 50–64.

167 Smith, J.D., Fisher, G.W., Waggoner, A.S. and Campbell, P.G. (2007) The use of quantum dots for analysis of chick CAM vasculature. *Microvascular Research*, **73**, 75–83.

168 Jayagopal, A., Russ, P.K. and Haselton, F.R. (2007) Surface engineering of quantum dots for *in vivo* vascular imaging. *Bioconjugate Chemistry*, **18**, 1424–33.

169 Cai, W. and Chen, X. (2008) Preparation of peptide-conjugated quantum dots for tumor vasculature-targeted imaging. *Nature Protocols*, **3**, 89–96.

170 Curnis, F., Sacchi, A., Gasparri, A., Longhi, R., Bachi, A., Doglioni, C., Bordignon, C., Traversari, C., Rizzardi, G.P. and Corti, A. (2008) Isoaspartate-glycine-arginine a new tumor vasculature-targeting motif. *Cancer Research*, **68**, 7073–82.

171 Schulze, T., Bembenek, A. and Schlag, P.M. (2004) Sentinel lymph node biopsy progress in surgical treatment of cancer. *Langenbeck's Archives of Surgery*, **389**, 532–50.

172 Morton, D.L., Wen, D.R., Wong, J.H., Economou, J.S., Cagle, L.A., Storm, F.K., Foshag, L.J. and Cochran, A.J. (1992) Technical details of intraoperative lymphatic mapping for early stage melanoma. *Archives of Surgery*, **127**, 392–9.

173 Kim, S., Lim, Y.T., Soltesz, E.G., De Grand, A.M., Lee, J., Nakayama, A., Parker, J.A., Mihaljevic, T., Laurence, R.G., Dor, D.M., Cohn, L.H., Bawendi, M.G. and Frangioni, J.V. (2004) Near-infrared fluorescent type II quantum dots for sentinel lymph node mapping. *Nature Biotechnology*, **22**, 93–7.

174 Soltesz, E.G., Kim, S., Laurence, R.G., De Grand, A.M., Parungo, C.P., Dor, D.M., Cohn, L.H., Bawendi, M.G., Frangioni, J.V. and Mihaljevic, T. (2005)

Intraoperative sentinel lymph node mapping of the lung using near-infrared fluorescent quantum dots. *The Annals of Thoracic Surgery*, **79**, 269–77.

175 Parungo, C.P., Ohnishi, S., Kim, S.W., Kim, S., Laurence, R.G., Soltesz, E.G., Chen, F.Y., Colson, Y.L., Cohn, L.H., Bawendi, M.G. and Frangioni, J.V. (2005) Intraoperative identification of esophageal sentinel lymph nodes with near-infrared fluorescence imaging. *The Journal of Thoracic and Cardiovascular Surgery*, **129**, 844–50.

176 Parungo, C.P., Colson, Y.L., Kim, S.W., Kim, S., Cohn, L.H., Bawendi, M.G. and Frangioni, J.V. (2005) Sentinel lymph node mapping of the pleural space. *Chest*, **127**, 1799–804.

177 Ohnishi, S., Lomnes, S.J., Laurence, R.G., Gogbashian, A., Mariani, G. and Frangioni, J.V. (2005) Organic alternatives to quantum dots for intraoperative near-infrared fluorescent sentinel lymph node mapping. *Molecular Imaging*, **4**, 172–81.

178 Soltesz, E.G., Kim, S., Kim, S.W., Laurence, R.G., De Grand, A.M., Parungo, C.P., Cohn, L.H., Bawendi, M.G. and Frangioni, J.V. (2006) Sentinel lymph node mapping of the gastrointestinal tract by using invisible light. *Annals of Surgical Oncology*, **13**, 386–96.

179 Robe, A., Pic, E., Lassalle, H.P., Bezdetnaya, L., Guillemin, F. and Marchal, F. (2008) Quantum dots in axillary lymph node mapping: biodistribution study in healthy mice. *BMC Cancer*, **8**, 111.

180 Maeda, H., Wu, J., Sawa, T., Matsumura, Y. and Hori, K. (2000) Tumor vascular permeability and the EPR effect in macromolecular therapeutics: a review. *Journal of Controlled Release*, **65**, 271–84.

181 Gao, X., Chung, L.W. and Nie, S. (2007) Quantum dots for *in vivo* molecular and cellular imaging. *Methods in Molecular Biology*, **374**, 135–45.

182 Inoue, Y., Izawa, K., Yoshikawa, K., Yamada, H., Tojo, A. and Ohtomo, K. (2007) *In vivo* fluorescence imaging of the reticuloendothelial system using quantum dots in combination with bioluminescent tumour monitoring. *European Journal of Nuclear Medicine and Molecular Imaging*, **34**, 2048–56.

183 Schipper, M.L., Iyer, G., Koh, A.L., Cheng, Z., Ebenstein, Y., Aharoni, A., Keren, S., Bentolila, L.A., Li, J., Rao, J., Chen, X., Banin, U., Wu, A.M., Sinclair, R., Weiss, S. and Gambhir, S.S. (2009) Particle size, surface coating, and PEGylation influence the biodistribution of quantum dots in living mice. *Small*, **5**, 126–34.

184 Corot, C., Petry, K.G., Trivedi, R., Saleh, A., Jonkmanns, C., Le Bas, J.F., Blezer, E., Rausch, M., Brochet, B., Foster-Gareau, P., Balériaux, D., Gaillard, S. and Dousset, V. (2004) Macrophage imaging in central nervous system and in carotid atherosclerotic plaque using ultrasmall superparamagnetic iron oxide in magnetic resonance imaging. *Investigative Radiology*, **39**, 619–25.

185 Devi, G.R. (2006) siRNA-based approaches in cancer therapy. *Cancer Gene Therapy*, **13**, 819–29.

186 Hoet, P.H., Brüske-Hohlfeld, I. and Salata, O.V. (2004) Nanoparticles – known and unknown health risks. *Journal of Nanobiotechnology*, **2**, 12.

187 Singh, S. and Nalwa, H.S. (2007) Nanotechnology and health safety – toxicity and risk assessments of nanostructured materials on human health. *Journal of Nanoscience and Nanotechnology*, **7**, 3048–70.

188 Hardman, R. (2006) A toxicologic review of quantum dots: toxicity depends on physicochemical and environmental factors. *Environmental Health Perspectives*, **114**, 165–72.

189 Derfus, A.M., Chan, W.C. and Bhatia, S.N. (2004) Probing the cytotoxicity of semiconductor quantum dots. *Nano Letters*, **4**, 11–18.

190 Pradhan, N., Goorskey, D., Thessing, J. and Peng, X. (2005) An alternative of CdSe nanocrystal emitters: pure and tunable impurity emissions in ZnSe nanocrystals. *Journal of the American Chemical Society*, **127**, 17586–7.

191 Yu, W.W., Chang, E., Drezek, R. and Colvin, V.L. (2006) Water-soluble quantum dots for biomedical applications. *Biochemical and Biophysical Research Communications*, **348**, 781–6.

192 Ballou, B., Lagerholm, B.C., Ernst, L.A., Bruchez, M.P. and Waggoner, A.S. (2004) Noninvasive imaging of quantum dots in mice. *Bioconjugate Chemistry*, **15**, 79–86.

193 Hoshino, A., Fujioka, K., Oku, T., Suga, M., Sasaki, Y.F., Ohta, T., Yasuhara, M., Suzuki, K. and Yamamoto, K. (2004) Physicochemical properties and cellular toxicity of nanocrystal quantum dots depend on their surface modification. *Nano Letters*, **4**, 2163–9.

194 Aldana, J., Wang, Y.A. and Peng, X. (2001) Photochemical instability of CdSe nanocrystals coated by hydrophilic thiols. *Journal of the American Chemical Society*, **123**, 8844–50.

195 Gerion, D., Pinaud, F., Williams, S.C., Parak, W.J., Zanchet, D., Weiss, S. and Alivisatos, A.P. (2001) Synthesis and properties of biocompatible water-soluble silica-coated CdSe/ZnS semiconductor quantum dots. *Journal of Physical Chemistry B*, **105**, 8861–71.

196 Zhang, T., Stilwell, J.L., Gerion, D., Ding, L., Elboudwarej, O., Cooke, P.A., Gray, J.W., Alivisatos, A.P. and Chen, F.F. (2006) Cellular effect of high doses of silica-coated quantum dot profiled with high throughput gene expression analysis and high content cellomics measurements. *Nano Letters*, **6**, 800–8.

197 Cho, S.J., Maysinger, D., Jain, M., Röder, B., Hackbarth, S. and Winnik, F.M. (2007) Long-term exposure to CdTe quantum dots causes functional impairments in live cells. *Langmuir*, **23**, 1974–80.

198 Lovri, J., Cho, S.J., Winnik, F.M. and Maysinger, D. (2005) Unmodified cadmium telluride quantum dots induce reactive oxygen species formation leading to multiple organelle damage and cell death. *Chemistry & Biology*, **12**, 1227–34.

199 Ipe, B.I., Lehnig, M. and Niemeyer, C.M. (2005) On the generation of free radical species from quantum dots. *Small*, **1**, 706–9.

200 Chan, W.H., Shiao, N.H. and Lu, P.Z. (2006) CdSe quantum dots induce apoptosis in human neuroblastoma cells via mitochondrial-dependent pathways and inhibition of survival signals. *Toxicology Letters*, **167**, 191–200.

201 Green, M. and Howman, E. (2005) Semiconductor quantum dots and free radical induced DNA nicking. *Chemical Communications (Cambridge, England)*, **1**, 121–3.

202 Ryman-Rasmussen, J.P., Riviere, J.E. and Monteiro-Riviere, N.A. (2007) Surface coatings determine cytotoxicity and irritation potential of quantum dot nanoparticles in epidermal keratinocytes. *Journal of Investigative Dermatology*, **127**, 143–53.

203 Wang, L., Nagesha, D.K., Selvarasah, S., Dokmeci, M.R. and Carrier, R.L. (2008) Toxicity of CdSe nanoparticles in Caco-2 cell cultures. *Journal of Nanobiotechnology*, **6**, 11.

204 Jaiswal, J.K., Goldman, E.R., Mattoussi, H. and Simon, S.M. (2004) Use of quantum dots for live cell imaging. *Nature Methods*, **1**, 73–8.

205 Shiohara, A., Hoshino, A., Hanaki, K., Suzuki, K. and Yamamoto, K. (2004) On the cyto-toxicity caused by quantum dots. *Microbiology and Immunology*, **48**, 669–75.

206 Fischer, H.C., Liu, L., Pang, K.S. and Chan, W.C.W. (2006) Pharmacokinetics of nanoscale quantum dots: *in vivo* distribution, sequestration, and clearance in the rat. *Advanced Functional Materials*, **16**, 1299–305.

207 Hsieh, S.C., Wang, F.F., Lin, C.S., Chen, Y.J., Hung, S.C. and Wang, Y.J. (2006) The inhibition of osteogenesis with human bone marrow mesenchymal stem cells by CdSe/ZnS quantum dot labels. *Biomaterials*, **27**, 1656–64.

208 Hsieh, S.C., Wang, F.F., Hung, S.C., Chen, Y.J. and Wang, Y.J. (2006) The internalized CdSe/ZnS quantum dots impair the chondrogenesis of bone marrow mesenchymal stem cells. *Journal of Biomedical Materials Research Part B, Applied Biomaterials*, **79**, 95–101.

209 Choi, A.O., Cho, S.J., Desbarats, J., Lovrić, J. and Maysinger, D. (2007) Quantum dot-induced cell death involves Fas upregulation and lipid peroxidation in human neuroblastoma cells. *Journal of Nanobiotechnology*, **5**, 1.

2
Quantum Dots for Targeted Tumor Imaging
Eue-Soon Jang and Xiaoyuan Chen

2.1
Introduction

Today, cancer has become one of the most serious global health threats with, according to the American Cancer Society, 7.6 million people having died from cancer worldwide during 2007 [1]. Cancer usually results from abnormal cell division caused by complex interactions between carcinogens and the host's genome, with such abnormalities in the cell division process often giving rise to the formation of tissue lumps called *tumors*. Tumors may be either benign (noncancerous) or malignant (cancerous). Malignant tumor tissue grows and invades all adjacent tissues and organs around the primary cancer site, thereby obstructing the normal physiological functions of the body. However, the cancerous cells gradually detach from the primary site and spread through the blood and lymphatic system to various parts of the body, forming new tumors; this process is known as cancer *metastasis* (see Figure 2.1) [2]. Eventually, the malignant tumors cause various organs of the body to start malfunctioning, leading to death.

The early diagnosis of cancer plays a crucial role in increasing the chance of survival, and thus a highly sensitive molecular imaging system is extremely important for diagnosing cancer. Among various imaging techniques, the resolution of fluorescence microscopy is much higher than that of positron emission tomography (PET), magnetic resonance imaging (MRI), and optical coherence tomography (OCT), as shown in Figure 2.2 [3]. In addition, fluorescence microscopy systems are rapid and the procedure is economic compared to other noninvasive imaging systems such as MRI, PET, computed tomography (CT), single photon emission CT (SPECT), and ultrasound scans, as shown in Table 2.1 [4]. In particular, fluorescence microscopic methods using quantum dots (QDs) have a high sensitivity in detecting abnormalities on the microscopic level.

In comparison with organic fluorophores, QDs have unique chemical and physical properties such as excellent photoresistance, chemical resistance, narrow symmetric fluorescence emission with broad absorption spectra, large effective excitation and emission Stokes shifts, and high quantum yield (as summarized in Tables 2.2 and 2.3 [5]).

Nanomaterials for the Life Sciences Vol.6: Semiconductor Nanomaterials.
Edited by Challa S. S. R. Kumar
Copyright © 2010 WILEY-VCH Verlag GmbH & Co. KGaA, Weinheim
ISBN: 978-3-527-32166-7

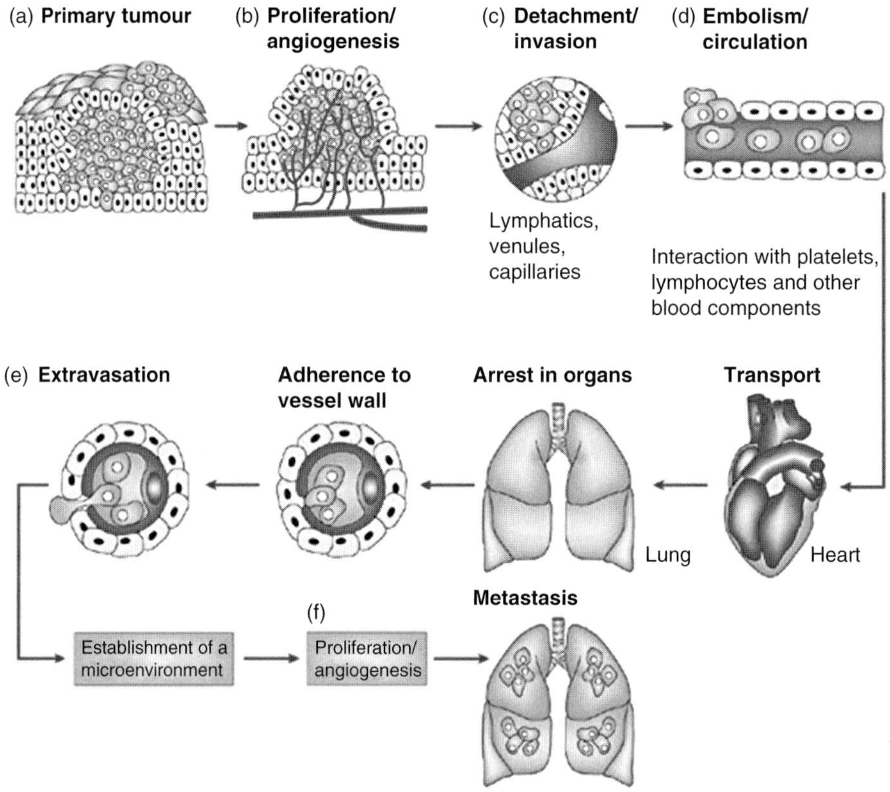

Figure 2.1 The main steps in the formation of a metastasis (adapted from Ref. [2]). (a) Cellular transformation and tumor growth; (b) Extensive vascularization must occur if a tumor mass is to exceed 1–2 mm in diameter; (c) Local invasion of the host stroma by some tumor cells occurs by several parallel mechanisms. Thin-walled venules, such as lymphatic channels, offer very little resistance to penetration by tumor cells and provide the most common route for tumor-cell entry into the circulation; (d) Detachment and embolization of single tumor cells or aggregates occurs next, with most circulating tumor cells being rapidly destroyed. After the tumor cells have survived the circulation, they become trapped in the capillary beds of distant organs by adhering either to capillary endothelial cells or to subendothelial basement membrane that might be exposed; (e) Extravasation occurs next, probably by mechanisms similar to those that operate during invasion; (f) Proliferation within the organ parenchyma completes the metastatic process. To continue growing, the micrometastasis must develop a vascular network and evade destruction by host defenses. The cells can then invade blood vessels, enter the circulation, and produce additional metastases. Reprinted with permission from Ref. [2]; © 2003, *Nature Reviews Cancer.*

2.2
Types of Quantum Dots (QDs)

When energy above a band gap is applied to semiconductors, electrons transfer from the valence to the conduction band, thereby creating electron-hole pairs known as *excitons*. The distance between the electron and the hole is called the

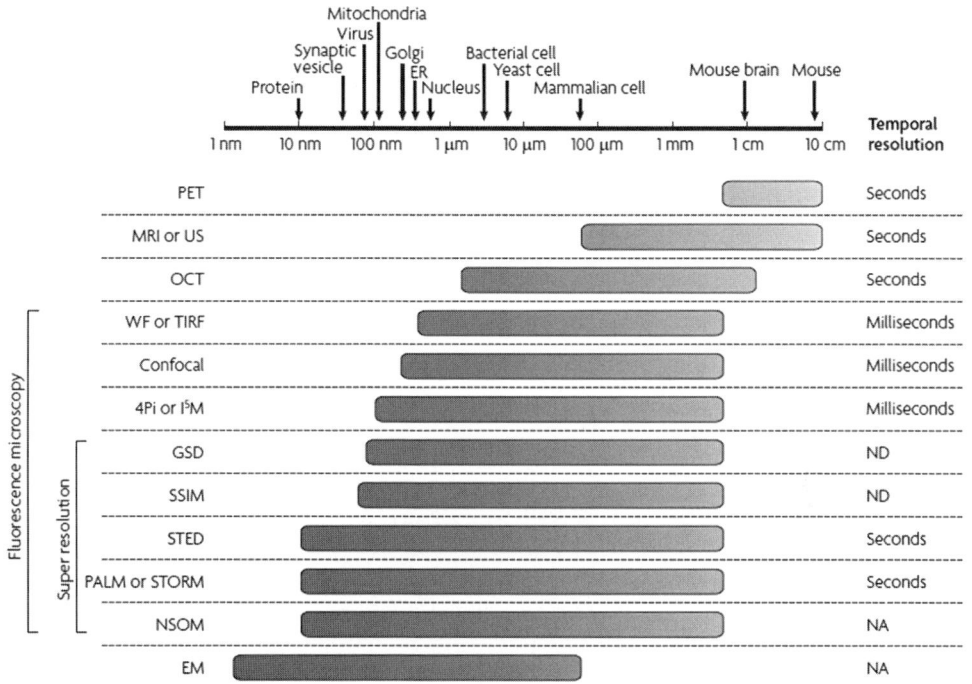

Technique	Spatial range	Temporal resolution
PET	1 mm – 10 cm	Seconds
MRI or US	100 μm – 10 cm	Seconds
OCT	1 μm – 10 cm	Seconds
WF or TIRF	100 nm – 1 cm	Milliseconds
Confocal	100 nm – 1 cm	Milliseconds
4Pi or I⁵M	100 nm – 1 cm	Milliseconds
GSD	10 nm – 1 cm	ND
SSIM	100 nm – 1 cm	ND
STED	10 nm – 1 cm	Seconds
PALM or STORM	10 nm – 1 cm	Seconds
NSOM	10 nm – 1 cm	NA
EM	1 nm – 1 mm	NA

Size references: Protein (1 nm), Synaptic vesicle, Virus, Mitochondria (10–100 nm), Golgi, ER, Nucleus (100 nm – 1 μm), Bacterial cell, Yeast cell, Mammalian cell (1–100 μm), Mouse brain (1 cm), Mouse (10 cm).

Figure 2.2 Comparison of the spatial and temporal resolutions of biological imaging techniques. The size scale is logarithmic. Average sizes of biological features are given; specific sizes vary widely among different species and cell lines. The spatial and temporal resolutions are estimates of current practices, and some were taken from Ref. [118]. The spatial resolution is given for the focal plane. The temporal resolution is not applicable (NA) for electron microscopy (EM) or near-field scanning optical microscopy (NSOM) because they image static samples. Ground-state depletion (GSD) and saturated structured-illumination microscopy (SSIM) have not been shown on biological samples, and thus their temporal resolutions are not determined (ND). ER, endoplasmic reticulum; MRI, magnetic resonance imaging; OCT, optical coherence tomography; PALM, photoactivated localization microscopy; PET, positron-emission microscopy; STED, stimulated emission depletion; STORM, stochastic optical reconstruction microscopy; TIRF, total internal reflection fluorescence; US, ultrasound; WF, wide-field microscopy. Reprinted with permission from Ref. [3]; © 2008, Nature Materials.

Bohr radius (a few nanometers) of the exciton. If the size of semiconductors is smaller than the Bohr exciton radius, then a quantum confinement effect is generated and the band gap is increased in inverse proportion to the radius of the nanocrystal. Semiconducting QDs are composed of a few hundred or a few thousand atoms, and therefore the energy band is quantized to discrete levels (see Figure 2.3a) [6]. Over the past two decades, the size-dependent optical properties of QDs have been studied extensively in a variety of semiconductor QDs, and the wavelengths of fluorescence emissions have been finely tuned from the ultraviolet (UV) to near-infrared (NIR) regions by size control (see Figure 2.3b) [7]. However, the ratio of emitted to absorbed photons (the quantum yield; QY) is seriously reduced by many surface defects of QDs resulting from trapping electrons or

Table 2.1 Overview of imaging systems.

Technique	Resolution[a]	Depth	Time[b]	Quantitative[c]	Multi-channel	Imaging agents	Target	Cost[a,d]	Main small-animal use	Clinical use
MRI	10–100 μm	No limit	Minutes to hours	Yes	No	Paramagnetic chelates, magnetic particles	Anatomical, physiological, molecular	$$$	Versatile imaging modality with high soft-tissue contrast	Yes
CT	50 μm	No limit	Minutes	Yes	No	Iodinated molecules	Anatomical, physiological	$$	Imaging lungs and bone	Yes
Ultrasound	50 μm	cm	Seconds to minutes	Yes	No	Microbubbles	Anatomical, physiological	$$	Vascular and interventional imaging[c]	Yes
PET	1–2 mm	No limit	Minutes to hours	Yes	No	^{18}F-, ^{64}CU- or ^{11}C-labeled compounds	Physiological, molecular	$$$	Versatile imaging modality with many tracers	Yes
SPECT	1–2 mm	No limit	Minutes to hours	Yes	No	^{99m}Tc- or ^{111}In-labeled compounds	Physiological, molecular	$$	Imaging labeled antibodies, proteins and peptides	Yes

Fluorescence reflectance imaging	2–3 mm	<1 cm	Seconds to minutes	No	Yes	Photoproteins, fluorochromes	Physiological, molecular	$	Rapid screening of molecular events in surface-based disease	Yes
FMT	1 mm	<10 cm	Minutes to hours	Yes	Yes	Near-infrared fluorochromes	Physiological, molecular	$$	Quantitative imaging of fluorochrome reporters	In development
Bioluminescence imaging	Several mm	cm	Minutes	No	Yes	Luciferins	Molecular	$$	Gene expression, cell and bacterium tracking	No
Intravital microscopy[f]	1 μm	<400–800 μm	Seconds to hours	No	Yes	Photoproteins, fluorochromes	Anatomical, physiological, molecular	$$$	All of the above at higher resolutions but limited depths and coverage	In development[g]

a For high-resolution, small-imaging systems. (Clinical imaging systems differ.).
b Time for image acquisition.
c Quantitative here means inherently quantitative. All approaches allow relative quantification.
d Cost is based on purchase price of imaging systems in the United States. $, <US$ 100 000; $$, US$ 100 000–300 000; $$$, >US$ 300 000.
e Interventional means used for interventional procedures such as biopsies or injection of cells under ultrasound guidance.
f Laser-scanning confocal or multiphoton microscopy.
g For microendoscopy and skin imaging.
Adapted from Ref. [4].

Table 2.2 Comparison of organic dyes and QDs.

Property	Organic dye	QD[a]
Absorption spectra	Discrete bands, FWHM[b] 35 nm[c] to 80–100 nm[d]	Steady increase toward UV wavelengths starting from absorption onset; enables free selection of excitation wavelength
Molar absorption coefficient	2.5×10^4–$2.5 \times 10^5\,M^{-1}\,cm^{-1}$ (at long-wavelength absorption maximum)	10^5–$10^6\,M^{-1}\,cm^{-1}$ at first excitonic absorption peak, increasing toward UV wavelengths; larger (longer wavelength) QDs generally have higher absorption
Emission spectra	Asymmetric, often tailing to long-wavelength side; FWHM, 35 nm[c] to 70–100 nm[d]	Symmetric, Gaussian profile; FWHM, 30–90 mm
Stokes' shift	Normally <50 nm[c], up to >150 nm[d]	Typically <50 nm for visible wavelength-emitting QDs
Quantum yield	0.5–1.0 (visible[c]), 0.05–0.25 (NIR[c])	0.1–0.8 (visible), 0.2–0.7 (NIR)
Fluorescence lifetimes	1–10 ns, mono-exponential decay	10–100 ns, typically multiexponential decay
Two-photon action cross-section	1×10^{-52}–$5 \times 10^{-48}\,cm^4\,s\,photon^{-1}$ (typically about $1 \times 10^{-49}\,cm^4\,s\,photon^{-1}$)	2×10^{-47}–$4.7 \times 10^{-46}\,cm^4\,s\,photon^{-1}$
Solubility or dispersibility	Control by substitution pattern	Control via surface chemistry (ligands)
Binding to biomolecules	Via functional groups following established protocols Often several dyes bind to a single biomolecule Labeling-induced effects on spectroscopic properties of reporter studied for many common dyes	Via ligand chemistry; few protocols available Several biomolecules bind to a single QD Very little information available on labeling-induced effects
Size	~0.5 nm; molecule	6–60 nm (hydrodynamic diameter); colloid
Thermal stability	Dependent on dye class; can be critical for NIR-wavelength dyes	High; depends on shell or ligands
Photochemical stability	Sufficient for many applications (visible wavelength), but can be insufficient for high-light flux applications; often problematic for NIR-wavelength dyes	High (visible and NIR wavelengths); orders of magnitude higher than that of organic dyes; can reveal photobrightening
Toxicity	From very low to high; dependent on dye	Little known yet (heavy metal leakage must be prevented, potential nanotoxicity)

Table 2.2 Continued

Property	Organic dye	QD[a]
Reproducibility of labels (optical, chemical properties)	Good, owing to defined molecular structure and established methods of characterization; available from commercial sources	Limited by complex structure and surface chemistry; limited data available; few commercial systems available
Applicability to single-molecule analysis	Moderate; limited by photobleaching	Good; limited by blinking
FRET	Well-described FRET pairs; mostly single-donor-single-acceptor configurations; enables optimization of reporter properties	Few examples; single-donor-multiple-acceptor configurations possible; limitation of FRET efficiency due to nanometer size of QD coating
Spectral multiplexing	Possible, three colors (MegaStokes dyes), four colors (energy-transfer cassettes)	Ideal for multicolor experiments; up to five colors demonstrated
Lifetime multiplexing	Possible	Lifetime discrimination between QDs and organic dyes
Signal amplification	Established techniques	Unsuitable for many enzyme-based techniques, other techniques remain to be adapted and/or established

Properties of organic dyes are dependent on dye class and are tunable via substitution pattern. Properties of QDs are dependent on material, size, size distribution and surface chemistry.

a Emission wavelength regions for QD materials (approximate): CdSe, 470–660 nm; CdTe, 520–750 nm; InP, 620–720 nm; PbS, >900 nm; and PbSe, >1000 nm.
b FWHM, full width at half height of the maximum.
c Dyes with resonant emission such as fluoresceins, rhodamines and cyanines.
d CT dyes.
e Definition of spectral regions used here: visible, 400–700 nm; and NIR, >700 nm.
Unless stated otherwise, all values were determined in water for organic dyes and in organic solvents for QDs, and refer to the free dye or QD.
Adapted from Ref. [5].

holes. One way to increase the QY is by surface passivation of the QDs through an overgrowth of a second semiconductor on core QDs, and this results in core–shell nanostructures. This shell can be finely controlled to obtain QY values close to 85% [8]. Moreover, this approach enhances both chemical and physical resistances by several orders of magnitude relative to organic dyes [5, 9–11].

Figure 2.4 and Table 2.4 show the band gaps and the relative position of electronic energy levels of the Group III–V, II–VI, and IV–VI bulk semiconductors that are commonly used for the core–shell QDs synthesis [12, 14]. Although tremendous core–shell systems have recently been developed, they are simply

Table 2.3 Comparison of the optical properties of selected organic dyes and QDs.

Dyes	Absorbance[a] (nm)	Fluorescence[b] (nm)	FWHM[c] (nm)	ε^d (M^{-1} cm^{-1})	ϕ^f		QDs	Absorbance[e] (nm)	Fluorescence[f] (nm)	FWHM[c] (nm)	ε^d (M^{-1} cm^{-1})	ϕ^f
Visible wavelengths (emission < 700 nm)												
Fluorescein	500 in basic ethanol	541 in basic ethanol	35 in basic ethanol	9.2×10^4 in basic ethanol	0.97 in basic ethanol		CdS	350–470	370–500	~30	1.0×10^5 (for 350-nm diameter) and 9.5×10^5 (for 450 nm diameter) in methanol	≤0.6
Cy3 (Cy3.18)	550 in PBS, 560 in ethanol	565 in PBS, 575 in ethanol	34 in PBS	1.5×10^5 in ethanol	0.04 in PBS, 0.09 in ethanol							
TAMRA	554 in water	573 in water	39 in methanol	~1×10^5 in methanol	0.28 in water	g	CdSe	450–640	470–660	~30	10×10^5 (for 500 nm diameter) and 7.0×10^5 (for 630 nm diameter) in methanol	0.65–0.85
Texas Red (sulforhodamine 101)	587 in methanol, 576 in ethanol	602 in methanol, 591 in ethanol	35 in methanol	9.6×10^4 in methanol, 1.4×10^5 in ethanol	0.93 in ethanol, 0.35 in water	g						
Nile Red	552 in methanol, 519 in dioxane	636 in methanol, 580 in dioxane	75 in dioxane	4.5×10^4 in methanol, ~4.0×10^4 in dioxane	0.7 in dioxane	g	CdTe	500–700	520–750	35–45	1.3×10^5 (for 570 nm diameter) and 6.0×10^5 (for 700 nm diameter) in methanol	0.3–0.75
Cy5 (Cy5.18)	650 in PBS, 658 in ethanol	667 in PBS, 677 in ethanol	39 in PBS	2.5×10^5 in ethanol	0.27 in PBS, 0.4 in ethanol		InP	550–650	620–720	50–90		0.1–0.6

2.2 Types of Quantum Dots (QDs)

NIR wavelengths (emission > 700 nm)

	λ_{abs} [a]	λ_{em} [b]	FWHM [c]	ε [d]	Φ
Atto740	740 in PBS	764 in PBS	43 in PBS	1.2×10^5 in PBS	0.10 in PBS
Cy7[i]	747 in water	774 in water	50 in water	2.0×10^5 in water	0.28 in water
Alexa 750	749 in phosphate buffer	775 in phosphate buffer	49 in phosphate buffer	2.4×10^5 in phosphate buffer	0.12 in phosphate buffer
IR125 (ICG)	781 in water/methanol 75/25%, 782 in methanol, 786 in ethanol	825 in methanol, 815 in water	58 in methanol	2.1×10^5 in water/methanol 75/25%, 1.95×10^5 in methanol, 1.94×10^5 in ethanol	0.02 in water/methanol 75/25%, 0.04 in methanol, 0.05 in ethanol, 0.01 in water
PbS[h]	800–3000 [e]	>900 [f]	80–90		0.26 in HEPES buffer, 0.70 in hexane
PbSe[g]	900–4000 [e]	>1000 [f]	80–90	1.23×10^5 in CHCl$_3$	0.4–0.5 in CHCl$_3$, 0.12–0.81 in hexane

a Longest wavelength absorption maximum.
b Shortest wavelength emission maximum.
c Full-width at half-height of the emission band.
d ε values provided for the main (longest-wavelength) absorption band (dyes) and the first excitonic absorption peak (QDs).
e Size-tunable position of the first excitonic absorption maximum.
f Size-tunable position of the emission maximum.
g Manufactured by Invitrogen.
h Manufactured by ATTOTech.
i Manufactured by Amersham. Adapted from Ref. [5].

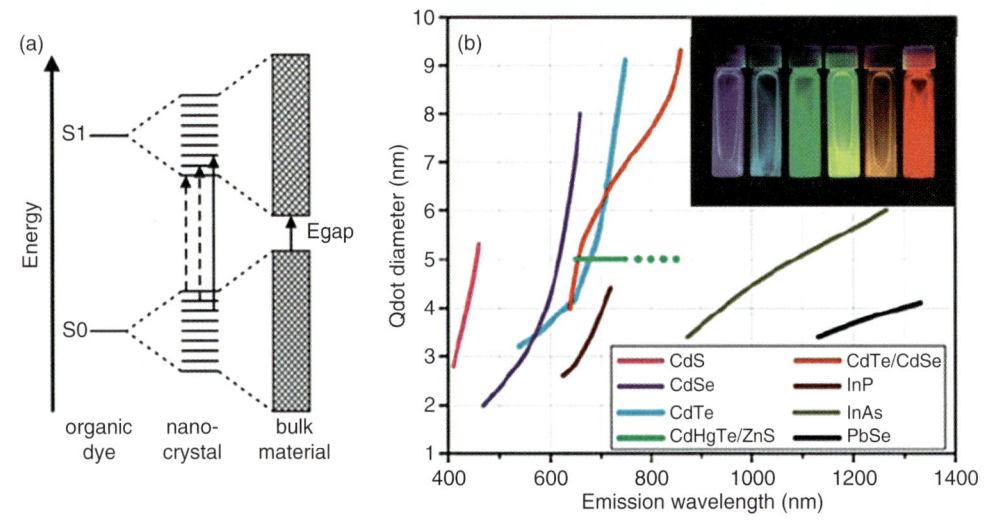

Figure 2.3 (a) A schematic diagram showing the development of electronic states on increasing length scales, from molecular (organic dye), to nanoscopic (nanocrystal), to macroscopic (bulk material). S_0 = ground state and S_1 = excited state of an organic dye. Adapted from Ref. [6]; (b) Emission maxima and sizes of semiconducting QDs with different composition. Adapted from Ref. [7]. Inset: Change of fluorescence emission by size variation (2–8 nm) of CdSe QDs under UV light. (Photography by Felice Frankel.)

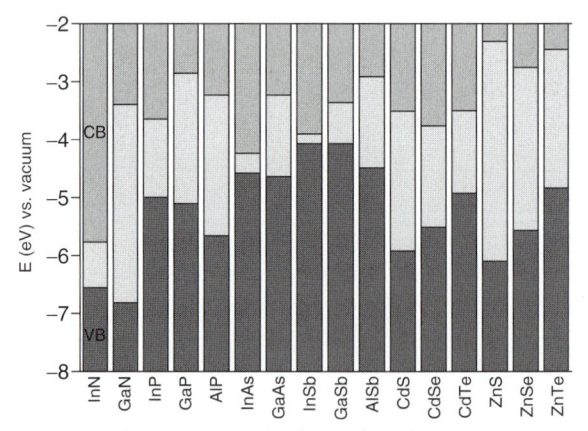

Figure 2.4 Electronic energy levels of selected III–V and II–VI semiconductors using the valence-band offsets. (VB = valence band; CB = conduction band). From Ref. [12].

classified in three types as type-I, reverse type-I, and type-II, depending on different band alignments (Figure 2.5) [12]. In type-I the band gap of the shell material is larger than that of the core; in reverse type-I the band gap of the shell is smaller than that of the core; and in type-II the valence-band edge of the shell material is located in the bandgap of the core. Such typical energy band alignments lead to a

2.2 Types of Quantum Dots (QDs)

Table 2.4 Material parameters of selected bulk semiconductors.

Material	Structure (300 K)	Type	E_{gap} (eV)	Lattice parameter (Å)	Density (kg m^{-3})
ZnS	Zinc blende	II–VI	3.61	5.41	4090
ZnSe	Zinc blende	II–VI	2.69	5.668	5266
ZnTe	Zinc blende	II–VI	2.39	6.104	5636
CdS	Wurtzite	II–VI	2.49	4.136/6.714	4820
CdSe	Wurtzite	II–VI	1.74	4.3/7.01	5810
CdTe	Zinc blende	II–VI	1.43	6.482	5870
GaN	Wurtzite	II–V	3.44	3.188/5.185	6095
GaP	Zinc-blende	II–V	2.27	5.45	4138
GaAs	Zinc-blende	II–V	1.42	5.653	5318
GaSb	Zinc-blende	II–V	0.75	6.096	5614
InN	Wurtzite	II–V	0.8	3.545/5.703	6810
InP	Zinc blende	II–V	1.35	5.869	4787
InAs	Zinc-blende	II–V	0.35	6.058	5667
InSb	Zinc-blende	II–V	0.23	6.479	5774
PbS	Rocksalt	IV–VI	0.41	5.936	7597
PbSe	Rocksalt	IV–VI	0.28	6.117	8260
PbTe	Rocksalt	IV–VI	0.31	6.462	8219

Adapted from Ref. [13].

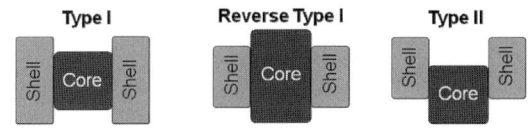

Figure 2.5 Schematic representation of the energy-level alignment in different core–shell QDs. The upper and lower edges of the rectangles correspond to the positions of the conduction- and valence-band edge of the core and shell materials, respectively. Adapted from Ref. [15].

different confinement phenomenon of the electron and the hole in different regions of the core–shell QDs.

2.2.1
Type I Core–Shell QDs

In type-I core–shell QDs, all charge carriers are confined in the core material, and thus the radiative recombination process occurs in the core QD (see Figure 2.6). Consequently, the wavelength of the fluorescence emission is mainly attributed by controlling the size of the core materials. One purpose of the shell overgrowth in type-I core–shell QDs is passivation of the core materials, so as to increase the fluorescence QY and photostability. Notably, the surface of the core is protected from chemical and physical environments by the growing shell layer. Moreover,

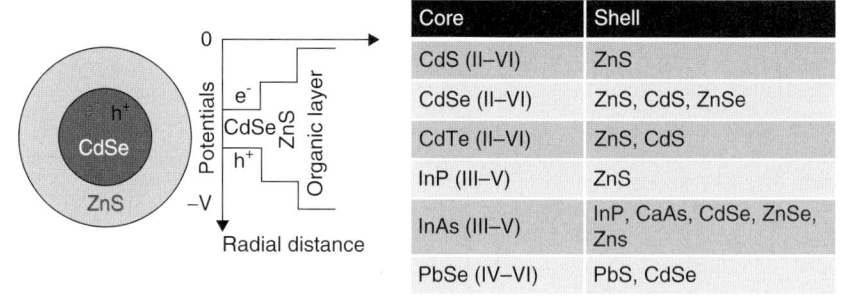

Figure 2.6 Schematic band structure and examples of type-I core–shell QDs.

Core	Shell
CdS (II–VI)	ZnS
CdSe (II–VI)	ZnS, CdS, ZnSe
CdTe (II–VI)	ZnS, CdS
InP (III–V)	ZnS
InAs (III–V)	InP, CaAs, CdSe, ZnSe, Zns
PbSe (IV–VI)	PbS, CdSe

the surface dangling bonds, which serve as trap sites for charge carriers, may be reduced such that the QY can be enhanced by overgrowth of the shell [13, 16, 17]. If the shell layer is too thick, however, then strain at the interface between the core and shell may be induced by the lattice mismatch and, as a result, the QY may generally deteriorate due to the generation of defect sites. In the opposite case, both the photostability and chemical resistance are reduced for the resultant core–shell QDs. Clearly, control of the shell thickness represents a key point in the core–shell system for effective passivation, and a successive ion layer adsorption and reaction (SILAR) method is usually adapted for growth of the shell layers [15]. This method is based on forming one monolayer at a time by alternating the injections of cationic and anionic precursors. In addition, the temperature for shell growth is generally lower than that used for the core QDs synthesis, so as to prevent nucleation of the shell material and an uncontrolled ripening of the core QDs.

In the type-I core–shell system, the Group II–VI, III–V, and IV–VI semiconductors are commonly adapted for the core materials. Type-I core–shell systems based on three different types of core material are discussed further in the following section.

2.2.1.1 Group II–VI Semiconductor Core–Shell QDs

The CdSe/ZnS QD is a representative of the type-I core–shell system, and the most intensively studied of the molecular imaging probes. In the CdSe/ZnS system, a QY above 50% is easily obtained by growing only one to two monolayers of ZnS [16, 18].

A general approach for the synthesis of core–shell QDs should consider not only an appropriate band alignment but also the lattice mismatch between the core and shell materials so as to obtain epitaxial growth. In particular, a large lattice mismatch will result in strain and the formation of defect states at the core–shell interface, or within the shell. Therefore, the QY could be reduced by generating defects as trap states of charge carriers [19]. The lattice mismatch between CdSe and CdS was only 3.9%, which was significantly lower than that of ZnS (10.3%) and ZnSe (6.3%) [8, 13]. Therefore, CdS may be the most suitable shell material for CdSe-based core–shell QDs with epitaxial growth. Peng *et al.* reported the

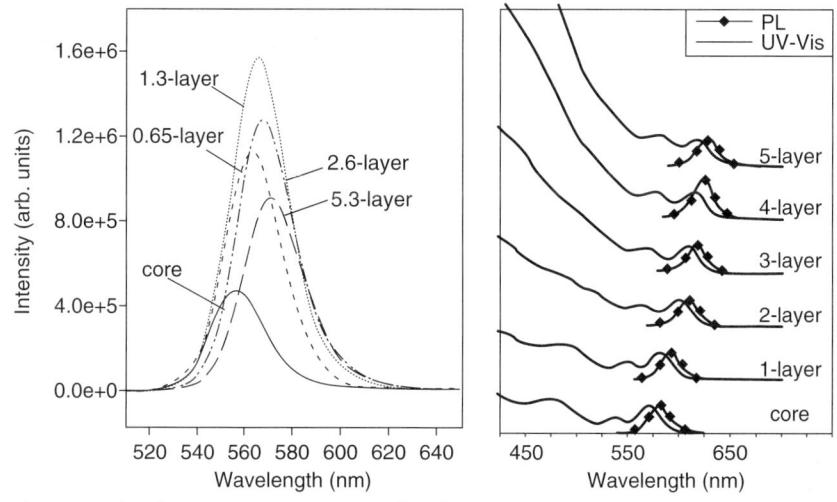

Figure 2.7 Photoluminescence (PL) spectra for CdSe/ZnS core–shell QDs (left panel) and UV-visible and PL spectra of CdSe/CdS core–shell QDs (right panel) upon growth of shell layers. Adapted from Refs [16] and [19].

synthesis and detailed characterization of a series of CdSe/CdS QDs with QYs above 50%. Moreover, the CdSe/CdS system has the same cation at the interface, which leads to a large offset in the valence band and a relatively small offset in the conduction band; this also indicates that the photogenerated electrons in the core might be easily transferred into the shell layer, in contrast to CdSe/ZnS core–shell system. As shown in Figure 2.7, a bathochromic shift (red shift) in the photoluminescence (PL) spectra of CdSe/CdS is approximately 50 nm upon growth of the CdS shell (five-layer), but is only 10 nm in the CdSe/ZnS system. This indicates that the photogenerated electrons for the CdSe/CdS system may be easily delocalized over the entire core–shell structure, and more so than for the CdSe/ZnS system.

In contrast to the CdSe/CdS core–shell system, CdSe/ZnSe has a common anion species in the core and shell materials, such that a small valence band offset and a large conduction band offset between the core and the shell is expected. This leads to an efficient confinement of the electrons in the core, and the holes may be leaked from the core to the shell by a relatively small barrier in the CdSe/ZnSe system.

As the band gap of CdTe (1.43 eV) in the bulk state is smaller than that for CdSe (1.74 eV) (see Table 2.4), it represents a good candidate for the design of NIR-emitting (700–900 nm) QDs. These NIR QDs play a key role in the imaging of various biological targets, because fluorescence imaging in the UV and visible regions is interfered with by the autofluorescence of blood, tissues, and proteins (see Figure 2.8) [20, 21]. Moreover, the penetration depth of NIR fluorescence is greater than that of both UV and visible fluorescence. Although NIR QDs

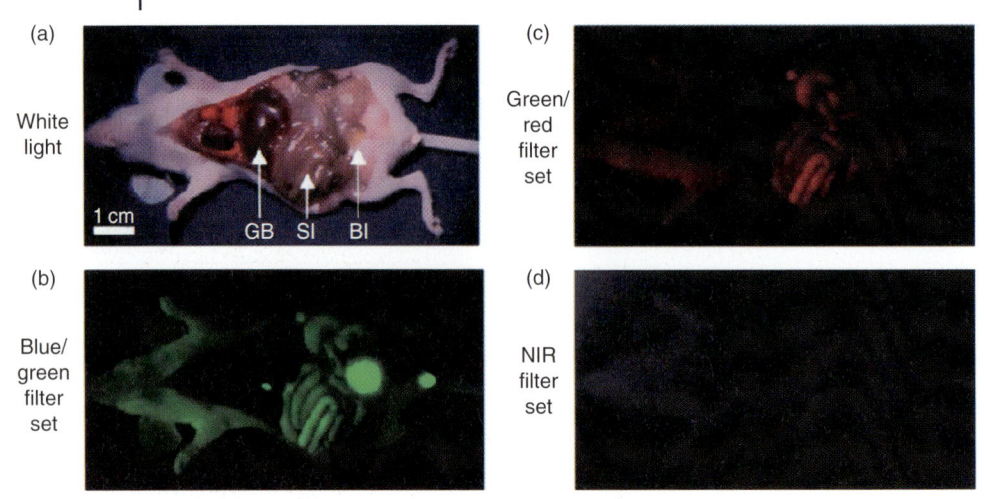

Figure 2.8 Photograph of autofluorescence observed in a mouse with various wavelength filter sets. GB = gallbladder; SI = small intestine; BI = bladder. Reprinted with permission from Ref. [20]; © 2002, Elsevier.

demonstrate a great potential for imaging in living subjects, very little success has been reported in the synthesis of CdTe-based core–shell systems, as most of shell materials with a type-I band alignment have a large lattice mismatch with respect to CdTe [22, 23]. However, the results of the CdTe/CdS core–shell system are worthy of notice because a high QY of 85% is realized via the aqueous synthetic approach [24]. Such high fluorescence emission is comparable to that of CdSe/CdS core–shell system with the small lattice mismatch between the core and the shell materials, as noted above.

2.2.1.2 Group III–V Semiconductor Core–Shell QDs

Recent research into Group III–V semiconductor-based QDs has increased intensively on the basis that they are less toxic than Cd-based II–VI semiconducting QDs. The small band gaps of III–V semiconductors are also suitable for NIR fluorescence applications (see Table 2.4). However, far fewer synthetic approaches are available to III–V semiconductor-based core–shell systems than to II–VI semiconductor core–shell QDs.

Indium phosphide (InP) is representative of III–V semiconducting QDs, and its fluorescence emission can be tuned from visible to NIR regions simply by controlling the crystal size. However, an early version of an InP QD exhibited poor optical properties, such as a broad PL peak width and band-edge emission related defect sites, as compared to CdSe QD; in particular, the QY was low, at less than 1% [24–26]. Subsequently, the QY has been increased remarkably, by up to 30–40%, by eliminating the surface phosphorus atoms lying at the origin of the trap states via a photoassisted etching process [27, 28]. Such success in preparing the InP

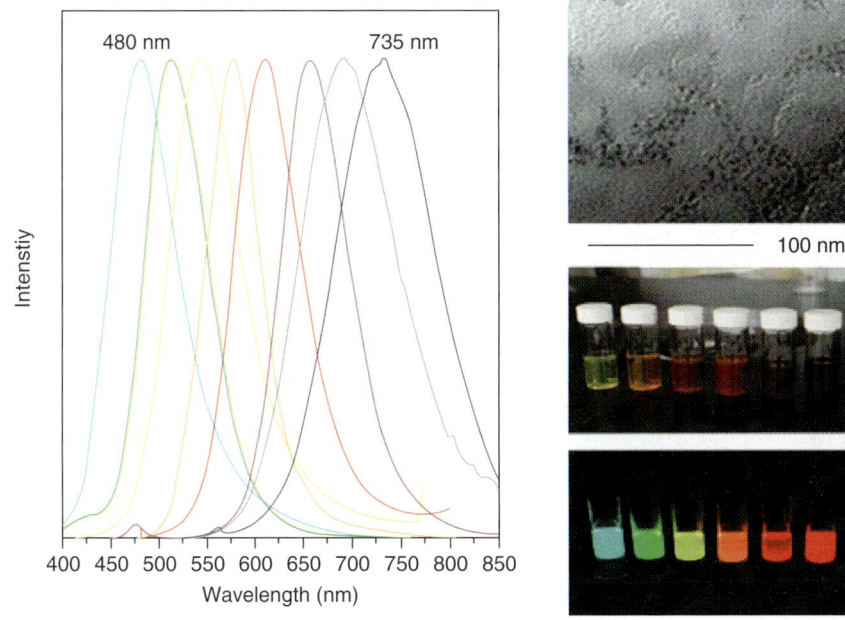

Figure 2.9 Left: Photoluminescence spectra of different-sized InP/ZnS CS NCs. Right: Upper panel: Transmission electron microscopy image (mean particle diameter 4.5 nm); middle panel: photograph of different-sized samples under room light; bottom panel: under UV light. Reprinted with permission from Ref. [29]; © 2008, Royal Society of Chemistry.

core QD with high optical properties subsequently allowed development of the InP/ZnS core–shell QD system with a high quantum yield of 60–70% and a wide fluorescence emission that ranged from UV to NIR (Figure 2.9) [29, 30]. This recent progress in InP/ZnS core–shell systems promises excellent applications for biological fluorescence marking, rather than CdSe-based core–shell systems.

Another III–V core–shell system is based on the InAs core and the InP, GaAs, CdSe, ZnSe, and ZnS shells [31, 32]. Here, the fluorescence emission is tunable between 800 and 1400 nm by controlling the InAs core size (1–4 nm). In addition, the QYs are changed on the shell materials (InAs/InP = 0; InAs/ZnS = 8%; InAs/CdSe and InAs/ZnSe = 20%) [33]. For the InAs/ZnSe and InAs/ZnS core–shell systems, the photogenerated carriers are strongly confined within the InAs core due to the large band offsets (see Figure 2.10), and thus absorption and emission onset are almost the same as that of the InAs core. In contrast, a bathochromic shift (red shift) of the emission is observed from the InAs/CdSe core–shell system as a consequence of a small band offset between the core and the shell (see Figure 2.10).

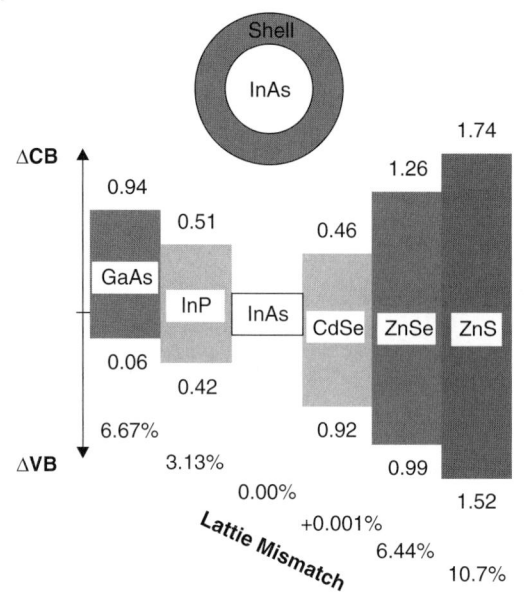

Figure 2.10 Schematic representation of the band offsets (in eV) and lattice mismatch (in %) between the core InAs and the III–V semiconductor shells (left side), and II–VI semiconductor shells (right side). CB = conduction band, VB = valence band. Reprinted with permission from Ref. [32]; © 2000, American Chemical Society.

2.2.1.3 Group IV–VI Semiconductor Core–Shell QDs

The bulk band gaps of Group IV–VI semiconductors are relatively small, which means that these materials are good candidates for the design of NIR core–shell systems. The majority of studies with the IV–VI core–shell system have focused on PbSe/PbS core–shell QD. For example, Lifshitz and coworkers reported the synthesis of PbSe/PbS and PbSe/PbSe$_x$S$_{1-x}$ core–shell QDs emitting in the range of 1–2 mm with QYs of 40–50% and 65%, respectively [33, 34]. To this end, Hollingsworth and coworkers developed PbSe/CdSe core–shell QDs by the exchange of surface lead ions with cadmium [35]. As a consequence, a greatly improved stability of the optical properties was observed, which could be further enhanced by subsequent overgrowth of the CS system with an additional ZnS shell.

2.2.2
Reverse Type I and Type II Core–Shell QDs

In the reverse type-I core–shell system, the band gap of the core material (ZnSe) is larger than that of the shell material (CdSe) [36], and consequently both the electrons and holes are localized into the shell layer (see Figure 2.11). Therefore, fine-tuning of the fluorescence emission over a broad spectral range (400–700 nm) is usually achieved by controlling the thickness of the shell layer (Figure 2.12)

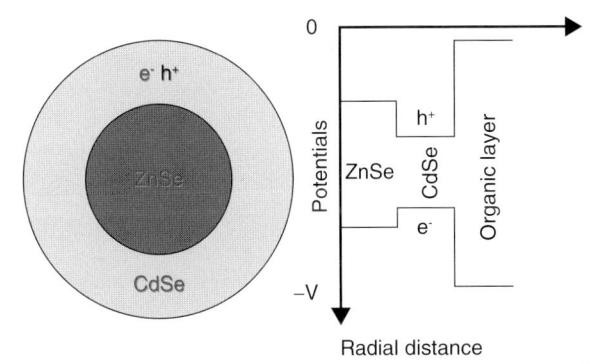

Figure 2.11 Schematic band structure of reverse type-I core–shell QDs.

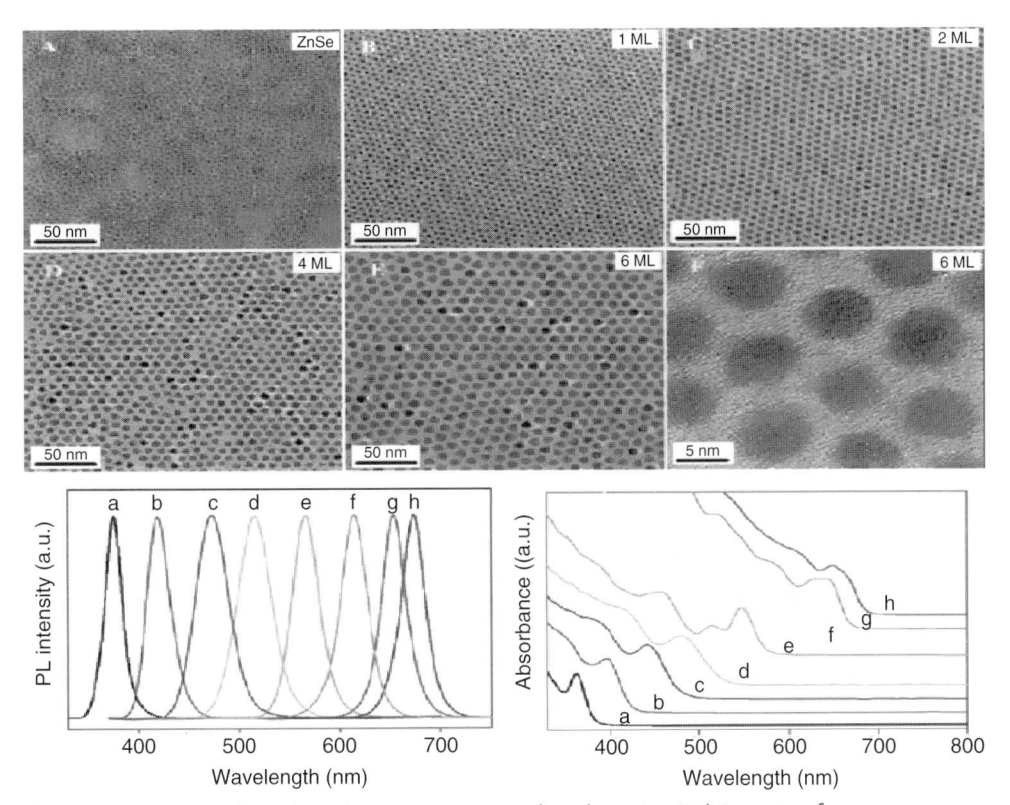

Figure 2.12 Upper panels: High-resolution (HR)-TEM images of ZnSe core QDs and the corresponding ZnSe/CdSe core–shell QDs with different shell thicknesses (expressed in CdSe monolayers, ML). Lower panels: Normalized photoluminescence (left) and corresponding absorption (right) spectra of ZnSe/CdSe core–shell QDs with different numbers of CdSe monolayers: (a) 0; (b) 0.1; (c) 0.2; (d) 0.5; (e) 1; (f) 2; (g) 4; (h) 6. Reprinted with permission from Ref. [37]; © 2005, American Chemical Society.

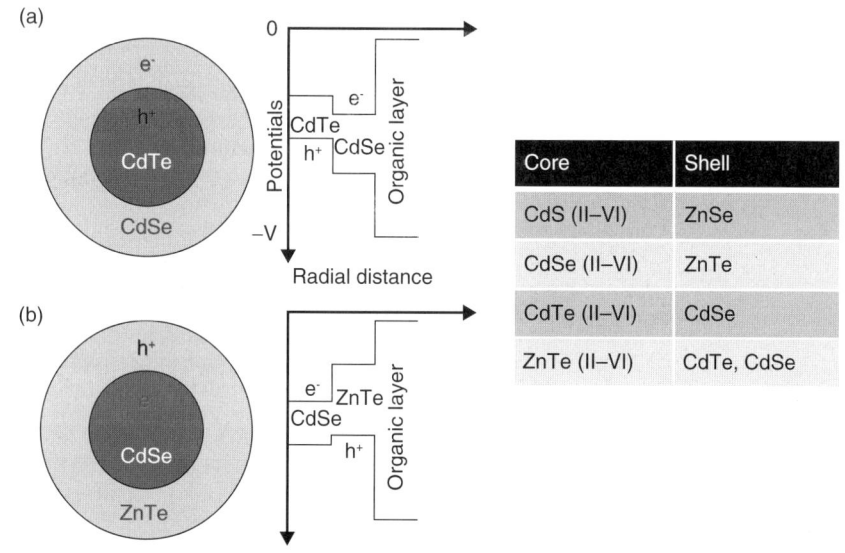

Core	Shell
CdS (II–VI)	ZnSe
CdSe (II–VI)	ZnTe
CdTe (II–VI)	CdSe
ZnTe (II–VI)	CdTe, CdSe

Figure 2.13 Schematic band structures and examples of type-II core–shell QDs.

[37]. Moreover, a QY of 45–80% is of a similar degree to that of the CdSe-based type-I core–shell system. An alternative inverted type-I core–shell system, namely CdS/CdSe core–shell QD, has an emission range of 520–650 nm and a QY of 20–40% [38].

In type-II core–shell systems, both the valence and conduction bands of the core materials are lower (or higher) than those of the shell materials (Figure 2.13), and therefore the electrons (or the holes) will be localized into the shell, and the holes (or electrons) confined into the core, respectively (see Figure 2.13). Consequently, it is to be expected that the optical properties of the type-II core–shell system would be quite different from those of the type-I and reverse type-I core–shell systems. The fluorescence emission of type-II core–shell systems can be changed by controlling the thickness of the core or the shell, or both. Figure 2.14a shows the room-temperature PL emission spectra of CdTe/CdSe QDs in which the emission range is changed dramatically, from 700 nm to 1000 nm, simply by controlling the core size and shell thickness [39]. In contrast, the observed mean decay lifetime (57 ns) of CdTe/CdSe QDs was significantly longer than that of the corresponding core CdTe QDs (9.6 ns). Moreover, the radiative and nanoradiative lifetimes of the CdTe/CdSe core–shell system were 120-fold and 3.6-fold larger than those of the CdTe core QDs, respectively. The longer radiative lifetime of type-II core–shell system derives from the slow electron-hole recombination of excitons as a consequence of charge carriers separated between the core and the shell. In addition, the longer nonradiative lifetime may result from the suppression of hole-trap-mediated nonradiative channels on the CdTe/CdSe QDs, because the holes are localized into the CdTe core [39].

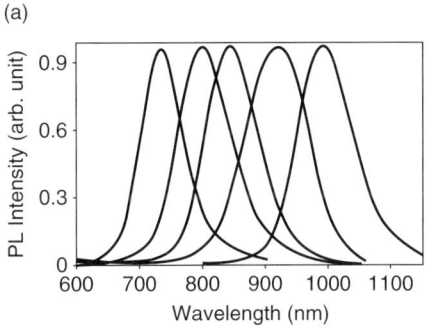

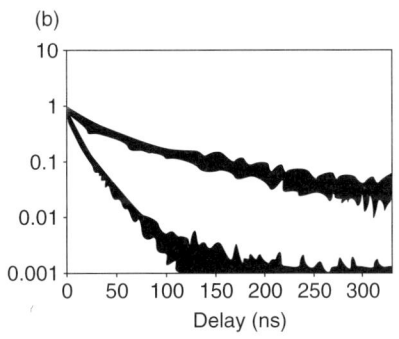

Figure 2.14 (a) Normalized room-temperature photoluminescence (PL) spectra of different CdTe/CdSe QDs with a core radius/shell thickness 16 Å/19 Å, 16 Å/32 Å, 32 Å/11 Å, 32 Å/24 Å, and 56 Å/19 Å (from left to the right in the spectra); (b) Normalized PL intensity decays of CdTe/CdSe (32 Å/11 Å) QD and CdTe core (32 Å) (dotted line). Reprinted with permission from Ref. [39]; © 2005, American Chemical Society.

By contrast, the ZnSe/CdSe type-II system is composed of the ZnSe core with a wide band gap, and the CdSe shell with a narrow bad gap; such ZnSe/CdSe core–shell QDs exhibit a high QY of 80–90% [40]. One interesting property of the ZnSe/CdSe system is the change in band edge alignment from type-I to type-II and reversible type-I by increasing the CdSe shell thickness (see Figure 2.15) [40]. Another property is the tunability between type-I and type-II regimes, which can be achieved by simply varying the shell thickness for a fixed core radius.

2.2.3
Core–Shell–Shell (CSS) QDs

The requirements for QDs with multiple passivating shells derive from an insufficiency of the core and the shell materials with appropriate band alignments and small lattice mismatch for the core–shell system. In particular, a large lattice mismatch between the core and shell materials will result in a low QY, and therefore the insertion of another shell with an intermediate lattice parameter into the core–shell system could lead to a reduction in lattice strain at the core–shell interface. In the InAs/CdSe/ZnS CSS system, in order to reduce the strain effect resulting from a large lattice mismatch (~7%) between InAs and ZnSe, the CdSe shell is adapted for the buffer layer because the lattice mismatch between InAs and CdSe is actually almost zero (see Figure 2.16). Normally, such CSS systems would have a higher photostability and QY than the core–shell systems. As another approach in the CSS system, the InAs QDs may be overcoated with a CdSe/ZnSe double shell, such that the resultant InAs/CdSe/ZnSe QDs would exhibit a high QY of 70% in the spectral region of 800–1600 nm [41].

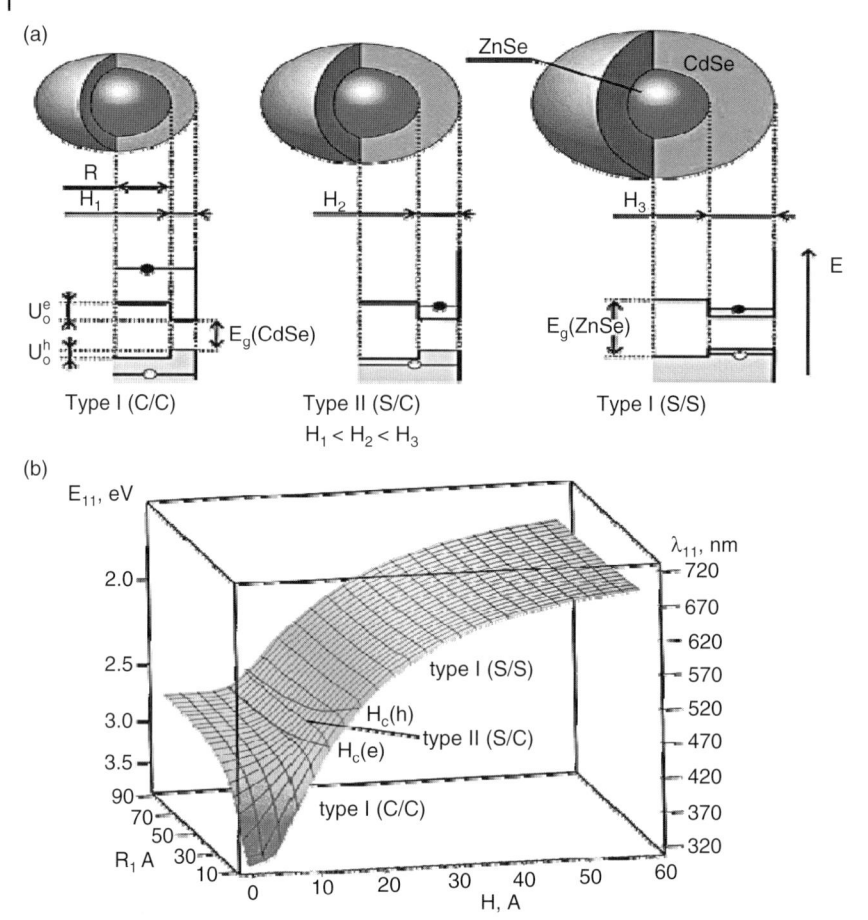

Figure 2.15 (a) Schematic illustration of band edge alignments for ZnSe/CdSe core–shell QDs with a fixed core radii and different shell thickness. R = core radius; H = shell thickness; U_0^e = conduction band energy; U_0^h = valence band energy. Thin shell (left): Both the electron and hole are delocalized over the core and the shell (type-I (C/C)). Intermediate shell (middle): The hole is still delocalized over the entire core–shell system, while the electron is confined primarily in the shell (type-II (S/C)). Thick shell (right): Both the electron and the hole are localized mostly in the shell (type-I (S/S) or reverse type-I); (b) The energy/wavelength diagram of the CdSe/ZnSe core–shell system as a function of core radius and shell thickness. The two red lines mark the energies corresponding to the transitions between different types. Reprinted with permission from Ref. [40]; © 2004, American Chemical Society.

The CdSe/ZnTe/ZnS and CdSe/CdTe/ZnTe CSS systems have also been acquired from type-II core–shell systems by the addition of a large-bandgap outer layer. In the case of the CdSe/CdTe/ZnTe CSS QDs, the intermediate CdTe shell serves as a barrier layer for electrons in the CdSe core and holes in the ZnTe outer shell; as a consequence, the radiative lifetime was increased to 10 ms [42].

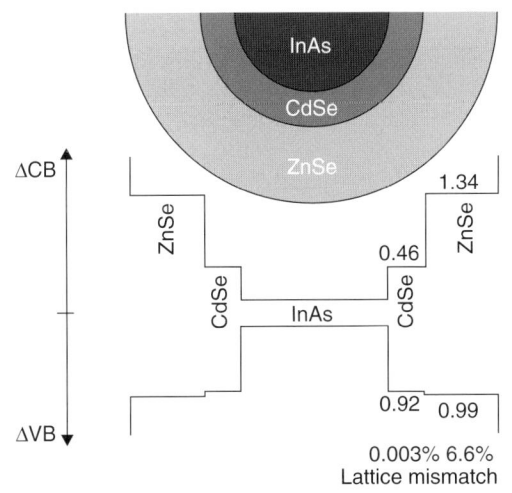

Figure 2.16 Schematic illustration of band structure for the InAs/CdSe/ZnSe CSS system. The band offset (eV) and lattice mismatch (%) between InAs and the two II–VI semiconductors are indicated. CB = conduction band; VB = valence band. Reprinted with permission from Ref. [41]; © 2006, American Chemical Society.

2.3
Surface Modifications of QDs for Bioapplications

2.3.1
Preparation of Water-Soluble QDs

Most QDs are synthesized in nonpolar organic solutions using aliphatic coordinating ligands such as trioctyl phosphine (TOP) and trioctyl phosphine oxide (TOPO). Consequently, they must be rendered water-dispersible for the bioapplications by either the exchange or functionalization of the inherently coordinated organic ligands on the surface of the QDs. Recently, a variety of methods has been developed to prevent aggregation or precipitation of the QDs in aqueous solution, the procedure being commonly classified by three strategies. The different approaches to achieve phase transfer of the QDs are illustrated in Figure 2.17 [43].

As in the first approach, the TOPOs on the surface of QDs are substituted by bifunctional ligands, in which one end group (e.g., thiol) has a high affinity with the QD surface, while the other end group is hydrophilic. Usually, thiolated or aminated water-soluble ligands, such as mercaptoacetic acid (MAA) [44], mercaptopropionic acid (MPA) [45, 46], mercaptoundecanoic acid (MUA), methoxysilane, cystamine, and cysteine, are used for the preparation of water-dispersible QDs. These amine- or carboxyl-terminated ligands create the surface charges of the QDs such that, consequently, the QDs may become electrostatically stabilized under an aqueous medium. This method is very simple and useful for biological

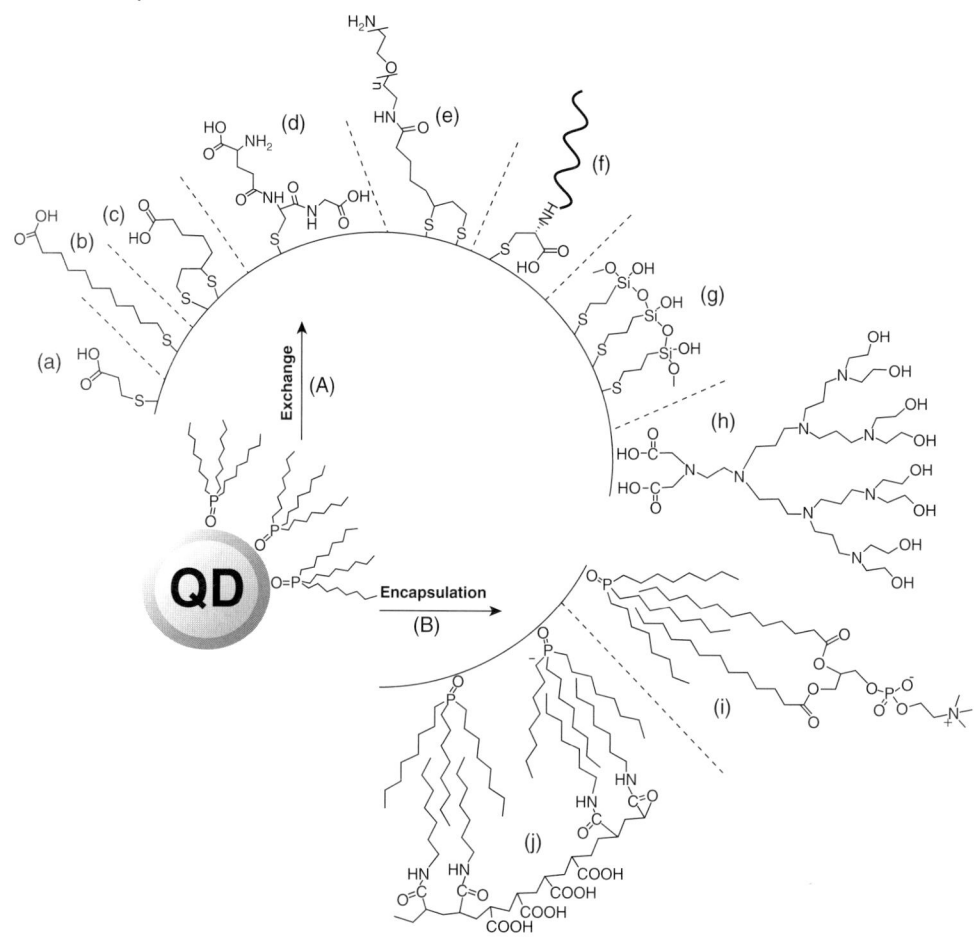

Figure 2.17 Overview of strategies to prepare water-dispersible QDs. (A) Exchange of the organic encapsulating layer with a water-soluble layer; (a–d) thiolated or dithiolated functional monolayers; (e) glutathione layer; (f) cysteine-terminated peptide; (g) thiolated siloxane; (h) carboxylic acid-functionalized dendrone; (B) Encapsulation of QDs stabilized with an organic encapsulating layer in functional bilayer films composed of (i) a phospholipid-encapsulating layer, and (j) a diblock copolymer. Reprinted with permission from Ref. [43]; © 2008, Wiley-VCH.

assays, except that the CdSe QDs coated by hydrophilic thiol ligands exhibit photochemical instability [47]. In addition, the MPA-bonded CdSe/ZnS QDs are unstable due to dynamic thiol–ZnS interactions which result in precipitation of the QDs. Therefore, the water-compatibility must be maintained generally for short periods, generally less than one week [48]. In the case of dihydrolipoic acid (DHLA) -substituted QDs, their water stability is increased remarkably up to a few years, owing to the bidentate chelate effect of the dithiol groups [49, 50].

A second method of preparing water-soluble QDs is via the encapsulation of polyethylene glycol (PEG) [51, 52], oligomeric phosphines [53], amphiphilic triblock copolymers [54], dendrimers, and phospholipids[55] through van der

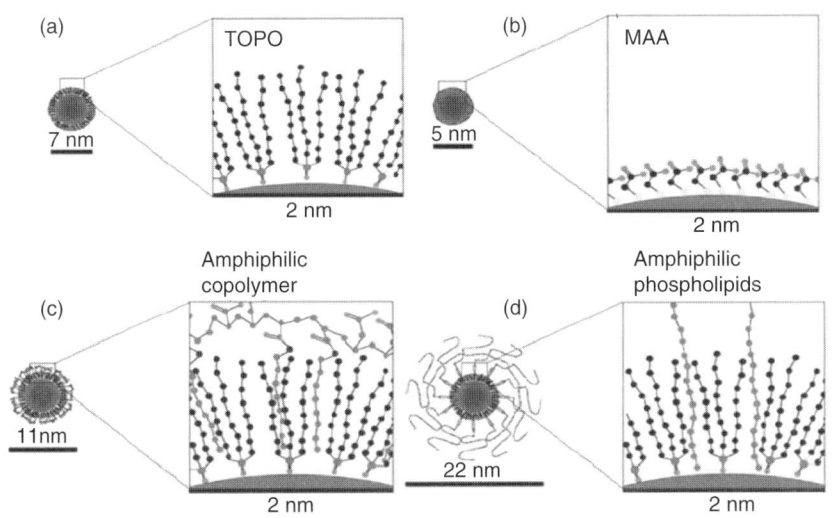

Figure 2.18 Schematic illustration of size variation of QDs (5.5 nm) coated with (a) TOPO, (b) MAA, (c) amphiphilic anionic polymer, (d) amphiphilic phospholipids. Reprinted with permission from Ref. [56]; © 2006, Biomedical Engineering Society.

Waals interactions with TOPO ligands on the surface of the QDs. However, the mean diameter of the resultant QDs is generally much larger than that of the mercaptopropionic acid or cystamine conjugated QDs (see Figure 2.18) [56]. Unfortunately, such an excessive size could hinder the widespread implementation of QDs for molecular imaging. Moreover, amphiphilic polymers often lead to an increase of the surface charge, resulting in nonspecific binding to the cell membranes. Although PEGylation of the polymer-coated QDs may reduce the nonspecific adsorption [57], the size of the resulting QDs would be further increased.

In contrast, a silica shell may be generated by the silanization of methoxysilane bonded to the QD surface. Such a silica coating can enhance the stability of the QDs against a broad pH range in various solvents. Moreover, the silica shells can be easily functionalized, thus allowing the QDs to be conjugated to other biomolecules or nanoparticles. However, the QY of silica-coated QDs in PBS buffer solution (~18%) is much less than that of inherent QDs in hydrophobic solvents, mainly because the amorphous silica layer is insufficient to exclude the external aqueous solvents. This problem may be solved by applying a surfactant-assisted microemulsion technique, as shown in Figure 2.19 [58, 59]. Namely, the hydrophobic polymers are sandwiched between the QDs, while the silica shell prevents permeation of the water molecules. This approach to creating silica-coated QDs may be extended to the synthesis of multimodal imaging probes. In general, two basic structures of the QD-based optical/MR dual modality probes have been developed (Figure 2.20). The first structure consists of the QD, the biocompatible ligands (e.g., PEG), and the paramagnetic lipids (Figure 2.20a) [60, 61]. Although

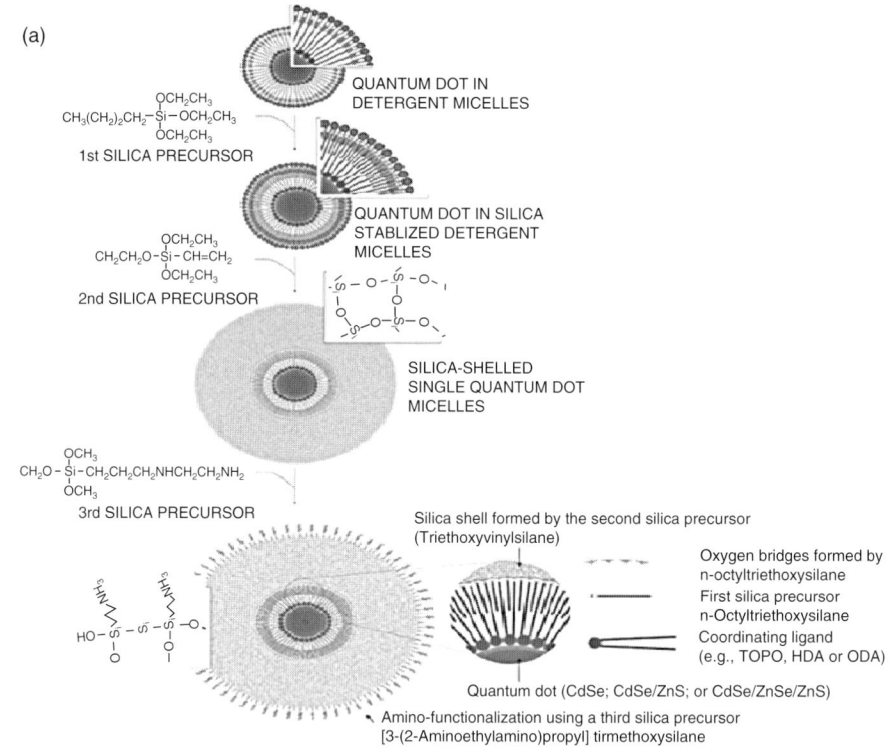

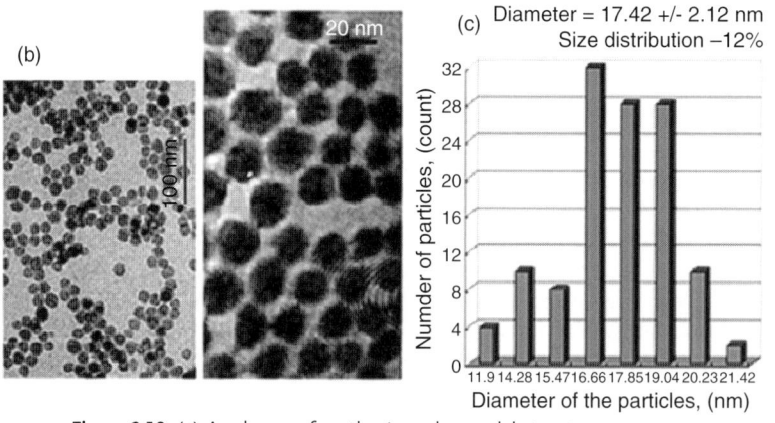

Figure 2.19 (a) A scheme of synthesis and a model structure of the silica-coated QD micelles; (b) TEM images of the silica-coated CdSe/ZnS QD micelles; (c) Size-histogram of the TEM result. The dark spots in the silica spheres indicate the QDs with heavy elements. Reprinted with permission from Refs [58] and [59]; © 2006, American Chemical Society.

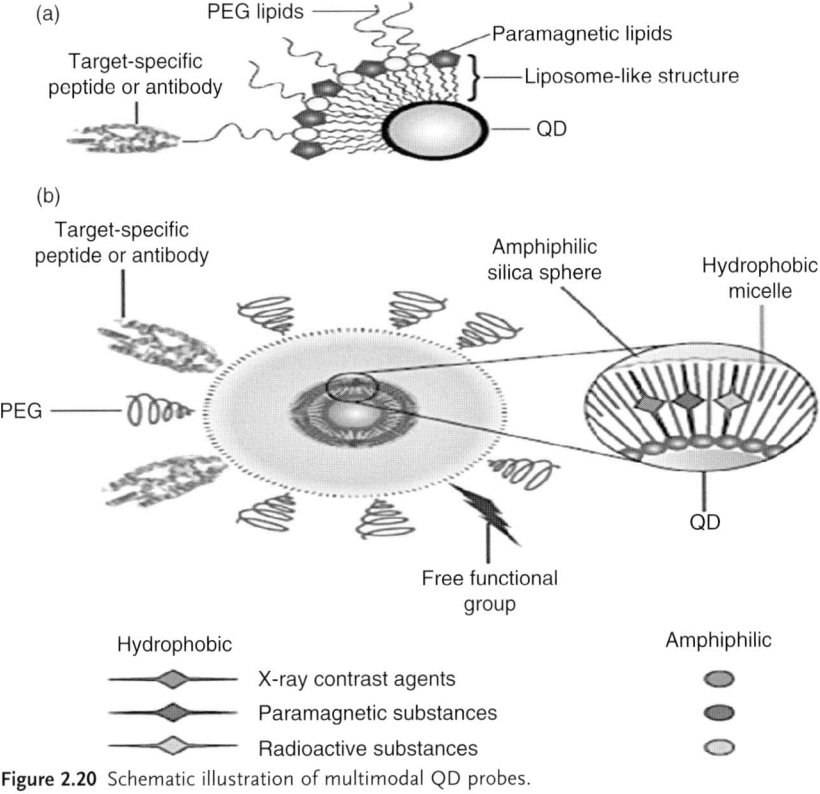

Figure 2.20 Schematic illustration of multimodal QD probes. (a) QD with target-specific molecules and paramagnetic lipids; (b) The structure of a multimodal QD probe, based on silica-coated QD micelles. Adapted from Ref. [60]; © 2007, Nature Publishing Group.

this structure is very compact, the paramagnetic molecules compete with the target-specific ligands for the limited number of active sites on the surfaces of the QDs. Therefore, the MR contrast effect might be reduced by a low concentration of the conjugated paramagnetic molecules through such a competitive process. The second structure was developed from the silica-coated QDs. For this, paramagnetic molecules are intercalated into the hydrophobic micelle layer on the surface of the QDs, after which the resultant structure is encapsulated by silica layer (see Figure 2.20b) [58]. The paramagnetic molecules can be substituted with either radioisotopic substances or X-ray contrast agents. This type of multimodal probe has a high QY of 30–50%, mainly because the QDs are effectively isolated from the aqueous solutions by the hydrophobic micelle. In addition, as the paramagnetic molecules are also protected by silica layer, this results in a MR contrast effect that is less affected by the external environment than the multimodal probes of the first structure. Most recently, this strategy was extended to a multimodal imaging probe with the CSS QDs and the gadolinium-based T_1 contrast agents

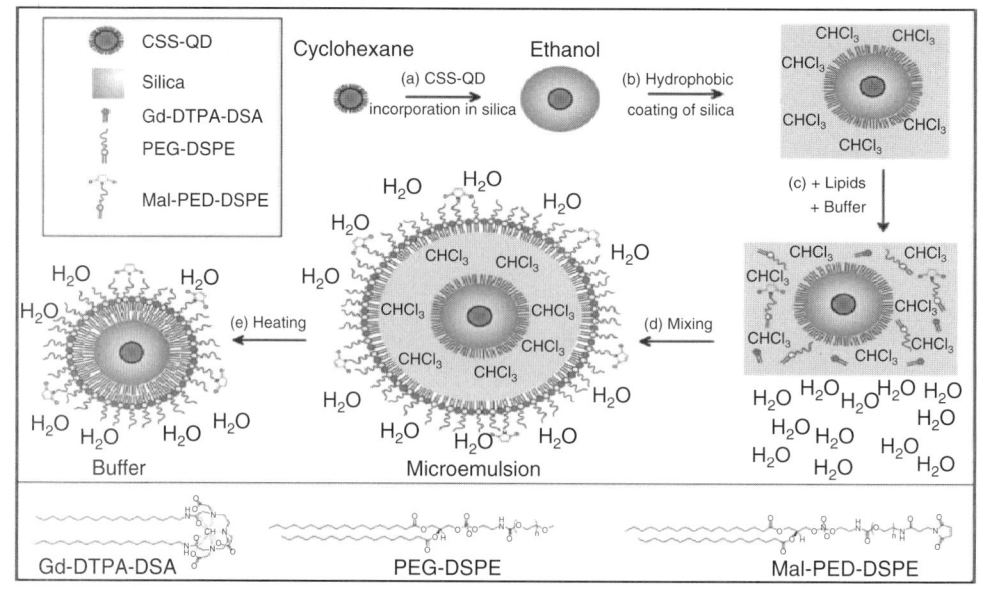

Figure 2.21 Overview of the synthesis of QD-SiO$_2$–Ga multimodal probe. (a) Incorporation of the CSS QDs into silica shell by the reverse micelle method; (b) Hydrophobic coating of silica by octadecanol, after which they can be dispersed in chloroform; (c) Addition of the different lipids to the QD/silica particles in chloroform, which is subsequently added to a HEPES (C$_8$H$_{18}$N$_2$O$_4$S) buffer; (d) Vigorous stirring results in an emulsion, with the chloroform and nanoparticles enclosed in a lipid monolayer; (e) Chloroform is evaporated by heating the mixture, resulting in water-soluble QD-SiO$_2$–Ga multimodal probe. The lower panel displays the structure formulas of the lipids. Reprinted with permission from Ref. [63]; © 2008, American Chemical Society.

(see Figure 2.21) [62]. In this structure, the Ga-DTPA-DSA molecules were incorporated onto the outer silica shell because a T$_1$ contrast effect could be generated by interaction with H$_2$O molecules in the biological environment.

Recently, functionalized PEGs with –OH, –NH$_2$, and –COOH have been developed, such that the effective reduction of nonspecific binding could be demonstrated by fluorescence observations of a human cell line incubated with ligand-coated QDs. As shown in Figure 2.22, the nonspecific binding event is minimized in the hydroxy-PEG-coated QDs and carboxy-PEG-coated QDs, whereas the amine-PEG-coated QDs showed a significant enhancement of nonspecific binding [52]. It is likely that the amine-PEG-coated QDs exhibited a highly positive charge, and this led to an increase in the electrostatic interaction with the cell membrane.

2.3.2
Bioconjugation of QDs

In order to improve the use of QDs as molecular imaging probes with specific targeting abilities, they must first be conjugated to a biological molecule, such as a peptides, protein, antibody or oligonucleotide, without disturbing the biological

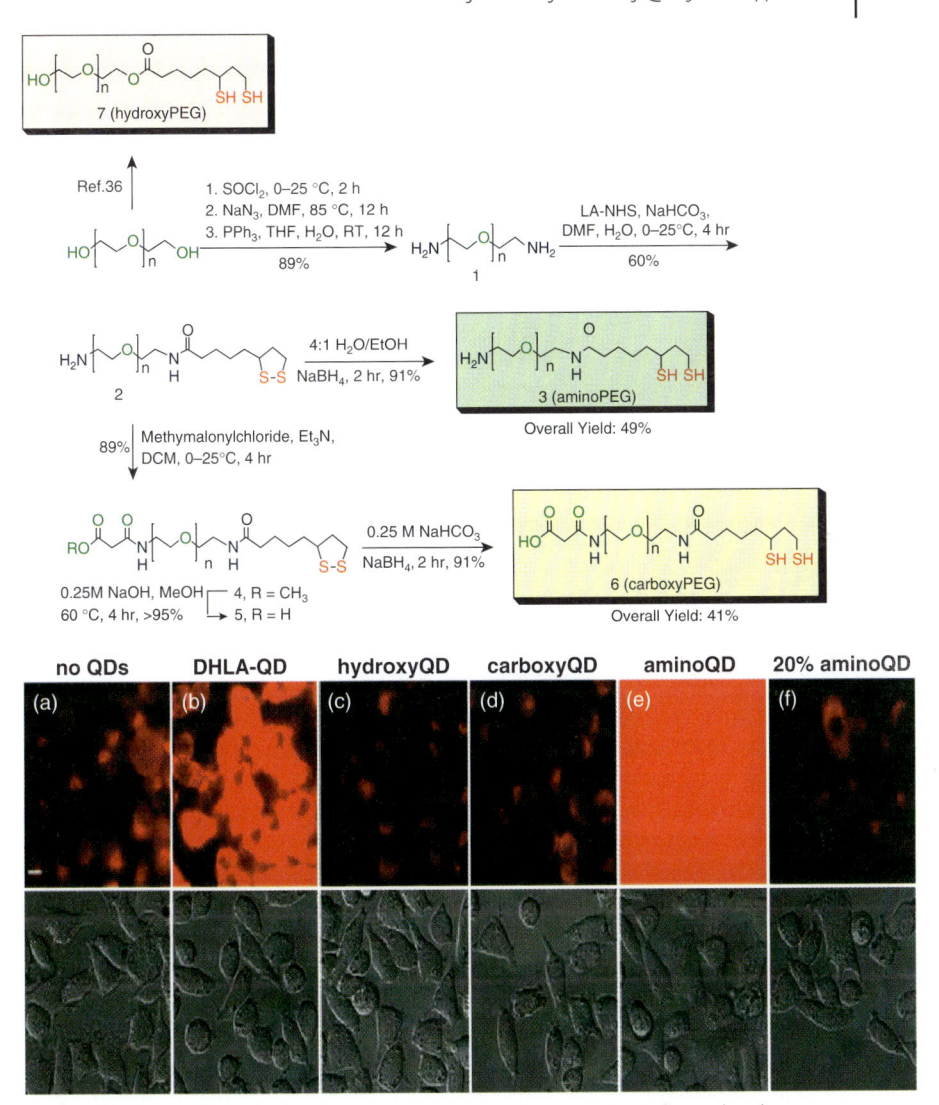

Figure 2.22 Schematic illustration of synthetic procedure of hydroxy-PEG, amino-PEG, and carboxy-PEG, and nonspecific binding of QD605 to HeLa cells as a function of ligand coating. Fluorescence images with 420 nm excitation/605 nm emission (top row). Differential interference contrast images (bottom row). (a) Control sample (showing cell autofluorescence), QDs ligand-exchanged with (b) DHLA, (c) hydroxyl-PEG, (d) carboxy-PEG, (e) amino-PEG, and (f) 20% amino-PEG (with 80% hydroxy-PEG). Scale bar = 10 μm. Reprinted with permission from Ref. [52]; © 2008, American Chemical Society.

activity of the latter molecules. As noted above, most water-soluble QDs have carboxyl, amine, and inherently hydrophobic groups on their outer surfaces, which allows the linking of biomolecules to the QD surface through three basic conjugation methods that can be categorized as covalent bonding, direct conjugation, and electrostatic interaction (Figure 2.23).

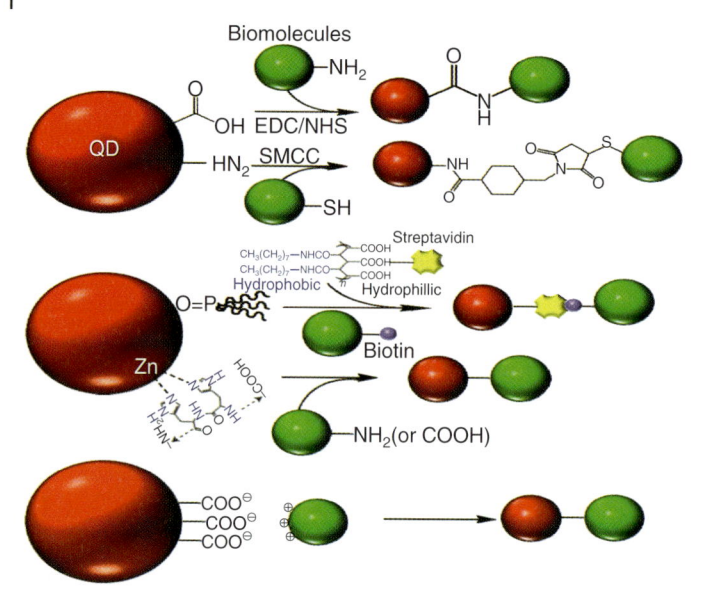

Figure 2.23 Schematic representation of conjugating methods of biomolecules to QDs. (a) Covalent bonding between carboxyl (or amine) groups on the QD surface and biomolecules; (b) Direct attachment of biomolecules to either Zn or S atoms on the QD surface; (c) Hydrophobic attachment between streptavidin-conjugated QDs and biotinylated biomolecules; (d) Electrostatic interactions between QD surfaces with negative surface charge and biomolecules with positive surface charge. Reprinted with permission from Ref. [68]; © 2007, Nature Protocols.

Currently, 1-ethyl-3-(3-dimethylaminopropyl) carbodiimide (EDC) is the most popular crosslinker for the covalent conjugation of a primary amine with a carboxyl group, whereas 4-(N-maleimidomethyl)-cyclohexanecarboxylic acid N-hydroxysuccinimide ester (SMCC) can be used to couple the primary amine with the thiol groups (Figure 2.23a). The use of these methods has led to numerous reports of QDs being conjugated with a variety of biological molecules, including biotin [62], oligonucleotides [63], peptides [64], and proteins including avidin/streptavidin [65], albumin [66], adapter proteins (e.g., protein A, protein G) and antibodies [65, 67]. However, it appears that QDs bioconjugated through the EDC crosslinker are more easily aggregated, presumably because the numerous surface functional sites can crosslink the many protein target sites.

A further bioconjugation method relies on direct bonding to the QD surface (see Figure 2.23b). For example, polyhistidine with a metal affinity could first be bonded to Zn atoms on the QD surface, after which the biomolecules could be coordinated to polyhistidine [68–74]. For those QDs with sulfur atoms on their surfaces, the thiol bonding of cysteine residues can be useful for conjugation with the QDs, and indeed the Weiss group has shown that phytochelatin peptides containing multiple cysteine residues can both "cap" a QD and impart subsequent biological activity [75–77].

By contrast, streptavidin-coated QDs may be used to conjugate with biotinylated proteins and antibodies (Figure 2.23c), and as a consequence have been shown to be useful for detecting Her2 cancer markers on the surface of SK-BR-3 human breast cancer cells [78]. Whilst the strong affinity of streptavidin–QDs towards biotinylated proteins and antibodies makes them an attractive proposition, they will unfortunately bind indiscriminately to all biotinylated proteins, such that only one targeted use is allowed. Moreover, the increased volume of the streptavidin–QD, when conjugated with a biotinylated antibody, may physically limit its attachment to some receptors.

The engineering of proteins to express positively charged domains allows them to self-assemble onto the surface of negatively charged QDs through an electrostatic assembly (Figure 2.23d). In fact, this approach has been proven useful for attaching a variety of engineered proteins to QDs, including maltose-binding protein (MBP), that express a positively charged leucine-zipper domain [48]. Following self-assembly, the resultant bioconjugate may be purified over an amylose resin. Further modification of this strategy allows both the attachment of antibodies to the QD, and purification of the bioconjugate for subsequent immunoassays [79–84]. Genetic engineering can also be used to introduce the positively charged leucine-zipper onto protein G, such that positively charged protein G can be immobilized on the QD and then used to bind the Fc region of antibodies [81, 84].

2.4 Methods of Delivering QDs into Tumors

2.4.1 Passive Targeting

In a fast-growing tumor, rapid vascularization leads to be an enhanced permeability and retention (EPR) effect [85–88]. Subsequently, the accumulation of QDs at the tumor site occurs by their leaking from the circulation, as tumors lack an effective lymphatic drainage system (Figure 2.24) [89]. This passive delivery of QDs into cancer cells relies on the inherent physico-chemical properties of the QDs, notably of the particle size and surface properties.

A variety of size-based clearance mechanisms has been shown to exist in the blood (Figure 2.25) [90–93]. For tumor targeting, the particle size should be less than 100 nm in diameter in order to maximize the circulation times, whereas particles larger than 500 nm can be easily phagocytosed by macrophages [90]. Chan et al. found the cell uptake for gold nanoparticles of 50 nm diameter to be much higher than for gold nanoparticles of 14 and 74 nm diameter [91]. In addition, Osaki et al. found the endocytosis of QDs to be highly size-dependent (50 nm > 15 nm > 5 nm), although the different surface properties of the QDs precluded any rigorous comparison [92]. When another group determined the size-dependent immunogenicity of polystyrene (PS) particles with diameters of 1 ~ 6 µm, the phagocytosis was maximal for particles of 2–3 µm [93]. Interestingly,

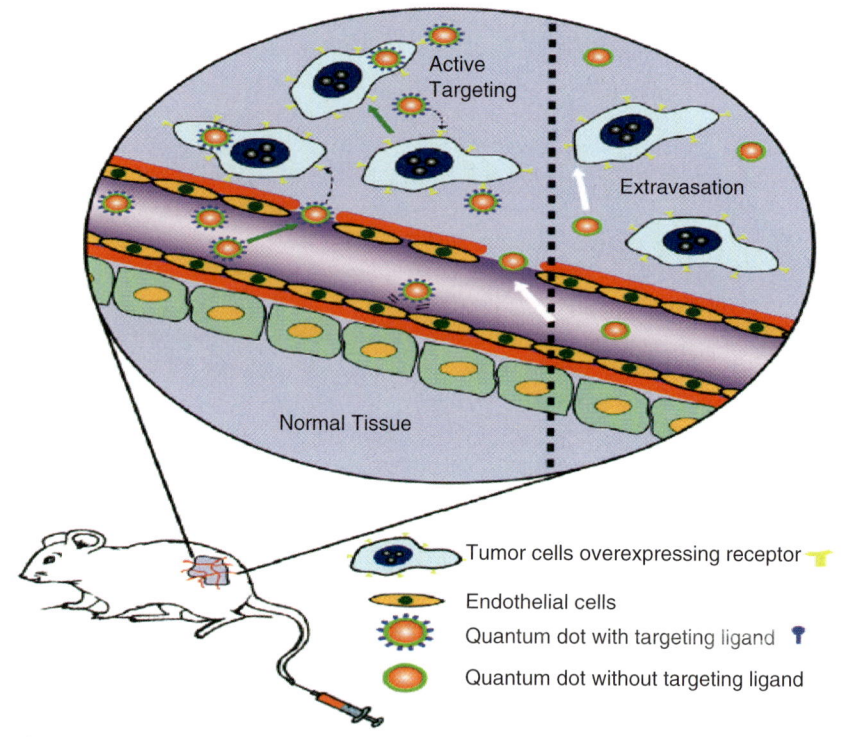

Figure 2.24 Schematic representation of QDs delivery process into the tumor site. In passive targeting, QDs are accumulated by an enhanced permeability and retention (EPR) effect. Active cellular targeting can be achieved by the bioconjugated QDs with peptides, protein, antibodies that promote cancer cell-specific binding. Reprinted with permission from Ref. [89]; © 2007, Elsevier.

this range was similar to the general size range of bacteria, which are the most common targets of macrophages.

The surface charges or ligands of the QDs must also be considered in the passive targeting of tumors. For example, when Gao et al. investigated the passive targeting of QDs with different surface ligands (Figure 2.26) [45], florescence emission was not observed from the tumor in the case of hydroxylated QDs, which indicated a rapid blood clearance by the reticuloendothelial system (RES). Such a low uptake of the carboxylated QDs derives from the negative charges on the QD surfaces due to carboxyl anions. In the case of PEGylated QDs, the organ uptake rate was reduced and the blood circulation time improved due to an antibiofouling effect of PEG, and this led in turn to a slow accumulation of QDs within the tumors. QDs conjugated with prostate-specific membrane antigen (PSMA)-specific monoclonal antibodies were also rapidly delivered by the tumor xenografts. This antibody-mediated active targeting is discussed further later in the chapter.

2.4 Methods of Delivering QDs into Tumors

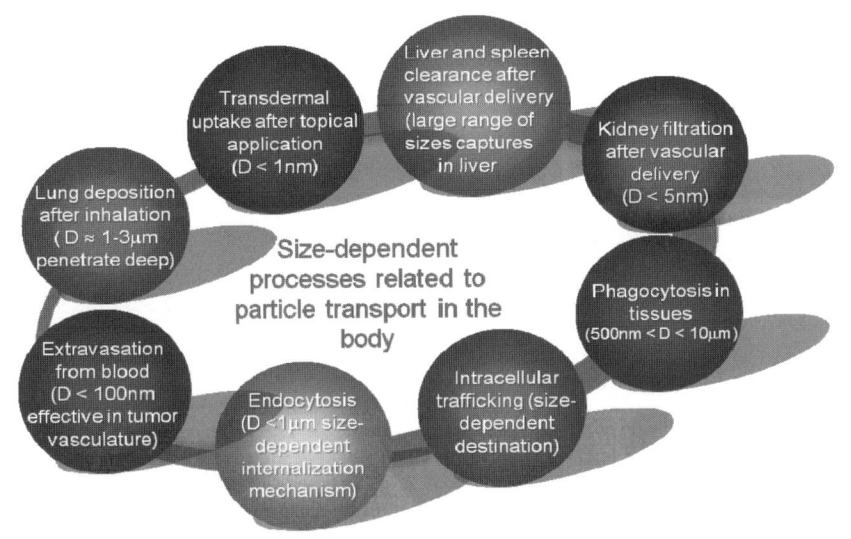

Figure 2.25 Size-dependent processes of particle transport in the human body. Particles can pass through biological barriers by a number of different processes. These include passive (diffusive) and active processes ranging from extravasation to transdermal uptake. Reprinted with permission from Ref. [90]; © 2009, *Nature Materials*.

Opsonization describes the mechanisms of the immune system whereby foreign particles (e.g., bacteria, viruses, or foreign bodies) are rendered more recognizable to those cells in the immune system which attack them. This usually involves the attachment of proteins (such as activated complement proteins or antibodies) to the foreign agent (such as a QD particle), such that the agent is "flagged" for attention by the attacking cells, such as macrophages. In passive targeting, the PEGylation of QDs represents a very useful means of increasing the blood circulation time, most likely by sterically hindering the adsorption of opsonizing proteins, which in turn delays the recognition and clearance of particles by the RES [94–99]. The blood circulation time will be changed by the functional group and chain length of the PEG coated onto the QDs. For example, carboxylated-PEG-coated QDs will be rapidly taken up by the RES, whereas amine-PEG-coated QDs will have varying half-lives within the circulation, depending on the molecular weight of the PEG. According to Ballou *et al.*, the circulating half-lives may be as short as 12 min for amphiphilic poly(acrylic acid), short-chain (750 Da) methoxy-PEG or long-chain (3400 Da) carboxy-PEG QDs, whereas the circulating half-life may be up to 70 min for long-chain (5000 Da) methoxy-PEG QDs (see Figure 2.27) [94–96].

The main advantage of passive delivery is simplicity. Although the QDs incubated with the cells are subsequently internalized by nonspecific endocytosis, high QD concentrations and long periods of time are generally required for passive targeting, and this may result in an enhancement of cytotoxicity.

(a)

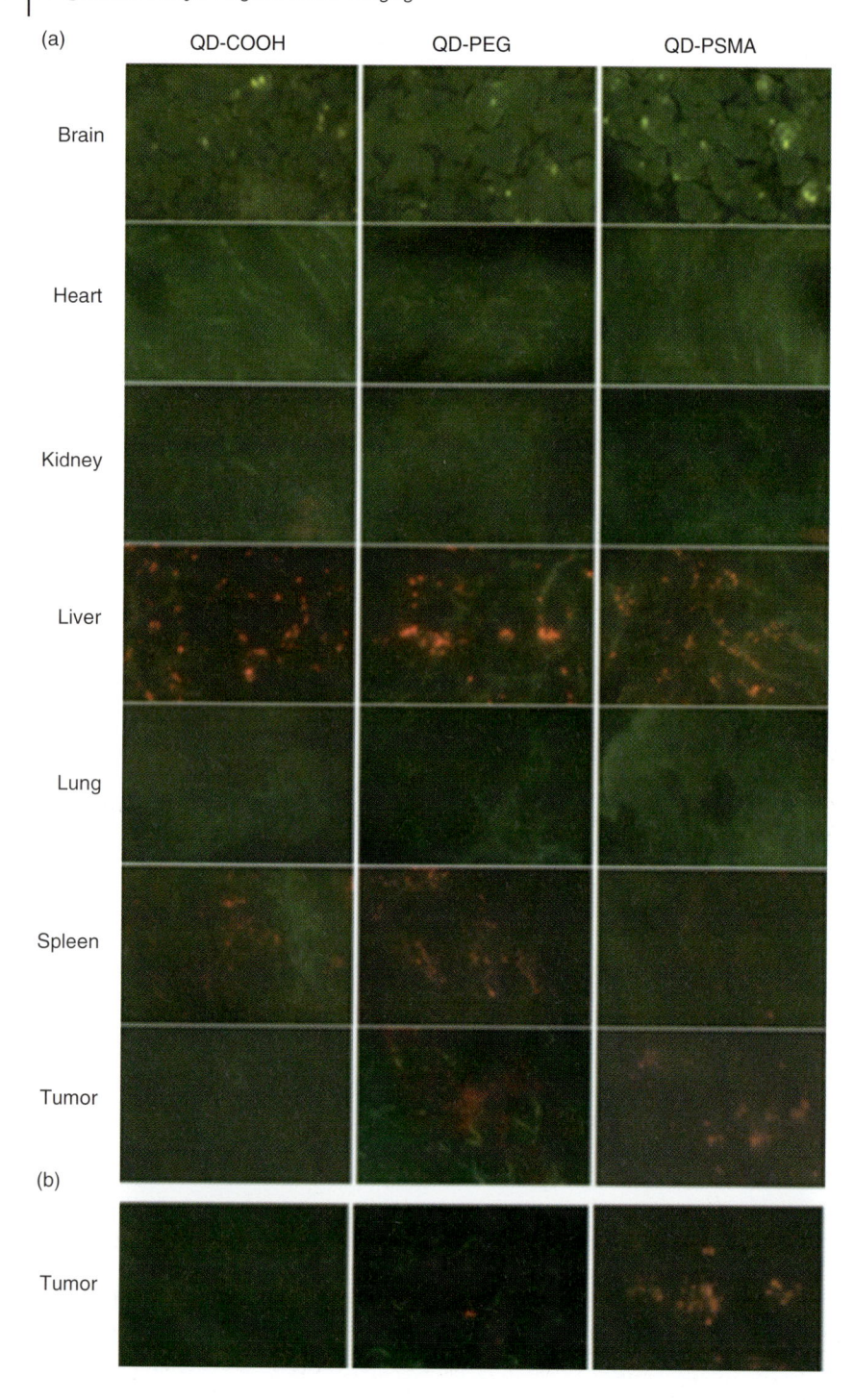

| | QD-COOH | QD-PEG | QD-PSMA |

Brain

Heart

Kidney

Liver

Lung

Spleen

Tumor

(b)

Tumor

Figure 2.26 Histological examination of three different QDs uptake, retention, and distribution in different normal host organs and in C4-2 tumor xenografts maintained in athymic nude mice. (a) Left column: QD coated with surface carboxylic acid groups (6.0 nmol and 6 h circulation). Middle column: QD with surface coated with PEG groups (6.0 nmol and 24 h circulation). Right column: QD with surface modified by PEG and bioconjugated with a prostate-specific membrane antigen (PSMA) antibody (0.4 nmol and 2 h circulation); (b) Same as panel (a), except that the amount of QD injection was reduced to 0.4 nmol and the circulation was reduced to 2 h. QDs were detected by their characteristic red-orange fluorescence; all other signals were due to background autofluorescence. Adapted from Ref. [45].

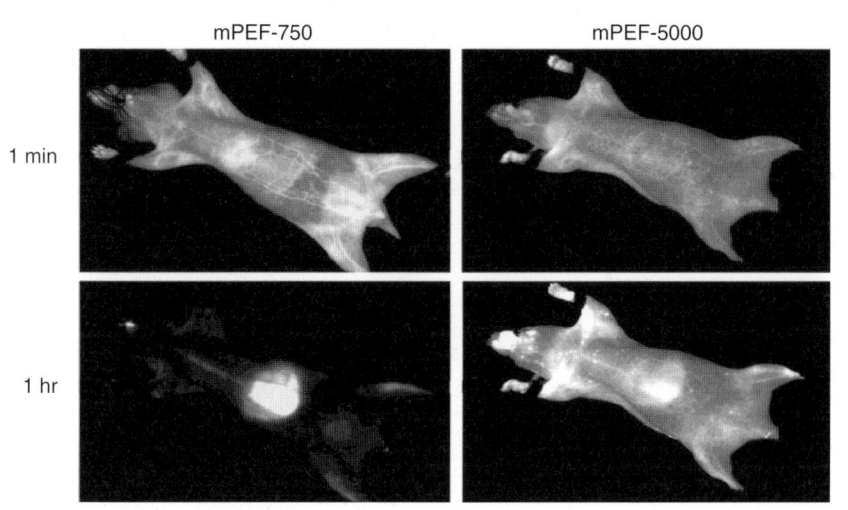

Figure 2.27 Different surface coatings of QDs (CdSe/ZnS) result in different *in vivo* kinetics. mPEG-750-coated QDs circulate much more quickly than mPEG-5000-coated QDs. Even at 1 min, a significant liver uptake of mPEG-750-coated QDs is visible. At 1 h, the mPEG-750-coated QDs were completely cleared from the circulation, while mPEG-5000-coated QDs persisted. Reprinted with permission from Ref. [94]; © 2004, American Chemical Society.

2.4.2
Active Targeting

2.4.2.1 Peptide-Mediated Tumor Uptake

Peptides or peptidomimetics are potentially better as active targeting ligands than antibodies, because many peptides (tens or even hundreds) can be combined to the surface of single QD compared to antibody–QD conjugates. Consequently, peptide–QD conjugates may exhibit a stronger binding affinity and a better targeting efficacy than antibody–QD conjugates because of the polyvalency effect [99].

Specific tumor targeting in living subjects was first reported by using peptide-conjugated QDs (CdSe/ZnS) [100] injected intravenously into MDA-MB-435 breast carcinoma xenograft-bearing nude mice. Three peptides were tested: (i) CGFECVRQCPERC (denoted as GFE), which binds to membrane dipeptidase on

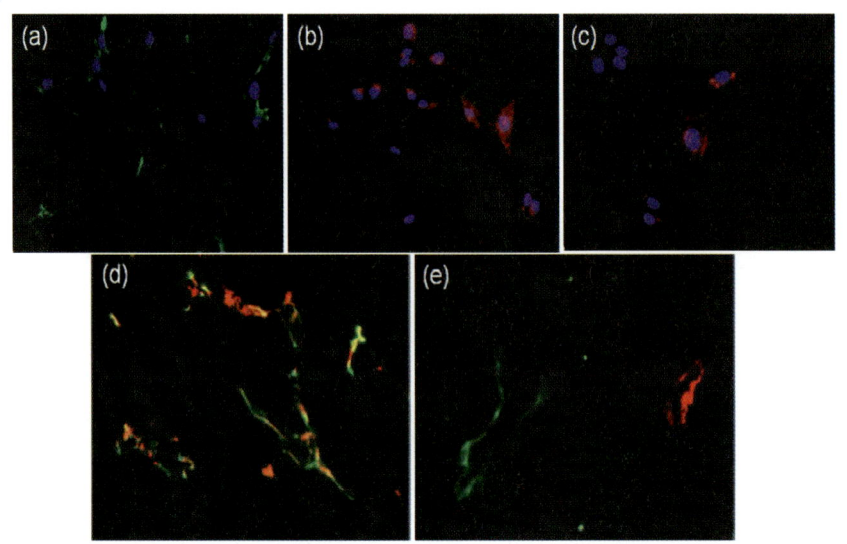

Figure 2.28 (a–c) In vitro targeting of peptide-conjugated QDs to endothelial cells and breast cancer cells. (a) Binding of GFE-conjugated QDs to lung endothelial cells that express membrane dipeptidase; Binding of (b) QD-F3 and (c) QD-LyP-1 to MDA-MB-435 breast carcinoma cells; (d,e) In vivo targeting of peptide-conjugated QDs to tumor vasculature. Red F3- or LyP-1-conjugated QDs, both PEG-coated, were injected into the tail vein of nude mice bearing MDA-MB-435 breast carcinoma xenograft tumors. Blood vessels were visualized by coinjecting tomato lectin (green); (d) QD-F3 colocalize with blood vessels in tumor tissue; (e) QD-LyP-1 also accumulate in tumor tissue, but do not colocalize with the blood vessel marker. Adapted from Ref. [100].

the endothelial cells [101, 102]; (ii) KDEPQRRSARLSAKPAPPKPEPKPKKAPAKK (denoted as F3), which binds preferentially to blood vessels and tumor cells in various tumors [103]; and (iii) CGNKRTRGC (denoted as LyP-1), which recognizes lymphatic vessels and tumor cells in certain tumors [104]. Since the QDs used in this study emit in the visible range (550 nm and 625 nm fluorescence maxima), and this was not optimal for in vivo imaging, a series of ex vivo histological analyses were carried out to show that the QDs had been specifically directed to the tumor vasculature and organ targets by the surface peptide molecules (see Figure 2.28). A high level of PEG substitution on the QDs was found to be important for reducing nonselective accumulation in the RES, thereby increasing the circulation half-life and targeting efficiency. QD-F3 was found to colocalize with blood vessels in the tumor tissue, while QD-LyP-1 also accumulated in the tumor tissue but did not colocalize with the blood vessel marker (Figure 2.28d,e). These results were the first to demonstrate the feasibility of specific targeting of QD in vivo, and opened up a new field of QD-based research.

Recently, the in vivo targeted imaging of tumor vasculature was reported using peptide-conjugated QDs [105]. In this situation, integrin $\alpha_v\beta_3$, which binds to the RGD-containing components of the interstitial matrix (such as vitronectin, fibronectin, and thrombospondin) is overexpressed on activated endothelial cells and tumor cells, but is not readily detectable in resting endothelial cells and most

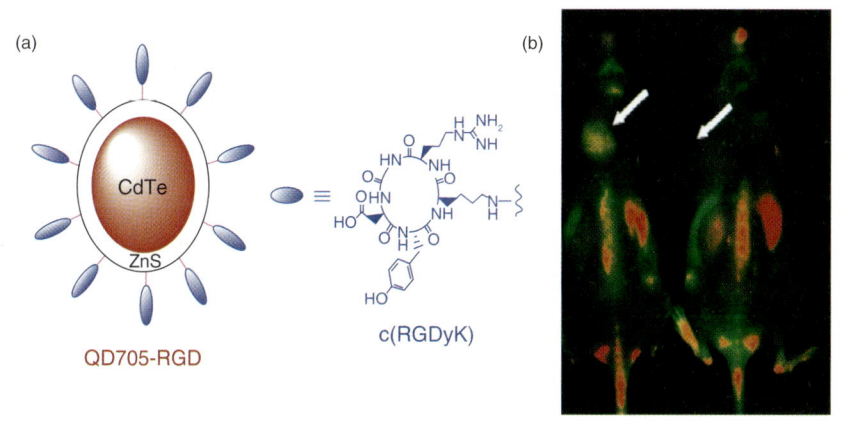

Figure 2.29 (a) A schematic illustration of RGD Peptide-conjugated QD705; (b) *In vivo* NIRF imaging of tumor vasculature in U87MG human glioblastoma-tumor-bearing mice. The mouse on the left was injected with QD705-RGD and the mouse on the right with QD705. The arrows indicate tumors. Adapted from Ref. [105].

normal organ systems [106]. Many previous studies have shown that, as a cell adhesion molecule, integrin $\alpha_v\beta_3$ may serve as excellent target for tumor imaging purposes because it plays a key role in tumor angiogenesis and metastasis [107–113]. In these pioneering studies, arginine-glycine-aspartic acid (RGD; a potent integrin $\alpha_v\beta_3$ antagonist) containing peptides was conjugated to QD705 (CdTe/ZnS core–shell QDs; emission maximum at 705 nm), such that QD705–RGD exhibited a high-affinity integrin $\alpha_v\beta_3$-specific binding in cell culture and ex vivo. *In vivo* NIR fluorescence (NIRF) imaging was successfully achieved in nude mice bearing subcutaneous integrin $\alpha_v\beta_3$-positive U87MG human glioblastoma tumors, where tumor fluorescence intensity reached a maximum at 6 h post injection (Figure 2.29) [105, 114, 115]. The size of the QD705–RGD (ca. 20 nm diameter) prevented efficient extravasation, and consequently the QD705–RGD mainly targeted the tumor vasculature rather than the tumor cells. This was directly confirmed using intravital microscopy with subcellular (ca. 0.5 μm) resolution, in which the binding of QD800–RGD conjugates to tumor blood vessels was directly observed in an SKOV-3 mouse ear tumor model [116]. Using this method, it was revealed that QD800–RGD does not extravasate in an SKOV-3 mouse ear tumor model, but specifically binds the target in the tumor neovasculature as aggregates rather than individually. As the sprouting neovasculature in many tumor types overexpresses integrin $\alpha_v\beta_3$, QD705–RGD appears to show great potential as a universal NIRF probe for detecting tumor vasculature in living subjects.

2.4.2.2 Antibody-Mediated Uptake

Angiogenic tumors produce vascular endothelial growth factor (VEGF) [117–119], which hyperpermeabilizes the tumor neovasculature and causes leakage of the circulating macromolecules and nanoparticles. Subsequent macromolecule or

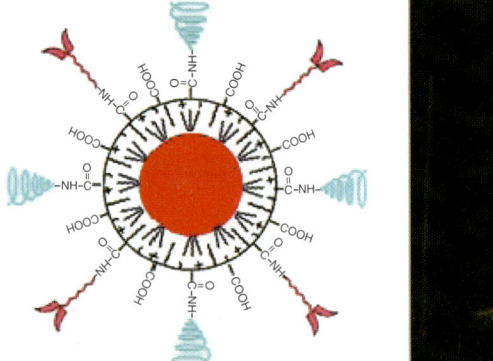

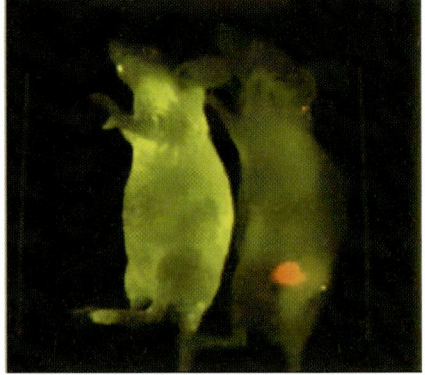

Figure 2.30 Antibody-conjugated QDs for *in vivo* cancer targeting and imaging. The mouse on the left was a control. Adapted from Ref. [45].

nanoparticle accumulation occurs as the tumor lacks any effective lymphatic drainage system. ABC triblock copolymer-coated QDs for prostate cancer targeting and imaging in live animals have been reported, as previously mentioned [45], and recent studies have identified PSMA as a cell-surface marker for both prostate epithelial cells and neovascular endothelial cells [120]. When polymer-coated QDs were conjugated to PSMA-specific monoclonal antibodies, it was estimated that there were five to six antibody molecules per QD. By using spectral imaging techniques where the fluorescence signals from QDs and mouse autofluorescence can be separated, based on the emission spectra [121], the intravenously injected probes were found to accumulate within the tumor site (Figure 2.30) [45]. Multiplexed imaging was also demonstrated in live animals using QD-labeled cancer cells. However, as no histological analysis was carried out to investigate the expression level of PSMA on the tumor cells and tumor vasculature, it was unclear whether these QD conjugates had targeted the tumor vasculature or tumor cells. In addition, the QDs used in these studies were not optimized for tissue penetration or imaging sensitivity, because the emission wavelength was in the visible region instead of the NIR region.

In a recent study, QDs were linked to anti-alpha-fetoprotein (AFP; a marker for hepatocellular carcinoma cell lines) antibody for *in vivo* tumor targeting and imaging [122]. Although no *in vitro* validation of the QD probe was carried out before the *in vivo* experiments, it was reported that active tumor targeting and spectroscopic hepatoma imaging was achieved using an integrated fluorescence imaging system. The heterogeneous distribution of the QD-based probe in the tumor was also evaluated. Unfortunately, one major flaw of this study was that it was unclear whether the anti-AFP antibody was actually attached to the QD, or not, and consequently there was insufficient experimental evidence to support the conclusion that the tumor contrast observed was from active, rather than passive, targeting.

The tracking of a single QD conjugated with a tumor-targeting antibody in tumors of living mice was achieved using a dorsal skinfold chamber and a high-speed confocal microscope fitted with a high-sensitivity camera [123]. For this, QDs labeled with anti-HER2 monoclonal antibodies were injected into mice bearing HER2-overexpressing breast cancer, so as to analyze the molecular processes of its tumor delivery. The movement of a single QD–antibody conjugate (the total number of QD particles injected was ca. 1.2×10^{14}) was observed at 30 frames per second inside the tumor, through the dorsal skinfold chamber. The QDs were observed during six processes of delivery: in a blood vessel; during extravasation; in the extracellular region; binding to HER2 on the cell membrane; moving from the cell membrane to the perinuclear region; and in the perinuclear region. At each stage, movement of the QD–antibody conjugate followed a "stop-and-go" pattern. Despite the technical difficulties of the experiment, no information was obtained with regards to the percentage of intravenously injected QDs that was extravasated, and therefore little can be concluded concerning the overall behavior of these QD–antibody conjugates *in vivo*. It was also unclear whether the "stop-and-go" pattern was typical for the majority of injected QD conjugates, or if it was limited to a small subset of QDs. It is likely that the majority of the QD conjugates were taken up by the RES shortly after injection, and that only certain QD conjugates, such as the smallest particles, were actually extravasated.

2.5
Recent Advances in QDs Technology

2.5.1
Bioluminescence Resonance Energy Transfer (BRET)

Recently, QDs have shown great potential for molecular imaging and cellular investigations of biological processes, although the need for external light excitation may partially offset the favorable tissue penetration properties of NIR QDs. This type of excitation also results in significantly increased background autofluorescence, but the use of direct bioluminescence light to excite the QDs has partially overcome this problem [124]. In the past, whilst luciferases have been widely used as reporter genes in biological research [125, 126], the bioluminescence activity of commonly used luciferases is too labile in serum. Specific mutations of *Renilla* luciferase, selected using a consensus sequence-driven strategy, were screened for their ability to confer stability of activity in serum, as well as their light output [127]. A mutant *Renilla* luciferase with eight mutations (RLuc8) was selected with a 200-fold increase in resistance to inactivation in murine serum, and a fourfold increase in light output. Multiple molecules of RLuc8 were covalently conjugated to a single fluorescent QD, thus forming a conjugate which was about 22 nm in terms of hydrodynamic diameter (Figure 2.31) [124]. When RLuc8 binds its substrate coelenterazine, it converts chemical energy into photon energy and emits broad-spectrum blue light that peaks at 480 nm. Due to a complete overlap of the

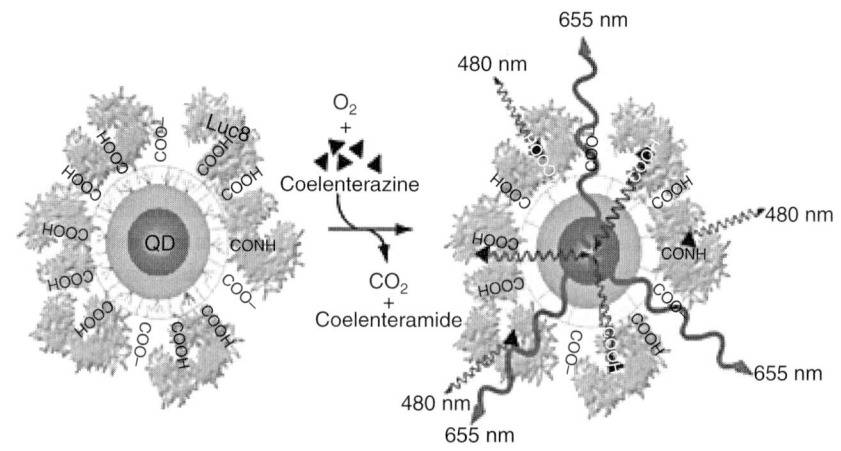

Figure 2.31 Schematic representation of self-illuminating QDs based on bioluminescence resonance energy transfer. Adapted from Ref. [124].

RLuc8 emission and QD absorption spectra, the QDs were efficiently excited in the absence of external light. *In vivo* imaging showed a greatly enhanced signal-to-background ratio after injection of the QD-RLuc8 conjugate into the bloodstream. RLuc8 can serve as a BRET donor for virtually any QD, and these probes can be used for multiplexed imaging. Moreover, BRET has the potential to greatly improve NIRF detection in living tissues, and similar QD conjugates can be obtained when RLuc8 is fused to other proteins, thus enabling new possibilities for imaging biological events [128]. One of the major goals that BRET will have to achieve before being widely used for *in vivo* imaging is that of targeting specificity. Since there are many RLuc8 molecules on the QD surface which cover the majority of the QD surface area, it remains to be tested whether there will be sufficient space remaining to which enough ligands can be attached for desirable targeting efficacy.

2.5.2
Non-Cd-Based QDs

Currently, many serious questions and concerns have been raised regarding the cytotoxicity of inorganic QDs containing Cd, Se, Zn, Te, Hg, and Pb [129–131]. These chemicals can be potent toxins, neurotoxins, and/or teratogens, depending on the dosage, complexation, and accumulation in the liver and nervous system. At very low doses, these metals are bound by metallothionein proteins and may be either excreted slowly or sequestered *in vivo* in adipose and other tissues [129, 132]. Cadmium, which has a half-life of about 20 years in humans, is a suspected carcinogen that can accumulate in the liver, kidney, and many other tissues, since there is no known active mechanism for cadmium to be excreted from the human body [133]. Although many studies have found no adverse effects of QDs on cell viability, morphology, function, or development over the duration of experiments (hours to days) at concentrations optimized for labeling efficiency [54, 65, 134, 135], the cellular toxicity of QDs under extreme conditions such as photo-oxidation and strong UV excitation has been clearly demonstrated [132, 136]. In general, the less protected the QD core or core–shell is, the sooner the appearance of signs of interference with cell viability or function as a result of Cd^{2+} and/or Se^{2-} release. Thick ZnS overcoating (four to six monolayers), in combination with efficient surface capping, has been shown to substantially reduce the desorption of core ions and to make the QDs more biologically inert [132]. Interestingly, the toxicity of QDs has been utilized for photodynamic therapy applications such as tumor ablation [137, 138]. As QD technology evolves and brighter probes are created with improved detection efficiency, the easiest way to decrease cytotoxic effects would be to use smaller quantities of QDs. In many cases, the amount of free Cd^{2+} ions released by QDs is far below the dose needed to cause cadmium poisoning in animal models. InAs-based QDs may serve as a substitute for Cd-based QDs, with a lower cytotoxicity [137–141]. The amount of As used is estimated to be several orders of magnitude lower than the dose of As_2O_3 used to treat human leukemia. Mn- or Cu-doped zinc chalcogenide QDs have been reported that can cover a similar emission window as that of CdSe QDs [142, 143]. Besides reducing the toxicity by replacing Cd with Zn, these QDs are also less sensitive to environmental changes, such as thermal, chemical, and photochemical disturbances. These doped QDs also have color-tunability with good quantum efficiency, and are promising candidates for future efforts to reduce QD-based cytotoxicities. They also have narrow emission spectra (45–65 nm full width at half maximum), and can cover most of the visible spectral window. Most recently, a high-quality $CuInS_2$/ZnS core–shell system with a QY of 30% was developed by using generic and air-stable chemicals in a noncoordinating solvent [144]. For $CuInS_2$/ZnS QDs, an emission peak is tunable from 500 to 950 nm by controlling the core size (Figure 2.32) [144]. In the near future, it is expected that such doped QDs that emit in the NIR region will be applied for *in vivo* molecular imaging.

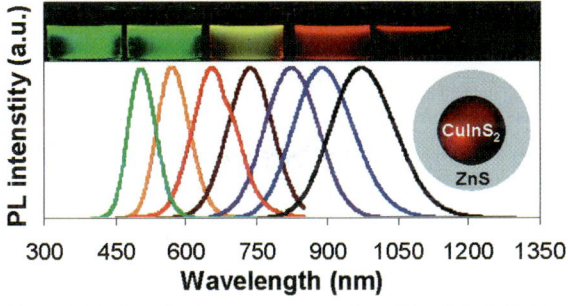

Figure 2.32 Photoluminescence properties of the $CuInS_2$/ZnS core–shell nanocrystals. Adapted from Ref. [144].

2.5.3
Moving Towards Smaller QDs

For inorganic nanoparticles such as QDs, the particle size and shape is relatively rigid compared to other organic nanoparticles, such as dendrimers. To date, most of the QDs evaluated *in vivo* have been 15 nm or more in hydrodynamic diameter. Although the tumor vasculature is typically quite leaky, such a small size does not permit efficient extravasation and it is expected that, with smaller sizes, QDs will extravasate more efficiently and provide a better *in vivo* targeting of both the tumor vasculature and tumor cells. Smaller-sized QDs are also expected to have a lower RES uptake, which will in turn translate into a better image quality. Different core–shell structures and thinner polymer coatings have been reported to reduce the overall size of QDs [143, 145], and unusually small, water-soluble QDs composed of InAs/ZnSe (core diameter <2 nm) have been developed [143]. Although these QDs have a lower QY (<10%), the smaller size is attractive for imaging applications. These unusually small QDs were not trapped in sentinel lymph nodes (SLNs), but rather migrated into the lymphatic system and the channels between the nodes. In addition, these small QDs were also able to migrate out of the blood vessels and into the interstitial fluid. Dendron-coated QDs have high stability, versatility, and chemical/biochemical proccessibility [146, 147]. Unlike the typical polymer coating, dendron-ligands are tight and small in radial dimension, which results in an overall smaller size of QDs. The surface density and length of the PEG units on the outer surface of the resulting dendron-coated QDs can be varied by synthesizing dendron ligands with different terminal structures. For this, a "peptide toolkit" has been reported which can provide a straightforward means of improving biocompatibility for cell biology and *in vivo* applications [64]. In the future, it is likely that small-molecule or peptide-coated QDs will provide better opportunities for the development and expansion of *in vivo* applications than will protein- or antibody-conjugated QDs.

2.5.4
Multifunctional Probes

Among all of the molecular imaging modalities currently available, no single modality is perfect and sufficient to obtain all necessary information [148]. Due to the current obstacles in fluorescence tomography [149–151], it is difficult to adequately quantify QD signals in living subjects based on fluorescence intensity alone, especially in deep tissues. However, a combination of QD-based imaging with 3-D tomography techniques such as PET, SPECT and MRI can permit the elucidation of targeting mechanisms, biodistribution, and dynamics in living animals with higher sensitivity and/or accuracy. One of the most promising applications for QDs is the development of multifunctional QD-based probes for multimodality molecular imaging *in vitro* and *in vivo*. A multimodality approach would make it possible to image targeted QDs at all scales, from whole-body down to nanometer resolution, using a single probe. A series of core–shell CdSe/Zn$_{1-x}$Mn$_x$S nanoparticles has been synthesized for use in both optical imaging and MRI [152]. In this case, the Mn^{2+} content was in the range of 0.6 to 6.2%, and varied with the thickness of the shell or amount of Mn^{2+} introduced to the reaction. The QY and Mn^{2+} concentration in the nanoparticles were sufficient to produce contrast for both modalities at a relatively low concentration. Bifunctional nanocomposite systems consisting of Fe$_2$O$_3$ magnetic nanoparticles and CdSe QDs have been synthesized [153]. QDs can be coated with paramagnetic and PEGylated lipids for use as detectable and targeted probes with MRI, as shown in Figure 2.20 [63, 154]. These QDs are useful as dual-modality contrast agents due to their high relaxivity and an ability to retain their optical properties. The details of several other QDs-based probes for both fluorescence imaging and MRI have also been reported [154–156]. For example, polymer-coated Fe$_2$O$_3$ cores overcoated with a CdSe–ZnS QD shell and functionalized with antibodies have been used to magnetically capture breast cancer cells and view them with fluorescence imaging [157]. Likewise, magnetic QDs composed of CdS–FePt have been synthesized [158]. The QDs have relatively large surface areas which can be conjugated with more than one targeting ligand. In this way, novel tumor-specific antibody fragments, growth factors, peptides, and small molecules can be attached to QDs for their delivery to tumors *in vivo* for the multiparameter imaging of biomarkers, with the ultimate goal of guiding therapy selection and predicting responses to therapy. This nanoplatform approach will enable the simultaneous detection and measurement of many biomarkers, which should in turn lead to a better signal/contrast than QDs modified with only one type of targeting ligand. The ability to accurately assess the pharmacokinetics and tumor targeting efficacy of biologically modified QDs is crucially important to assess future studies that involve multi-targeting (i.e., targeting multiple targets with one QD) and, eventually, multiplexing (i.e., targeting multiple targets simultaneously using QDs of different emission wavelengths). Dual-modality PET/NIRF imaging probes offer synergistic advantages over single-modality imaging probes by overcoming the difficulty of quantifying fluorescence intensity both *in vivo* and *ex vivo*. Subsequently, the tumor-targeting efficacy of dual-functional QD-based probes was evaluated quantitatively using both NIRF and PET imaging (Figure 2.33) [159], when both RGD peptides and the macrocyclic chelator, DOTA, were conjugated to QD705. Here, the RGD peptides allowed for integrin $\alpha_v\beta_3$ targeting, while the DOTA was able to complex ^{64}Cu (a positron emitter with 12.7 h half-life) so as to enable PET imaging [108, 160]. Noninvasive PET imaging using radiolabeled QD conjugates may provide a robust and reliable measure of the *in vivo* biodistribution of QDs. Yet, with further improvements in QD technology, it is expected that the accurate evaluation of *in vivo* tumor targeting efficacy using quantitative imaging modalities (e.g., PET) will greatly facilitate the future biomedical applications of QDs. Such information will also be critical for fluorescence-guided surgery via the sensitive, specific, and real-time intraoperative visualization of molecular features of normal and disease processes.

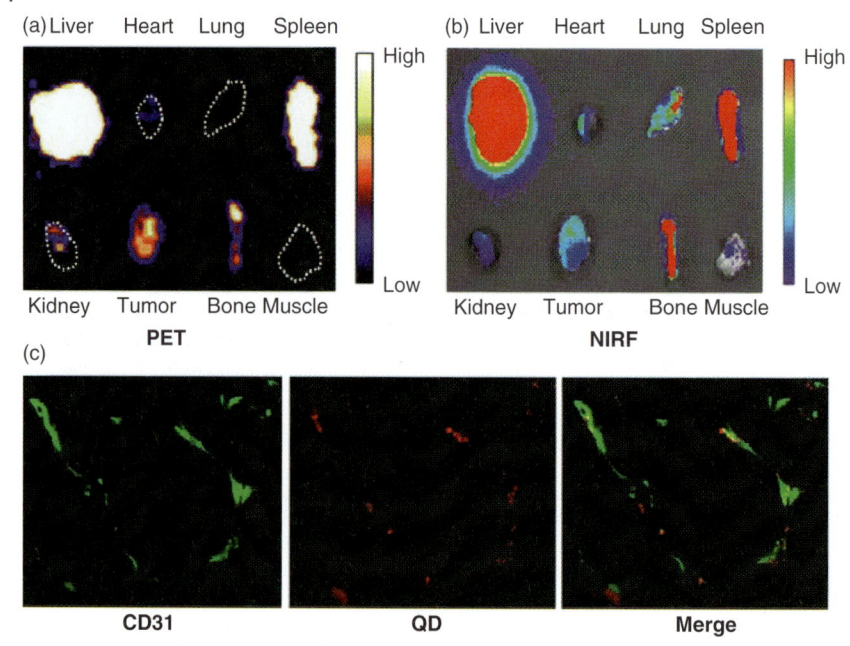

Figure 2.33 Dual-functional QDs based probe for both PET and NIRF imaging. (a) PET image of harvested major organs/tissues at 5 h post-injection of the dual-functional probe; (b) NIRF image of harvested major organs/tissues at 5 h post-injection of the probe; (c) Immunofluorescence staining of the tumor tissue revealed that QDs are targeting the tumor vasculature. Adapted from Ref. [159].

2.6
Conclusion and Perspectives

Since the first demonstration of QDs for biological applications [1, 2], numerous breakthroughs in QD technology have led to the recent success of *in vivo* tumor targeted imaging of QDs in live animals. The future development of improved QD-based biological probes for *in vivo* optical imaging shows great promise for both basic science and clinical applications. Today, QDs-based molecular imaging has the potential to significantly impact cancer diagnosis and cancer patient management. It is expected that *ex vivo* diagnostics, in combination with *in vivo* diagnostics, can markedly impact future cancer patient management by providing a synergistic approach that neither strategy can provide alone. After further development and validation, QD-based approaches (both *ex vivo* nanosensors and *in vivo* imaging) will eventually be able to predict which patients will likely respond to a specific anticancer therapy, and to monitor their response to personalized therapy (Figure 2.34). With their capacity to provide enormous sensitivity, throughput, and flexibility, QDs have the potential to profoundly impact cancer patient management in the future. Unfortunately, QD-based tumor imaging in mice cannot be

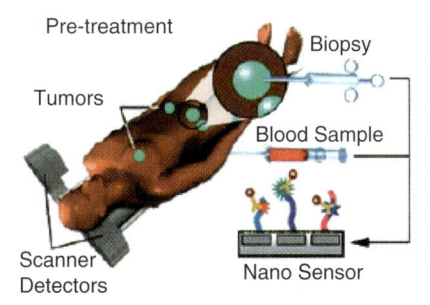

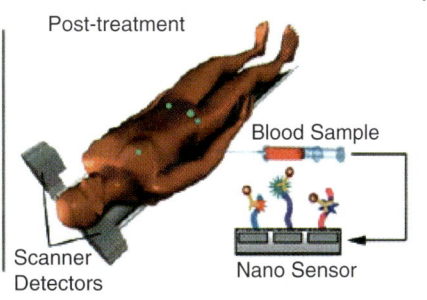

Figure 2.34 Patients can have their tumors biopsied and blood samples drawn for protein profiling by ex vivo nanosensors to predict their response to a given therapy. In addition, they will also be imaged with molecular imaging probes of different types to predict their response. Post-treatment and potentially during treatment, patient response will be evaluated by blood analysis and molecular imaging to ensure the accurate differentiation of responders from non-responders. Adapted from ref. [115].

directly scaled up to *in vivo* imaging in human applications, due to limited optical signal penetration depth. However, within clinical settings optical imaging is relevant for tissues close to the surface of the skin, to tissues accessible by endoscopy, and to intraoperative visualization. Currently, NIR optical imaging devices to detect and diagnose breast cancer are undergoing tests in patients, and the initial results have been encouraging [160, 161]. Multiwavelength QDs emitting in the NIR region can allow for the multiplexed imaging of deeper tissues, thus significantly extending potential human applications. QD-based multitarget imaging might also play an important role in optically guided surgery in the future. Overall, the major roadblocks for the clinical translation of QDs are inefficient delivery, toxicity, and a lack of quantification. However, with the development of smaller, non-Cd-based multifunctional QDs, and further improvements in conjugation strategy, it is expected that QDs will achieve optimal tumor targeting efficacy with an acceptable toxicity profile for clinical translation in the near future, using either NIRF imaging alone or multimodality imaging.

Acknowledgments

These studies were supported in part by grants NCI R21 CA121842, P50 CA114747, and U54 CA119367. Eue-Soon Jang would like to acknowledge partial support from the Korea Research Foundation (KRF) to join Stanford University.

References

1 Dunhan, W. (2007) Report sees 7.6 million global 2007 cancer deaths. American Cancer Society, 17 December 2007.

2 Fidler, I.J. (2003) The pathogenesis of cancer metastasis: the "seed and soil" hypothesis revisited. *Nature Reviews Cancer*, **3**, 453–8.

3 Weissleder, R. and Pittet, M.J. (2008) Imaging in the era of molecular oncology. *Nature*, **452**, 580–9.
4 Resch-Genger, U., Grabolle, M., Cavaliere-Jaricot, S., Nitschke, R. and Nann, T. (2008) Quantum dots versus organic dyes as fluorescent labels. *Nature Methods*, **5**, 763–75.
5 Alivisatos, A.P. (1997) Nanocrystals: building blocks for modern materials design. *Endeavour*, **21**, 56–60.
6 Michalet, X., Pinaud, F.F., Bentolila, L.A., Tsay, J.M., Doose, S., Li, J.J., Sundaresan, G., Wu, A.M., Gambhir, S.S. and Weiss, S. (2005) Quantum dots for live cells, in vivo imaging, and diagnostics. *Science*, **307**, 538–44.
7 Reiss, P., Bleuse, J. and Pron, A. (2002) Highly luminescent CdSe/ZnSe core/shell nanocrystals of low size dispersion. *Nano Letters*, **2**, 781–4.
8 Sun, Y.H. et al. (2006) Photostability and pH sensitivity of CdSe/ZnSe/ZnS quantum dots in living cells. *Nanotechnology*, **17**, 4469–76.
9 Ziegler, J., Merkulov, A., Grabolle, M., Resch-Genger, U. and Nann, T. (2007) High quality ZnS shells for CdSe nanoparticles-a rapid, low toxic microwave synthesis. *Langmuir*, **23**, 7751–9.
10 Nida, D.L., Nitin, N., Yu, W.W., Colvin, V.L. and Richards-Kortum, R. (2008) Photostability of quantum dots with amphiphilic polymer-based passivation. *Nanotechnology*, **19**, 035701.
11 Reiss, P., Protière, M. and Li, L. (2009) Core/shell semiconductor nanocrystals. *Small*, **5**, 154–68.
12 Wei, S.H. and Zunger, A. (1998) Calculated natural band offsets of all II–VI and III–V semiconductors: chemical trends and the role of cation d orbitals. *Applied Physics Letters*, **72**, 2011–13.
13 Li, J.J., Wang, Y.A., Guo, W.Z., Keay, J.C., Mishima, T.D., Johnson, M.B. and Peng, X.G. (2003) Large-scale synthesis of nearly monodisperse CeSe/CdS core/shell nanocrystals using air-stable reagents via successive ion layer adsorption and reaction. *Journal of the American Chemical Society*, **125**, 12567–75.
14 Hines, M.A. and Guyot-Sionnest, P. (1996) Synthesis and characterization of strongly luminescing ZnS-Capped CdSe nanocrystals. *Journal of Physical Chemistry*, **100**, 468–71.
15 Talapin, D.V., Rogach, A.L., Kornowski, A., Haase, M. and Weller, H. (2001) Highly luminescent monodisperse CdSe and CdSe/ZnS nanocrystals synthesized in a hexadecylamine-trioctylphosphine oxide-trioctylphospine mixture. *Nano Letters*, **1**, 207–11.
16 Dabbousi, B.O., RodriguezViejo, J., Mikulec, F.V., Heine, J.R., Mattoussi, H., Ober, R., Jensen, K.F. and Bawendi, M.G. (1997) (CdSe)ZnS core-shell quantum dots: synthesis and characterization of a size series of highly luminescent nanocrystallites. *Journal of Physical Chemistry, B*, **101**, 9463–75.
17 Peng, X., Schlamp, M.C., Kadavanich, A.V. and Alivisatos, A.P. (1997) Epitaxial growth of highly luminescent CdSe/CdS core/shell nanocrystals with photostability and electronic accessibility. *Journal of the American Chemical Society*, **119**, 7019–29.
18 Chen, X., Lou, Y., Samia, A.C. and Burda, C. (2003) Coherency strain effects on the optical response of core/shell heteronanostructures. *Nano Letters*, **3**, 799–803.
19 Frangioni, J.V. (2003) In vivo near-infrared fluorescence imaging. *Current Opinion in Chemical Biology*, **7**, 626–34.
20 Cheng, Z., Wu, Z., Xiong, Z., Gambhir, S.S. and Chen, X. (2005) Near-infrared fluorescent RGD peptides for optical imaging of integrin $\alpha_v\beta_3$ expression in living mice. *Bioconjugate Chemistry*, **16**, 1433–41.
21 Tsay, J.M., Pflughoefft, M., Bentolila, L.A. and Weiss, S. (2004) Hybrid approach to the synthesis of highly luminescent CdTe/ZnS and CdHgTe/ZnS nanocrystals. *Journal of the American Chemical Society*, **126**, 1926–7.
22 Rogach, A.L., Franzl, T., Klar, T.A., Feldmann, J., Gaponik, N., Lesnyak, V., Shavel, A., Eychmüller, A., Rakovich, Y.P. and Donegan, J.F. (2007) Aqueous synthesis of thiol-capped CdTe nanocrystals: state-of-the-art. *Journal of Physical Chemistry, C*, **111**, 14628–37.

23 Mićić, O.I., Curtis, C.J., Jones, K.M., Sprague, J.R. and Nozik, A.J. (1994) Synthesis and characterization of InP quantum dots. *Journal of Physical Chemistry*, **98**, 4966–9.

24 Guzelian, A.A., Katari, J.E.B., Kadavanich, A.V., Banin, U., Hamad, K., Juban, E., Alivisatos, A.P., Wolters, R.H., Arnold, C.C. and Heath, J.R. (1996) Synthesis of size-selected, surface-passivated InP nanocrystals. *Journal of Physical Chemistry*, **100**, 7212–19.

25 Mićić, O.I., Cheong, H.M., Fu, H., Zunger, A., Sprague, J., Mascarenhas, A. and Nozik, A.J. (1997) Size-dependent spectroscopy of InP quantum dots. *Journal of Physical Chemistry, B*, **101**, 4904–12.

26 Mićić, O.I., Sprague, J., Lu, Z.H. and Nozik, A.J. (1996) Highly efficient band-edge emission from InP quantum dots. *Applied Physics Letters*, **68**, 3150–2.

27 Talapin, D.V., Gaponik, N., Borchert, H., Rogach, A.L., Haase, M. and Weller, H. (2002) Etching of colloidal InP nanocrystals with fluorides: photochemical nature of the process resulting in high photoluminescence efficiency. *Journal of Physical Chemistry, B*, **106**, 12 659–63.

28 Xu, S., Ziegler, J. and Nann, T. (2008) Rapid synthesis of highly luminescent InP and InP/ZnS nanocrystals. *Journal of Material Chemistry*, **18**, 2653–6.

29 Li, L. and Reiss, P. (2008) One-pot synthesis of highly luminescent InP/ZnS nanocrystals without precursor injection. *Journal of the American Chemical Society*, **130**, 11 588–9.

30 Cao, Y.W. and Banin, U. (1999) Synthesis and characterization of InAs/InP and InAs/CdSe core/shell nanocrystals. *Angewandte Chemie, International Edition*, **38**, 3692–4.

31 Cao, Y.W. and Banin, U. (2000) Growth and properties of semiconductor core/shell nanocrystals with InAs cores. *Journal of the American Chemical Society*, **122**, 9692–702.

32 Brumer, M., Kigel, A., Amirav, L., Sashchiuk, A., Solomesch, O., Tessler, N. and Lifshitz, E. (2005) PbSe/PbS and PbSe/PbSe$_x$S$_{1-x}$ core/shell nanocrystals. *Advanced Functional Materials*, **15**, 1111–16.

33 Lifshitz, E., Brumer, M., Kigel, A., Sashchiuk, A., Bashouti, M., Sirota, M., Galun, E., Burshtein, Z., Le Quang, A.Q., Ledoux-Rak, I. and Zyss, J. (2006) Air-stable PbSe/PbS and PbSe/PbSe$_x$S$_{1-x}$ core-shell nanocrystal quantum dots and their applications. *Journal of Physical Chemistry, B*, **110**, 25 356–65.

34 Pietryga, J.M., Werder, D.J., Williams, D.J., Casson, J.L., Schaller, R.D., Klimov, V.I. and Hollingsworth, J.A. (2008) Utilizing the lability of lead selenide to produce heterostructured nanocrystals with bright, stable infrared emission. *Journal of the American Chemical Society*, **130**, 4879–85.

35 Balet, L.P., Ivanov, S.A., Piryatinski, A., Achermann, M. and Klimov, V.I. (2004) Inverted core/shell nanocrystals continuously tunable between type-I and type-II localization regimes. *Nano Letters*, **4**, 1485–8.

36 Zhong, X.H., Xie, R.G., Zhang, Y., Basché, T. and Knoll, W. (2005) High-quality violet- to red-emitting ZnSe/CdSe core/shell nanocrystals. *Chemistry of Materials*, **17**, 4038–42.

37 Battaglia, D., Li, J.J., Wang, Y.J. and Peng, X.G. (2003) Colloidal two-dimensional systems: CdSe quantum shells and wells. *Angewandte Chemie, International Edition*, **42**, 5035–9.

38 Kim, S., Fisher, B., Eisler, H.J. and Bawendi, M. (2003) Type-II core/shell CdS/ZnSe nanocrystals: synthesis, electronic structures, and spectroscopic properties. *Journal of the American Chemical Society*, **125**, 11 466–7.

39 Ivanov, S.A., Piryatinski, A., Nanda, J., Tretiak, S., Zavadil, K.R., Wallace, W.O., Werder, D. and Klimov, V.I. (2007) Light amplification using inverted core/shell nanocrystals: towards lasing in the single-exciton regime. *Journal of the American Chemical Society*, **129**, 11 708–19.

40 Aharoni, A., Mokari, T., Popov, I. and Banin, U. (2006) Synthesis of InAs/CdSe/ZnSe core/shell1/shell2 structures with bright and stable near-infrared fluorescence. *Journal of the American Chemical Society*, **128**, 257–64.

41 Chen, C.Y., Cheng, C.T., Lai, C.W., Hu, Y.H., Chou, P.T., Chou, Y.H. and Chiu, H.T. (2005) Type-II CdSe/CdTe/ZnTe (core-shell-shell) quantum dots with cascade band edges: the separation of electron (at CdSe) and hole (at ZnTe) by the CdTe layer. *Small*, **1**, 1215–20.

42 Gill, R., Zayats, M. and Willner, I. (2008) Semiconductor quantum dots for bioanalysis. *Angewandte Chemie, International Edition*, **47**, 7602–25.

43 Chan, W.C.W. and Nie, S. (1998) Quantum dot bioconjugates for ultrasensitive nonisotopic detection. *Science*, **281**, 2016–18.

44 Parak, W.J., Gerion, D., Zanchet, D., Woerz, A.S., Pellegrino, T., Micheel, C., Williams, S.C., Seitz, M., Bruehl, R.E., Bryant, Z., Bustamante, C., Bertozzi, C.R. and Alivisatos, A.P. (2002) Conjugation of DNA to silanized colloidal semiconductor nanocrystalline quantum dots. *Chemistry of Materials*, **14**, 2113–19.

45 Gao, X., Cui, Y., Levenson, R.M., Chung, L.W.K. and Nie, S. (2004) In vivo cancer targeting and imaging with semiconductor quantum dots. *Nature Biotechnology*, **22**, 969–76.

46 Aldana, J., Wang, Y.A. and Peng, X. (2001) Photochemical instability of CdSe nanocrystals coated by hydrophilic thiols. *Journal of the American Chemical Society*, **123**, 8844–50.

47 Parak, W.J., Gerion, D., Pellegrino, T., Zanchet, D., Micheel, C., Williams, S.C., Boudreau, B., Gros, M.A.L., Larabell, C.A. and Alivisatos, A.P. (2003) Biological applications of colloidal nanocrystals. *Nanotechnology*, **14**, R15–27.

48 Mattoussi, H., Mauro, J.M., Goldman, E.R., Anderson, G.P., Sundar, V.C., Mikulec, F.V. and Bawendi, M.G. (2000) Self-assembly of CdSe ZnS quantum dot bioconjugates using an engineered recombinant protein. *Journal of the American Chemical Society*, **122**, 12142–50.

49 Medintz, I.L., Clapp, A.R., Mattoussi, H., Goldman, E.R., Fisher, B. and Mauro, J.M. (2003) Self-assembled nanoscale biosensors based on quantum dot FRET donors. *Nature Materials*, **2**, 630–8.

50 Yu, W.W., Chang, E., Falkner, J.C., Zhang, J., Al-Somali, A.M., Sayes, C.M., Johns, J., Drezek, R. and Colvin, V.L. (2007) Forming biocompatible and nonaggregated nanocrystals in water using amphiphilic polymers. *Journal of the American Chemical Society*, **129**, 2871–9.

51 Liu, W., Howarth, M., Greytak, A.B., Zheng, Y., Nocera, D.G., Ting, A.Y. and Bawendi, M.G. (2008) Compact biocompatible quantum dots functionalized for cellular imaging. *Journal of the American Chemical Society*, **130**, 1274–84.

52 Kim, S. and Bawendi, M.G. (2003) Oligomeric ligands for luminescent and stable nanocrystal quantum dots. *Journal of the American Chemical Society*, **125**, 14652–3.

53 Pellegrino, T., Manna, L., Kudera, S., Liedl, T., Koktysh, D., Rogach, A.L., Keller, S., Radler, J., Natile, G. and Parak, W.J. (2004) Hydrophobic nanocrystals coated with an amphiphilic polymer shell: a general route to water soluble nanocrystals. *Nano Letters*, **4**, 703–7.

54 Dubertret, B., Skourides, P., Norris, D.J., Noireaux, V., Brivanlou, A.H. and Libchaber, A. (2002) In Vivo imaging of quantum dots encapsulated in phospholipid micelles. *Science*, **298**, 1759–62.

55 Smith, A.M., Ruan, G., Rhyner, M.N. and Nie, S. (2006) Engineering luminescent quantum dots for in vivo molecular and cellular imaging. *Annals of Biomedical Engineering*, **34**, 3–14.

56 Bentzen, E.L., Tomlinson, I.D., Mason, J., Gresch, P., Warnement, M.R., Wright, D., Bush, S.E., Blakely, R., and Rosenthal, S.J. (2005) Surface modification to reduce nonspecific binding of quantum dots in live cell assays. *Bioconjugate Chemistry*, **16**, 1488–94.

57 Zhelev, Z., Ohba, H. and Bakalova, R. (2006) Single quantum dot-micelles coated with silica shell as potentially non-cytotoxic fluorescent cell tracers. *Journal of the American Chemical Society*, **128**, 6324–5.

58 Bakalova, R., Zhelev, Z., Aoki, I., Ohba, H., Imai, Y. and Kanno, I. (2006) Single-shelled single quantum dot micelles as imaging probes with dual or multimodality. *Analytical Chemistry*, **78**, 5925–32.

59 Mulder, W.J.M., Koole, R., Brandwijk, R.J., Storm, G., Chin, P.T.K., Strijkers, G.J., Donegá, C.M., Nicolay, K. and Griffioen, A.W. (2006) Quantum dots with a paramagnetic coating as a bimodal molecular imaging probe. *Nano Letters*, **6**, 1–6.

60 Koole, R., van Schooneveld, M.M., Hilhorst, J., Castermans, K., Cormode, D.P., Strijkers, G.J., Donega, C.M., Vanmaekelbergh, D., Griffioen, A.W., Nicolay, K., Fayad, Z.A., Meijerink, A. and Mulder, W.J.M. (2008) Paramagnetic lipid-coated silica nanoparticles with a fluorescent quantum dot core: a new contrast agent platform for multimodality imaging. *Bioconjugate Chemistry*, **19**, 2471–9.

61 Bakalova, R., Zhelev, Z., Aoki, I. and Kanno, I. (2007) Designing quantum-dot probes. *Nature Photonics*, **1**, 487–9.

62 Bruchez, M.J., Moronne, M., Gin, P., Weiss, S. and Alivisatos, A.P. (1998) Semiconductor nanocrystals as fluorescent biological labels. *Science*, **281**, 2013–16.

63 Xiao, Y. and Barker, P.E. (2004) Semiconductor nanocrystal probes for human metaphase chromosomes. *Nucleic Acids Research*, **32**, e28.

64 Pinaud, F., King, D., Moore, H.P. and Weiss, S. (2004) Bioactivation and cell targeting of semiconductor CdSe/ZnS nanocrystals with phytochelatin-related peptides. *Journal of the American Chemical Society*, **126**, 6115–23.

65 Wu, X., Liu, H., Liu, J., Haley, K.N., Treadway, J.A., Larson, J.P., Ge, N., Peale, F. and Bruchez, M.P. (2003) Immunofluorescent labeling of cancer marker Her2 and other cellular targets with semiconductor quantum dots. *Nature Biotechnology*, **21**, 41–6.

66 Gao, X., Chan, W.C. and Nie, S. (2002) Quantum dot nanocrystals for ultrasensitive biological labeling and multicolor optical encoding. *Journal of Biomedical Optics*, **7**, 532–7.

67 Xing, Y., Chaudry, Q., Shen, C., Kong, K.Y., Zhau, H.E., Chung, L.W., Petros, J.A., O'Regan, R.M., Yezhelyev, M.V., Simons, J.W., Wang, M.D. and Nie, S. (2007) Bioconjugated quantum dots for multiplexed and quantitative immunohistochemistry. *Nature Protocols*, **2**, 1152–65.

68 Zhang, P. (2006) Investigation of novel quantum dots/proteins/cellulose bioconjugate using NSOM and fluorescence. *Journal of Fluorescence*, **16**, 349–53.

69 Slocik, J.M., Moore, J.T. and Wright, D.W. (2002) Monoclonal antibody recognition of histidine-rich peptide encapsulated nanoclusters. *Nano Letters*, **2**, 169–73.

70 Sandros, M.G., Gao, D. and Benson, D.E. (2005) A modular nanoparticle-based system for reagentless small molecule biosensing. *Journal of the American Chemical Society*, **127**, 12198–9.

71 Sandros, M.G., Gao, D., Gokdemir, C. and Benson, D.E. (2005) General high-affinity approach for the synthesis of fluorophore appended protein nanoparticle assemblies. *Chemical Communications*, **22**, 2832–4.

72 Sandros, M.G., Shete, V. and Benson, D.E. (2006) Selective, reversible, reagentless maltose biosensing with core-shell semiconducting nanoparticles. *The Analyst*, **131**, 229–35.

73 Ding, S.Y., Rumbles, G., Jones, M., Tucker, M.P., Nedeljkovic, J., Simon, M.N., Wall, J.S. and Himmel, M.E. (2004) Bioconjugation of (CdSe)ZnS quantum dots using a genetically engineered multiple polyhistidine tagged cohesin/dockerin protein polymer. *Macromolecular Materials and Engineering*, **289**, 622–8.

74 Ding, S.Y., Jones, M., Tucker, M.P., Nedeljkovic, J.M., Wall, J., Simon, M.N., Rumbles, G. and Himmel, M.E. (2003) Quantum dot molecules assembled with genetically engineered proteins. *Nano Letters*, **3**, 1581–5.

75 Roullier, V., Clarke, S., You, C., Pinaud, F., Gouzer, G.G., Schaible, D., Marchi-Artzner, V., Piehler, J. and Dahan, M. (2009) High-affinity labeling and

tracking of individual histidine-tagged proteins in live cells using Ni^{2+} tris-nitrilotriacetic acid quantum dot conjugates. *Nano Letters*, **9**, 1228–34.

76 Tsay, J.M., Doose, S. and Weiss, S. (2006) Rotational and translational diffusion of peptide-coated CdSe/CdS/ZnS nanorods studied by fluorescence correlation spectroscopy. *Journal of the American Chemical Society*, **128**, 1639–47.

77 Tsay, J.M., Doose, S., Pinaud, F. and Weiss, S. (2005) Enhancing the photoluminescence of peptide-coated nanocrystals with shell composition and UV irradiation. *Journal of Physical Chemistry, B*, **109**, 1669–74.

78 Xiao, Y. and Barker, P.E. (2004) Semiconductor nanocrystal probes for human metaphase chromosomes. *Nucleic Acids Research*, **32**, e161.

79 Goldman, E.R., Medintz, I.L. and Mattoussi, H. (2006) Luminescent quantum dots in immunoassays. *Analytical and Bioanalytical Chemistry*, **384**, 560–3.

80 Goldman, E.R., Mattoussi, H.M., Anderson, G.P., Medintz, I.L. and Mauro, J.M. (2005) Fluoroimmunoassays using antibody-conjugated quantum dots. *Methods in Molecular Biology*, **303**, 19–34.

81 Goldman, E.R., Anderson, G.P., Tran, P.T., Mattoussi, H., Charles, P.T. and Mauro, J.M. (2002) Conjugation of luminescent quantum dots with antibodies using an engineered adaptor protein to provide new reagents for fluoroimmunoassays. *Analytical Chemistry*, **74**, 841–7.

82 Goldman, E.R., Balighian, E.D., Kuno, M.K., Labrenz, S., Tran, P.T., Anderson, G.P., Mauro, J.M. and Mattoussi, H. (2002) Luminescent quantum dot-adaptor protein-antibody conjugates for use in fluoroimmunoassays. *Physica Status Solidi B*, **229**, 407–14.

83 Goldman, E.R., Balighian, E.D., Mattoussi, H., Kuno, M.K., Mauro, J.M., Tran, P.T. and Anderson, G.P. (2002) Avidin: a natural bridge for quantum dot-antibody conjugates. *Journal of the American Chemical Society*, **124**, 6378–82.

84 Goldman, E.R., Clapp, A.R., Anderson, G.P., Uyeda, H.T., Mauro, J.M., Medintz, I.L. and Mattoussi, H. (2004) Multiplexed toxin analysis using four colors of quantum dot fluororeagents. *Analytical Chemistry*, **76**, 684–8.

85 Matsumura, Y. and Maeda, H. (1986) A new concept for macromolecular therapeutics in cancer chemotherapy: mechanism of tumoritropic accumulation of proteins and the antitumor agent SMANCS. *Cancer Research*, **46**, 6387–92.

86 Duncan, R. (2003) The dawning era of polymer therapeutics. *Nature Reviews. Drug Discovery*, **2**, 347–60.

87 Jain, R.K. (2001) Delivery of molecular medicine to solid tumors: lessons from *in vivo* imaging of gene expression and function. *Journal of Controlled Release*, **74**, 7–25.

88 Jain, R.K. (1999) Understanding barriers to drug delivery: high resolution *in vivo* imaging is key. *Clinical Cancer Research*, **5**, 1605–6.

89 Smith, A.M., Duan, H., Mohs, A.M. and Nie, S. (2008) Bioconjugated quantum dots for *in vivo* molecular and cellular imaging. *Advanced Drug Delivery Reviews*, **60**, 1226–40.

90 Mitragotri, S. and Joerg, L. (2008) Physical approaches to biomaterial design. *Nature Materials*, **8**, 15–23.

91 Chithrani, B.D., Ghazani, A.A. and Chan, W.C. (2006) Determining the size and shape dependence of gold nanoparticle uptake into mammalian cells. *Nano Letters*, **6**, 662–8.

92 Osaki, F., Kanamori, T., Sando, S., Sera, T. and Aoyama, Y. (2004) A quantum dot conjugated sugar ball and its cellular uptake. On the size effects of endocytosis in the subviral region. *Journal of the American Chemical Society*, **126**, 6520–1.

93 Champion, J.A., Walker, A. and Mitragotri, S. (2008) Role of particle size in phagocytosis of polymeric microspheres. *Pharmaceutical Research*, **25**, 1815–21.

94 Ballou, B., Lagerholm, B.C., Ernst, L.A., Bruchez, M.P. and Waggoner, A.S. (2004) Noninvasive imaging of quantum dots in mice. *Bioconjugate Chemistry*, **15**, 79–86.

95 Ballou, B., Ernst, L.A., Andreko, S., Bruchez, M.P., Lagerholm, B.C. and Waggoner, A.S. (2008) Long-term retention of fluorescent quantum dots in vivo, in Nanomaterials for Application in Medicine and Biology (eds M. Giersig and G.B. Khomutov), Springer Press, Netherlands, pp. 127–37.

96 Ballou, B. (2005) Quantum dot surfaces for use in vivo and in vitro. Current Topics in Developmental Biology, **70**, 103–20.

97 Akerman, M.E., Chan, W.C.W., Laakkonen, P., Bhatia, S.N. and Ruoslahti, E. (2002) Proceedings of the National Academy of Sciences of the United States of America, **99**, 12617.

98 Allen, C., Santos, N.D., Gallagher, R., Chiu, G.N., Shu, Y., Li, W.M., Johnstone, S.A., Janoff, A.S., Mayer, L.D., Webb, M.S. and Bally, M.B. (2002) Controlling the physical behavior and biological performance of liposome formulations through use of surface grafted poly(ethylene glycol). Bioscience Reports, **22**, 225–50.

99 (a) Santos, N.D., Allen, C., Doppen, A.-M., Anantha, M., Cox, K.A.K., Gallagher, R.C., Karlsson, G., Edwards, K., Kenner, G., Samuels, L., Webb, M.S. and Bally, M.B. (2007) Influence of poly(ethylene glycol) grafting density and polymer length on liposomes: relating plasma circulation lifetimes to protein binding. Biochimica et Biophysica Acta – Biomembranes, **1768**, 1367–77. (b) Mammen, M., Chio, S. and Whitesides, G.M. (1998) Polyvalent interactions in biological systems: implications for design and use of multivalent ligands and inhibitors. Angewandte Chemie, International Edition in English, **37**, 2755–94.

100 Akerman, M.E., Chan, W.C.W., Laakkonen, P., Bhatia, S.N. and Ruoslahti, E. (2002) Nanocrystal targeting in vivo. Proceedings of the National Academy of Sciences of the United States of America, **99**, 12617–21.

101 Rajotte, D. and Ruoslahti, E. (1999) Membrane dipeptidase is the receptor for a lung-targeting peptide identified by in vivo phage display. Journal of Biological Chemistry, **274**, 11593–8.

102 Rajotte, D., Arap, W., Hagedorn, M., Koivunen, E., Pasqualini, R. and Ruoslahti, E. (1998) Molecular heterogeneity of the vascular endothelium revealed by in vivo phage display. Journal of Clinical Investigation, **102**, 430–7.

103 Porkka, K., Laakkonen, P., Hoffman, J.A., Bernasconi, M. and Ruoslahti, E. (2002) A fragment of the HMGN2 protein homes to the nuclei of tumor cells and tumor endothelial cells in vivo. Proceedings of the National Academy of Sciences of the United States of America, **99**, 7444–9.

104 Laakkonen, P., Porkka, K., Hoffman, J.A. and Ruoslahti, E. (2002) A tumor-homing peptide with a targeting specificity related to lymphatic vessels. Nature Medicine, **8**, 751–5.

105 Cai, W., Shin, D.W., Chen, K., Gheysens, O., Cao, Q., Wang, S.X., Gambhir, S.S. and Chen, X. (2006) Peptide-labeled near-infrared quantum dots for imaging tumor vasculature in living subjects. Nano Letters, **6**, 669–76.

106 Cai, W. and Chen, X. (2006) Anti-angiogenic cancer therapy based on integrin $\alpha_v\beta_3$ antagonism. Anti-Cancer Agents Medicinal Chemistry, **6**, 407–28.

107 Cai, W., Rao, J., Gambhir, S.S. and Chen, X. (2006) How molecular imaging is speeding up antiangiogenic drug development. Molecular Cancer Therapeutics, **5**, 2624–33.

108 Cai, W., Wu, Y., Chen, K., Cao, Q., Tice, D.A. and Chen, X. (2006) In vitro and in vivo characterization of 64Cu-labeled Abegrin, a humanized monoclonal antibody against integrin $\alpha v\beta_3$. Cancer Research, **66**, 9673–81.

109 Cai, W., Zhang, X., Wu, Y. and Chen, X. (2006) A thiol-reactive 18F-labeling agent, N-[2-(4-18F-fluorobenzamido)ethyl]maleimide, and synthesis of RGD peptide-based tracer for PET imaging of $\alpha_v\beta_3$ integrin expression. Journal of Nuclear Medicine, **47**, 1172–80.

110 Chen, X., Conti, P.S. and Moats, R.A. (2004) In vivo near-infrared fluorescence imaging of integrin $\alpha_v\beta_3$ in brain tumor xenografts. Cancer Research, **64**, 8009–14.

111 Wu, Y., Zhang, X., Xiong, Z., Cheng, Z., Fisher, D.R., Liu, S. and Chen, X. (2005)

microPET imaging of glioma integrin $α_vβ_3$ expression using (64)Cu-labeled tetrameric RGD peptide. *Journal of Nuclear Medicine*, **46**, 1707–18.

112 Xiong, Z., Cheng, Z., Zhang, X., Patel, M., Wu, J.C., Gambhir, S.S. and Chen, X. (2006) Imaging chemically modified adenovirus for targeting tumors expressing integrin alphavbeta3 in living mice with mutant herpes simplex virus type 1 thymidine kinase PET reporter gene. *Journal of Nuclear Medicine*, **47**, 130–9.

113 Zhang, X., Xiong, Z., Wu, Y., Cai, W., Tseng, J.R., Gambhir, S.S. and Chen, X. (2006) Quantitative PET imaging of tumor integrin $α_vβ_3$ expression with 18F-FRGD2. *Journal of Nuclear Medicine*, **47**, 113–21.

114 Cai, W. and Chen, X. (2007) Nanoplatforms for targeted molecular imaging in living subjects. *Small*, **3**, 1840–54.

115 Cai, W., Hsu, A.R., Li, Z.B. and Chen, X. (2007) Are quantum dots ready for *in vivo* imaging in human subjects? *Nanoscale Research Letters*, **2**, 265–81.

116 Smith, B.R., Cheng, Z., De, A., Koh, A.L., Sinclair, R. and Gambhir, S.S. (2008) Real-time intravital imaging of RGD-quantum dot binding to luminal endothelium in mouse tumor neovasculature. *Nano Letters*, **8**, 2599–606.

117 Cai, W., Chen, K., Mohamedali, K.A., Cao, Q., Gambhir, S.S., Rosenblum, M.G. and Chen, X. (2006) PET of vascular endothelial growth factor receptor expression. *Journal of Nuclear Medicine*, **47**, 2048–56.

118 Cai, W. and Chen, X. (2007) Multimodality imaging of vascular endothelial growth factor and vascular endothelial growth factor receptor expression. *Frontiers in Bioscience*, **12**, 4267–79.

119 Ferrara, N. (2004) Vascular endothelial growth factor: basic science and clinical progress. *Endocrine Reviews*, **25**, 581–611.

120 Chang, S.S., Reuter, V.E., Heston, W.D. and Gaudin, P.B. (2001) Metastatic renal cell carcinoma neovasculature expresses prostate-specific membrane antigen. *Urology*, **57**, 801–5.

121 Mansfield, J.R., Gossage, K.W., Hoyt, C.C. and Levenson, R.M. (2005) Autofluorescence removal, multiplexing, and automated analysis methods for in-vivo fluorescence imaging. *Journal of Biomedical Optics*, **10**, 41207.

122 Yu, X., Chen, L., Li, K., Li, Y., Xiao, S., Luo, X., Liu, J., Zhou, L., Deng, Y., Pang, D. and Wang, Q. (2007) Immunofluorescence detection with quantum dot bioconjugates for hepatoma *in vivo*. *Journal of Biomedical Optics*, **12**, 014008.

123 Tada, H., Higuchi, H., Wanatabe, T.M. and Ohuchi, N. (2007) *In vivo* real-time tracking of single quantum dots conjugated with monoclonal anti-HER2 antibody in tumors of mice. *Cancer Research*, **67**, 1138–44.

124 So, M.K., Xu, C., Loening, A.M., Gambhir, S.S. and Rao, J. (2006) Self-illuminating quantum dot conjugates for *in vivo* imaging. *Nature Biotechnology*, **24**, 339–43.

125 Contag, C.H. and Bachmann, M.H. (2002) Advances in *in vivo* bioluminescence imaging of gene expression. *Annual Review of Biomedical Engineering*, **4**, 235–60.

126 Negrin, R.S. and Contag, C.H. (2006) *In vivo* imaging using bioluminescence: a tool for probing graft-versus-host disease. *Nature Reviews. Immunology*, **6**, 484–90.

127 Loening, A.M., Fenn, T.D., Wu, A.M. and Gambhir, S.S. (2006) Consensus guided mutagenesis of Renilla luciferase yields enhanced stability and light output. *Protein Engineering, Design & Selection*, **19**, 391–400.

128 Zhang, Y., So, M.K., Loening, A.M., Yao, H., Gambhir, S.S. and Rao, J. (2006) HaloTag protein-mediated site-specific conjugation of bioluminescent proteins to quantum dots. *Angewandte Chemie, International Edition*, **45**, 4936–40.

129 Colvin, V.L. (2003) The potential environmental impact of engineered nanomaterials. *Nature Biotechnology*, **21**, 1166–70.

130 Hoet, P.H., Bruske-Hohlfeld, I. and Salata, O.V. (2004) Nanoparticles-known and unknown health risks. *Journal of Nanobiotechnology*, **2**, 12.

131 Tsay, J.M. and Michalet, X. (2005) New light on quantum dot cytotoxicity. *Chemistry & Biology*, **12**, 1159–61.
132 Derfus, A.M., Chan, W.C.W. and Bhatia, S.N. (2004) Probing the cytotoxicity of semiconductor quantum dots. *Nano Letters*, **4**, 11–18.
133 Nath, R., Prasad, R., Palinal, V.K. and Chopra, R.K. (1984) Molecular basis of cadmium toxicity. *Progress in Food & Nutrition Science*, **8**, 109–63.
134 Jaiswal, J.K., Mattoussi, H., Mauro, J.M. and Simon, S.M. (2003) Long-term multiple color imaging of live cells using quantum dot bioconjugates. *Nature Biotechnology*, **21**, 47–51.
135 Voura, E.B., Jaiswal, J.K., Mattoussi, H. and Simon, S.M. (2004) Tracking metastatic tumor cell extravasation with quantum dot nanocrystals and fluorescence emission-scanning microscopy. *Nature Medicine*, **10**, 993–8.
136 Kirchner, C., Liedl, T., Kudera, S., Pellegrino, T., Javier, A.M., Gaub, H.E., Stoelzle, S., Fertig, N. and Parak, W.J. (2005) Cytotoxicity of colloidal CdSe and CdSe/ZnS nanoparticles. *Nano Letters*, **5**, 331–8.
137 Kim, S.W., Zimmer, J.P., Ohnishi, S., Tracy, J.B., Frangioni, J.V. and Bawendi, M.G. (2005) Engineering InAs(x)P(1-x)/InP/ZnSe III–V alloyed core/shell quantum dots for the near-infrared. *Journal of the American Chemical Society*, **127**, 10526–32.
138 Bakalova, R., Ohba, H., Zhelev, Z., Ishikawa, M. and Baba, Y. (2004) Quantum dots as photosensitizers? *Nature Biotechnology*, **22**, 1360–1.
139 Samia, A.C., Chen, X. and Burda, C. (2003) Semiconductor quantum dots for photodynamic therapy. *Journal of the American Chemical Society*, **125**, 15736–7.
140 Landin, L., Miller, M.S., Pistol, M., Pryor, C.E. and Samuelson, L. (1998) Optical studies of individual InAs quantum dots in GaAs: few-particle effects. *Science*, **280**, 262–4.
141 Tanaka, N., Yamasaki, J., Fuchi, S. and Takeda, Y. (2004) First observation of In(x)Ga(1-x)As quantum dots in GaP by spherical-aberration-corrected HRTEM in comparison with ADF-STEM and conventional HRTEM. *Microscopy and Microanalysis*, **10**, 139–45.
142 Pradhan, N., Goorskey, D., Thessing, J. and Peng, X. (2005) An alternative of CdSe nanocrystal emitters: pure and tunable impurity emissions in ZnSe nanocrystals. *Journal of the American Chemical Society*, **127**, 17586–5787.
143 Pradhan, N., Battaglia, D.M., Liu, Y. and Peng, X. (2007) Efficient, stable, small, and water-soluble doped ZnSe nanocrystal emitters as non-cadmium biomedical labels. *Nano Letters*, **7**, 312–17.
144 Xie, R., Rutherford, M. and Peng, X. (2009) Formation of high-quality I–III–VI semiconductor nanocrystals by tuning relative reactivity of cationic precursors. *Journal of the American Chemical Society*, **131**, 5691–7.
145 Zimmer, J.P., Kim, S.W., Ohnishi, S., Tanaka, E., Frangioni, J.V. and Bawendi, M.G. (2006) *Journal of the American Chemical Society*, **128**, 2526.
146 Guo, W., Li, J.J., Wang, Y.A. and Peng, X. (2003) Luminescent CdSe/CdS core/shell nanocrystals in dendron boxes: superior chemical, photochemical and thermal stability. *Journal of the American Chemical Society*, **125**, 3901–9.
147 Liu, Y., Kim, M., Wang, Y., Wang, Y.A. and Peng, X. (2006) Highly luminescent, stable, and water-soluble CdSe/CdS core-shell dendron nanocrystals with carboxylate anchoring groups. *Langmuir*, **22**, 6341–5.
148 Massoud, T.F. and Gambhir, S.S. (2003) Molecular imaging in living subjects: seeing fundamental biological processes in a new light. *Genes and Development*, **17**, 545–80.
149 Montet, X., Ntziachristos, V., Grimm, J. and Weissleder, R. (2005) Tomographic fluorescence mapping of tumor targets. *Cancer Research*, **65**, 6330–6.
150 Montet, X., Figueiredo, J.L., Alencar, H., Ntziachristos, V., Mahmood, U. and Weissleder, R. (2007) Tomographic fluorescence imaging of tumor vascular volume in mice. *Radiology*, **242**, 751–8.
151 Ntziachristos, V., Tung, C.H., Bremer, C. and Weissleder, R. (2002) Fluorescence molecular tomography

resolves protease activity *in vivo*. *Nature Medicine*, **8**, 757–60.

152 Wang, S., Jarrett, B.R., Kauzlarich, S.M. and Louie, A.Y. (2007) Core/shell quantum dots with high relaxivity and photoluminescence for multimodality imaging. *Journal of the American Chemical Society*, **129**, 3848–56.

153 Selvan, S.T., Patra, P.K., Ang, C.Y. and Ying, J.Y. (2007) Synthesis of silica-coated semiconductor and magnetic quantum dots and their use in the imaging of live cells. *Angewandte Chemie, International Edition*, **46**, 2448–52.

154 van Schooneveld, M.M., Vucic, E., Koole, R., Zhou, Y., Stocks, J., Cormode, D.P., Tang, C.Y., Gordon, R.E., Nicolay, K., Meijerink, A., Fayad, Z.A. and Mulder, W.J. (2008) Improved biocompatibility and pharmacokinetics of silica nanoparticles by means of a lipid coating: a multimodality investigation. *Nano Letters*, **8**, 2517–25.

155 van Tilborg, G.A., Mulder, W.J., Chin, P.T., Storm, G., Reutelingsperger, C.P., Nicolay, K. and Strijkers, G.J. (2006) Annexin A5-conjugated quantum dots with a paramagnetic lipidic coating for the multimodal detection of apoptotic cells. *Bioconjugate Chemistry*, **17**, 865–8.

156 Prinzen, L., Miserus, R.J., Dirksen, A., Hackeng, T.M., Deckers, N., Bitsch, N.J., Megens, R.T., Douma, K., Heemskerk, J.W., Kooi, M.E., Frederik, P.M., Slaaf, D.W., van Zandvoort, M.A. and Reutelingsperger, C.P. (2007) Optical and magnetic resonance imaging of cell death and platelet activation using annexin a5-functionalized quantum dots. *Nano Letters*, **7**, 93–100.

157 Wang, D., He, J., Rosensweig, N. and Rosenzweig, Z. (2004) *Nano Letters*, **4**, 409.

158 Gu, H., Zheng, R., Zhang, X. and Xu, B. (2004) Facile one-pot synthesis of bifunctional heterodimers of nanoparticles: a conjugate of quantum dot and magnetic nanoparticles. *Journal of the American Chemical Society*, **126**, 5664–5.

159 Cai, W., Chen, K., Li, Z.B., Gambhir, S.S. and Chen, X. (2007) Dual-function probe for PET and near-infrared fluorescence imaging of tumor vasculature. *Journal of Nuclear Medicine*, **48**, 1862–70.

160 (a) Hsu, A.R., Cai, W., Veeravagu, A., Mohamedali, K.A., Chen, K., Kim, S., Vogel, H., Hou, L.C., Tse, V., Rosenblum, M.G. and Chen, X. (2007) Multimodality molecular imaging of glioblastoma growth inhibition with vasculature-targeting fusion toxin VEGF121/rGel. *Journal of Nuclear Medicine*, **48**, 445–54.
(b) Taroni, P., Danesini, G., Torricelli, A., Pifferi, A., Spinelli, L. and Cubeddu, R. (2004) Clinical trial of time-resolved scanning optical mammography at 4 wavelengths between 683 and 975 nm. *Journal of Biomedical Optics*, **9**, 464–73.

161 Intes, X. (2005) Time-domain optical mammography SoftScan: initial results. *Academic Radiology*, **12**, 934–47.

3
Multiplexed Bioimaging Using Quantum Dots
Richard Byers and Eleni Tholouli

3.1
Introduction

During recent years, several reviews have detailed the use of quantum dots (QDs) in biological imaging. These have tended to concentrate on the underlying technology, and often included substantial details on the fabrication of QDs and methods of modification for biological use, rather than on examples of QD use. Where such examples are discussed in detail they usually include use in both *in vivo* and *in vitro* imaging, but with a relative lack of detail for the latter. Most reviews have also focused on uses for biological rather than biomedical imaging, and very few – if any – have paid attention to the issues of standardization, quantification, and workflow required to take a new technology to clinical use. Finally, multiplex imaging, in which QDs really "shine," requires sophisticated image analysis for maximum effect, and this aspect is often brushed over. The aim of this chapter is to address these shortcomings by providing a brief overview of the nature of QDs, with particular reference to their value in multiplex imaging. Attention is also focused on seminal and more recent examples of their use in immunohistochemistry (IHC) and *in situ* hybridization (ISH), both in single and multiplex experiments. The use of QDs in combination with electron microscopy is also discussed. The pros and cons of different attachment chemistries are described with relevance to practicality and robustness in a range of different studies, and their use as multiplex detectors in liquid (flow cytometry) and solid (microarray, either protein or genomic) state systems are briefly outlined. Issues regarding standardization and quantification are detailed, as is spectral imaging. The chapter concludes with some details of the use of QDs in clinical biomarker measurements, and some suggestions for their future perspective.

3.2
The Need for Novel Multiplex Imaging Systems

Today, the Human Genome Project has cataloged the majority of gene sequences, such that biological/biomedical investigations are now principally focused on the analysis of function and interaction of these genes, and of their protein products in cells, organ systems, and organisms. There is, therefore, a need for a methodology that is capable of detecting several genes or proteins at the same time, and of providing good spatial and cellular resolution at the morphological level [1]. In the field of cancer, the above-described approach has led to global gene expression profiling experiments that have identified several gene signatures as markers, either of tumor categories with distinct behaviors, or predictive of disease progression and response to therapy. The application of such signatures to individual patients in a clinical setting holds great potential for improving diagnosis, for guiding tailored molecular therapy [2–5], and for making informed therapeutic decisions. The generation and use of such signatures in routine clinical practice is presently limited by the need for relatively large amounts of high-quality frozen material for microarray gene expression profiling [1, 6]. In contrast, nearly all clinical material is routinely formalin-fixed and processed to paraffin [6], a practice that is likely to continue for the foreseeable future, especially outside large academic centers. In addition, gene signatures have largely been derived from tissue homogenates, despite recent studies having demonstrated the importance of localization of the genes that comprise these signatures not only to the tumor cells, but also to adjacent cells, such as immune cells and stromal cells [7]. There is, therefore, a clear need for a generic method for the simultaneous visualization of multiple genes in formalin-fixed, paraffin-embedded tissue (FFPET), that will allow investigations to be made of the spatial localization of microarray gene signatures. Ideally, this process should be carried out by an automated or semi-automated system that will enable it to be applied not only to the clinical environment but also to high-throughput studies [8].

Fluorescence-based detection systems, which would be ideal for the *in situ* detection of multiple probes, are hampered by low sensitivity [9, 10], the intrinsic high autofluorescence of paraffin-embedded tissue [11, 12], and the low resultant signal-to-noise ratio (SNR). In addition, the relatively small numbers of fluorophores with broad overlapping spectra [13], and which also require different excitation wavelengths, make the deconvolution of multiple signals extremely challenging. Furthermore, fluorescence-based techniques are nonpermanent and fade with time, rendering them suboptimal for a clinical test in which a permanent record is best for clinical governance. QDs are relatively newly identified fluorescent semiconductor nanocrystals that possess an extremely high fluorescence efficiency and photostability, which makes them near-optimal for many fluorescent applications [14–22]. Consequently, QDs have been used for bioimaging by immunofluorescence [14], for molecule [14, 19] and cell labeling [20, 21, 23] and, more recently, in human clinical material [24]. In addition, while their excitation wavelength is constant, their emission wavelength is sharp, symmetrical and tunable (depending on the crystal diameter), with potential for multicolor staining [18]. Taken together, these properties indicate that QDs may indeed be useful for application to clinical samples.

Multiplex imaging using a range of probes, including antibody and/or DNA probes, has resulted in the generation of multicolor images, from which has arisen the problem of color resolution to extract meaningful data. The simultaneous detection of multiple fluorescent signals requires spectral deconvolution to resolve individual signals, and spectral imaging achieves this by generating a complete optical profile for each pixel in the image field. From this, multiple spectral distributions can be reconstructed via a least squares fitting linear unmixing approach [25]. The spectral information in the acquired datasets can then be used to discriminate between autofluorescence and a true fluorescent signal, and between different fluorescent signals [21].

In this review of multiplex imaging using QDs, both of these aspects will be addressed, namely the multiplex use of QDs and spectral imaging, each of which is required for reliable multiplex biological imaging.

3.3 Quantum Dots

3.3.1 Optical Properties

Quantum dots are near-spherical semiconductor nanocrystals that are composed of elements from Groups II–VI (CdSe) or III–V (InP) of the Periodic Table [26–28] (Figure 3.1a). Initially developed for use in the semiconductor industry, QDs were subsequently found to be highly fluorescent, which indicated their possible use as fluorophores. Key to this was the fact that their excitation states/band gaps were spatially confined, and this resulted in physical–particularly optical–properties that were intermediate between those of bulk compounds and discrete molecules. This quantum confinement underlies the ability of QDs to emit light at different wavelengths, depending upon the size of the core diameter [29] (Figure 3.1b). Essentially, the size of the electron "orbit" band gap is dependent upon the size of the QDs, with larger QDs having smaller band gaps that results in the emission of light at the red end of the spectrum, while smaller QDs emit a higher energy, and therefore blue light. Furthermore, due to their size, the entire crystal acts as a single molecule with all constituent atoms being excited and emitting light together, with a high resultant signal intensity. Because QD crystals can be manufactured to tight diametric tolerances, the emission spectra of a given amount of QDs is tight and, by virtue of Gaussian size distribution, is also symmetrical [18, 30]. These features result in QDs being tunable, extremely bright, and also amenable to multiplex detection. Most importantly, however, their relative brightness is dependent upon their diameter, and therefore, the wavelength of emission. When Xing et al. [31] investigated the relative brightness of different QDs, they reported that the integrated signal intensity of green QDs (525 nm) was 17-fold lower than that of red QDs (655 nm), and almost 32-fold lower than that of near-infrared (NIR) QDs (705 nm). It is not surprising, therefore, that most users employ red QDs, which are also further from the peak of autofluorescence (ca. 560 nm) that is abundant in formalin-fixed, paraffin-embedded tissues. This has

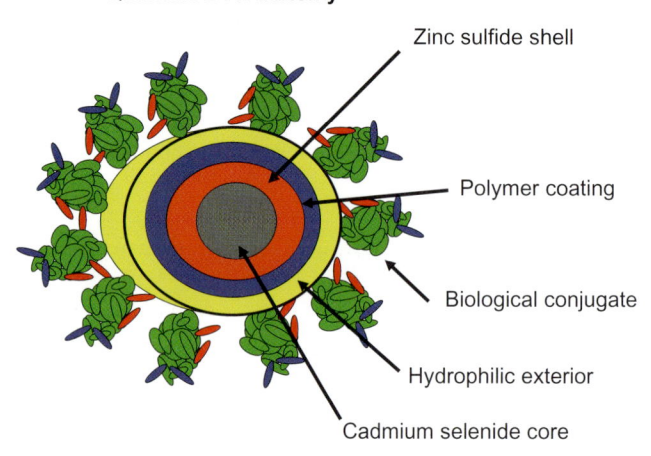

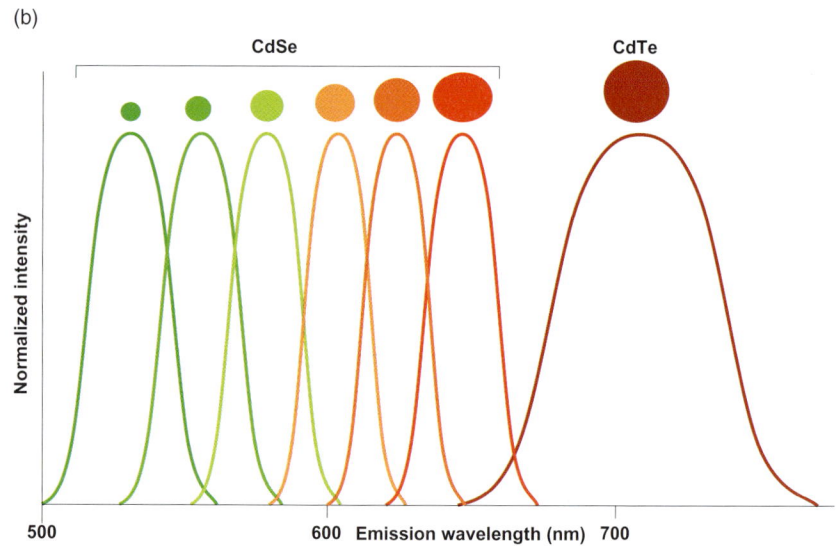

Figure 3.1 Schematic representation of a QD. Quantum dots are composed of a cadmium selenide (CdSe) core, around which a zinc sulfide (ZnS) shell is added to increase the quantum yield. The addition of polymer and hydrophilic coverings is required for stability, and the outside of the assembly is coated with biomolecules such as streptavidin or amine to enable attachment to other molecules; (b) Emission wavelengths of QDs. Following excitation, their emission wavelength is narrow, symmetrical and reaches into the near-infrared. Reproduced with permission from Ref. [44].

implications for the comparison of expression levels between different elements if detected with different-sized QDs, and consequently data normalization is required. It also effectively reduces the number of available QDs for multiplexing, although the use of QDs and image detection systems into the infrared range [23] can help to overcome this problem.

3.3.2
Manufacture

Quantum dots are usually prepared by injecting liquid precursors into hot (300 °C) organic solvents, such as trioctyl phosphine oxide (TOPO) and hexadecylamine; this enables nanocrystals of different size to be manufactured by altering the amount of precursors and crystal growth time [26, 27, 30]. The process produces a heavy metal core comprising CdSe, CdS or CdTe, but has a relatively low quantum yield (brightness) of typically less than 10%. A shell of a high bandgap semiconductor, such as ZnS, is then grown epitaxially around the core, increasing the quantum yield up to 80%. The ZnS layer also protects the core from oxidation and prevents leaching of the Cd/Se into the surrounding solution. By themselves, however, QDs are neither water-soluble nor biocompatible, which precludes their use in biological imaging. Bawendi et al. first reported a modification of the QD surface chemistry by using high-temperature growth solvents such as a mixture of trioctyl phosphine (TOP) and TOPO. The QDs prepared using high-temperature solvents are not intrinsically soluble in aqueous solution, but are amenable to surface fictionalization with hydrophilic ligands, either by "cap exchange" or by encapsulation with an organic coating. This has made them more biocompatible [18]; typically, QDs with surface TOPO can also be used to add hydrophilic ligands or other bioconjugates. Here, TOPO acts as a coordinating ligand, and is important for shielding the core from contact with the outside environment.

These modifications have enabled the use of QDs in aqueous solutions, as first reported by Bruchez et al. [15] and Chan et al. [16]. Bruchez et al. [15] used two different sized CdSe–CdS core–shell QDs to visualize the nucleus and cytoplasmic actin filaments in mouse fibroblasts, although there was nonspecific staining of the nuclear membrane, whilst Chan et al. [16], demonstrated an antibody-induced agglutination of QDs that had been labeled with human immunoglobulin G (IgG). Wu et al. [14] improved the surface functionalization of QDs by linking them with streptavidin and IgGs, which facilitated their use in the labeling of a range of cellular targets (cell-surface receptors, cytoskeletal proteins and nuclear antigens) at different subcellular locations (surface, intracellular and intranuclear), and in different specimen types (cultured live cells, fixed cells and tissue sections). For this, Wu and colleagues used QDs to visualize HER2 in fixed cells from a breast cancer cell line, using both direct QD–IgG conjugates, and streptavidin-coated QDs to detect primary biotinylated antibodies. The same group also used streptavidin-coated QDs to detect HER2 in fixed-tissue sections, and went on to demonstrate duplex staining for nuclear antigen and tubulin using streptavidin-coated QDs to disclose antibody localization. They also performed duplex staining for HER2 and nuclear antigen using a combination of directly conjugated QD–IgG against HER2 and a streptavidin-coated QD to detect biotinylated anti-nuclear antibody, thus demonstrating that QDs conjugated to different secondary detection reagents are effective in common fluorescence labeling. This was the first report of their mature use and, in particular for widespread application, the first of their use in fixed-tissue sections. Since then, QDs have been conjugated with many biological molecules, such as proteins, antibodies, oligonucleotides and streptavidin [32]. Their powerful optical properties, combined with their ability to be conjugated with different biomolecules, has rendered QDs ideal for multiplex labeling experiments.

3.4
Bioimaging Applications of Quantum Dots

3.4.1
Quantum Dot Use for Immunohistochemistry

For proof of principal, Bruchez et al. [15] were the first to demonstrate the use of QDs in immunofluorescence for the detection of actin filaments in mouse fibroblasts. Wu et al. [14] were the first to apply this method, which subsequently has been used to detect a vast array of proteins [33–35] in tissue sections, in fluoroimmunoassays [36, 37], and in flow cytometry [38]. Zahavy et al. [39] used two QD colors for the dual-labeling of B and T cells in mouse spleen. For this, a combination of biotinylated primary antibody and streptavidin-coated QDa was used to detect B cells, together with a construct of sequential rat anti-human primary CD3 antibody, biotinylated rabbit anti-rat secondary antibody and QD-conjugated goat anti-rabbit antibody to detect T cells. Image acquisition was performed using a single filter cube, which made interpretation of the images difficult. This report illustrated two problems generic to multiplex studies: (i) the difficulty of antibody cross-reactivity requiring use of different antibody species or strategies (i.e., the use of streptavidin-coated QDs or direct QD–antibody conjugates) for the detection of different antigens (notably, Wu et al. [14] used a similar approach); and (ii) the image analysis required to achieve a good signal separation for each of the QDs used. The real value of QDs lies in higher plex studies, in which these problems – which are generic to multiplex staining in general and not germane to QDs – are compounded [31, 40–42]. However, the fluorescent properties of QDs – namely high brightness, symmetrical emission spectra, common excitation wavelength and tunability – mitigate against these generic problems of multiplex imaging, thus facilitating higher-plex imaging in general. The multiplex capability of QDs has been demonstrated most impressively by Fountaine et al. [40], who reported their use in the simultaneous measurement of five markers. Relying on sequential staining for each antibody, CD20, IgD, MIB-1, CD3 and CD68 were correctly detected in human lymph nodes (Figure 3.2a,b). An avidin–biotin block was applied for 10 min between the end of each antibody detection step (i.e., primary antibody, secondary antibody and streptavidin-coated QDs) and the start of the next sequential primary. Standard mouse nonconjugated primary antibodies were used for each antigen; each of these was disclosed by using a biotinylated secondary antibody followed by a streptavidin-coated QDs. This simple approach could be applied readily to any combination of antibodies, but was seen to be time-consuming. It also demonstrated a potential for the transfer of streptavidin-coated QDs between different secondary antibodies, a problem first identified by Sweeney et al. [8], although the data acquired by Fountaine et al. [40] did not support the existence of this possible problem. In these studies, the images were acquired with a range of filters which, although potentially restrictive, was overcome by others by using spectral imaging.

Other groups have investigated the use of different conjugation chemistries to enable simultaneous marker application. Schwock et al. [43] extended this multiplex approach to measure STAT signaling pathways in needle core biopsies as a possible clinical tool for informing tailored treatment. The aim was to develop a method, validated against parallel Western blotting, to measure the expression levels of phosphorylated proteins (specifically STAT3) in needle core biopsies as a clinical tool for treatment stratification. Standard IHC had disclosed the use of 3,3'-diaminobenzidine (DAB) to be effective in such measurements. However, the small size of such biopsies and the need to measure several different markers – a situation which would be compounded as tailored treatment increased – raised the need for a method capable of quantitative measurement of several markers at the same time. To this end, and as a proof-of-principle, QDs were used for the detection and measurement of the levels of STAT3. Of note, these authors also emphasized that this approach may be capable of multiplexing.

A range of attachment methods have been reported to link QDs to biological molecules. This area was studied systematically by Xing et al. [31], who described several different conjugation strategies, citing a lack of robust protocols and experimental procedures for the low level of success and adoption of QDs in clinical/medical applications. In particular, previous studies used a range of different attachment methods, with no emerging consensus on optimal methods of QD–antibody bioconjugation, tissue preparation, image analysis, multiplex methods, and data quantitation. Comparisons were made between sulfhydryl, amide, Fc-sugar, His-tag or biotin–avidin binding, among which Fc-sugar and His-tag provided the best results. In practical terms, however, the commercially available streptavidin-coated QDs and biotinylated antibodies represent the most feasible conjugation strategy, as they are easily applicable to already established protocols that require only minor optimization with equally good results. A vast number of already optimized protocols are accessible via the internet, and a variety of QD–antibody kits are commercially available (http://probes.invitrogen.com/products/qdot/manuals.html).

Xing et al. [31] applied their methods to the detection of four tumor biomarkers of the epithelial–mesenchymal transition process that is known to be important

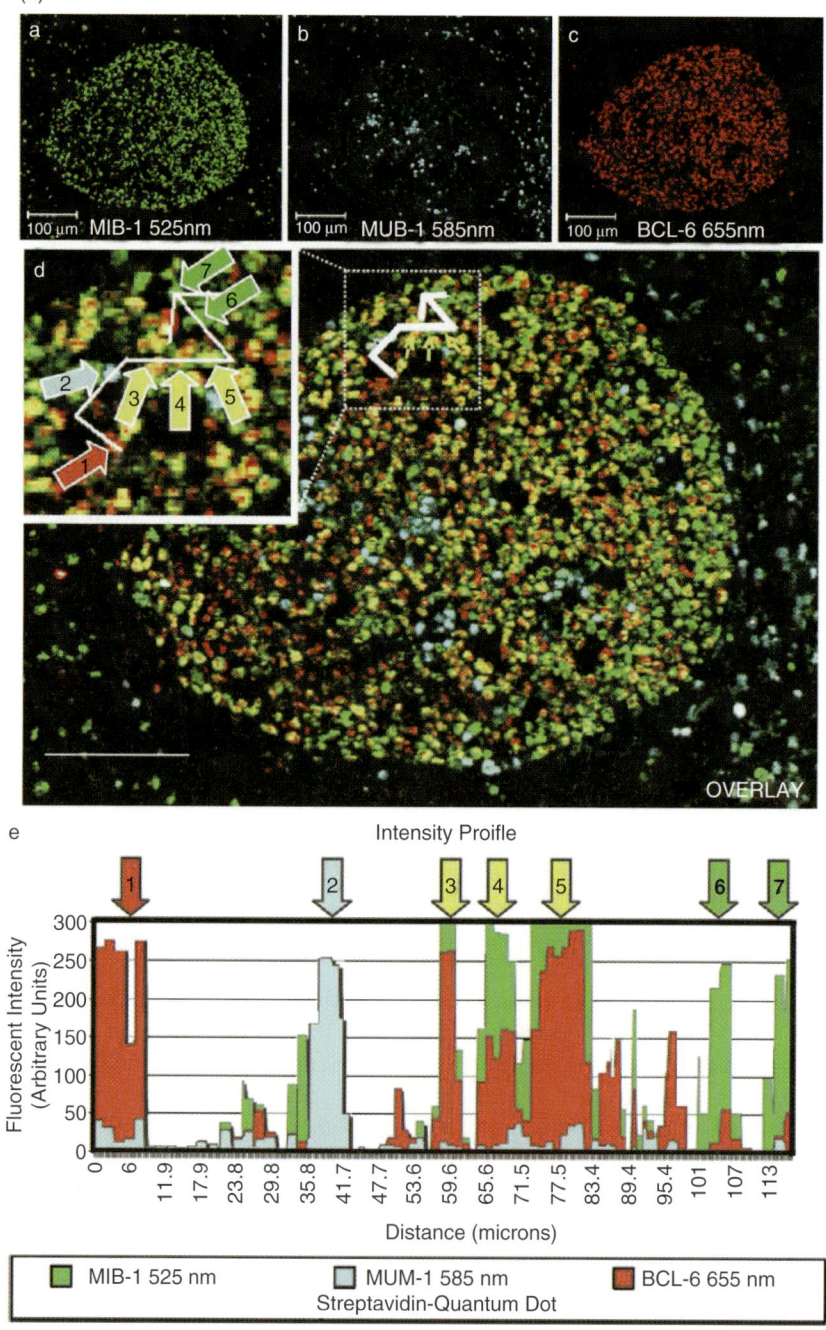

Figure 3.2 (a) Triplex immunofluorescence using QDs to detect MIB-1, MUM-1, and Bcl-6 in lymph node germinal center cells; single staining is shown for each antibody in panels a–c, with triplex staining in panel d (the amount of each antibody is shown in panel e along the inset line in panel d); (b) Quintuplet immunofluorescence using QDs to detect CD20, IgD, MIB-1, CD3, and CD68 in a lymph node germinal center. Reproduced with permission from Ref. [40].

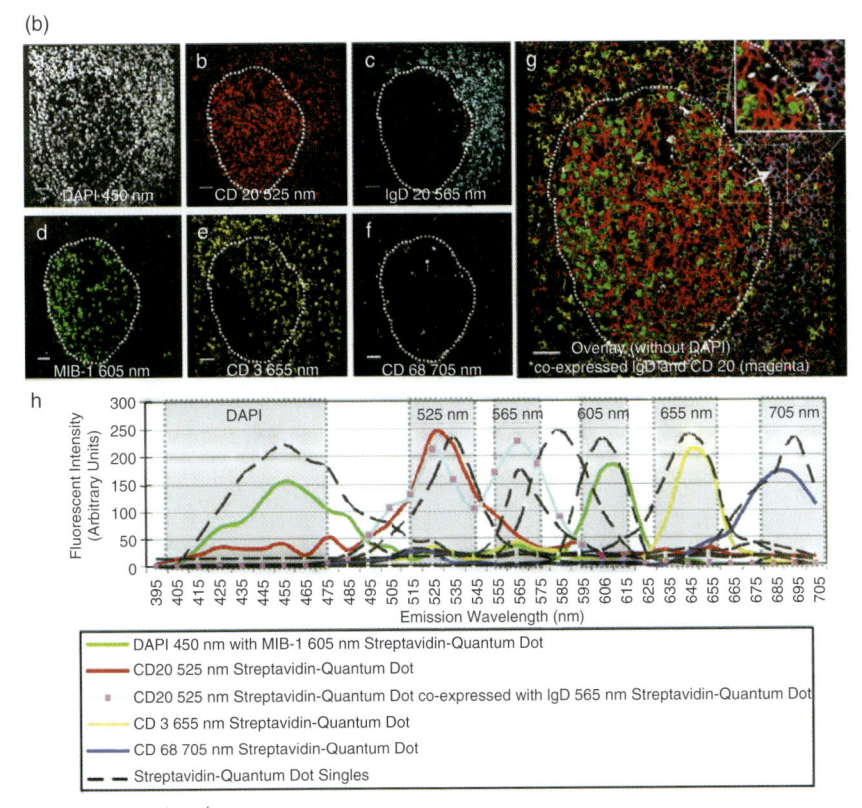

Figure 3.2 *Continued*

in the progression and metastasis of prostate cancer to the bone (namely, N-cadherin, elongation factor-1 alpha, E-cadherin and vimentin), and showed an ability to detect each separately in four-plex staining. The simultaneous detection of four other markers (mdm-3, p53, EGR-1, p21) in archival formalin-fixed paraffin-embedded prostate cancer tissue samples was also demonstrated, although it was not clear which of the different QD conjugation methods was used in these two multiplex experiments.

The same group also proposed a workflow for the use of multiplex QD imaging in clinical practice, which closely matched that described by Bostick et al. [41] and Tholouli et al. [24] (Figure 3.3). Robust protocols and a validation of each step in such a bench to bed workline are essential if QDs are to become more useful in a clinical setting. Xing et al. [31] also described a dedicated software package which was developed as part of such a workline, called Q-IHC, and which enabled integrated image processing and bioinformatic analysis of both traditional and QD-based IHC. In order to measure the distribution of different antigens, multiple images were collected, following which image processing was used to carry out automatic boundary identification, semi-automatic

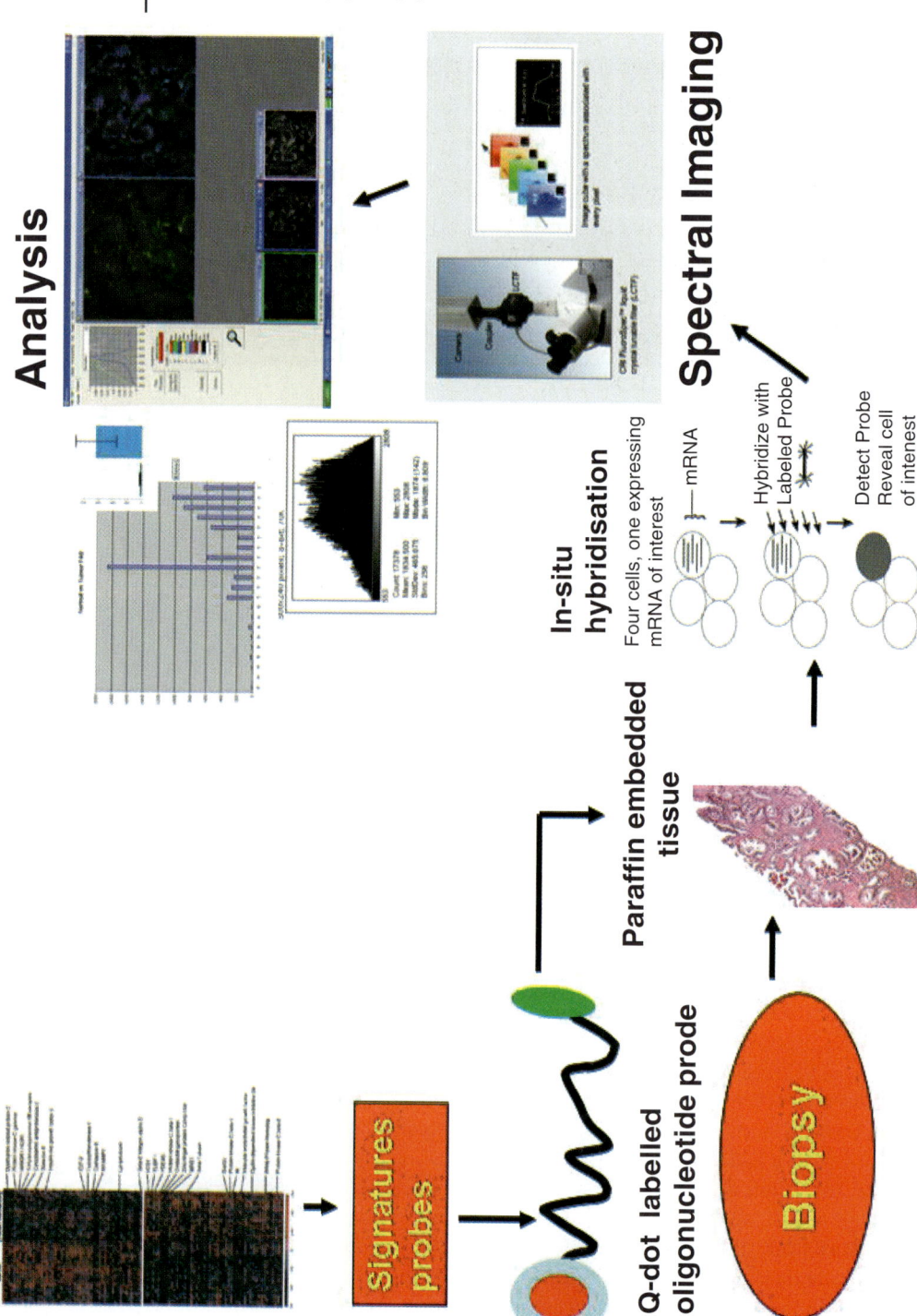

Figure 3.3 Overview of the process for multiplex *in situ* gene expression measurement. Diagnostic or prognostic gene signatures are used to select informative genes for *in situ* detection. Probes to these are labeled with QDs and the resultant hybridization images analyzed using spectral imaging and downstream image analysis. Reproduced with permission from Byers, E. and Tholouli, R. (2008) *Diagnost. Histopath.*, 14, 223–35.

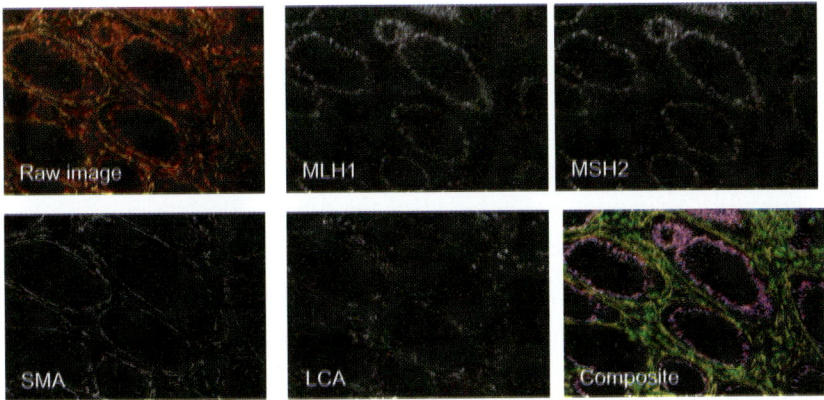

Figure 3.4 Multiplex immunohistochemistry (IHC) for mismatch repair genes in the normal colon. Quadruplex QD-IHC was performed to localize the mismatch repair genes MLH1 and MSH2, leukocyte common antigen (LCA) and smooth muscle actin (SMA) in normal human colon. Monoclonal antibodies to MLH1 (labeled with a 605 nm QD), MSH2 (labeled with a 655 nm QD), SMA (labeled with a 605 nm QD) and LCA (labeled with a 705 nm QD) were applied sequentially, and the resultant raw image subjected to spectral analysis to resolve the component signal intensity maps for each antigen. A composite image showing all four antigen distributions was generated from these, using false color with MLH1 in red, MSH2 in blue, SMA in yellow, and LCA in green (the colocalization of MLH1 and MSH2 resulted in a purple false color). Reproduced with permission from Ref. [44].

image segmentation, and color-based tissue classification based on biomarker staining. An image analysis module was then used to quantify the numerical expression of each biomarker. These values, it was proposed, might then be used in clinical decision making.

Subsequently, these established IHC protocols were adapted and successfully multiplexed up to four QD markers [44] (Figure 3.4); some practical tips found to be useful for the application of QDs have been provided by Tholouli et al. [44].

3.4.2
Quantum Dot Use for *In Situ* Hybridization

A significant focus of current biomedical research is the understanding of the molecular mechanisms of cells, their function, and activity. As the amount of data acquired and the number of biomarkers identified continue to increase, it is becoming particularly important to carry out *in situ* expression studies utilizing more than one marker. QDs have enabled this for protein expression by immunofluorescence and, more recently, they have also been used to detect nucleic acids by ISH. One of the first attempts to use QD-labeled oligonucleotide probes was to identify mutations in human sperm by standard fluorescence *in situ* hybridization (FISH), in 2001 [45]. In this study, however, the successful binding of oligonucleotides to the QD surface using carboxylic acid groups proved difficult due to inefficient loading, poor stability and high nonspecific binding when used for FISH. Consequently, the QD surfaces were modified using carboxyl groups so as to increase the solubility and stability of the conjugates [46]. Since then, further improvements in the water solubility and bioconjugation of QDs have allowed their widespread use in biological applications. Presently, a wide range of modified biomolecules can be directly conjugated to QDs in a single step, and the conjugates used for ISH or IHC, without compromising their properties. Streptavidin-conjugated QDs have also been used successfully to quantify FISH signals on human metaphase [47], mouse [48] and plant chromosomes [49], as well as in microorganisms such as *Escherichia coli* [50]. Xiao and Barker [47] used a biotinylated BAC DNA probe to detect HER2 in metaphase spreads, whilst in fresh animal tissue Matsuno et al. [51] combined QD-based ISH with IHC. For this, rat pituitary was fixed in 4% paraformaldehyde and combined IHC and ISH used to for the simultaneous detection of mRNA and protein for growth hormone (GH) or prolactin (PRL). For this, a QD–IgG secondary antibody was used to detect the primary antibody, and a streptavidin–QD to detect a biotinylated oligonucleotide cDNA mRNA probe. The first successful duplex ISH was described by Chan et al. [52], who reported the simultaneous detection of two nucleic acid targets by utilizing directly labeled 54-mer oligonucleotide probes to QDs in fresh-frozen mouse brain tissue (Figure 3.5a). In this case, an *in vitro* conjugation of oligonucleotide probes with streptavidin-coated QDs was carried out, followed by superdex gel filtration, to produce prehybridization QD–probe conjugates. The initial biocytin blocking of excess QD-bound streptavidin sites was used to avoid a too-heavy loading of the QDs by the probe, which would have interfered with hybridization. Image acquisition was performed using a confocal microscope with different filters for each QD, with image analysis in the NIH ImageJ program (http://rsbweb.nih.gov/ij/). In this way, up to four different mRNAs could be simultaneously detected, using a combination of Alexa fluorophores and QDs (Figure 3.5b). These authors also carried out a simultaneous detection of Vmat 2 mRNA and tyrosine hydroxylase (TH) protein using a QD-conjugated oligonucleotide probe for mRNA detection and a streptavidin-coated QD for protein detection.

Although, unfortunately, the problems encountered with FFPET materials were not addressed, a protocol which successfully applied QD-labeled oligonucleotide probes for ISH in routinely processed FFPET samples was shortly made available [24] (Figure 3.6a,b). By applying the same principle, it was possible to demonstrate duplex ISH with QD-labeled ribonucleotide probes, and to combine this with IHC in a semi-automated system, thus allowing a high throughput of samples [42]. Subsequently, Matsuno et al. [51, 53] reported the details of a QD-based ISH and IHC by using electron microscopy that allowed the subcellular (three-dimensional; 3-D) localization of pituitary hormones (GH and PRL) and their corresponding mRNAs.

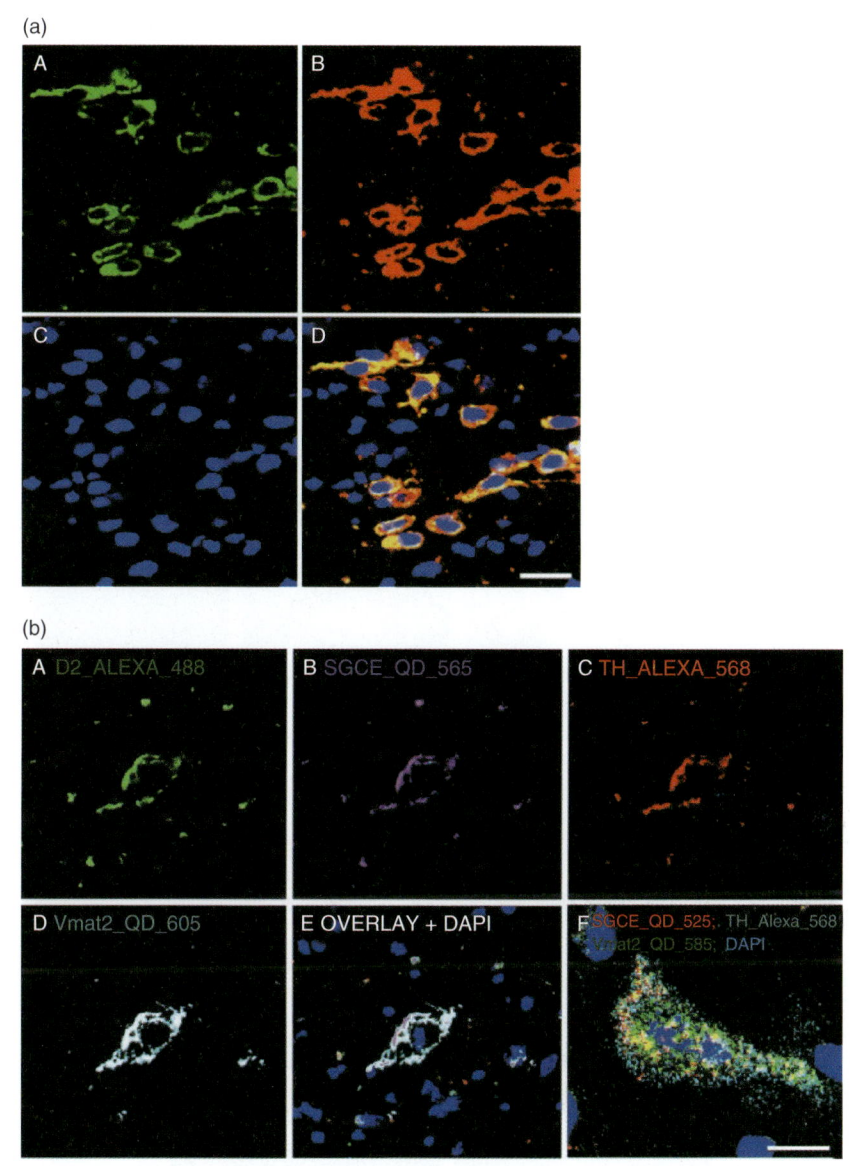

Figure 3.5 (a) Double-labeled FISH using QD-labeled oligonucleotide probes to detect Vmat2 mRNA in murine neurons in panels A and B, with overlap in panel D; (b) Combined multiplex QD and Alexafluor FISH to detect four different mRNA targets in murine neurons; mRNA targets are labeled in panels A to D, with overlay in panel F. Reproduced with permission from Ref. [52].

| | H&E | IHC for MPO | QD-ISH for MPO |

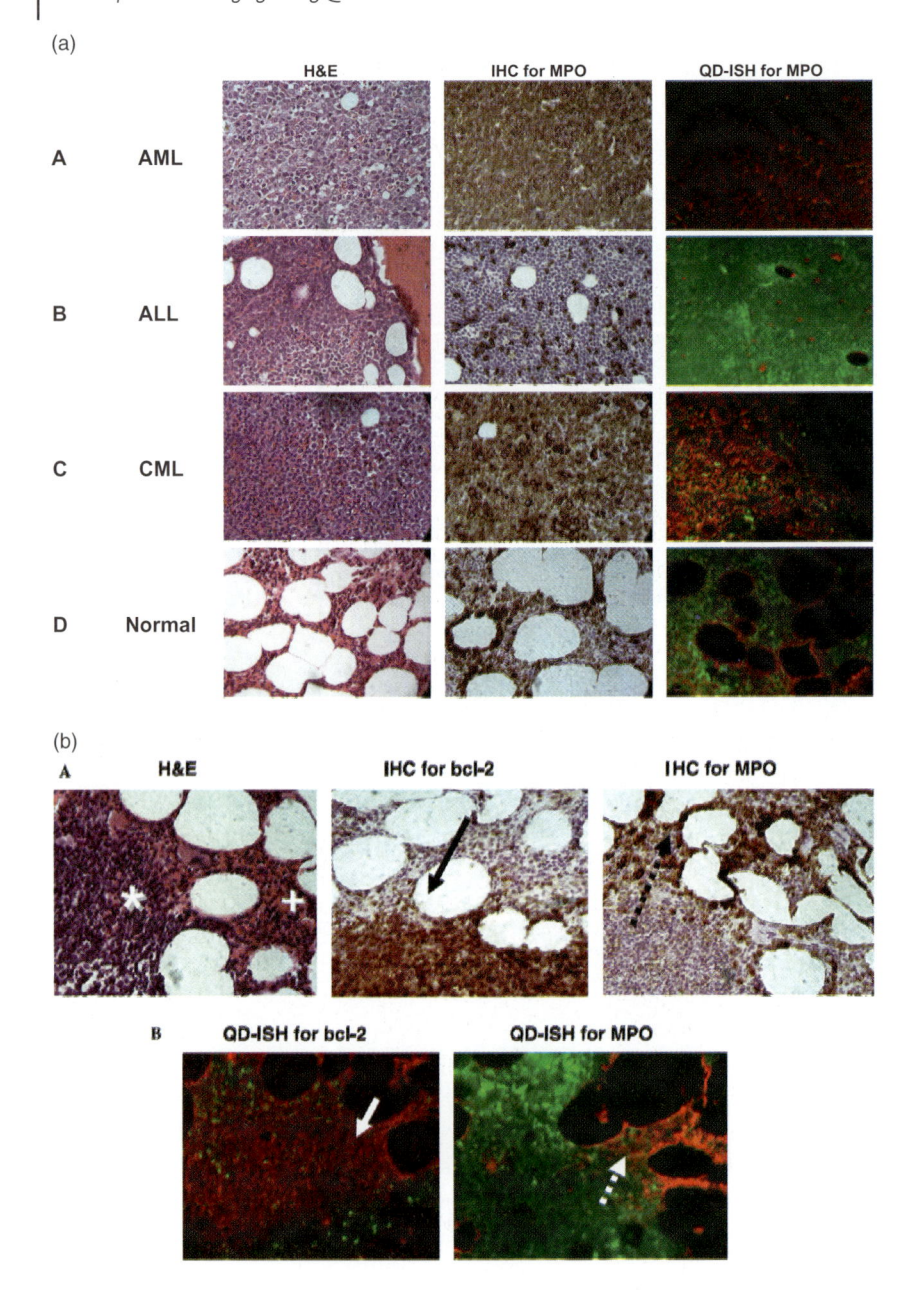

(a)

A AML

B ALL

C CML

D Normal

(b)

A H&E   IHC for bcl-2   IHC for MPO

B QD-ISH for bcl-2   QD-ISH for MPO

Figure 3.6 (a) Single QD-based *in-situ* hybridization in human clinical tissue. A: Bone marrow infiltrated by myeloblasts replacing normal hematopoiesis (acute myeloid leukemia – FAB M3). The cells were strongly positive for myeloperoxidase (MPO) by IHC, disclosed with DAB, and single-gene ISH using a QD-labeled probe for MPO in acute myeloid leukemia demonstrated strong cytoplasmic hybridization signals in all myeloblasts. B: Bone marrow infiltrated by lymphoblasts (acute lymphoblastic leukemia – FAB L1) which were negative for MPO by IHC. Single-gene ISH using QD-labeled probe for MPO showed a restriction of hybridization signal to occasional residual myeloid cells and absence in the leukemic blasts. C: Bone marrow infiltrated by chronic myeloid leukemia was positive for MPO by IHC and by ISH using a QD-labeled probe for MPO, though the signal intensity lower than that in AML (shown in A, above), whilst residual nonmyeloid cells were negative. D: Normal bone marrow showed only patchy positivity for MPO, by both IHC and QD-ISH, consistent with the normal distribution of myeloid precursor cells. All images were falsely colored after spectral analysis to display hybridization signal in red and tissue autofluorescence in green; (b) Visualization of two mRNA targets by QD-ISH. A: Bone marrow infiltrated by FL showing a lymphoid aggregate of small lymphoid cells (*) with an area of normal hematopoiesis (+). The same area was examined by standard IHC for bcl-2 and MPO showing positivity for bcl-2 in the lymphoma cells (solid arrow) and MPO in the adjacent normal marrow (broken arrow). B: QD-ISH for (ii) bcl-2 and (iii) MPO in serial sections from the same biopsy showed hybridization signal in the same distribution as that of IHC positivity, with bcl-2 upregulated in an area of lymphoma cells (solid arrow) and MPO in adjacent myeloid cells (broken arrow); the MPO probe was labeled with a 605 nm QD, the bcl-2 probe with a 655 nm QD; a signal for each is shown after spectral analysis with hybridization signal in red and tissue autofluorescence in green. Reproduced with permission from Ref. [24].

3.4.3
Quantum Dot Use in Solid- and Liquid-State Multiplex Detection Platforms

Whilst the most biologically exciting use has been application to multiplex imaging, QDs have also been used extensively to facilitate multiplex detection in solid- and liquid-phase experiments. Hahn *et al.* [54] reported the use of QDs in the multiplex flow cytometric detection of bacterial pathogens, citing their advantages of brighter fluorescence intensity, lower detection thresholds and improved accuracy compared to conventional organic fluorophores, with a net one order magnitude increase in brightness compared to fluorescein isothiocyanate (FITC). Most reports of the use of QDs in flow cytometry have focused on the detection of pathogens, with mixed success; some have reported significant nonspecific binding to water-borne detritus which resulted in poor performance compared to conventional methods. Recently, Chattopadhyay *et al.* [55] reported the use of QDs for the multiplex detection of T-cell antigens, combining seven-color QD detection with 10 other fluorochromes to yield a 17-color staining panel. A significant advantage here was the ability of the QDs to be excited by a single laser, in contrast to conventional multicolor methods which require multiple lasers. While the resultant highly multiplexed method is likely to aid a number of applications [56], it is particularly useful for studies of T-cell differentiation, which require the use of multiple markers for cellular characterization. This method provided the single largest increment in the number of measurable fluorophores at the time, with particular impact in the area of polychromatic flow cytometry, though their application is expected to extend to standard flow cytometry systems [55].

QDs have also been used for multiplex protein detection in protein arrays [57, 58]. Gokharna et al. [59] used PEGylated (PEG) QDs directly conjugated to antibodies against prostate-specific antigen (PSA) to detect the antigen itself. The use of a PEG QD allowed direct conjugation to the PSA antibody, without using biotin–streptavidin interactions, thus minimizing any nonspecific staining whilst retaining a high affinity. Although only one antigen was detected in this study, it provided a microarray platform for protein detection that could easily be adapted to the use of multiple QD–antibody conjugates. Geho et al. [60] instead used PEGylated, streptavidin-conjugated QDs as detection agents in a reverse-phase protein microarray, in which heterogeneous protein mixtures from cellular extracts were directly spotted onto a protein biochip. This method has the potential for multiplexed high-throughput reverse-phase protein microarray analysis, in which numerous analytes can be measured in parallel in a single spot.

QDs have also been used in cDNA arrays, either solid, or bead-based. For this, Han et al. [61] used QD-tagged polymeric microbeads to detect nucleic acid and protein sequences, using different-sized QDs to yield 10 intensity levels and six colors that, in theory, was capable or coding for one million different combinations, although practical considerations would be expected to reduce this to approximately 10 000 to 40 000 recognizable codes. The utility of the approach was tested in DNA hybridization assays using oligonucleotide-conjugated multicolored beads which were capable of correctly measuring the relative amounts of four different DNA sequences. Furthermore, it was shown that the coding and target signals could be read simultaneously at the single-bead level, with potential for high-throughput use. Eastman et al. [62] used four different sizes of QDs (525, 545, 565, and 585 nm), with 12 intensity levels, mixed with a polymer and coated onto magnetic microbeads, to generate nanobarcoded beads (QBeads). Gene-specific oligonucleotide probes were then conjugated to each spectrally nanobarcoded bead to create a multiplexed panel, to which biotinylated cRNAs generated from test RNA were hybridized; a fifth streptavidin QD (655 nm or infrared QD) binding to biotin on the cRNA was used as a quantification reporter. This system had a high sensitivity for target molecules detection, approaching that of quantitative PCR.

Conversely, relatively few reports have been of the use of QDs for RNA or DNA detection in solid-phase platforms. Liang et al. [63] fabricated a microRNA microarray using complementary oligo-DNA probes immobilized on glass slides to capture 3′ biotinylated miRNAs, followed by detection using streptavidin-coated QDs. The device was used to analyze 11 microRNAs, and showed high sensitivity and consistence with Northern blot results. Kalrin-Neumann et al. [64] used QDs instead of organic dyes for four-color genotyping to improve the SNR and enable multiplex genomic analysis, quoting the capability of >20k plex detection; however, such advances have yet to be translated into commercial use.

3.5
Translation to Clinical Biomarker Measurement

3.5.1
Quantitation

Meaningful gene expression profiling, at either mRNA or protein level, is critically dependent upon robust methods for expression quantitation. Flow cytometry routinely uses reference microbeads for this purpose, and several groups have developed protocols for quantitation using QDs. Smith *et al.* [65] generated a surface-pegylated QD construct capable of multivalent targeted binding providing a modular platform for quantitation of cell-surface receptors. Specifically a QD–PEG–NGR construct was made by the conjugation of PEGylated peptide with carboxylated QDs; the NGR tripeptide is a CD13-targeting molecule identified as a tumor homing sequence that selectively targets the tumor vasculature *in vivo*. Ligand–cell interactions were successfully tested using the QD–PEG–NGR in solution with cell samples, and QD probe binding was quantified by flow cytometry using reference microspheres. Specifically R-phycoerythrin (R-PE) calibration microspheres were used to correlate fluorescence measurements producing a standard curve relating measured relative fluorescent intensity (MFI) to the number of equivalent R-PE molecules. The standard curve of calculated intensities was then used to determine the number of R-PE molecules per QD probe, from which a new standard curve was produced relating MFI measured by flow cytometry to the number of QD probes. This standard curve enabled the direct conversion of fluorescence measured by flow cytometry to the number of bound QD probes per cell. Conversely, Wu *et al.* [38] developed a simple assay for the production of QD calibration beads using commercially available streptavidin-coated QDs which were incorporated into an assembly of biotinylated M2 antiFLAG antibodies, biotinylated FLAG peptides and streptavidin-functionalized beads. The law of mass action was used to define the site density of dots on each bead, and the fluorescence intensity of the QD-bead assemblies was tested against commercial fluorescein calibration beads. The utility of the calibration beads was then tested by measuring the surface density of QD585 dots attached to the ligand of the epidermal growth factor receptor (EGFR) on A431 cells. A streptavidin-coated QD was bound using biotinylated FLAG peptide to a biotinylated antiFLAG Ab, which was in turn bound to a streptavidin-coated microbead.

Xiao *et al.* [66] quantified the expression of HER2 and telomerase using monospecific polyclonal chicken IgY antibodies against human HER2 and telomerase in Western blots and IHC of tumor and normal cells, using fluorescent microspheres as a fluorescence standard. The IgY antibodies used have the advantages of a lack of complement activation, and a lack of binding to protein A and G, to rheumatoid factor, or to the cell-surface Fc receptor, thus potentially eliminating false positives and reducing background. However, in common with other approaches, the quantitation was relative than absolute and to date there have been no reports of quantitative methods calibrated in amounts of protein or mRNA; hence, the development of such methods remains a major challenge. Zajac et al. [67] reported the development of QD protein microarrays for the detection of cancer markers, detecting up to six different cytokines in protein solution down to picomolar concentrations. A comparison between use of QDs directly conjugated with antibody against a selected marker and the use of streptavidin-coated QDs and biotinylated detector antibody, demonstrated better performance of the latter, which was also cheaper and technically simpler. Other studies have confirmed the greater utility of this simple approach over more complex attachment and conjugation strategies. Here, a computer analysis was used to quantify the amount of protein detected by the array, thus demonstrating its use for monitoring changes in biomarker concentrations in the physiological range. The translation of these techniques from protein measurements to tissue analysis would be invaluable for the analysis of *in situ* expression.

Xing et al. [31] noted the need to validate QD staining against other available methods, and carried out a simple semi-quantitation of the amount of three breast cancer biomarkers, namely estrogen receptor (ER), progesterone receptor (PR) and HER2, in paraffin-embedded tissue samples. For this, the fluorescence intensity values for each IHC method were compared by using a scoring system that employed a scale of 1+ to 3+. A 3+ score for ER, PR or HER2 using traditional IHC corresponded to a 85–100% relative antigen expression when using QDs, whilst a 1+ or 2+ score corresponded to a 11–48% expression. From these findings, it was suggested that the quantitative nature of QDs could simplify and standardize the categorization of antigens at low levels. This procedure has been investigated further by comparing the ability of QDs and conventional IHC to measure mismatch repair gene mutation status in hereditary nonpolyposis colorectal cancer, in which conventional IHC scoring by an expert pathologist outperformed a QD-based quantified status determination (E. Barrow et al., unpublished results).

3.5.2
Imaging Analysis

Whilst QDs allow multiple label-staining there remains the problem of color resolution, and therefore – despite the improved fluorescence properties of QDs over conventional fluorophores – effective multiplexed and quantitative imaging using QDs still requires sophisticated image analysis [25, 68, 69]. Crucially, the data quality from highly multiplex imaging using QDs is compromised without such sophisticated image analysis. Conventional RGB (red, green, blue) cameras summate the intensity of each color channel, reducing the complexity of the spectral information to a three-digit readout, but are incapable of:

- distinguishing pure from colocalized color mixtures; that is, pure yellow from a red and green mixture;

- removing tissue autofluorescence, which is often high in formalin-fixed paraffin-embedded tissue;
- distinguishing spectrally overlapping fluorescent signals; and
- producing a numerical readout of signal distribution and intensity [70].

Spectral imaging (also known as "hyperspectral imaging") is a branch of spectroscopy in which a complete spectrum or some spectral information (such as the Doppler shift or Zeeman splitting of a spectral line) is collected at every location in an image plane. The process has been used extensively in astronomy, solar physics and Earth remote sensing and, more recently, in biological imaging [69]. For the analysis of fluorescence imaging it differs from conventional methods, in which different filters or "cubes" are used to isolate and visualize each fluorophore, by collection of the entire spectral information for each pixel, from which individual spectral components are resolved digitally.

When using a fluorescence microscope combined with a CCD camera and a liquid crystal tunable filter (LCTF), a series of images can be captured along a specific wavelength range. The Nuance spectral imaging system supplied by Cambridge Research Instruments (CRI, Woburn, MA, USA), which the present authors have used extensively, achieves this by using stacked liquid crystal filters to produce a solid-state tunable Lyot filter that enables fluorescent image files to be collected at serially stepped wavelength intervals from 450 to 720 nm The resultant image files each comprise the concatenated stack of images at each wavelength interval per pixel, and can be used to reconstruct multiple spectral distributions via a maximum likelihood method [71]. Specifically, the maximum likelihood distributions at each pixel are determined for spectral distributions obtained from autofluorescence and for the QDs used in a given experiment. These distributions represent signal intensity at each pixel for the defined spectra, and can be converted to composite false color images to visualize staining distribution and intensity for each QD. This system therefore offers the possibility to separate electronically the different spectra or signals, and in particular the tissue autofluorescence, from other fluorescent markers.

These advances in image processing and computer technologies have provided significant improvements in SNRs and an accurate separation of multiple colors, simultaneously capturing signal intensity and enabling signal quantitation. Gao *et al.* [21] used spectral imaging to visualize fluorescent probes targeting prostate cancer. They were able to successfully remove the background noise, such that multiple fluorescent signals were identified within the acquired images in a live mouse. This application has subsequently gained popularity due to its ease of use. Xiao *et al.* [66] have used QDs to quantify HER2 and telomerase, demonstrating the feasibility of their use for the relative quantitation of cancer biomarkers. Alternatively, 3-D and *in vivo* analyses of small animals can be performed with confocal laser scanning microscopy [72], although the underlying principle utilizes a similar approach as the spectral imaging analysis described above. In this case, the scanners and detectors are arranged around a sample so as to ensure optimal illumination, and a spectral signature is collected at regular wavelength intervals for each pixel, thus creating a stack.

The possibility of quantitative fluorescence gene expression measurement with QDs requires the development of fluorescent standards, as described above. Wu et al. [38] reported the production of calibrated QDs for a range of emission spectra, of importance in multiplex studies. One possible confounding factor in the use of QDs for quantitative measurement is the characteristic, of single QDs, of fluorescence intermittency ("blinking"). This may hamper their use as single-photon sources, either for quantitation or for real-time monitoring of single biomolecules. However, the problem may be obviated when quantifying over an area, in which multiple reporter QDs will be present, and when imaging is undertaken over a prolonged period relative to the frequency of blinking.

3.5.3
Combinational Microscopy Methods

A few groups have combined light and electron microscopic imaging. Giepmans et al. [73] exploited the high electron-density of QDs to visualize them using electron microscopy. Streptavidin-coated QDs were used to detect alpha-tubulin, via a biotinylated secondary antibody, in fixed sections of rat lung fibroblasts, by using light microscopy. Electron microscopy was used to visualize QDs on carbon films, demonstrating differences in the shape of different-sized QDs, with larger dots having a rhomboid shape. Streptavidin-coated QDs were then used for the disclosure of anti-Cx43, as visualized by electron microscopy, which showed a localization of staining to the gap junctions of cells (Figure 3.7a). Next, triplex staining for GFAP, Cx43 and IP3R was performed in mouse cerebellum, with visualization of staining by both light and electron microscopy; the size and shape difference of the different QDs used for disclosure of each antibody was then used to distinguish label identity in the electron micrographs (Figure 3.7b). Deerinck et al. [74] also showed simultaneous light/fluorescent and electron microscopic detection of antibody staining, for beta-tubulin, in HeLa cells. In particular, they noted the extreme durability of QD-fluorescence, which was preserved through processing for electron microscopy, as long as post-fixation with osmium tetroxide was omitted. Typically, specimens can be embedded in either epoxy resin or in London Resin, and visualized first by fluorescence microscopy, followed by electron microscopic visualization of selected areas. This enables an initial relatively wide-field survey to be followed by directly correlated imaging using electron microscopy.

3.5.4
Clinical and Mechanistic Biological Applications

Most of the applications noted above have used QDs to visualize multiple markers in test models, either as proof of principle or for method development and validation. However, in some cases the QDs have been used in particularly imaginative

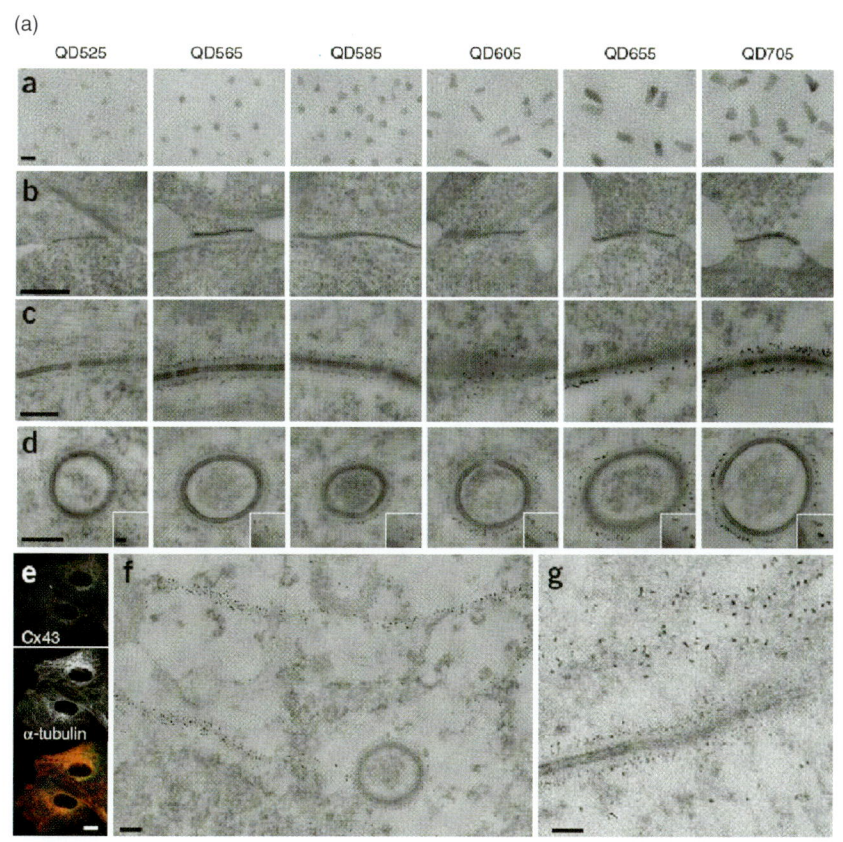

Figure 3.7 (a) Multiplex electron microscopy using QDs. Different sized streptavidin–QD conjugates (shown on carbon films in panel a) were used for the disclosure of Cx43 antibody staining in fibroblasts, showing localization to gap junctions (panel b, higher magnification in panel c and annular junctions in panel d). Double staining for Cx42 and alpha-tubulin is shown in panels e–g, with clearly identifiable localization of different-sized QDs (smaller labeling Cx43 and larger alpha-tubulin) in panel g; (b) Double- and triple-labeling of proteins for electron microscopic imaging. (Panels a,b) Mouse cerebellum probed for IP3R and GFAP with 565 nm and 655 nm QDs, respectively [imaged by light (a) and electron (b) microscopy]; (panels c–e) Mouse cerebellum probed for GFAP, Cx43 and NFP68 using 525 nm, 565 nm and 655 nm QDs, respectively (imaged by light [singly in c and as overlay in d] and electron (e) microscopy). Reproduced with permission from Ref. [73].

ways to study complex or difficult biological problems. For example, Matsumo et al. [51, 53] used confocal laser scanning microscopy and combined QD ISH and IHC to visualize, in 3-D form, the relationship between GH mRNA and protein in the rat pituitary. This was especially useful for the analysis of protein and mRNA localization and interaction in subcellular organelles, in which the 3-D structure of, and localization of biomolecules to, are important. This method may therefore facilitate a 3-D understanding of protein–protein and protein–

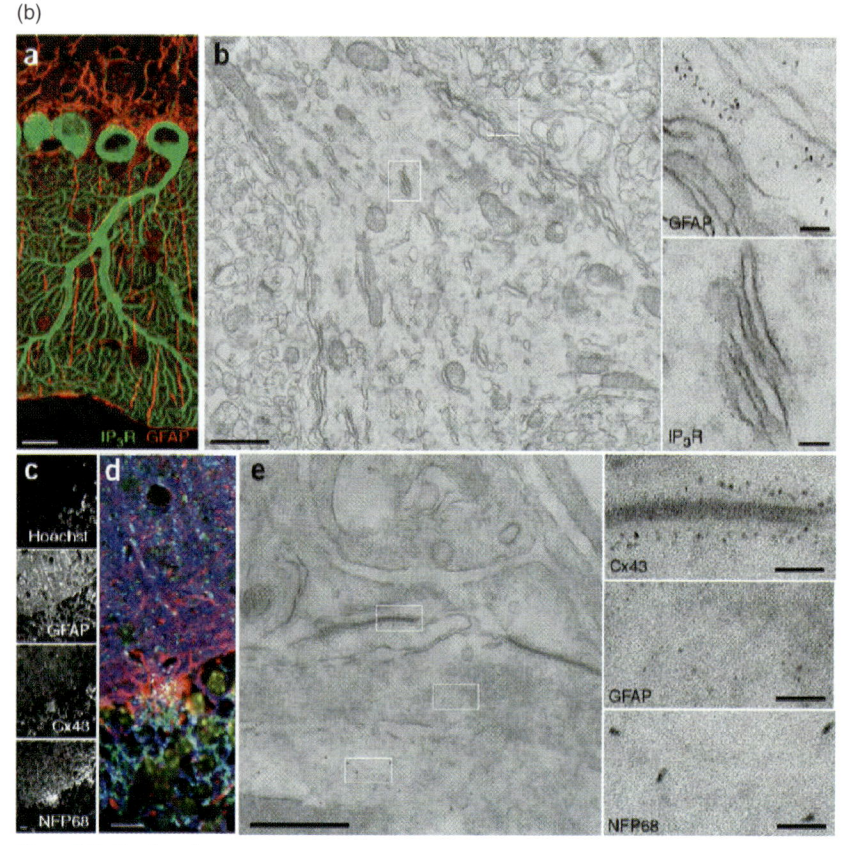

Figure 3.7 *Continued*

mRNA interactions at the subcellular level. Specifically, for GH and PRL, as studied by Matsumo et al. [53], the results suggested that PRL was being transported to the plasma membrane and secreted more rapidly than GH. Lidke et al. [19] used QDs to study the cellular localization of EGF, using QDs bearing EGF (QD-EGF) that had been prepared by incubating streptavidin-coated QDs with biotinylated EGF; in this case, erbB1, which binds EGF, was labeled with green fluorescent protein (GFP). In live cells there was rapid colocalization of QD-EGF and ErbB1-GFP, followed by endocytosis, which was shown by using transferrin-Alexa633 to occur via clathrin-coated pits. Subsequently, it was shown that the uptake of EGF-QDs occurred by a previously unreported retrograde transport mechanism, which could not have been detected without the use of QD-EGF. Finally, the heterodimerization of erbB2 (labeled with yellow fluorescent protein; YFP), but not erbB3 with erbB1, was demonstrated after EGF stimulation. The main impact of this report was its demonstration of the power of QD-ligands to visualize complex protein interactions and cellular processes down to the single-molecule level.

3.5.5
Cancer Molecular Profiling

Cancer diagnostics increasingly relies on the measurement of multiple biomarkers at either the genotypic, mRNA, or protein level. Despite the simultaneous measurement of multiple markers being difficult when using conventional techniques, there is an increasing need for methods to perform such measurements in clinical samples as the number of informative markers required for prediction continues to increase while the size of many diagnostic samples decrease. This is particularly important with regards to improvements in imaging, which have resulted in the radiologically guided sampling of deep tumors. Consequently, much interest has been expressed in the possibility of using QDs for this purpose.

Caldwell et al. [75] used spectral imaging to measure the average intensity of QD–antibody staining for just two proteins (MDM-2 and B-actin) in a tissue microarray of tissue samples from a renal cell carcinoma, thus demonstrating the ability of the method to distinguish cancer from normal adjacent tissue. Bostick et al. [41] proposed the use of QDs to detect up to five biomarkers per slide, from which more biomarkers could be measured using multiple slides each stained with five different biomarkers. A similar approach, using QD-ISH, was taken by Tholouli et al. [24] to measure nine prognostic genes in acute myeloid leukemia (E. Tholouli et al., unpublished results). Here, a custom-built image analysis method was used to quantify the expression of each biomarker, and a workflow for the analysis, similar to that proposed by Byers et al. [42] and Tholouli et al. [24]. In future, it will be important for clinical applications that such systems are robust, standardized, streamlined, fast, easy to use, and, ideally, automatable; as an example, the system described by Bostick et al. [41] took seven hours to analyze six biomarkers.

Byers [42], Tholouli [24], Sweeney [8], and colleagues have extensively explored the use of QDs for the measurement of biomarkers in clinical tissue. In two related reports – one detailing a manual and the other an automated system – Byers et al. [42] and Tholouli et al. [24] demonstrated multiplex QD-ISH in archival clinical tissue samples. Byers et al. [42] demonstrated the photostability of QDs over a period of 18 months, and also (in a preliminary study) the use of QD fluorescence intensity to measure Fas mRNA expression in fixed LNCap cells in a semi-quantitative manner; the results obtained correlated well with parallel real-time PCR measurements of mRNA expression. The use for mRNA detection, again for Fas, was also demonstrated in 30-year-old archival prostatic tissue in a tissue microarray, and also as combined ISH and IHC in murine lung (Figure 3.8a–c). Tholouli et al. [24] undertook a comprehensive test of the method in EDTA-decalcified formalin-fixed bone marrow trephine samples, applying strict ISH controls, and demonstrated triplex ISH for X-linked inhibitor of apoptosis protein (XIAP), survivin, and Bcl2; a comparison of the expression values obtained by single and triplex ISH showed good concordance (Figure 3.9).

Each of these studies employed spectral imaging using a CRI Nuance system (Woburn, MA, USA) to unmix raw images and produce intensity maps for each

(a)

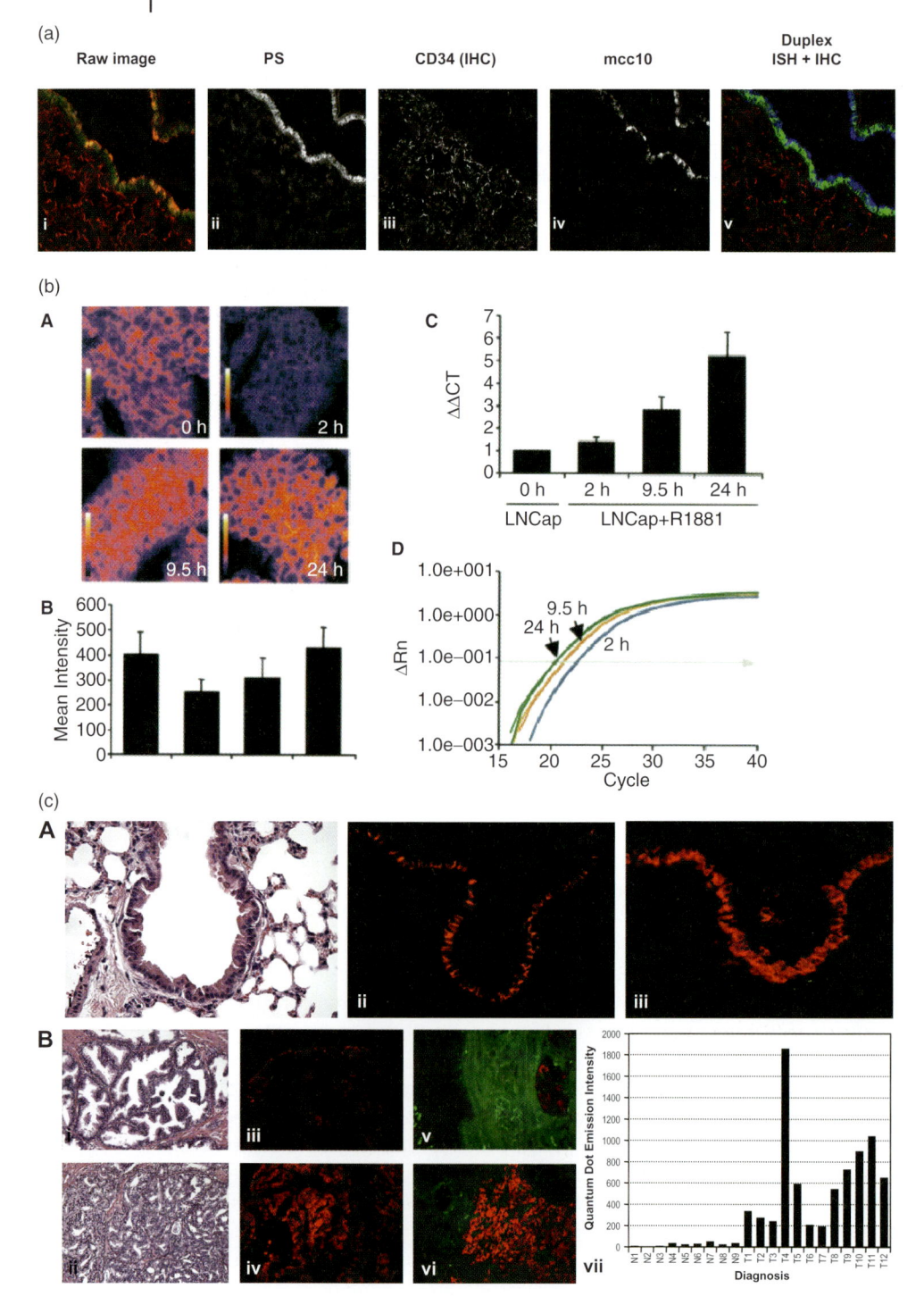

Raw image | PS | CD34 (IHC) | mcc10 | Duplex ISH + IHC
i | ii | iii | iv | v

(b)

A
0 h | 2 h
9.5 h | 24 h

B
Mean Intensity (0–600)

C
$\Delta\Delta CT$ (0–7)
0 h — LNCap
2 h, 9.5 h, 24 h — LNCap+R1881

D
ΔRn vs Cycle (15–40)
24 h, 9.5 h, 2 h

(c)

A
i | ii | iii

B
i | iii | v
ii | iv | vi

vii — Quantum Dot Emission Intensity (0–2000) vs Diagnosis (N1–N9, T1–T12)

Figure 3.8 (a) Combined duplex Q-ISH and IHC was performed in mouse lung for mcc10, pulmonary surfactant (PS) and CD34 using QD-labeled riboprobes for mcc10 and PS with the addition of IHC for CD34, which highlights the alveolar capillary network, but is not present in the bronchiolar epithelium. The resultant raw image (i) was subjected to spectral analysis using the spectra for 525 nm (PS), 605 nm (mcc10) and 655 nm (CD34) QDs, from which distributions of PS (ii), CD34 (iii) and mcc10 (iv) were resolved; a composite false color image of resolved images for PS, CD34 and mcc10 is shown in (v) (PS in green, CD34 in red, mcc10 in blue; all ×40 original magnification); (b) Quantification of QD-ISH. Expression FAS was measured at a series of time points in LNCap cells following stimulation of expression with the synthetic androgen R1881 by both QD-ISH (panels A and B) and quantitative RT-PCR (panels C and D), and showed a high correlation between FAS levels measured by QD-ISH (panel B) and quantitative RT-PCR (panel C); (c) QD-based *in-situ* hybridization is photostable and quantifiable. A: Signal intensity for the same slide of mouse lung, hybridized with DIG-labeled antisense mcc10 riboprobe visualized twice over an 18-month interval, demonstrating the stability of QD fluorescence. There was no significant loss of intensity between the first (ii) and second (iii) visualizations; hematoxylin and eosin-stained section shown in (i) (all ×40 original magnification). B: FAS hybridization in 30-year (i) benign and (ii) malignant prostatic core biopsies in a tissue microarray demonstrating (iii) low signal in benign tissue, and (iv) high signal in prostatic adenocarcinoma; 605 nm QD disclosed IHC for FAS is shown in (v) benign and (vi) malignant prostatic tissue (all ×20 original magnification). Signal intensity was determined by spectral analysis over nine benign (N) and 12 malignant (T) cores (vii), and measured using IPLab, demonstrating a higher expression of FAS in the malignant cores. Reproduced with permission from Ref. [42].

of the QDs used, from which areas of positivity, colocalization and intensity could be calculated. Sweeney *et al.* [8] have published a modified method using *in vitro* QD–antibody conjugation as a possible high-throughput method for biomarker validation and measurement in clinical samples, demonstrating triplex staining for cytokeratin, CD34, and cleaved caspase 3. The utility of the CRI imaging system for the colocalization, and measurement of QD area and intensity of staining was highlighted in this study (Figure 3.10a,b).

3.6
Summary and Future Perspectives

QDs are relatively novel and near-ideal fluorophores that, during the relatively short time since their first use, have been employed in many biological and biomedical imaging applications. They can be tagged or conjugated to many different biological probes, including antibodies and nucleic acids. Their unique spectral qualities enable not only a significantly improved signal detection compared to previous detection systems but also the multiplexing of several different probes, including different types of probe such as DNA and protein. When combined with

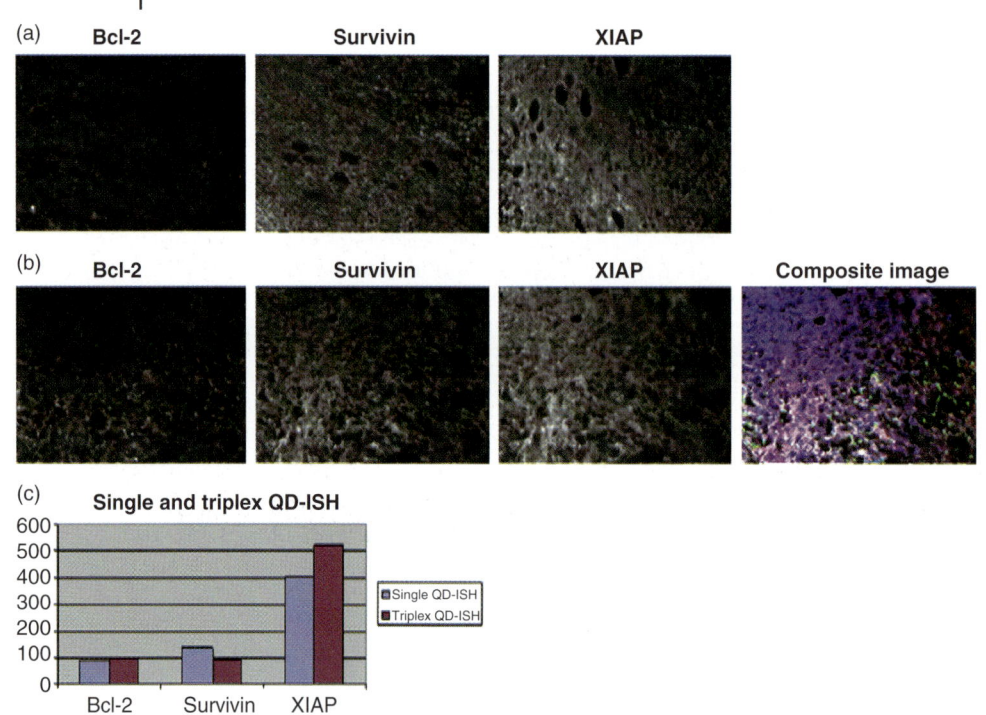

Figure 3.9 Triplex QD-ISH. (a) Single-gene QD-ISH was performed in bone marrow infiltrated by AML. Antisense oligonucleotide probes to bcl-2 (labeled with 605 nm QDs), survivin (labeled with 655 nm QDs) and XIAP (labeled with 705 nm QDs), all known to be expressed in AML at different levels, were hybridized separately. The resultant raw images were subjected to spectral analysis to resolve hybridization signal intensity maps for bcl-2, survivin and XIAP, shown as gray scale; (b) Triplex QD-ISH was performed in bone marrow infiltrated by AML. Antisense oligonucleotide probes to bcl-2 (labeled with 605 nm QDs), survivin (labeled with 655 nm QDs) and XIAP (labeled with 705 nm QDs) were hybridized, and the resultant raw images subjected to spectral analysis to resolve component signal intensity maps for bcl-2, survivin and XIAP, shown as gray scale. A composite image showing all three gene expression distributions was generated from these, using false color with bcl-2 in red, survivin in green and XIAP in blue; (c) Quantification of hybridization signal intensity for bcl-2, survivin and XIAP in single and triplex QD-ISH; signal intensity for each gene was measured in IPLab from gray scale signal intensity maps (shown in a and b, above) generated after spectral analysis of raw images; arbitrary intensity units on y-axis. Reproduced with permission from Ref. [24].

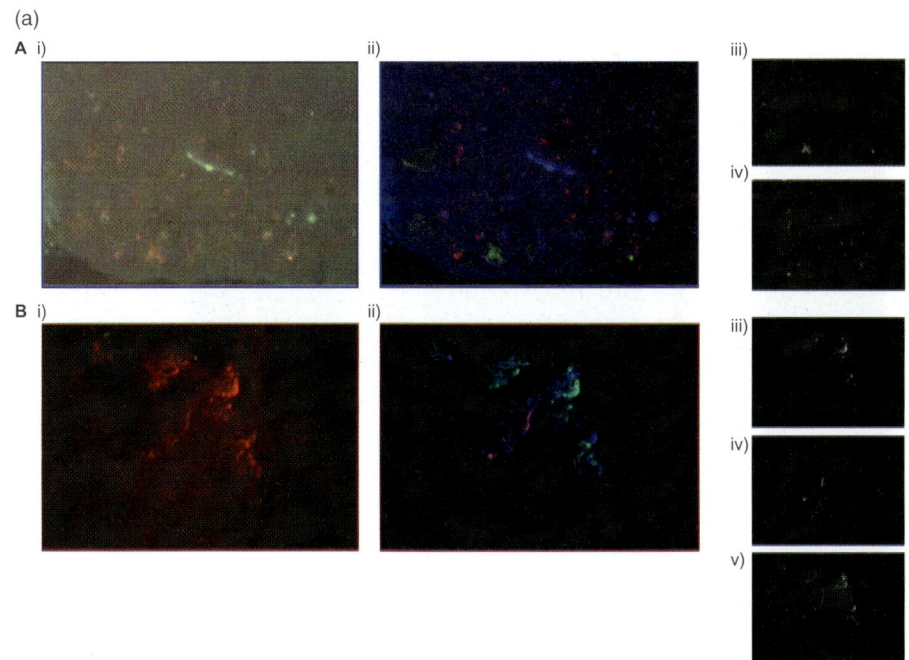

Figure 3.10 (a) One-step multiplexed staining using QDots conjugated to biotinylated primary antibodies on tonsil tissue. A: Raw (i) and false-color composite images (ii) of dual staining with Cytokeratin 18-QD605 (green), cleaved Caspase 3-QD655 (red) and autofluorescence (blue); the individual unmixed images are shown as thumbnails: (iii) Cytokeratin 18 and (iv) cleaved Caspase 3. B: Raw (i) and false-color composite images (ii) of triple staining with CD34-QD605 (red), Cytokeratin 18-QD655 (green) and cleaved Caspase 3-QD705 (blue), with autofluorescence removed; the individual unmixed images are (iii) Cytokeratin 18, (iv) CD34 and (v) cleaved Caspase 3; (b) Spectral imaging and signal quantitation. The raw image (i) was unmixed into individual intensity maps for (ii) Cytokeratin 18 and (v) CD34, and combined to form a falsely colored composite image (iv). Nuance software was used to quantify Cytokeratin 18 (iii) and (vi) CD34, and the average signal intensity/pixel for each antibody was calculated (viii). The percentage colocalization of CK18 and CD34 was also measured using Nuance software (vii) and (ix). Reproduced with permission from Ref. [8].

spectral imaging, QDs represent a particularly powerful bioimaging platform capable of generating quantitative *in situ* protein and mRNA expression data. This aspect makes them particularly promising for multiplex *in situ* gene expression measurements, at either the protein or mRNA level, in clinical tissues. This will facilitate the investigation of predictive cancer gene signatures in clinical tissues, and also enable more sophisticated investigations of the interaction of the elements of cancer gene signatures *in situ*. Today, the use of QDs is relatively straightforward; indeed, it is expected that they will become of increasing interest to research groups involved in pathological and allied biomedical studies.

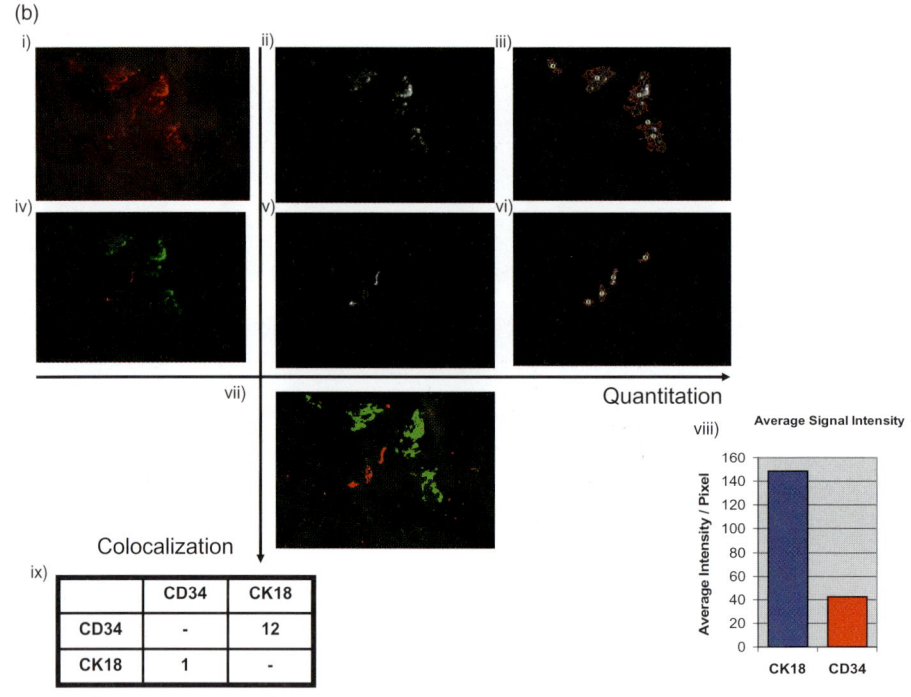

Figure 3.10 Continued

References

1 Ebert, B.L. and Golub, T.R. (2004) Genomic approaches to hematologic malignancies. *Blood*, **104**, 923–32.
2 Wadlow, R. and Ramaswamy, S. (2005) DNA microarrays in clinical cancer research. *Current Molecular Medicine*, **5**, 111–20.
3 Chin, K.V., Alabanza, L., Fujii, K., Kudoh, K., Kita, T., Kikuchi, Y., Selvanayagam, Z.E., Wong, Y.F., Lin, Y. and Shih, W.C. (2005) Application of expression genomics for predicting treatment response in cancer. *Annals of the New York Academy of Sciences*, **1058**, 186–95.
4 Miller, L.D. and Liu, E.T. (2007) Expression genomics in breast cancer research: microarrays at the crossroads of biology and medicine. *Breast Cancer Research*, **9**, 206.
5 Ramaswamy, S. and Golub, T.R. (2002) DNA microarrays in clinical oncology. *Journal of Clinical Oncology*, **20**, 1932–41.
6 Byers, R., Roebuck, J., Sakhinia, E. and Hoyland, J. (2004) PolyA PCR amplification of cDNA from RNA extracted from formalin-fixed paraffin-embedded tissue. *Diagnostic Molecular Pathology*, **13**, 144–50.
7 Dave, S.S., Wright, G., Tan, B., Rosenwald, A., Gascoyne, R.D., Chan, W.C., Fisher, R.I., Braziel, R.M., Rimsza, L.M., Grogan, T.M., Miller, T.P., LeBlanc, M., Greiner, T.C., Weisenburger, D.D., Lynch, J.C., Vose, J., Armitage, J.O., Smeland, E.B., Kvaloy, S., Holte, H., Delabie, J., Connors, J.M., Lansdorp, P.M., Ouyang, Q., Lister, T.A., Davies, A.J., Norton, A.J., Muller-Hermelink, H.K., Ott, G., Campo, E., Montserrat, E., Wilson, W.H., Jaffe, E.S., Simon, R., Yang, L., Powell, J., Zhao, H., Goldschmidt, N., Chiorazzi, M. and Staudt, L.M. (2004) Prediction of survival in follicular lymphoma based on

molecular features of tumor-infiltrating immune cells. *The New England Journal of Medicine*, **351**, 2159–69.

8 Sweeney, E., Ward, T.H., Gray, N., Womack, C., Jayson, G., Hughes, A., Dive, C. and Byers, R. (2008) Quantitative multiplexed quantum dot immunohistochemistry. *Biochemical and Biophysical Research Communications*, **374**, 181–6.

9 Zwirglmaier, K. (2005) Fluorescence in situ hybridisation (FISH) – the next generation. *FEMS Microbiology Letters*, **246**, 151–8.

10 Wilcox, J.N. (1993) Fundamental principles of in situ hybridization. *Journal of Histochemistry and Cytochemistry*, **41**, 1725–33.

11 Baschong, W., Suetterlin, R. and Laeng, R.H. (2001) Control of autofluorescence of archival formaldehyde-fixed, paraffin-embedded tissue in confocal laser scanning microscopy (CLSM). *Journal of Histochemistry and Cytochemistry*, **49**, 1565–72.

12 Del Castillo, P., Llorente, A.R. and Stockert, J.C. (1989) Influence of fixation, exciting light and section thickness on the primary fluorescence of samples for microfluorometric analysis. *Basic and Applied Histochemistry*, **33**, 251–7.

13 Fu, A., Gu, W., Larabell, C. and Alivisatos, A.P. (2005) Semiconductor nanocrystals for biological imaging. *Current Opinion in Neurobiology*, **15**, 568–75.

14 Wu, X., Liu, H., Liu, J., Haley, K.N., Treadway, J.A., Larson, J.P., Ge, N., Peale, F. and Bruchez, M.P. (2003) Immunofluorescent labeling of cancer marker Her2 and other cellular targets with semiconductor quantum dots. *Nature Biotechnology*, **21**, 41–6.

15 Bruchez, M. Jr., Moronne, M., Gin, P., Weiss, S. and Alivisatos, A.P. (1998) Semiconductor nanocrystals as fluorescent biological labels. *Science*, **281**, 2013–16.

16 Chan, W.C. and Nie, S. (1998) Quantum dot bioconjugates for ultrasensitive nonisotopic detection. *Science*, **281**, 2016–18.

17 Dahan, M., Levi, S., Luccardini, C., Rostaing, P., Riveau, B. and Triller, A. (2003) Diffusion dynamics of glycine receptors revealed by single-quantum dot tracking. *Science*, **302**, 442–5.

18 Medintz, I.L., Uyeda, H.T., Goldman, E.R. and Mattoussi, H. (2005) Quantum dot bioconjugates for imaging, labelling and sensing. *Nature Materials*, **4**, 435–46.

19 Lidke, D.S., Nagy, P., Heintzmann, R., Arndt-Jovin, D.J., Post, J.N., Grecco, H.E., Jares-Erijman, E.A. and Jovin, T.M. (2004) Quantum dot ligands provide new insights into erbB/HER receptor-mediated signal transduction. *Nature Biotechnology*, **22**, 198–203.

20 Jaiswal, J.K., Mattoussi, H., Mauro, J.M. and Simon, S.M. (2003) Long-term multiple color imaging of live cells using quantum dot bioconjugates. *Nature Biotechnology*, **21**, 47–51.

21 Gao, X., Cui, Y., Levenson, R.M., Chung, L.W. and Nie, S. (2004) In vivo cancer targeting and imaging with semiconductor quantum dots. *Nature Biotechnology*, **22**, 969–76.

22 Dubertret, B., Skourides, P., Norris, D.J., Noireaux, V., Brivanlou, A.H. and Libchaber, A. (2002) In vivo imaging of quantum dots encapsulated in phospholipid micelles. *Science*, **298**, 1759–62.

23 Kim, S., Lim, Y.T., Soltesz, E.G., De Grand, A.M., Lee, J., Nakayama, A., Parker, J.A., Mihaljevic, T., Laurence, R.G., Dor, D.M., Cohn, L.H., Bawendi, M.G. and Frangioni, J.V. (2004) Near-infrared fluorescent type II quantum dots for sentinel lymph node mapping. *Nature Biotechnology*, **22**, 93–7.

24 Tholouli, E., Hoyland, J.A., Di Vizio, D., O'Connell, F., Macdermott, S.A., Twomey, D., Levenson, R., Yin, J.A., Golub, T.R., Loda, M. and Byers, R. (2006) Imaging of multiple mRNA targets using quantum dot based in situ hybridization and spectral deconvolution in clinical biopsies. *Biochemical and Biophysical Research Communications*, **348**, 628–36.

25 Farkas, D.L., Du, C., Fisher, G.W., Lau, C., Niu, W., Wachman, E.S. and Levenson, R.M. (1998) Non-invasive image acquisition and advanced processing in optical bioimaging.

Computerized Medical Imaging and Graphics, **22**, 89–102.

26 Peng, X., Manna, L., Yang, W., Wickham, J., Scher, E., Kadavanich, A. and Alivisatos, A.P. (2000) Shape control of CdSe nanocrystals. *Nature*, **404**, 59–61.

27 Peng, Z.A. and Peng, X. (2001) Formation of high-quality CdTe, CdSe, and CdS nanocrystals using CdO as precursor. *Journal of the American Chemical Society*, **123**, 183–4.

28 Schmidt, M.E., Blanton, S.A., Hines, M.A. and Guyot-Sionnest, P. (1996) Size-dependent two-photon excitation spectroscopy of CdSe nanocrystals. *Physical Review B, Condensed Matter*, **53**, 12629–32.

29 Wilson, W.L., Szajowski, P.F. and Brus, L.E. (1993) Quantum confinement in size-selected, surface-oxidized silicon nanocrystals. *Science*, **262**, 1242–4.

30 Norris, D.J., Efros, A.L., Rosen, M. and Bawendi, M.G. (1996) Size dependence of exciton fine structure in CdSe quantum dots. *Physical Review B, Condensed Matter*, **53**, 16347–54.

31 Xing, Y., Chaudry, Q., Shen, C., Kong, K.Y., Zhau, H.E., Chung, L.W., Petros, J.A., O'Regan, R.M., Yezhelyev, M.V., Simons, J.W., Wang, M.D. and Nie, S. (2007) Bioconjugated quantum dots for multiplexed and quantitative immunohistochemistry. *Nature Protocols*, **2**, 1152–65.

32 Pinaud, F., King, D., Moore, H.P. and Weiss, S. (2004) Bioactivation and cell targeting of semiconductor CdSe/ZnS nanocrystals with phytochelatin-related peptides. *Journal of the American Chemical Society*, **126**, 6115–23.

33 Akhtar, R.S., Latham, C.B., Siniscalco, D., Fuccio, C. and Roth, K.A. (2007) Immunohistochemical detection with quantum dots. *Methods in Molecular Biology*, **374**, 11–28.

34 Zhou, M., Nakatani, E., Gronenberg, L.S., Tokimoto, T., Wirth, M.J., Hruby, V.J., Roberts, A., Lynch, R.M. and Ghosh, I. (2007) Peptide-labeled quantum dots for imaging GPCRs in whole cells and as single molecules. *Bioconjugate Chemistry*, **18**, 323–32.

35 Ornberg, R.L. and Liu, H. (2007) Immunofluorescent labeling of proteins in cultured cells with quantum dot secondary antibody conjugates. *Methods in Molecular Biology*, **374**, 3–10.

36 Wei, Q., Lee, M., Yu, X., Lee, E.K., Seong, G.H., Choo, J. and Cho, Y.W. (2006) Development of an open sandwich fluoroimmunoassay based on fluorescence resonance energy transfer. *Analytical Biochemistry*, **358**, 31–7.

37 Shen, J., Xu, F., Jiang, H., Wang, Z., Tong, J., Guo, P. and Ding, S. (2007) Characterization and application of quantum dot nanocrystal-monoclonal antibody conjugates for the determination of sulfamethazine in milk by fluoroimmunoassay. *Analytical and Bioanalytical Chemistry*, **389**, 2243–50.

38 Wu, Y., Campos, S.K., Lopez, G.P., Ozbun, M.A., Sklar, L.A. and Buranda, T. (2007) The development of quantum dot calibration beads and quantitative multicolor bioassays in flow cytometry and microscopy. *Analytical Biochemistry*, **364**, 180–92.

39 Zahavy, E., Freeman, E., Lustig, S., Keysary, A. and Yitzhaki, S. (2005) Double labeling and simultaneous detection of B- and T cells using fluorescent nano-crystal (q-dots) in paraffin-embedded tissues. *Journal of Fluorescence*, **15**, 661–5.

40 Fountaine, T.J., Wincovitch, S.M., Geho, D.H., Garfield, S.H. and Pittaluga, S. (2006) Multispectral imaging of clinically relevant cellular targets in tonsil and lymphoid tissue using semiconductor quantum dots. *Modern Pathology*, **19**, 1181–91.

41 Bostick, R.M., Kong, K.Y., Ahearn, T.U., Chaudry, Q., Cohen, V. and Wang, M.D. (2006) Detecting and quantifying biomarkers of risk for colorectal cancer using quantum dots and novel image analysis algorithms. *Conference Proceedings IEEE Engineering in Medicine and Biology Society*, **1**, 3313–16.

42 Byers, R.J., Di Vizio, D., O'Connell, F., Tholouli, E., Levenson, R.M., Gossage, K., Twomey, D., Yang, Y., Benedettini, E., Rose, J., Ligon, K.L., Finn, S.P., Golub, T.R. and Loda, M. (2007) Semiautomated multiplexed quantum dot-based in situ hybridization and spectral deconvolution. *Journal of Molecular Diagnostics*, **9**, 20–9.

43 Schwock, J., Ho, J.C., Luther, E., Hedley, D.W. and Geddie, W.R. (2007) Measurement of signaling pathway activities in solid tumor fine-needle biopsies by slide-based cytometry. *Diagnostic Molecular Pathology*, **16**, 130–40.

44 Tholouli, E., Sweeney, E., Barrow, E., Clay, V., Hoyland, J.A. and Byers, R.J. (2008) Quantum dots light up pathology. *Journal of Pathology*, **216**, 275–85.

45 Pathak, S., Choi, S.K., Arnheim, N. and Thompson, M.E. (2001) Hydroxylated quantum dots as luminescent probes for in situ hybridization. *Journal of the American Chemical Society*, **123**, 4103–4.

46 Wang, Q., Xu, Y., Zhao, X., Chang, Y., Liu, Y., Jiang, L., Sharma, J., Seo, D.K. and Yan, H. (2007) A facile one-step in situ functionalization of quantum dots with preserved photoluminescence for bioconjugation. *Journal of the American Chemical Society*, **129**, 6380–1.

47 Xiao, Y. and Barker, P.E. (2004) Semiconductor nanocrystal probes for human metaphase chromosomes. *Nucleic Acids Research*, **32**, e28.

48 Bentolila, L.A. and Weiss, S. (2006) Single-step multicolor fluorescence in situ hybridization using semiconductor quantum dot-DNA conjugates. *Cell Biochemistry and Biophysics*, **45**, 59–70.

49 Ma, L., Wu, S.M., Huang, J., Ding, Y., Pang, D.W. and Li, L. (2008) Fluorescence in situ hybridization (FISH) on maize metaphase chromosomes with quantum dot-labeled DNA conjugates. *Chromosoma*, **117**, 181–7.

50 Wu, S.M., Zhao, X., Zhang, Z.L., Xie, H.Y., Tian, Z.Q., Peng, J., Lu, Z.X., Pang, D.W. and Xie, Z.X. (2006) Quantum-dot-labeled DNA probes for fluorescence in situ hybridization (FISH) in the microorganism *Escherichia coli*. *ChemPhysChem*, **7**, 1062–7.

51 Matsuno, A., Itoh, J., Takekoshi, S., Nagashima, T. and Osamura, R.Y. (2005) Three-dimensional imaging of the intracellular localization of growth hormone and prolactin and their mRNA using nanocrystal (Quantum dot) and confocal laser scanning microscopy techniques. *Journal of Histochemistry and Cytochemistry*, **53**, 833–8.

52 Chan, P., Yuen, T., Ruf, F., Gonzalez-Maeso, J. and Sealfon, S.C. (2005) Method for multiplex cellular detection of mRNAs using quantum dot fluorescent in situ hybridization. *Nucleic Acids Research*, **33**, e161.

53 Matsuno, A., Mizutani, A., Takekoshi, S., Itoh, J., Okinaga, H., Nishina, Y., Takano, K., Nagashima, T., Osamura, R.Y. and Teramoto, A. (2006) Analyses of the mechanism of intracellular transport and secretion of pituitary hormone, with an insight of the subcellular localization of pituitary hormone and its mRNA. *Brain Tumor Pathology*, **23**, 1–5.

54 Hahn, M.A., Keng, P.C. and Krauss, T.D. (2008) Flow cytometric analysis to detect pathogens in bacterial cell mixtures using semiconductor quantum dots. *Analytical Chemistry*, **80**, 864–72.

55 Chattopadhyay, P.K., Price, D.A., Harper, T.F., Betts, M.R., Yu, J., Gostick, E., Perfetto, S.P., Goepfert, P., Koup, R.A., De Rosa, S.C., Bruchez, M.P. and Roederer, M. (2006) Quantum dot semiconductor nanocrystals for immunophenotyping by polychromatic flow cytometry. *Nature Medicine*, **12**, 972–7.

56 Hotz, C.Z. (2005) Applications of quantum dots in biology: an overview. *Methods in Molecular Biology*, **303**, 1–17.

57 Shingyoji, M., Gerion, D., Pinkel, D., Gray, J.W. and Chen, F. (2005) Quantum dots-based reverse phase protein microarray. *Talanta*, **67**, 472–8.

58 Kerman, K., Endo, T., Tsukamoto, M., Chikae, M., Takamura, Y. and Tamiya, E. (2007) Quantum dot-based immunosensor for the detection of prostate-specific antigen using fluorescence microscopy. *Talanta*, **71**, 1494–9.

59 Gokarna, A., Jin, L.H., Hwang, J.S., Cho, Y.H., Lim, Y.T., Chung, B.H., Youn, S.H., Choi, D.S. and Lim, J.H. (2008) Quantum dot-based protein micro- and nanoarrays for detection of prostate cancer biomarkers. *Proteomics*, **8**, 1809–18.

60 Geho, D., Lahar, N., Gurnani, P., Huebschman, M., Herrmann, P., Espina, V., Shi, A., Wulfkuhle, J., Garner, H., Petricoin, E. 3rd, Liotta, L.A. and Rosenblatt, K.P. (2005) Pegylated,

streptavidin-conjugated quantum dots are effective detection elements for reverse-phase protein microarrays. *Bioconjugate Chemistry*, **16**, 559–66.
61 Han, M., Gao, X., Su, J.Z. and Nie, S. (2001) Quantum-dot-tagged microbeads for multiplexed optical coding of biomolecules. *Nature Biotechnology*, **19**, 631–5.
62 Eastman, P.S., Ruan, W., Doctolero, M., Nuttall, R., de Feo, G., Park, J.S., Chu, J.S., Cooke, P., Gray, J.W., Li, S. and Chen, F.F. (2006) Qdot nanobarcodes for multiplexed gene expression analysis. *Nano Letters*, **6**, 1059–64.
63 Liang, R.Q., Li, W., Li, Y., Tan, C.Y., Li, J.X., Jin, Y.X. and Ruan, K.C. (2005) An oligonucleotide microarray for microRNA expression analysis based on labeling RNA with quantum dot and nanogold probe. *Nucleic Acids Research*, **33**, e17.
64 Karlin-Neumann, G., Sedova, M., Falkowski, M., Wang, Z., Lin, S. and Jain, M. (2007) Application of quantum dots to multicolor microarray experiments: four-color genotyping. *Methods in Molecular Biology*, **374**, 239–51.
65 Smith, R.A. and Giorgio, T.D. (2008) Quantitative measurement of multifunctional quantum dot binding to cellular targets using flow cytometry. *Cytometry Part A*, **75**, 465–74.
66 Xiao, Y., Gao, X., Gannot, G., Emmert-Buck, M.R., Srivastava, S., Wagner, P.D., Amos, M.D. and Barker, P.E. (2008) Quantitation of HER2 and telomerase biomarkers in solid tumors with IgY antibodies and nanocrystal detection. *International Journal of Cancer*, **122**, 2178–86.
67 Zajac, A., Song, D., Qian, W. and Zhukov, T. (2007) Protein microarrays and quantum dot probes for early cancer detection. *Colloids and Surfaces B, Biointerfaces*, **58**, 309–14.

68 Mansfield, J.R., Hoyt, C. and Levenson, R.M. (2008) Visualization of microscopy-based spectral imaging data from multi-label tissue sections. *Current Protocols Molecular in Biology*, Chapter 14, Unit 14.19.
69 Levenson, R.M. and Mansfield, J.R. (2006) Multispectral imaging in biology and medicine: slices of life. *Cytometry Part A*, **69**, 748–58.
70 True, L.D. and Gao, X. (2007) Quantum dots for molecular pathology: their time has arrived. *Journal of Molecular Diagnostics*, **9**, 7–11.
71 Mansfield, J.R., Gossage, K.W., Hoyt, C.C. and Levenson, R.M. (2005) Autofluorescence removal, multiplexing, and automated analysis methods for in-vivo fluorescence imaging. *Journal of Biomedical Optics*, **10**, 41207.
72 Itoh, J., Yasumura, K., Takeshita, T., Ishikawa, H., Kobayashi, H., Ogawa, K., Kawai, K., Serizawa, A. and Osamura, R.Y. (2000) Three-dimensional imaging of tumor angiogenesis. *Analytical and Quantitative Cytology and Histology*, **22**, 85–90.
73 Giepmans, B.N., Deerinck, T.J., Smarr, B.L., Jones, Y.Z. and Ellisman, M.H. (2005) Correlated light and electron microscopic imaging of multiple endogenous proteins using Quantum dots. *Nature Methods*, **2**, 743–9.
74 Deerinck, T.J. (2008) The application of fluorescent quantum dots to confocal, multiphoton, and electron microscopic imaging. *Toxicologic Pathology*, **36**, 112–16.
75 Caldwell, M.L., Moffitt, R.A., Liu, J., Parry, R., Sharma, Y. and Wang, M.D. (2008) Simple quantification of multiplexed quantum dot staining in clinical tissue samples. *Conference Proceedings IEEE Engineering in Medicine and Biology Society*, **2008**, 1907–10.

4
Multiplexed Detection Using Quantum Dots

Young-Pil Kim, Zuyong Xia and Jianghong Rao

4.1
Introduction

In recent years, rapid and sensitive detection systems, when used in a multiplexed manner, have attracted much interest in the areas of clinical diagnostics, drug screening, and bioimaging applications. With today's ever-increasing knowledge of genomic and transcriptomic information covering a wide range of diseases, much greater numbers of biomarkers are becoming available for the development of bioanalytical assays for disease diagnostics, treatment monitoring, and drug development. Most complex human diseases such as cancer and atherosclerosis involve a number of genes and proteins, rather than single units. Thus, the ability to track a panel of molecular markers at the same time would allow not only a better understanding of complex human diseases, but also the ability to classify and differentiate among them, rather than to use a single biomarker on each occasion. The additional advantages of a multiplexed assay include time-savings with high-throughput screening, a low consumption of reagents and samples, and a reduction in errors between inter-sampling. Although many powerful analytical methods are currently available, fluorescent probes have long been used in these assays for multiplexed purposes, due to their high sensitivity with fluorescence detection [1]. However, the narrow absorption and wide crosstalk of organic dyes and fluorescent proteins represents a major hurdle for the simultaneous detection of multiple targets, due to requirements for the elaborate excitation of each probe and complicated analyses of the acquired data. Moreover, the low resistance of fluorescent probes to chemical and photodegradation may significantly limit their use in multiplexed monitoring.

Recently, a new class of fluorophores, semiconductor fluorescence nanocrystals (or quantum dots; QDs) has begun to emerge as an attractive alternate to organic dyes for bioanalytical and bioimaging applications. These QDs are composed of inorganic semiconductors, and have novel optical properties that can be used to optimize the signal-to-background ratio in fluorescence detection and imaging [2, 3]. In comparison with organic dyes and fluorescent proteins, QDs have several advantages and unique applications.

First, they have very large molar extinction coefficients in the order of 0.5–5 × 10^6 M^{-1} cm^{-1} [4], which is about 10- to 50-fold larger than that of organic dyes (5–10 × 10^4 M^{-1} cm^{-1}). Therefore, QDs are able to absorb 10 to 50 times more photons than organic dyes at the same excitation photon flux (i.e., the number of incident photons per unit area), and this leads to a significant improvement in probe brightness. This, in turn, allows for brighter emissions under photon-limited *in vivo* conditions, where light intensities are severely attenuated by scattering and absorption. In theory, the lifetime-limited emission rates for single QDs are five- to tenfold lower than those of small organic dyes, because of their longer excited state lifetimes (20–50 ns versus <10 ns). In practice, however, fluorescence imaging usually operates under absorption-limited conditions, where the rate of absorption is the main limiting factor of fluorescence emission (versus the emission rate of the fluorophore). As a result, individual QDs have been found to be 10- to 20-fold brighter than organic dyes [5].

Second, QDs are several thousand times more stable against photobleaching (the loss of fluorescence due to photoinduced chemical damages) than organic dyes, and are thus well-suited for continuous tracking studies over long periods of time. In addition, the relatively longer excited state lifetimes of QDs can be used to separate the QD fluorescence from background fluorescence, in a technique known as "time-domain imaging" [6]; QDs emit light sufficiently slowly that most of the background autofluorescence emission is complete by the time that QD emission occurs.

Third, the large Stokes shifts of QDs (measured by the distance between the excitation and emission peaks) can be used to further improve detection sensitivity. This factor becomes especially important for *in vivo* molecular imaging due to the high autofluorescence background that is often seen in complex biomedical specimens. The Stokes shifts of semiconductor QDs may be as large as 300–400 nm, depending on the wavelength of the excitation light. Organic dye signals with a small Stokes shift are often buried by strong tissue autofluorescence, whereas QD signals with a large Stokes shift are clearly recognizable above the background. This degree of "color contrast" is only available to QD probes, as the signals and background can be easily separated by wavelength-resolved or spectral imaging [5]. A further advantage of QDs is that multicolor QD probes can be used to image and track multiple molecular targets simultaneously. By tuning the size and composition, QDs can emit light from the blue to the near-infrared (NIR). For example, CdS and ZnSe dots emit blue to near-UV light, different-sized CdSe dots emit light across the visible spectrum, and InP and InAs QDs emit in the far-red and NIR regions [7]. This is a very desirable feature for multiplex detection and imaging because the broad absorption profiles of QDs allow the simultaneous excitation of multiple colors, while their emission wavelengths can be continuously tuned by varying the particle size and chemical composition (Figure 4.1).

These superior optical properties of QDs have led to them finding a wide range of applications, from bioanalytical assays, live-cell imaging, fixed-cell and tissue labeling, to biosensing and *in vivo* animal imaging, that have been reviewed extensively [7–10]. In order to avoid repetition, this chapter will focus primarily on the multiplex

4.1 In Vitro Multiplexed Analysis Using QDs

(a)

(b)

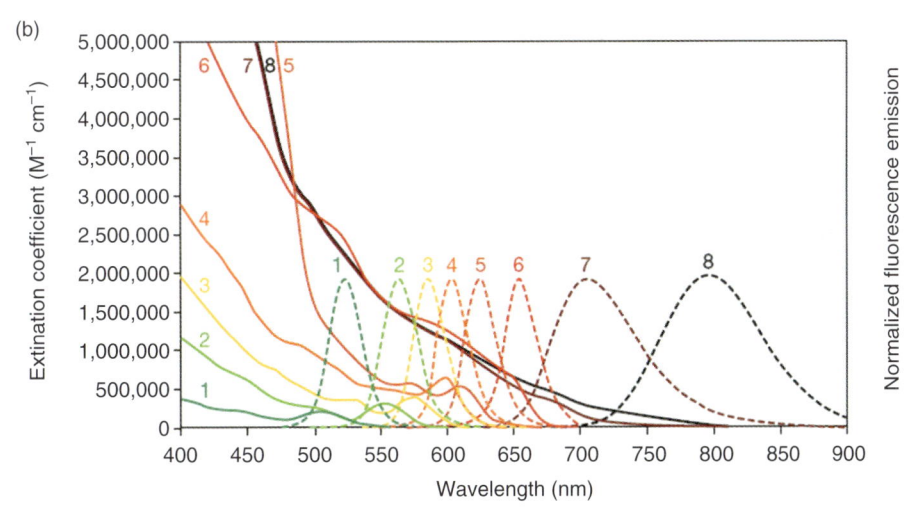

— 1. Qodt® 525 conjugate excitation
— 2. Qodt® 565 conjugate excitation
— 3. Qodt® 585 conjugate excitation
— 4. Qodt® 606 conjugate excitation
— 5. Qodt® 625 conjugate excitation
— 6. Qodt® 655 conjugate excitation
— 7. Qodt® 705 conjugate excitation
— 8. Qodt® 800 conjugate excitation

---- 1. Qodt® 525 conjugate emission
---- 2. Qodt® 565 conjugate emission
---- 3. Qodt® 585 conjugate emission
---- 4. Qodt® 606 conjugate emission
---- 5. Qodt® 625 conjugate emission
---- 6. Qodt® 655 conjugate emission
---- 7. Qodt® 705 conjugate emission
---- 8. Qodt® 800 conjugate emission

Figure 4.1 (a) Image of ten distinguishable emission colors of ZnS-capped CdSe QDs excited with a near-UV lamp. From left to right (blue to red), the emission maxima are located at 443, 473, 481, 500, 518, 543, 565, 587, 610, and 655 nm [7]; (b) Excitation (solid line) and size-tunable emission (solid line) spectra of QDs (http:www.invitrogen.com).

features of QDs for biological applications in terms of multiplexed detection and imaging. A brief survey of the QD-based *in vitro* assays, such as DNA hybridization, immunoassays, and enzyme activity detection, is followed by some highlights of the applications of QDs for multiplexed imaging in cells and in small live animals.

4.2
In Vitro Multiplexed Analysis Using QDs

4.2.1
DNA Hydridization

Analyses of gene expression profiling and single nucleotide polymorphisms (SNPs) have been recognized as a hallmark to studies relevant to distinct genetic malfunction, as well as in the prediction of disease progression [11, 12]. In order to improve the diagnosis, prognosis and tailored molecular therapy by targeting disease-causing genes or genetic variations, reliable methods with multiplexed and rapid screening are essential. Thus, over the past few decades, electrophoresis and "lab-on-a-chip" systems have been successfully applied to explore single nucleotide differences by the hybridization of labeled targets with a wide variety of probe designs, including molecular beacons, peptide nucleic acids, and nanoparticle-labeled oligonucleotides. However, the challenge for these methods is to assay, simultaneously, multiple different nucleic acids. It was shown subsequently by Han *et al.* [13], however, that this limitation could be overcome by using the optical properties of semiconductor QDs, and by generating multicolor-coded QDs by embedding different-sized QDs into polymeric microbeads. With different ratios of QDs of different colors, a very large number of multiplexed codes could be achieved with high selectivity and sensitivity and, even though there may be spectral overlapping and fluorescence intensity variations among the recognition codes, it would still be possible to distinguish 10 000 to 40 000 different codes if five to six colors with six intensity levels were to be used. Notably, in comparison to microspheres containing organic dyes, which would require multiple excitations, the use of QDs has significant advantages by being able to use a single wavelength for simultaneous excitation. Indeed, it was shown possible to detect DNA hybridization when different probe oligonucleotides were labeled with multicolor encoded beads (Figure 4.2). Since these assay methods are uniform, reproducible, and highly accurate (up to 99.99%), a combination of encoded beads and oligonucleotide probes would be expected to allow for practical genomic applications in a multiplexed manner. This resulted in the QD-encoded microsphere-based assay being commercialized as the Qbead™ system by the Invitrogen Corporation [14], and demonstrated for SNP genotyping of the cytochrome P450 family. Whilst this was highly relevant to modern drug discovery, it proved difficult for multiplexed analyses due to the high degree of homology (i.e., high cross-reactivities) required. Notably, the Qbead™ system enabled an accurate analysis with very small quantities of DNA in real samples, thus validating its high accuracy

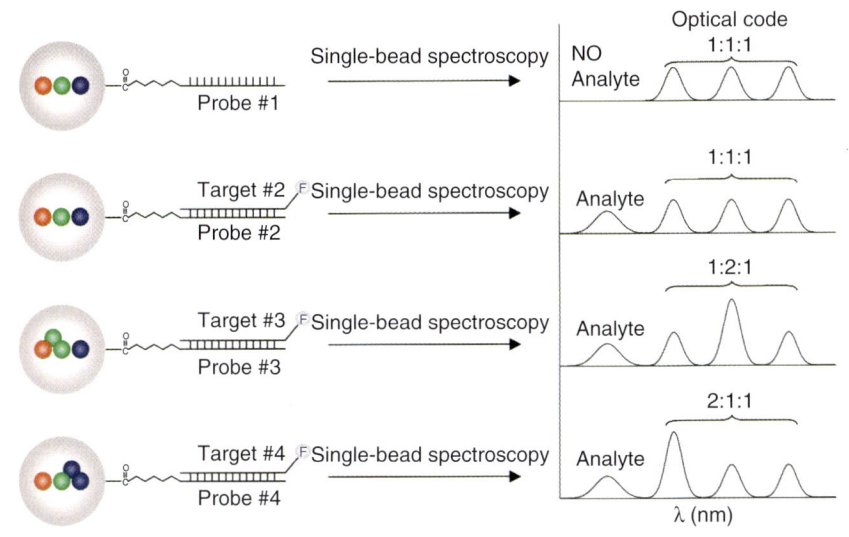

Figure 4.2 Schematic illustration of DNA hybridization assays using QD-tagged beads. Probe oligos (#1–4) were conjugated to the beads by crosslinking, and target oligos (#1–4) were detected with a blue fluorescent dye such as Cascade Blue. After hybridization, nonspecific molecules and excess reagents were removed by washing. For multiplexed assays, the oligonucleotide lengths and sequences were optimized so that all probes had similar melting temperatures (T_m = 66–99 °C) and hybridization kinetics (30 min) [13].

and reliability for multiplexed SNP genotyping. It was also reported that carboxyl-modified polystyrene beads embedding the QDs could be used to detect DNA targets effectively, and with a detection limit as low as 0.01 ~ 0.2 µg ml^{-1} in complex samples such as a denatured calf thymus DNA solution [15, 16].

The multiplexed characteristic of QDs has also been successfully utilized in DNA microarray technology. For example, the group of Alivisatos reported a microarray-based multi-allele detection through the direct conjugation of nanocrystals with DNA [17]. The developed system not only detected SNP mutations in the human p53 tumor suppressor gene (which is mutated in over 50% of all known human cancers), but also identified hepatitis B and C genotypes (75-mers) with two different QDs in the presence of a background of human genes. The most interesting aspects of these DNA–nanocrystal conjugates included their stability, their ability to maintain a stable activity for more than six months, their rapid hybridization, and their room-temperature incubation requirements. A simultaneous qualitative analysis of QD-conjugated DNA strands was also demonstrated using surface plasmon-enhanced fluorescence microscopy and spectrometry as multicolor images [18]. This concept was based on a combination of surface plasmon field enhancement and fluorescence spectroscopy, by which approach the resonant excitation of an evanescent surface plasmon mode (plasmon surface polariton, PSP) can be used to excite the fluorophores that are chemically attached to the target molecules. Upon binding to the probe DNA strand at the metal/solution interface, the fluorophore reaches the strong optical fields that can be obtained in PSP resonance, giving rise to significant enhancement factors. In particular, the use of a fluorescence detection scheme in combination with the resonant excitation of surface plasmons, has been shown to provide a considerable increase in detection sensitivity. Moreover, the use of QDs makes it possible to excite several QD populations simultaneously with a single light source at a single angle of incidence for resonance surface plasmon excitation. In this assay, 5′-biotin-tagged single-stranded DNA (ssDNA) sequences were attached to streptavidin-coupled CdSe/ZnS core–shell QDs, after which the specific hybridization of QD-conjugated DNA single strands to the sensor-attached complementary sequences was detected by a substantial shift in the angular reflectivity spectrum of the SPR, as well as by a large fluorescence signal. In another example of the DNA chip-based system, a multiplexed hybridization detection method based on the multicolor colocalization of oligonucleotide-functionalized QD probes was reported [19]. For this, two QD nanoprobes with different emission wavelengths were designed to bind in juxtaposition to the same target DNA of interest so as to form a "sandwiched" nanoassembly (Figure 4.3). Then, by measuring the color intensity of the nanoassemblies, the presence of a specific target nucleotide could be determined and a multiplexed sequence identification achieved by using mul-

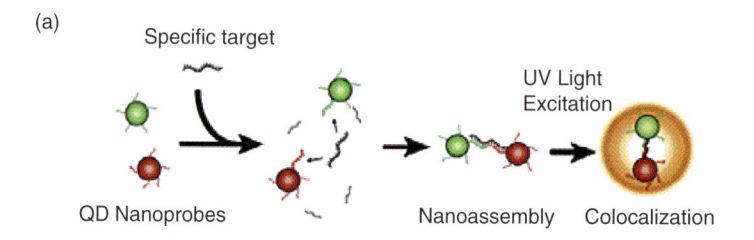

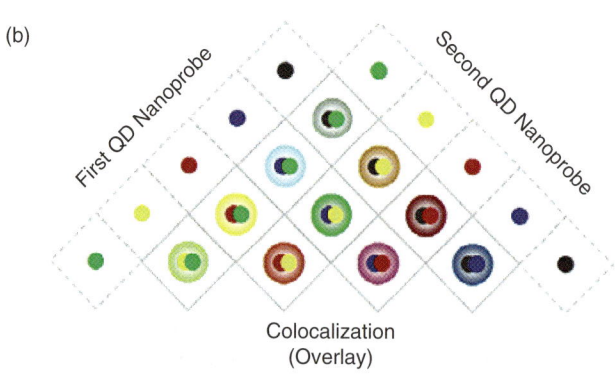

Figure 4.3 (a) QD nanoprobes prepared by surface-functionalizing QDs with target-specific oligonucleotide probes. Two target-specific QD nanoprobes with different emission wavelengths sandwich a target, forming a QD probe-target nanoassembly. The nanoassembly is detected as a blended color (orange) due to the colocalization of the both QD nanoprobes; (b) The color combination scheme for multiplexed colocalization detection [19].

tiple combinations of different-color QD nanoprobes. The use of QD-based probes on a fiber-optic microarray platform was recently described to perform the polychromatic (termed multiplexed) detection of eight different *Bacillus anthracis* subspecies [20]. Here, beads coated with ssDNA probes were localized into the etched wells of fiber-optic arrays so as to facilitate the use of virtually optical reporters such as QDs. Multiplex detection with an assortment of eight fluorescent reporters, including five different QDs, was demonstrated in a fiber-optic microarray platform, which provided a fourfold increase in throughput over standard two-color assays.

QDs have also been used for the electrochemical detection of oligonucleotide targets. For example, Wang *et al.* first reported that multiple nucleic acids could be easily detected by using a sandwich-type DNA hybridization and subsequently a stripping voltammetry method based on probe-attached magnetic beads and QD tags [21]. Three different nucleic acids were immobilized on different magnetic particles, which permitted a favorable and straightforward separation and efficient purification of the complementary target nucleic acids. As a detection probe, three encoding QDs (ZnS, CdS, and PbS) were functionalized with nucleic acids complementary to the target nucleic acids associated with the magnetic particles. When the target DNA was present, the DNA was linked between the QDs and magnetic beads by a sandwich-type hybridization, after which stripping voltammetry of the respective semiconductor nanoparticles yielded well-defined and resolved stripping waves, without any overlapping. As a result, this method enabled the simultaneous electrochemical analysis of several DNA analytes (Figure 4.4a). This principle could be further extended for the coding of individual SNPs [22], whereby nanocrystal QDs of four different compositions (ZnS, CdS, PbS, and CuS QDs) were linked to adenosine, cytidine, guanosine, and thymidine mononucleotides, respectively. According to the voltammetric codes by nanocrystal-nucleotide tags, the degree of mismatch could be identified within a single DNA target in a single voltammetric run. A distinct multipotential readout would, therefore, allow for a rapid and high-throughput analysis and fingerprinting of specific SNPs (Figure 4.4b).

4.2.2
Immunoassay

Antibody-based detection systems represent powerful tools for a variety of molecular and cellular analyses, as well as clinical diagnostics. For example, the Western analysis of protein expression in cells and tissues has served as a major tool in biochemistry and molecular biology for protein immunodetection. In comparison to multiple repeats of single target blotting, a multiplexed analysis would be much less time-consuming and labor-intensive, and also avoid the loss of immobilized proteins from the blot. Although the use of small organic fluorescent molecules has been investigated for this purpose, QDs came to be the fluorophore of choice [23–25]. By conjugating streptavidin-coated QDs with a biotinylated Z domain derived from *Staphylococcus aureus* protein A, Bello *et al.* constructed different

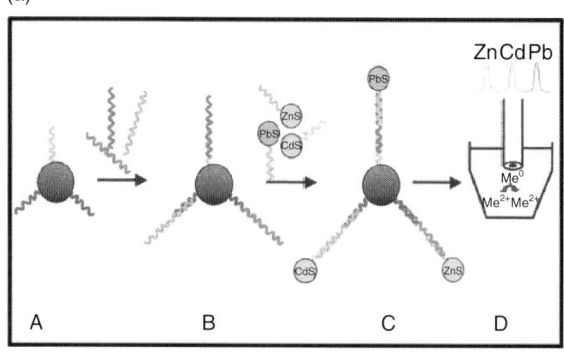

154 4 Multiplexed Detection Using Quantum Dots

Figure 4.4 (a) Multitarget electrical DNA detection protocol based on different inorganic colloid nanocrystal tracers. (A) Introduction of probe-modified magnetic beads; (B) Hybridization with the DNA targets; (C) Second hybridization with the QD-labeled probes; (D) Dissolution of QDs and electrochemical detection [21]; (b) Electrochemical coding of all eight possible one-base mismatches using inorganic nanocrystal tracers. Use of mismatch-containing hybrids (captured on magnetic beads) followed by sequential additions of ZnS-linked adenosine-5′ monophosphate, CdS-linked cytidine-5′ monophosphate, PbS-linked guanosine-5′ monophosphate, and CuS-linked thymidine-5′ monophosphate. Also shown (right) are the corresponding assemblies of nanocrystal-linked DNA/magnetic beads. (Note: the relative sizes of the magnetic beads and nanocrystal tags are not to scale.) [22].

---

probes with antibody [24]. For the simultaneous detection of two different types of protein, two probes – one with anti-apoAI (human apolipoprotein AI)–QD565 and one with anti-luciferase–QD655 – were easily visualized on the same blot [24]. Today, Western blot kits using QD–antibody are now commercially available. In fact, it has been reported that a multiplex Western blot with the commercial kit enabled the simultaneous analysis of p42 MAPK phosphorylation in platelet-derived growth factor (PDGF)-treated NIH-3T3 cells [25].

A QD-based multiplexed immunoassay was also developed by Goldman *et al.* to detect multiple toxins [26]. In this case, a dual fluoroimmunossay was initially executed using two-color QDs [27], essentially because the detection of more than two toxins was limited as the filters for the plate reader were incapable of fine color resolution. This problem was solved by employing a scanning emission spectrum and using a deconvolution method that assumed a superposition of independent QD spectra. Consequently, in an extension of these studies, four different toxins (cholera toxin, ricin, shiga-like toxin, and staphylococcal enterotoxin B) could be analyzed in the single wells of a microtiter plate (Figure 4.5). Despite the cross-reactivity of antibody and nonspecific interaction, four analytes were detected by a mix of four QD–antibody pairs in both low and high concentration ranges (30 and 1000 ng ml^{-1}).

Liu *et al.* demonstrated an effective and inexpensive multitarget electrochemical immunoassay based on the use of different colloidal nanocrystal tags for detecting specific proteins. [28]. First, β_2-microglobulin, immunoglobulin G (IgG), bovine serum albumin, and C-reactive protein were conjugated with ZnS, CdS, PbS, and CuS colloidal crystals, respectively. The multiprotein electrical detection was combined with the electrochemical stripping transduction, which led to very low detection limits (on the femtomolar level). This system was also operated together with an efficient magnetic separation so as to minimize any nonspecific binding effects. By capturing the target antigens using antibody-conjugated magnetic beads, the bound antigens could be detected by their reaction with a pool of nanocrystal–antibody pairs and a stripping voltammetric measurement of the corresponding metals. Each individual protein recognition event thus yielded a distinct voltammetric peak. A similar attempt has been made using an aptamer–QD conjugate [29], where the coupling of aptamers with the coding and amplification features

(a)

510 nm 555 nm 590 nm 610 nm

CT   ricin   SLT   SEB

Microtiter plate

(b)

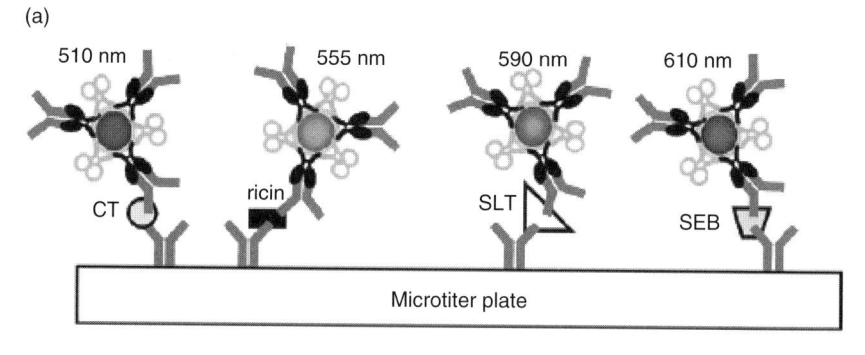

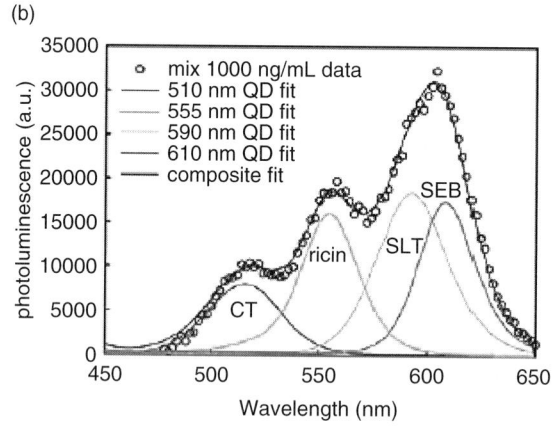

Figure 4.5 Multiplex immunoassays with QDs. (a) Schematic of the four-color multiplex assay. The indicated colors of QDs were prepared with antibodies against the four-indicated toxins and simultaneously incubated in microtiter-well plates containing the four-toxins immobilized by capture antibodies on the surface; (b) Multitoxin assay examining mixes of all four indicated toxins at 1000 ng ml^{-1}, each probed with a mix of QD–detection antibody conjugates. The measured values are shown as circles. Both, the composite fit and the fit from each of the four individual QD components are displayed. Reproduced with permission from Ref. [26]; ©, American Chemical Society.

of inorganic nanocrystals enabled a highly sensitive and selective simultaneous bioelectronic detection of lysozyme and thrombin (detection limit 20 ng ml^{-1}), along with the coimmobilization of the corresponding aptamers. With negligible cross-interferences, it is possible to measure five to six protein targets simultaneously in a single run. Nanoparticle-based electrochemical sensing system could be readily multiplexed, and should be capable of scaling-up in multiwell microtiter plates so as to allow the simultaneous parallel detection of numerous proteins or samples. A lab-on-a-chip immunoassay using the double detection of microsphere light scattering and QD emission was demonstrated by Lucas et al. [30]. For this, highly carboxylated polystyrene/polyacrylic acid (PS/PAA) submicron latex microspheres were coated with QDs, and an immunoassay then carried out using two types of nano-on-micro (NOM) combinations. In this case, one batch of microspheres was coated with QDs emitting at 655 nm and mouse IgG (mIgG), while the other batch was coated with QDs emitting at 605 nm and bovine serum albumin (BSA). A mixture of these two NOMs was used to identify either anti-mIgG or anti-BSA. When a positive antibody match occurred, the light scattering was increased at 380 nm while the QD emission was attenuated (605 or 655 nm), depending on which antibody was present.

4.2.3
Assaying the Activity of Enzymes

As an important class of proteins, enzymes carry out crucial biological functions in all cellular and physiological processes and, indeed, are often the targets for disease detection, diagnostics, drug discovery, and therapeutics [31, 32]. *Proteases*, which represent the second largest enzyme family in humans, are implicated in both normal physiological processes such as immunity [33–35], development [36], blood clotting and wound healing [37], and also in cardiovascular, oncologic, neurodegenerative, and inflammatory diseases [38–41]. Their importance in basic and pharmaceutical research demands sensitive assays that are capable of monitoring their activity both *in vitro* and *in vivo*.

Most protease assays are designed to detect the cleavage or digestion of an appropriate substrate, such as a fluorescent dye-labeled peptide or a protein, by the target protease [42–47]. Cleavage of the fluorogenic substrate by the protease results in changes in the fluorescence emission of the reporter molecule. The majority of fluorescent probes are quenched fluorogenic substrates, based on the principle of fluorescence resonance energy transfer (FRET). In a FRET-based probe, two fluorophores are attached to the ends of the substrate with a distance between them of <10 nm, such that the emission wavelength of the donor fluorophores overlaps with the excitation wavelength of acceptor fluorophores for nonradiative energy transfer. Cleavage of the substrate between the two fluorophores disrupts the energy transfer, and this results in an increase in the donor emission (Figure 4.6) [48, 49].

Quantum dots offer several advantages over organic dyes in FRET probes:

- Many fluorophores are available to form FRET pairs, with optimized spectral overlap.
- An optimal excitation wavelength, away from the absorption peak of paired fluorescent molecule, will minimize crosstalk between the FRET pair.
- The immobilization of multiple copies of fluorescent molecules on a single QD will increase the overall FRET efficiency.

Although QDs are frequently used as donors in FRET due to their broad excitation spectra and long excited-state life times [53], they may also act as energy acceptors in bioluminescence resonance energy transfer (BRET), with a bioluminescent protein as the energy donor [54]. The detection of protease activity (based on either FRET or BRET) between QDs and dyes, gold nanoparticles and bioluminescent proteins, is outlined in the following sections.

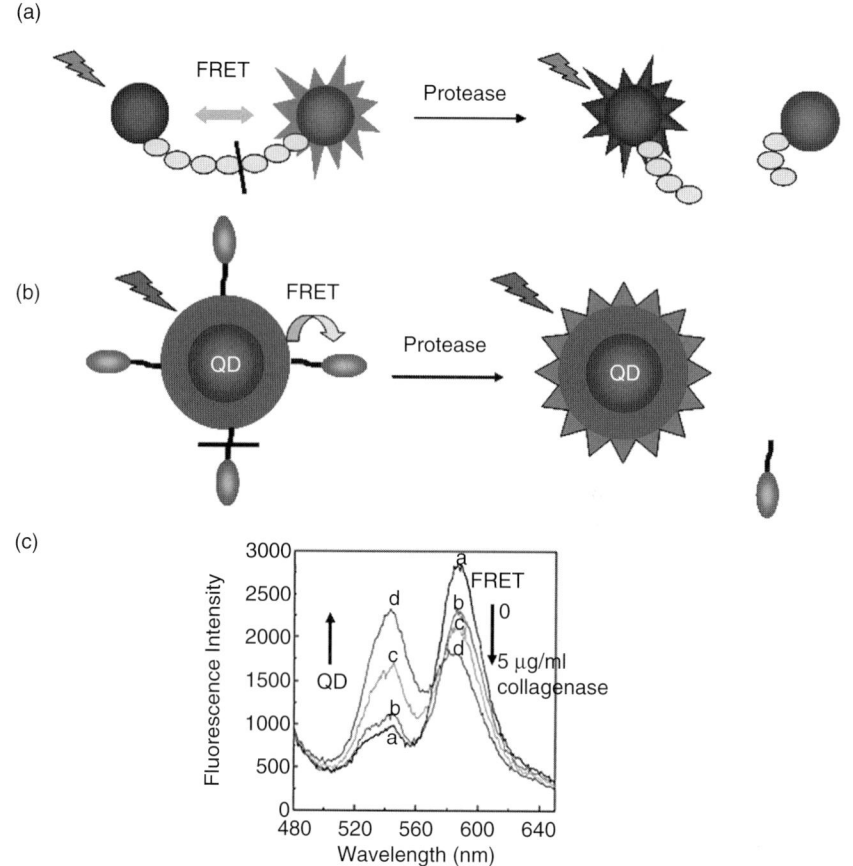

Figure 4.6 Fluorescence resonance energy transfer (FRET)-based probes for the detection of protease activity. (a) A FRET-based protease probe has donor and acceptor fluorophores linked by a substrate sequence. Proteolytic cleavage of the linker interrupts the FRET process that causes fluorescence changes; (b) Protease probes using QDs have multiple acceptors per QD for efficient energy transfer. The acceptor can be a quencher (e.g., gold nanoparticle [50, 51]) or a fluorescent dye [52]; (c) A ratiometric protease assay eliminates the artifacts from the variation of concentrations [52]. Upon enzymatic cleavage of the substrate, the QD fluorescence increases and emission from the acceptor dye decreases. The ratio changes can be used to quantify the enzyme activity.

4.2.3.1 FRET-Based Protease Detection with QDs as the Donor

In order to detect protease activity, quenching groups (whether organic quenchers or gold nanoparticles [50–52, 55, 56]) were first bound to the surface of a QD through a peptide sequence. This close proximity allowed for quenching of the QD emission via FRET, while subsequent cleavage of the peptide sequence by the corresponding protease led to a recovery of the QD fluorescence. The first report of protease detection with a QD-based probe utilized gold nanoparticles (AuNP) as the quenching moieties [50]; these were linked to a CdSe/CdS QD surface by a collagenase-degradable peptide sequence (Figure 4.6b). When the AuNP level was six per QD, a quenching efficiency of 71% was observed; however, following incubation of the probe with collagenase (0.2 mg ml^{-1}) a 51% increase in QD fluorescence was observed. A similar approach was developed using the organic dark quencher QXL-520 [55], in which peptides with a hexahistidine tag at the N-terminus were used for self-assembly on a dihydrolipoic acid (DHLA)-coated CdSe/ZnS QD. In this way, a rigid helical region could also be incorporated so as to separate the hexahistidine residues from the protease recognition and cleavage sites. A C-terminal cysteine thiol was also incorporated for the quencher attachment. In this case, the target enzymes were caspase-1, thrombin, collagenase, and chymotrypsin, while between eight and ten peptides were used per QD in order to avoid any potential steric crowding. The proteolytic assays were carried out with excesses of both enzyme and substrate, and not only allowed a quantitative monitoring of protease activity but also provided an insight into the mechanisms of enzymatic inhibition. Subsequently, when several inhibitory compounds were tested against the QD-thrombin-specific peptide substrate in a pharmaceutical screening assay, the K_m values for collagenase, chymotrypsin and thrombin were all in reasonable agreement with published values, but that for caspase-1 was 50% lower than had been reported previously. Assays with protein as the protease substrate were also developed [56], in which the proteolytic activity of proteinase K and papain was monitored using QD–protein conjugates as substrates. In this case, the digestion data were analyzed in such a way as to provide estimated minimal values of enzymatic activity, and the reaction velocities of the enzymes used. Mechanisms of enzymatic inhibition were also inferred from assays performed in the presence of specific inhibitors.

In a ratiometric FRET protease probe, QDs were conjugated with a rhodamine-labeled peptide [52] such that, in the resultant assembly, local excitation of the QDs was followed by an efficient energy transfer to the adjacent rhodamine dye. As a result, the corresponding emission spectrum showed one band at 545 nm for the QDs, and another at 590 nm for the rhodamine acceptor (Figure 4.6c). The addition of collagenase resulted in enzymatic cleavage of the peptide linkage, with separation of the energy donor and acceptor. In this way the energy transfer process was effectively suppressed, as the luminescence intensity at 545 nm was increased and that at 590 nm decreased. Such a ratiometric probe would be superior to a simple intensity-based probe because it would eliminate any artifacts caused by variations in the probe concentration.

Kim et al. recently reported a chip-based energy transfer system using a QD–AuNP conjugate on a surface in order to detect the activity and inhibitory effects of several proteases [matrix metalloproteinases (MMPs), thrombin, and caspase-3] [51]. For this, a biotinylated peptide substrate for the protease was conjugated to a monomaleimide-functionalized AuNP, and the resultant AuNPs (Pep-AuNPs) were associated with streptavidin (SA)-bound QDs deposited on a glass slide to form AuNPs–QDs conjugates, the photoluminescence of which was quenched. The addition of a protease to cleave the peptide substrate on the AuNP–QD conjugates led to a regeneration of the photoluminescence emission of the QDs.

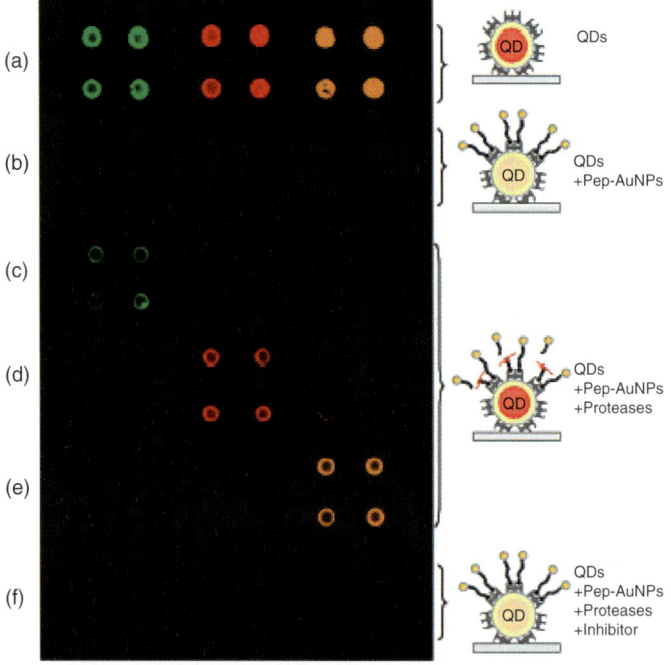

Figure 4.7 Multiplexed assay of proteases by using QDs with different colors. SA-QD525, SA-QD605, and SA-QD655 were used (from left to right). Biotinylated peptide substrates for MMP-7, caspase-3, and thrombin were conjugated to the AuNPs, after which the resulting Pep-AuNPs were associated with SA-QD525, SA-QD605, and SA-QD655, respectively. (a) SA-QDs only. (b) SA-QDs + respective Pep-AuNPs. (c) SA-QDs + Pep-AuNPs + MMP-7. (d) SA-QDs + Pep-AuNPs + caspase-3, (e) SA-QDs + Pep-AuNPs + thrombin. (f) QDs + Pep-AuNPs + mixture of the respective protease and its inhibitor [51].

Protease inhibitors also prevented any recovery of the photoluminescence of QDs by inhibiting the protease activity. When three types of SA-QD (SA-QD525, SA-QD605, and SA-QD655) were complexed with AuNPs presenting the corresponding peptide substrates, and then spotted separately onto a glass slide, a specific reaction of the protease induced a strong photoluminescence intensity from each spot, at a specified wavelength (Figure 4.7). Any cross-reactional images of the protease against other peptide substrates were found to be negligible, thus confirming the multiplexed feasibility of this assay system. This energy transfer-based assay system offers certain advantages over conventional FRET-based approaches. As the AuNPs can be employed as a common energy acceptor, a variety of QDs with different colors could be used as the energy donor, thus enabling a multiplexed assay. Moreover, the high quenching efficiency of AuNPs would allow a longer FRET distance between the donor and acceptor. Such a chip-based system might also overcome some of the drawbacks resulting from a solution-based format, including the aggregation of nanoparticles, fluctuations in photoluminescence, and the consumption of large amounts of reagents.

This FRET-based multiplexed assay has also been applied to enzymes other than proteases. For example, Suzuki et al. [57] recently reported the use of QD-based nanoprobes to detect multiple cellular signaling events, including the activities of protease (trypsin), deoxyribonuclease and DNA polymerase, as well as changes in the pH. This system was designed based on a FRET with QD as donor and an appended fluorophore as acceptor. Subsequently, both protease and deoxyribonuclease (DNase) induced changes in FRET efficiency between the donor (the QD) and the acceptor [green fluorescent protein (GFP) or a fluorophore-modified, double-stranded DNA (dsDNA)]. In contrast, DNA polymerase brought fluorescently labeled nucleotides to the surface of the QD, while pH-sensitive fluorophores that conjugated on the QD surface produced pH-dependent changes in FRET efficiency (Figure 4.8). Notably, this mixture of modified QDs showed distinct changes in emission peaks before and after enzyme treatment by simultaneous-wavelength excitation in the same tube.

4.2.3.2 BRET-Based Protease Detection with QDs as Acceptor

It appears that QDs cannot act as effective FRET acceptors when used with organic fluorophore donors [53]. This failure to observe FRET is attributed to both the efficient and unavoidable direct excitation of the QD acceptor, and the large difference in excited-state lifetimes between organic fluorophores and QDs. A direct excitation of the QD acceptor results from a strong QD absorption at the wavelength shorter than its emission wavelength. However, recent studies conducted by the present authors have demonstrated the feasibility of using QDs as the acceptor in a BRET system [54]. BRET is analogous to FRET, except that the energy derives from a chemical reaction catalyzed by the donor enzyme (e.g., a *Renilla* luciferase-mediated oxidation of its substrate, coelenterazine) rather than the absorption of excitation photons. In this system, the QDs are linked covalently to a bioluminescent protein, and act as the energy acceptor while the protein acts as the donor. The protein emits blue light with a peak at 480 nm upon addition of the substrate, coelenterazine; then, if the QDs are in close proximity to the protein, they may become excited and emit at the emission maximum. This QD-BRET-based probe has been applied to the detection of MMP-2 activity [58]. For this, a peptide sequence (GGPLGVRGGHHHHHH), which contains the MMP-2 substrate (PLGVR) and a six-histidine tag, was genetically fused to the C terminus of the BRET donor, a mutant of *Renilla* luciferase (Luc8). The QD-BRET system was established by the simultaneous coordination of carboxyl groups on the QD surface and the Luc8 His tag with Ni^{2+}. In the presence of Ni^{2+}, the carboxylic acids on the QDs were bound to the metal ions to form complexes with the 6× His tag on the Luc8 fusion protein, and the bioluminescence energy of Luc8 was efficiently transferred to the QDs in the formed complexes. Upon cleavage of the MMP-2 substrate sequence by MMP-2, the His tag was released from the fusion Luc8 and the BRET signal decreased. In comparison to FRET-based QD sensors, BRET-based QD biosensors offer several attractive features. Notably, the spectral

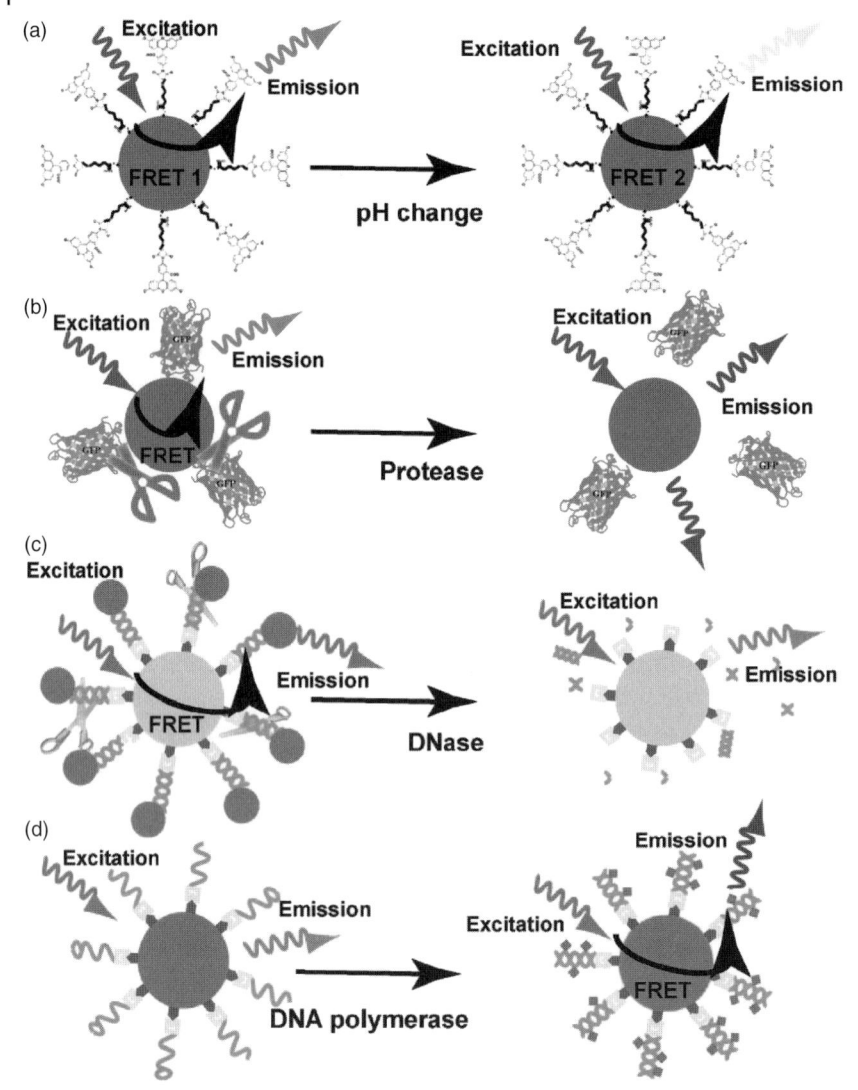

Figure 4.8 FRET-based QD bioprobes designed to give FRET changes on (from top to bottom): (a) pH change via pH-sensing dyes attached to a QD; (b) cleavage of a GFP variant with an inserted sequence recognized by a protease (e.g., trypsin) to release GFP from the QD surface; (c) digestion by DNase of dsDNA (labeled with fluorescent dUTP) bound to a QD; (d) incorporation of fluorescently labeled dUTPs into ssDNA on a QD by extension with DNA polymerase [57].

separation between the BRET donor and acceptor emissions is large, which makes it easy to detect both emissions for ratiometric measurements. In addition, the sensitivity is high due to a low background emission. One drawback of this approach, however, is that it cannot be used in complex biological media such as serum, because of the nonspecific/instable nature of an electrostatic interaction.

More recently, a QD-BRET nanosensor with a stable covalent linkage has been developed that allows the detection of protease activity in mouse sera and tumor lysates [59]. In this approach, the luciferase–protease substrate recombinant protein was genetically modified with an additional intein segment. Inteins catalyze the splicing reaction through the formation of an active thioester intermediate, and have been widely applied to protein conjugation and immobilization. Initially, the carboxylated QDs were functionalized with adipic dihydrazide (because hydrazides are excellent nucleophiles to attack the thioester intermediate of inteins). The reaction proceeded rapidly when the two components were mixed together, and resulted in cleavage of the intein and ligation of the C terminus of the recombinant protein to the QDs. Recently, this method has been applied successfully to the synthesis of a series of nanosensors for the sensitive detection of MMP-2, MMP-7, and urokinase-type plasminogen activator (uPA) (Figure 4.9a). Most importantly, the nanosensors were not only capable of detecting these proteases in complex biological media (e.g., mouse serum and tumor lysates) with a sensitivity down to $1\,ng\,ml^{-1}$, but could also detect multiple proteases present in one sample.

As a result of the wide absorption spectra of QDs, one bioluminescent protein (such as *Renilla* luciferase) can efficiently excite multiple QDs with different emissions [54], which in turn makes possible the multiplex detection of biological analytes. By conjugating *Renilla* luciferase to different emission QDs with the insertion of different protease substrates, it was possible to detect several proteases simultaneously [59]. For this, QD705 was used to prepare the sensor QD705-uPA-Luc8, and this was mixed with QD655-MMP-2-Luc8 for the simultaneous detection of MMP-2 and uPA. As shown in Figure 4.9b, both BRET emissions from QD655 and QD705 were observed in the spectrum. When MMP-2 was added to the mixture, only the BRET signals at 655 nm decreased, but when uPA was added there was a decrease in the BRET signals at 705 nm. When both MMP-2 and uPA were present, the BRET emissions at 655 and 705 nm were each decreased. Taken together, these results confirmed the ability of the QD-BRET detection platform to assay multiple targets simultaneously.

Unlike fluorescence-based methods, which generally encounter high-background signals from interfering species present in biological samples, bioluminescence-based detection provides great sensitivity due to its extremely low background. In particular, the BRET ratio (i.e., the ratio of donor and acceptor emissions) between the QD and the luciferase is modulated by the degree of protease activity, which makes it more reliable when compared to other assays. Most importantly, the elimination of physical light excitation results in a high sensitivity of the QD-BRET system; this allows it to function in complex biological media, including serum and tumor samples, in a multiplexed manner.

4.2.4
Other *In Vitro* Multiplexed Detection Systems

The QD-based energy transfer system was implemented to detect small cationic ions such as zinc and manganese [60]. Here, the surface of the CdSe/ZnS

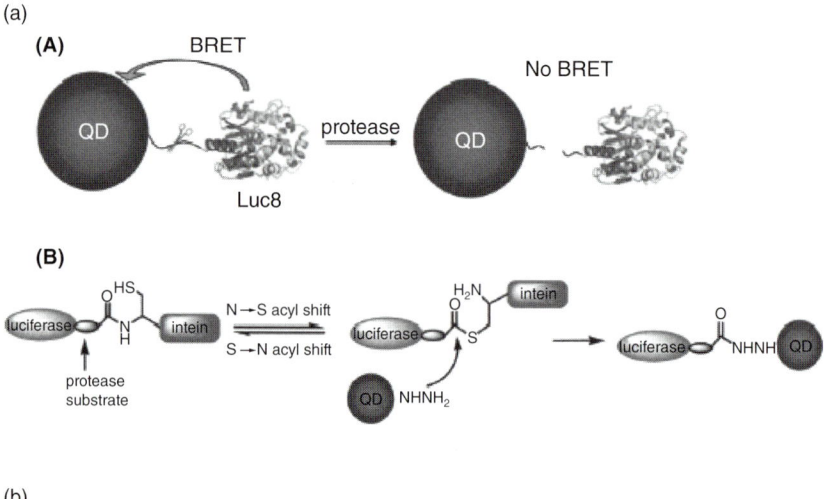

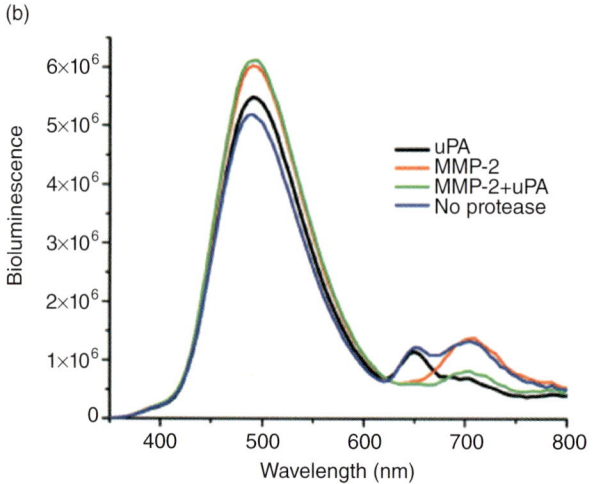

Figure 4.9 (a) (A) Schematic of the nanosensor comprising of a QD and luciferase proteins (Luc8) that are linked to the QD through an MMP peptide substrate and (B) intein-mediated site-specific conjugation of Luc8 fusion proteins to QDs; (b) Simultaneous detection of MMP-2 and uPA: A mixture of QD655-MMP-2-Luc8 (0.15 µM) and QD705-uPA-Luc8 (0.2 µM) were incubated with MMP-2 (1 µg ml^{-1}, red); uPA (10 µg ml^{-1}, black); MMP-2 (1 µg ml^{-1}) + uPA (10 µg ml^{-1}) (green); or no enzyme (blue) at room temperature for 1 h in 20 mM Tris buffer (pH 7.5) [59].

core–shell QD nanoparticles was modified with the nonfluorogenic reagent zincon (2-carboxyl-2-hydroxy-5-sulfoformazylbenzene), to develop new metal ion nanosensors with fluorescence detection. For this, the QD was first coated with negatively charged 3-mercaptopropionic acid (MPA) and a positive polyelectrolyte, after which poly(allylamine hydrochloride) (PAH$^+$) was distributed on the QD-MPA, using a layer-by-layer modification. Following the subsequent immobilization of zincon (a chromogenic reagent used widely for metal ion determination) on the positively charged QD-MPA-PAH$^+$, quenching of the CdSe/ZnS QD nanoparticle luminescence was observed, as zincon could act as a charge carrier of electrons on the QD surface. In the presence of Zn^{2+}, the fluorescence signal of the QD-zincon was further reduced as a result of RET between the QD emission and Zn^{2+}-zincon absorbance. However, for the Mn^{2+}-zincon complex there was no overlap of QD emission and Mn^{2+}-zincon absorbance, and this resulted in a "turning-on" of the emission of QD-zincon. When using these different reactions, the QD-zincon conjugates showed very good linearity over the range of 10 to 1000 mM and 5 to 500 mM for the Zn^{2+} and Mn^{2+} nanosensors, respectively.

The detection of protein glycosylation has been achieved with the energy transfer between QDs and AuNPs on a chip surface [61]. The rationale of this is based on the fact that energy transfer efficiency between lectin-modified QDs and carbohydrate-conjugated AuNPs can be modulated by a competitive inhibition of unknown glycoproteins. The potential of multiplexed detection of glycoproteins with different QDs has also been demonstrated in this study. Compared to other chip-based carbohydrate detection systems employing single fluorophores, the QD/AuNP system offered several advantages, including less photobleaching, a low background noise, and multiplexed analysis. The coupling of different QD nanoparticles with lectins and carbohydrates would enable this system to be used in the study of glycomics in a multiplexed manner.

4.3
Multiplexed Imaging Using QDs

4.3.1
Multiplexed Cellular Imaging

The application of QDs in cellular imaging has attracted much interest because of their unique optical properties and the possibility of multicolor imaging. Indeed, many reports have been made regarding the application of QDs for multiplexed imaging in cells *in vitro* and in small animals *in vivo*, some examples of which are detailed in the following sections.

Although QD labeling allows cells to be visualized under continuous illumination, and multiple targets to be monitored simultaneously [62–67], the major challenge for live-cell imaging with QDs is how to deliver them to the target cells. Normally, QDs can be taken up by cells in random fashion through endocytosis, but they may also be delivered into cells by microinjection, electroporation, and chemical-mediated transfection (e.g., lipofectamine) [68, 69]. Likewise, certain peptide sequences, such as TAT or poly-arginine, can assist the conjugated QDs in crossing the cell membrane. Moreover, the wide selection of excitation and emission wavelengths makes QDs suitable for the simultaneous tracking of multiple proteins and live cells for long periods of time and, therefore, for monitoring a range of biological processes in cells.

A simple method was reported for coding mammalian cells with multiple-emission QDs [70], whereby nine arginine peptides were conjugated to streptavidin-coated QDs for cell penetration and delivery. Cell coding was achieved by

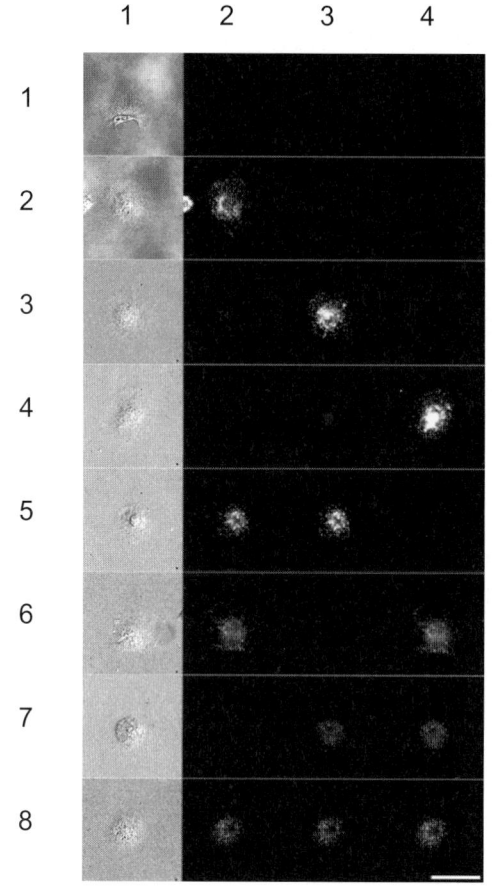

Figure 4.10 Qualitative analysis of cell multicolor coding by fluorescence microscopy. Images shown are: Column 1, differential interference contrast (DIC); column 2, fluorescence with 565/20 nm emission filter; column 3, fluorescence with 610/20 nm emission filter; column 4, fluorescence with 654/24 nm emission filter. The cells were labeled with (row 1) blank control, (row 2) sAv-565 QDs, (row 3) sAv-605 QDs, (row 4) sAv-655 QDs, (row 5) sAv-565 and sAv-605 QDs, (row 6) sAv-565 and sAv-655 QDs, (row 7) sAv-605 and sAv-655 QDs, (row 8) sAv-565, sAv-605, and sAv-655 QDs [70].

separately incubating cells with a combination of QDs such that, by varying the colors and intensities, eight optical codes (three colors and two intensities) were detected by appropriately chosen color channels by using flow cytometry or fluorescence microscopy. Following cell uptake, the QD codes were read quantitatively by measuring the fluorescence intensities in each of three appropriately chosen detector color channels on a cell-to-cell basis, by flow cytometry (Figure 4.10). The number of optical codes may be expanded by the use of additional spectrally separated QDs, and by the use of more than two intensities.

By conjugating QDs that emit at different wavelengths with IgG and streptavidin, two cellular targets could be detected simultaneously in fixed cancer cells,

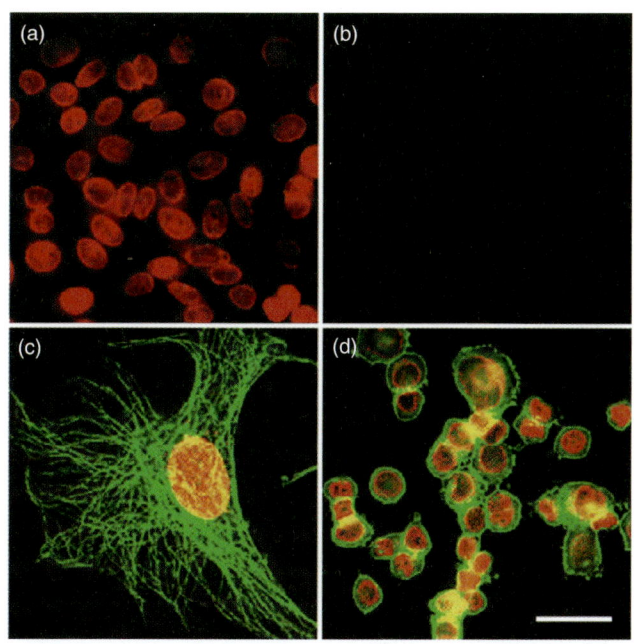

Figure 4.11 Detection of nuclear antigens and double labeling. (a) Nuclear antigens in the nuclei of human epithelial cells were labeled with polyclonal anti-nuclear antigen antibodies (ANA), anti-human IgG–biotin and QD 630–streptavidin; (b) When normal human IgGs were used in place of ANA, no detectable stain was observed; (c) The nucleus of a 3T3 cell was stained with ANA, anti-human IgG–biotin, and QD 630–streptavidin (red). The microtubules were labeled with mouse anti-α-tubulin antibody, anti-mouse IgG–biotin, and QD 535–streptavidin (green); (d) Her2 on the surface of SK-BR-3 cells was stained green with mouse anti-Her2 antibody and QD 535–IgG (green). Nuclear antigens were labeled with ANA, anti-human IgG–biotin and QD 630–streptavidin (red) [71].

using one single excitation wavelength [71]. The double labeling of cytosolic microtubules and nuclear antigens was carried out by the sequential incubation of specific antibodies, biotinylated IgG and streptavidin-conjugated QDs (Figure 4.11). When examined by epifluorescence microscopy with a 460 nm short-pass excitation filter and a 500 nm long-pass emission filter, both colors were clearly visible and spectrally resolved to the eye: the nuclear antigens appeared to be labeled with the red QD630–streptavidin, and the microtubules in the cytoplasm with the green QD535–streptavidin (Figure 4.11c). The combination of QD535–IgG and QD630–streptavidin was also used to detect, respectively, Her2 on the cell surface and nuclear antigens in the nucleus of SK-BR-3 cells. Under fluorescence microscopy, both QD630-labeled (red) nuclear antigens and QD535-labeled (green) membrane-associated Her2 were visible simultaneously (Figure 4.11d). These results showed that the QD–streptavidin conjugates were specific for their intended targets, and that they could be used effectively in the two-color fluorescence labeling of distinct cellular components.

This multiplex cell labeling may also allow the identification of specific cancer cells. For example, Prasad *et al.* conjugated targeting molecules such as transferrin and folic acid with quantum rods (QRs; these are QD-like nanoparticles) with different emission wavelengths (610 nm and 657 nm) to successfully multiplexed labeling human nasopharyngeal epidermal carcinoma cells [72]. The anti-Claudin 4- and anti-mesothelin-conjugated QRs were used as optical probes for the detection of pancreatic cancer [72], with the cells remaining stably labeled for over a week as they grew and developed. These approaches should permit the simultaneous study of multiple cells over long periods of time as they proceed through growth and development. The QD labeling of modified oligonucleotide probes was also used for sensitive mRNA detection with a low tissue background. Here, a QDs-based multiplex FISH technique proved to be ultrasensitive in the simultaneous study of multiple mRNA in tissue cultures and histological sections [73].

4.3.2
Multiplexed Imaging in Small Animals

Typically, the size of QDs has been suitable for lymphatic imaging following direct interstitial injection, and indeed, the multiplex imaging of lymph nodes has been demonstrated using QDs [74]. Figure 4.12 shows simultaneous multicolor *in vivo* wavelength-resolved spectral fluorescence lymphangiography using five QDs with similar physical sizes but different emission spectra. This allows the noninvasive and simultaneous visualization of five separate lymphatic flows and drainages, and may have implications for predicting the route of cancer metastasis into the lymph nodes. The practical implication of this study is that it is possible to separately study the drainage patterns and mixing of five adjacent lymphatic basins *in vivo* using high-resolution imaging within the laboratory environment. At least potentially, by using this method it may be possible to analyze five independent parameters in a single pass during an *in vivo* small animal image.

In order to track embryonic stem cells (ESC) *in vivo*, commercially available Qtracker QDs were used to label ESC for noninvasive imaging in living mice [75]. The QDs were found not to affect the viability and proliferation of the ESC, and had no profound effects on their differentiation capacity within the sensitivities of the screening assays used. Multiplex imaging *in vivo*, using the Maestro system, showed that QD525-, QD565-, QD605-, QD655-, QD705-, and QD800- labeled ESC could be detected *in vivo* using a single excitation wavelength (465 nm). At 24 h after labeling the ESC with QDs, 72% of them were positively stained, but by day 4 this percentage had fallen to 4%. Such a dramatic decrease may have been due either to the rapid division of ESC (doubling time 12–15 h), or to the QDs diffusing out of the dividing cells over time, causing a dilution of the QD signal.

Since BRET-based *in vivo* imaging does not require an external excitation light source, the issue of autofluorescence, which significantly reduces the signal-to-noise ratios (SNR) in fluorescence-based *in vivo* imaging, is not of concern with

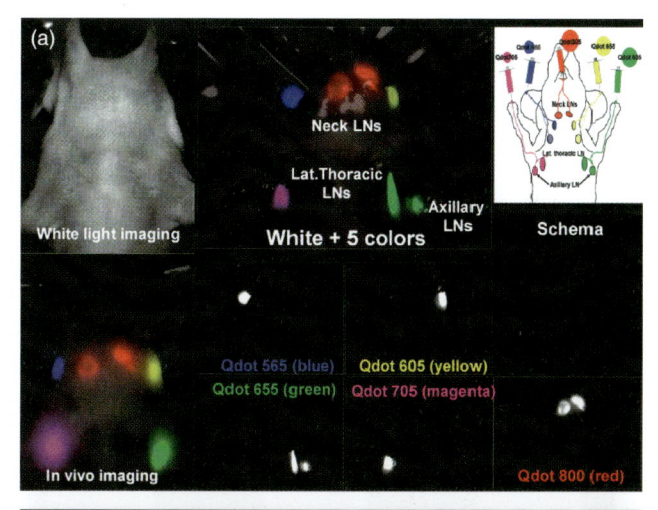

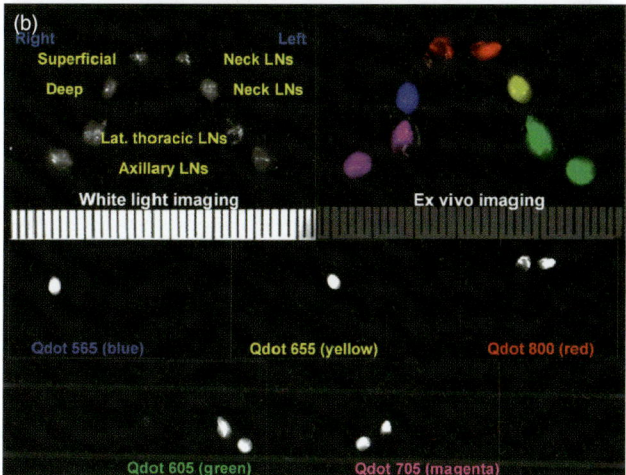

Figure 4.12 *In vivo* five-color lymphatic drainage imaging was able to visualize five distinct lymphatic drainages. (a) *In vivo* and intrasurgical spectral fluorescence imaging of a mouse injected with five carboxyl QDs (565, blue; 605, green; 655, yellow; 705, magenta; 800, red) intracutaneously into the middle digits of the bilateral upper extremities, the bilateral ears, and at the median chin, as shown in the schema. Five primary draining lymph nodes were simultaneously visualized with different colors through the skin in the *in vivo* image and are more clearly seen in the image taken at the surgery; (b) *Ex vivo* spectral fluorescence imaging of the eight draining lymph nodes after surgical resection [74].

QD-BRET probes. However, light scattering by tissues and absorption by hemoglobin still exist, and may significantly affect short-wavelength (<600 nm) emissions [76]. BRET configurations with QDs as acceptors have the potential to overcome this obstacle, due to the significantly longer wavelengths of emission.

In the case of QDs that emit at wavelengths longer than 600 nm, it has been shown that the absorbance of whole blood does not significantly affect BRET

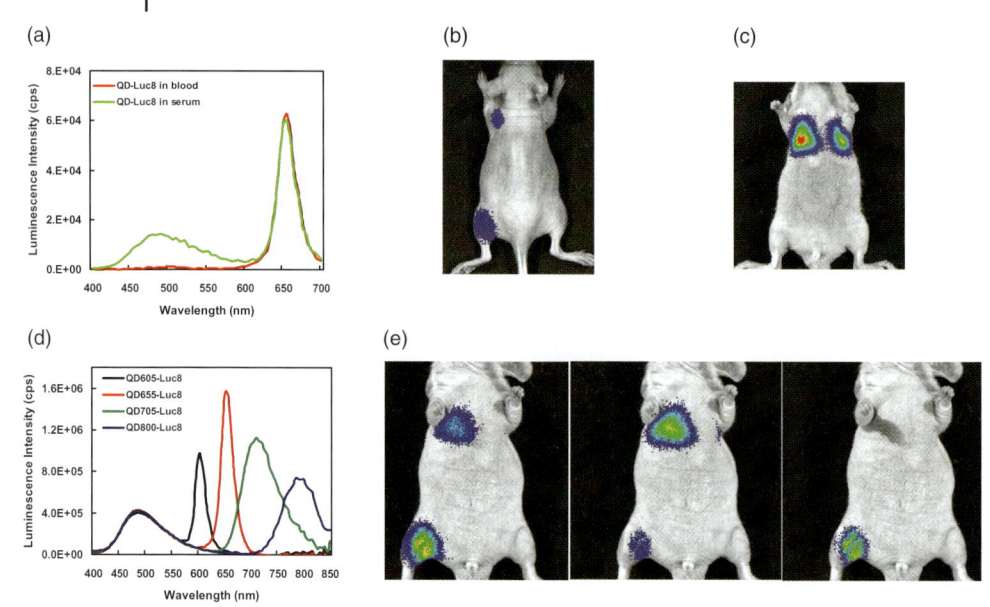

Figure 4.13 *In vivo* imaging applications with QD-BRET. (a) Bioluminescence emission spectrum of the QD-BRET conjugate (QD655-Luc8) in the mouse serum and whole blood; (b) Bioluminescence imaging of a live mouse injected with QD655-Luc8 subcutaneously (top) and intramuscularly (bottom) with an emission filter (650–660 nm); (c) Imaging C6 glioma cells labeled with the QD-BRET conjugate *in vitro* and injected into a living mouse via tail vein, acquired with a filter (575–650 nm); (d) Bioluminescence emission spectra of four QD-BRET conjugates (QD605-Luc8, QD655-Luc8, QD705-Luc8, and QD800-Luc8); (e) Multiplexed imaging of two groups of C6 glioma cells labeled by QD655-Luc8 conjugate and injected via tail vein, or labeled by QD800-Luc8 conjugate and injected intramuscularly; the left image was collected without any filter, the middle image with a 575–650 nm filter, and the right image with a 760–790 nm filter [77].

emission from QDs [54]. In contrast, the emission of donor bioluminescent proteins was almost eliminated by the absorption of their short wavelength emission due to hemoglobin (Figure 4.13a) [54]. The *in vivo* detection of QD-BRET signals has been demonstrated with conjugates at superficial sites and at deep tissue locations. By using a conjugate prepared by the direct coupling of QD655 with Luc8, it was shown that the BRET emission from QD655 could be detected from conjugates at both subcutaneous and intramuscular sites (Figure 4.13b). When C6 glioma cells were labeled by the QD-BRET conjugates and injected through the tail vein into a nude mouse, their trafficking into the lungs was readily imaged by using BRET emission from the QDs (Figure 4.13c). These successful examples highlight the advantage of QDs as BRET acceptors, due largely to the longer wavelengths of QD emission.

The features of broad excitation spectra and size-tunable emission of QDs enable the creation of many possible BRET pairs. For example, the same bioluminescent protein Luc8 can pair with QD605, QD655, QD705, and QD800 for multiplexed BRET imaging (Figure 4.13d). Each of these has displayed efficient BRET upon conjugation, and all can be imaged *in vivo* upon injection. When two groups of cells were labeled with different QD-BRET conjugates (e.g., QD655 and QD800) and introduced into the same mouse, both could be imaged and differentiated from their QD-BRET emission (Figure 4.13e). With the capability of tuning the emission of QDs by controlling their size and composition, additional BRET pairs may become available that would allow more interactions and events to be imaged simultaneously in the same animal.

4.4 Summary

Quantum dots as a novel fluorescent probes have, since their invention, proved to be tremendously useful in many areas of biological and medical research. In this chapter, attention has been focused on their applications for multiplexed detection, tissue/cell labeling, and imaging in cells and in small animals. Due to the generally wide absorption spectra of QDs, and their narrow emission spectra, QDs of different sizes and with different emission spectra can be excited at one single wavelength so as to emit at their individual emission wavelengths. This interesting property renders QDs as an excellent probe for a multiplexed assay, and this feature has been used to best advantage in many bioanalytical assays, ranging from DNA detection, immunoblotting for protein detection, detection of enzymatic activity, and small molecule analytes such as metal ions, pH value, to live-cell labeling, cancer cell detection, and *in vivo* cell trafficking and detection. Adapting these multiplexed assays with the platform of microfluidics and microarrays will further enhance their utilities. The further development of QDs for multiplexed detection and imaging will benefit from the more robust and more specific conjugation chemistries of biomolecules to QDs, and additional choices of QDs with a wide range of emission wavelengths and even more narrow emission peaks. Yet, for *in vivo* multiplexed imaging applications, the issues of potential cytotoxicity and unfavorable pharmacokinetics with current QDs [78–80] must first be addressed before they can be fully exploited in this arena and applied to human subjects.

References

1 Tong, A.K., Li, Z., Jones, G.S., Russo, J.J. and Ju, J. (2001) Combinatorial fluorescence energy transfer tags for multiplex biological assays. *Nature Biotechnology*, **19**, 756–9.
2 Bruchez, M. Jr, Moronne, M., Gin, P., Weiss, S. and Alivisatos, A.P. (1998) Semiconductor nanocrystals as fluorescent biological labels. *Science*, **281**, 2013–16.
3 Chan, W.C. and Nie, S. (1998) Quantum dot bioconjugates for ultrasensitive nonisotopic detection. *Science*, **281**, 2016–18.

4 Leatherdale, C.A., Woo, W.K., Mikulec, F.V. and Bawendi, M.G. (2002) On the absorption cross section of CdSe nanocrystal quantum dots. *Journal of Physical Chemistry, B*, **106**, 7619–22.

5 Gao, X., Cui, Y., Levenson, R.M., Chung, L.W. and Nie, S. (2004) In vivo cancer targeting and imaging with semiconductor quantum dots. *Nature Biotechnology*, **22**, 969–76.

6 Dahan, M., Laurence, T., Pinaud, F., Chemla, D.S., Alivisatos, A.P., Sauer, M. and Weiss, S. (2001) Time-gated biological imaging by use of colloidal quantum dots. *Optics Letters*, **26**, 825–7.

7 Nie, S., Xing, Y., Kim, G.J. and Simons, J.W. (2007) Nanotechnology applications in cancer. *Annual Reviews of Biomedical Engineering*, **9**, 257–88.

8 Michalet, X., Pinaud, F.F., Bentolila, L.A., Tsay, J.M., Doose, S., Li, J.J., Sundaresan, G., Wu, A.M., Gambhir, S.S. and Weiss, S. (2005) Quantum dots for live cells, in vivo imaging, and diagnostics. *Science*, **307**, 538–44.

9 Medintz, I., Uyeda, E.L., Goodman, E.R. and Mattousi, H. (2005) Quantum dot bioconjugates for imaging, labelling and sensing. *Nature Materials*, **4**, 435–46.

10 Alivisatos, P. (2004) The use of nanocrystals in biological detection. *Nature Biotechnology*, **22**, 47–52.

11 Golub, T.R., Slonim, D.K., Tamayo, P., Huard, C., Gaasenbeek, M., Mesirov, J.P., Coller, H., Loh, M.L., Downing, J.R., Caligiuri, M.A., Bloomfield, C.D. and Lander, E.S. (1999) Molecular classification of cancer: class discovery and class prediction by gene expression monitoring. *Science*, **286**, 531–7.

12 Gilles, P.N., Wu, D.J., Foster, C.B., Dillon, P.J. and Chanock, S.J. (1999) Single nucleotide polymorphic discrimination by an electronic dot blot assay on semiconductor microchips. *Nature Biotechnology*, **17**, 365–70.

13 Han, M., Gao, X., Su, J.Z. and Nie, S. (2001) Quantum-dot-tagged microbeads for multiplexed optical coding of biomolecules. *Nature Biotechnology*, **19**, 631–5.

14 Xu, H., Sha, M.Y., Wong, E.Y., Uphoff, J., Xu, Y., Treadway, J.A., Truong, A., O'Brien, E., Asquith, S., Stubbins, M., Spurr, N.K., Lai, E.H. and Mahoney, W. (2003) Multiplexed SNP genotyping using the Qbead system: a quantum dot-encoded microsphere-based assay. *Nucleic Acids Research*, **31**, e43.

15 Cao, Y.C., Huang, Z.L., Liu, T.C., Wang, H.Q., Zhu, X.X., Wang, Z., Zhao, Y.D., Liu, M.X. and Luo, Q.M. (2006) Preparation of silica encapsulated quantum dot encoded beads for multiplex assay and its properties. *Analytical Biochemistry*, **351**, 193–200.

16 Cao, Y.C., Liu, T.C., Hua, X.F., Zhu, X.X., Wang, H.Q., Huang, Z.L., Zhao, Y.D., Liu, M.X. and Luo, Q.M. (2006) Quantum dot optical encoded polystyrene beads for DNA detection. *Journal of Biomedical Optics*, **11**, 054025.

17 Gerion, D., Chen, F., Kannan, B., Fu, A., Parak, W.J., Chen, D.J., Majumdar, A. and Alivisatos, A.P. (2003) Room-temperature single-nucleotide polymorphism and multiallele DNA detection using fluorescent nanocrystals and microarrays. *Analytical Chemistry*, **75**, 4766–72.

18 Robelek, R., Niu, L., Schmid, E.L. and Knoll, W. (2004) Multiplexed hybridization detection of quantum dot-conjugated DNA sequences using surface plasmon enhanced fluorescence microscopy and spectrometry. *Analytical Chemistry*, **76**, 6160–5.

19 Ho, Y.P., Kung, M.C., Yang, S. and Wang, T.H. (2005) Multiplexed hybridization detection with multicolor colocalization of quantum dot nanoprobes. *Nano Letters*, **5**, 1693–7.

20 Shepard, J.R. (2006) Polychromatic microarrays: simultaneous multicolor array hybridization of eight samples. *Analytical Chemistry*, **78**, 2478–86.

21 Wang, J., Liu, G. and Merkoci, A. (2003) Electrochemical coding technology for simultaneous detection of multiple DNA targets. *Journal of the American Chemical Society*, **125**, 3214–15.

22 Liu, G., Lee, T.M. and Wang, J. (2005) Nanocrystal-based bioelectronic coding of single nucleotide polymorphisms. *Journal of the American Chemical Society*, **127**, 38–9.

23 Ornberg, R.L. and Liu, H. (2007) Immunofluorescent labeling of proteins in cultured cells with quantum dot

secondary antibody conjugates. *Methods in Molecular Biology*, **374**, 3–10.
24 Makrides, S.C., Gasbarro, C. and Bello, J.M. (2005) Bioconjugation of quantum dot luminescent probes for Western blot analysis. *Biotechniques*, **39**, 501–6.
25 Ornberg, R.L., Harper, T.F. and Liu, H. (2005) Western blot analysis with quantum dot fluorescence technology: a sensitive and quantitative method for multiplexed proteomics. *Nature Methods*, **2**, 79–81.
26 Goldman, E.R., Clapp, A.R., Anderson, G.P., Uyeda, H.T., Mauro, J.M., Medintz, I.L. and Mattoussi, H. (2004) Multiplexed toxin analysis using four colors of quantum dot fluororeagents. *Analytical Chemistry*, **76**, 684–8.
27 Goldman, E.R., Balighian, E.D., Mattoussi, H., Kuno, M.K., Mauro, J.M., Tran, P.T. and Anderson, G.P. (2002) Avidin: a natural bridge for quantum dot-antibody conjugates. *Journal of the American Chemical Society*, **124**, 6378–82.
28 Liu, G., Wang, J., Kim, J., Jan, M.R. and Collins, G.E. (2004) Electrochemical coding for multiplexed immunoassays of proteins. *Analytical Chemistry*, **76**, 7126–30.
29 Hansen, J.A., Wang, J., Kawde, A.N., Xiang, Y., Gothelf, K.V. and Collins, G. (2006) Quantum-dot/aptamer-based ultrasensitive multi-analyte electrochemical biosensor. *Journal of the American Chemical Society*, **128**, 2228–9.
30 Lucas, L.J., Chesler, J.N. and Yoon, J.Y. (2007) Lab-on-a-chip immunoassay for multiple antibodies using microsphere light scattering and quantum dot emission. *Biosensors & Bioelectronics*, **23**, 675–81.
31 Kozarich, J.W. (2003) Activity-based proteomics: enzyme chemistry redux. *Current Opinion in Chemical Biology*, **7**, 78–83.
32 Drews, J. (2000) Drug discovery: a historical perspective. *Science*, **287**, 1960–196.
33 Belaaouaj, A. (2002) Neutrophil elastase-mediated killing of bacteria: lessons from targeted mutagenesis. *Microbes and Infection*, **4**, 1259–64.
34 Packard, B.Z., Telford, W.G., Komoriya, A. and Henkart, P.A. (2007) Granzyme B activity in target cells detects attack by cytotoxic lymphocytes. *Journal of Immunology*, **179**, 3812–20.
35 Parks, W.C., Wilson, C.L. and Lopez-Boado, Y.S. (2004) Matrix metalloproteinases as modulators of inflammation and innate immunity. *Nature Reviews. Immunology*, **4**, 617–29.
36 Ge, G. and Greenspan, D.S. (2006) Developmental roles of the BMP1/TLD metalloproteinases. *Birth Defects Research. Part C, Embryo Today: Reviews*, **78**, 47–68.
37 Page-McCaw, A., Ewald, A.J. and Werb, Z. (2007) Matrix metalloproteinases and the regulation of tissue remodeling. *Nature Reviews. Molecular Cell Biology*, **8**, 221–33.
38 Puente, X.S., Sánchez, L.M., Overall, C.M. and López-Otín, C. (2003) Human and mouse proteases: a comparative genomic approach. *Nature Reviews. Genetics*, **4**, 544–58.
39 Vihinen, P., Ala-Aho, R. and Kähäri, V.-M. (2005) Matrix metalloproteinases as therapeutic targets in cancer. *Current Cancer Drug Targets*, **5**, 203–20.
40 Ala-Aho, R. and Kähäri, V.-M. (2005) Collagenases in cancer. *Biochimie*, **87**, 273–86.
41 Handsley, M.M. and Edwards, D.R. (2005) Metalloproteinases and their inhibitors in tumor angiogenesis. *International Journal of Cancer*, **115**, 849–60.
42 Nagai, T. and Miyawaki, A. (2004) A high-throughput method for development of FRET based indicators for proteolysis. *Biochemical and Biophysical Research Communications*, **319**, 72–7.
43 Mahajan, N.P., Harrison-Shostak, D.C., Michaux, J. and Herman, B. (1999) Novel mutant green fluorescent protein protease substrates reveal the activation of specific caspases during apoptosis. *Chemistry & Biology*, **6**, 401–9.
44 Rodems, S.M., Hamman, B.D., Lin, C., Zhao, J., Shah, S., Heidary, D., Makings, L., Stack, J.H. and Pollok, B.A. (2002) A FRET-based assay platform for ultra-high density drug screening of protein kinases and phosphatases. *ASSAY and Drug Development Technologies*, **1**, 9–19.
45 Rehm, M., Dussmann, H., Janicke, R.U., Tavare, J.M., Kogel, D. and Prehn, J.H.M. (2002) Single-cell fluorescence resonance energy transfer analysis demonstrates that

caspase activation during apoptosis is a rapid process. Role of caspase-3. *Journal of Biological Chemistry*, **277**, 24506–14.

46 Mizukami, S., Kikuchi, K., Higuchi, T., Urano, Y., Mashima, T., Tsuruo, T. and Nagano, T. (1999) Imaging of caspase-3 activation in HeLa cells stimulated with etoposide using a novel fluorescent probe. *FEBS Letters*, **453**, 356–60.

47 Luo, K.Q., Yu, V.C., Pu, Y.M. and Chang, D.C. (2003) Measuring dynamics of caspase-8 activation in a single living HeLa cell during TNFalpha-induced apoptosis. *Biochemical and Biophysical Research Communications*, **304**, 217–22.

48 Gurtu, V., Kain, S.R. and Zhang, G.H. (1997) Fluorometric and colorimetric detection of caspase activity associated with apoptosis. *Analytical Biochemistry*, **251**, 98–102.

49 Harris, J.L., Backes, B.J. and Leonetti, F. (2000) Rapid and general profiling of protease specificity by using combinatorial fluorogenic substrate libraries. *Proceeding of the National Academy of Sciences of the United States of America*, **97**, 7754–9.

50 Chang, E., Miller, J.S., Sun, J., Yu, W.W., Colvin, V.L., Drezek, R. and West, J.R. (2005) Protease-activated quantum dot probes. *Biochemical and Biophysical Research Communications*, **334**, 1317–21.

51 Kim, Y.P., Oh, Y.H., Oh, E., Ko, S., Han, M.K. and Kim, H.S. (2008) Energy transfer-based multiplexed assay of proteases by using gold nanoparticle and quantum dot conjugates on a surface. *Analytical Chemistry*, **80**, 4634–41.

52 Shi, L., De Paoli, V., Rosenzweig, N. and Rosenzweig, Z. (2006) Synthesis and application of quantum dots FRET-based protease sensors. *Journal of the American Chemical Society*, **128**, 10378–9.

53 Clapp, A.R., Medintz, I.L., Fisher, B.R., Anderson, G.P. and Mattoussi, H. (2005) Can luminescent quantum dots be efficient energy acceptors with organic dye donors? *Journal of the American Chemical Society*, **127**, 1242–50.

54 So, M.K., Xu, C., Loening, A.M., Gambhir, S.S. and Rao, J. (2006) Self-illuminating quantum dot conjugates for *in vivo* imaging. *Nature Biotechnology*, **24**, 339–43.

55 Medintz, I.L., Clapp, A.R., Brunel, F.M., Tiefenbrunn, T., Uyeda, H.T., Chang, E.L., Deschamps, J.R., Dawson, P.E. and Mattoussi, H. (2006) Proteolytic activity monitored by fluorescence resonance energy transfer through quantum-dot-peptide conjugates. *Nature Materials*, **5**, 581–9.

56 Clapp, A.R., Goldman, R.E., Uyeda, H.T., Chang, E.L., Whitley, J.L. and Medintz, I.L. (2008) Monitoring of enzymatic proteolysis using self-assembled quantum dot-protein substrate sensors. *Journal of Sensors*, **2008**, Article ID 797436, 1–10.

57 Suzuki, M., Husimi, Y., Komatsu, H., Suzuki, K. and Douglas, K.T. (2008) Quantum dot FRET biosensors that respond to pH, to proteolytic or nucleolytic cleavage, to DNA synthesis, or to a multiplexing combination. *Journal of the American Chemical Society*, **130**, 5720–5.

58 Yao, H., Zhang, Y., Xiao, F., Xia, Z. and Rao, J. (2007) Quantum dot/bioluminescence resonance energy transfer based highly sensitive detection of proteases. *Angewandte Chemie, International Edition*, **119**, 4424–7.

59 Xia, Z., Xing, Y., So, M.K., Koh, A.L., Sinclair, R. and Rao, J. (2008) Multiplex detection of protease activity with quantum dot nanosensors prepared by intein-mediated specific bioconjugation. *Analytical Chemistry*, **80**, 8649–55.

60 Ruedas-Rama, M.J. and Hall, E.A. (2009) Multiplexed energy transfer mechanisms in a dual-function quantum dot for zinc and manganese. *Analyst*, **134**, 159–69.

61 Kim, Y.P., Park, S., Oh, E., Oh, Y.H. and Kim, H.S. (2009) On-chip detection of protein glycosylation based on energy transfer between nanoparticles. *Biosensors & Bioelectronics*, **24**, 1189–94.

62 Hanaki, K., Momo, A., Oku, T., Komoto, A., Maenosono, S., Yamaguchi, Y. and Yamamoto, K. (2003) Semiconductor quantum dot/albumin complex is a long-life and highly photostable endosome marker. *Biochemical and Biophysical Research Communications*, **302**, 496–501.

63 Sukhanova, A., Devy, J., Venteo, L., Kaplan, H., Artemyev, M., Oleinikov, V., Klinov, D., Pluot, M., Cohen, J.H.

and Nabiev, I. (2004) Biocompatible fluorescent nanocrystals for immunolabeling of membrane proteins and cells. *Analytical Biochemistry*, 324, 60–7.

64 Kaul, Z., Yaguchi, T., Kaul, S.C., Hirano, T., Wadhwa, R. and Taira, K. (2003) Mortalin imaging in normal and cancer cells with quantum dot immunoconjugates. *Cell Research*, 13, 503–7.

65 Hoshino, A., Hanaki, K., Suzuki, K. and Yamamoto, K. (2004) Applications of T-lymphoma labeled with fluorescent quantum dots to cell tracing markers in mouse body. *Biochemical and Biophysical Research Communications*, 314, 46–53.

66 Chen, F. and Gerion, D. (2004) Fluorescent CdSe/ZnS nanocrystal-peptide conjugates for long-term, nontoxic imaging and nuclear targeting in living cells. *Nano Letters*, 4, 1827–32.

67 Jaiswal, J.K., Mattoussi, H., Mauro, J.M. and Simon, S.M. (2003) Long-term multiple color imaging of live cells using quantum dot bioconjugates. *Nature Biotechnology*, 21, 47–51.

68 Derfus, A.M., Chan, W.C.W. and Bhatia, S.N. (2004) Intracellular delivery of quantum dots for live cell labeling and organelle tracking. *Advanced Materials*, 16, 961–6.

69 Voura, E.B., Jaiswal, J.K., Mattoussi, H. and Simon, S.M. (2004) Tracking metastatic tumor cell extravasation with quantum dot nanocrystals and fluorescence emission-scanning microscopy. *Nature Medicine*, 10, 993–8.

70 Lagerholm, B.C., Wang, M., Ernst, L.A., Ly, D.H., Liu, H., Bruchez, M.P. and Waggoner, A.S. (2004) Multicolor coding of cells with cationic peptide coated quantum dots. *Nano Letters*, 4, 2019–22.

71 Wu, X., Liu, H., Liu, J., Haley, K.N., Treadway, J.A., Larson, J.P., Ge, N., Peale, F. and Bruchez, M.P. (2003) Immunofluorescent labeling of cancer marker Her2 and other cellular targets with semiconductor quantum dots. *Nature Biotechnology*, 21, 41–6.

72 Yong, K.-T., Roy, I., Pudavar, H.E., Bergey, E.J., Tramposch, K.M., Swihart, M.T. and Prasad, P.N. (2008) Multiplex imaging of pancreatic cancer cells by using functionalized quantum rods. *Advanced Materials*, 20, 1412–17.

73 Chan, P., Yuen, T., Ruf, F., Gonzalez-Maeso, J. and Sealfon, S.C. (2005) Method for multiplex cellular detection of mRNAs using quantum dot fluorescent in situ hybridization. *Nucleic Acids Research*, 33, e161.

74 Kobayashi, H., Hama, Y., Koyama, Y., Barrett, T., Regino, C.A., Urano, Y. and Choyke, P.L. (2007) Simultaneous multicolor imaging of five different lymphatic basins using quantum dots. *Nano Letters*, 7, 1711–16.

75 Lin, S., Xie, X., Patel, M.R., Yang, Y.H., Li, Z., Cao, F., Gheysens, O., Zhang, Y., Gambhir, S.S., Rao, J.H. and Wu, J.C. (2007) Quantum dot imaging for embryonic stem cells. *BMC Biotechnology*, 7, 67.

76 Zhao, H., Doyle, T.C., Coquoz, O., Kalish, F., Rice, B.W. and Contag, C.H. (2005) Emission spectra of bioluminescent reporters and interaction with mammalian tissue determine the sensitivity of detection in vivo. *Journal of Biomedical Optics*, 10, 041210.

77 Xia, Z. and Rao, J. (2009) Biosensing and imaging based on bioluminescence resonance energy transfer. *Current Opinion in Biotechnology*, 20, 37–44.

78 Schipper, M.L., Cheng, Z., Lee, S.-W., Keren, S., Bentolila, L.A., Sundaresan, G., Iyer, G., Gheysens, O., Ebenstein, Y., Li, J., Rao, J., Chen, X., Wu, A.M., Weiss, S.S. and Gambhir, S.S. (2007) MicroPET-based biodistribution of quantum dots in living mice. *Journal of Nuclear Medicine*, 48, 1511–18.

79 Schipper, M.L., Iyer, G., Koh, A.L., Cheng, Z., Ebenstein, Y., Aharoni, A., Keren, S., Bentolila, L.A., Li, J., Rao, J., Chen, X., Banin, U., Wu, A.M., Sinclair, R., Weiss, S.S. and Gambhir, S.S. (2009) Particle size, surface coating, and PEGylation influence the biodistribution of quantum dots in living mice. *Small*, 5, 126–34.

80 Rao, J. (2008) Shedding light on tumors using nanoparticles. *ACS Nano*, 2, 1984–6.

Part II
Therapy

5
Medical Diagnostics of Quantum Dot-Based Protein Micro- and Nanoarrays

Anisha Gokarna and Yong-Hoon Cho

5.1
Introduction

Microarray technology, which today is a well-known technique, has proven to be a powerful tool for the rapid, simultaneous detection and high-throughput analysis of large number of biological samples. This technology also incorporates a high level of sensitivity and precision. The term "microarray" is often used interchangeably with the term "biochip" or, more specifically, DNA microarrays with DNA chip, protein microarrays with protein chip, and so on. Microarrays are comprised of a micropatterned array of biosensing capture agents for the rapid and simultaneous probing of a large number of DNA, proteins, cells, or tissue fragments.

The benefits of using microarrays relate to the miniaturization of the arrays, followed by a multiplexing of protein interactions which, together, form a very attractive feature of microarray-based diagnostics. Moreover, in both microarrays and nanoarrays, the reagent consumption is several orders of magnitude less than that of microplate assays, both in terms of the amount of deposited probes and the sample per probe volume. Another advantage is that many microarray formats function equally well for different types of molecule, including nucleic acids, proteins, or small molecules.

Since many previous reports have contributed to the development of microarray technology, this chapter will include only the latest, multifaceted aspects such as the fabrication of micro- and nanoarrays of proteins, and their optical readout by using nanotechnology-based colloidal quantum dots (QDs) as labels. The chapter provides a comprehensive "snapshot" of the current state of clinical proteomics, from the vantage points of antibody and protein micro- and nanoarray technologies, and the labeling techniques used therein for their optical readout.

The chapter first provides an introduction to the foundation of protein microarrays, from its invention in 1961 to the compact protein arrays available today. A comparative view is provided of the traditional methods used to label protein arrays (using organic dyes) for optical readout, and of the latest techniques, using inorganic colloidal QDs. The various types of existing protein microarray, and differences in their fabrication, functioning, and applications, are then described. The next section introduces nanoarrays, the different methods available for immobilizing proteins at the nanoscale, and draws attention to the key issues of nanoarrays, notably that of "multiplexing." Subsequent sections describe the surface functionalization of QDs, in order to attach them to proteins or other biomolecules, and the potential problems that may arise during such conjugation. Details are then provided of the experimental aspects of protein microarrays or nanoarray fabrication with QDs as labels, and of the clinical applications of QD-conjugated protein microarrays, an example being their use as biomarkers in the detection of cancer. This section incorporates the detection of protein-protein interactions in cancer biomarkers, the non-specific binding effects that can occur in QDs-conjugated protein arrays due to the presence of QDs, and multiplexing in protein arrays using QDs. A comparison between organic dye-labeled protein microarrays and colloidal QD-labeled microarrays is also provided, noting in particular those issues relating to photostability and sensitivity. Finally, a discussion is provided of the fabrication and readout of various types of protein nanoarrays, ranging from organic dye-labeled protein nanoarrays to QDs-labeled protein arrays. Protein nanoarrays fabricated using parallel scanning probe microscopy (SPM), and rewritable/erasable chips fabricated using nanolithographic techniques, are also discussed. The chapter concludes with an outline of the future prospects for QDs-labeled protein arrays.

5.1.1
Invention of Protein Microarrays

The invention of the antibody microarray technique dates back to 1961, when Joseph G. Feinberg, working at the Beecham Research Laboratories (Betchworth, Surrey, UK), prepared a thin agar film on a glass coverslip. The agar had been impregnated with serum from a patient with autoimmune thyroid disease. However, when 1 µl samples of thyroglobulin antigen were spotted onto the agar surface, a series of micro immunoprecipitation reactions, made visible by staining with Ponceau S, was seen to develop. Ponceau S is the sodium salt of a diazo dye that can be used as a stain for the rapid reversible detection of protein bands on nitrocellulose or polyvinylidene difluoride (PVDF) membranes (Western blotting), as well as on cellulose acetate membranes. A Ponceau S stain is easily reversed with water washes, and this facilitates subsequent immunological detection. As these findings signaled an interaction between thyroglobulin and the anti-thyroglobulin antibody in the patient's serum, thus was born Feinberg's microspot test for antigens and antibodies [1].

Some 30 years later, Ekins and Chu [2] and Ekins *et al.* [3] further developed this system into a true array of spotted antibodies, and rhetorically called it the equivalent of a "CD (compact disc) for clinical chemistry". The present operational

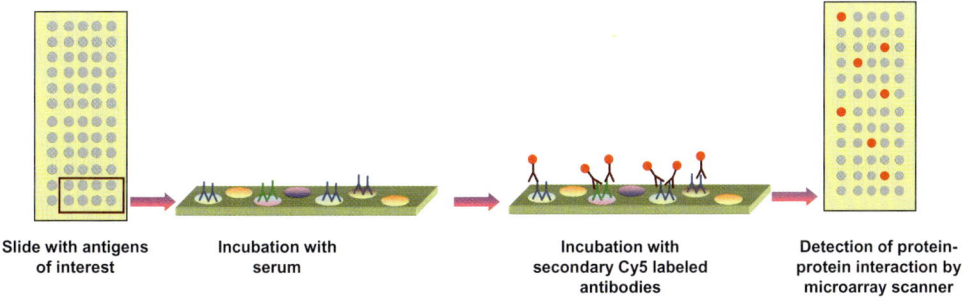

Figure 5.1 Protein–protein interaction occurring in a protein biochip and its optical readout using Cy5 organic dye as a label. Reproduced with permission from http://eye-research.org/joomla/images/stories/microarrayworkflow8.JPG) permission to be obtained!

version of a high-density protein or antibody microarray on a glass slide was finally brought to physical reality by MacBeath and Schreiber [4] and Zhu et al. [5]. Only very recently has Ekins and Chu's original rhetorical vision of a CD for an antibody microarray platform actually become a physical reality. The BioCD® from Quadraspec [6] is a spinning disk, label-free platform based on a self-referencing interferometric optical biosensor. On a similar basis, Gyros AB [7] have developed a CD-shaped microfluidics device in which samples are driven through the system by centrifugal force.

A general scheme of a protein microarray experiment is shown in Figure 5.1. In this set-up, a large set of capture proteins or antigens is arrayed on a suitable solid substrate and, after washing and blocking the unreacted sites, the array is probed with a sample containing the counterparts of the molecular recognition events under study. These counterparts are sometimes labeled by fluorescent molecules, such as organic dyes or colloidal QDs. If an interaction occurs, a signal is detected by using a variety of detection techniques (e.g., a microarray chip scanner or a confocal microscope).

The DNA microarray (also termed "DNA chips" or "gene chips") technology is quite similar to the protein microarray technology, the only difference being that DNAs are used in these types of microarrays in place of proteins. DNA spots ranging from 5 to 150 μm in size, with picomolar amounts of a specific DNA sequence, are fabricated on a solid support consisting of single-stranded DNA (ssDNA) with 20 to 1000 or even more number of bases attached. These spotted arrays help in the identification of DNA sequences of a gene present in a sample. For this purpose, fluorescently labeled mRNA or cDNA are hybridized with the immobilized DNA fragments present in the microarray spots.

5.1.2
Optical Readout of Protein Arrays: Traditional versus New Labels

Traditionally, during the fabrication of biochips, organic fluorophores (either genetically encoded fluorescent proteins or chemically synthesized fluorescent dyes) are used as labels. These act as fluorescent markers during the process of optical imaging of these biological entities. However, such fluorophores are highly sensitive to the environment and produce a low quantum yield in aqueous environments. Moreover, they exhibit narrow absorption spectra and broad emission spectra, such that each fluorophore can be excited by light of only a specific wavelength. Their broad emission spectrum often causes the signals from different fluorophores to overlap, and consequently appropriate cut-off filters must often be utilized for removal of the spectral "crosstalk" phenomena. These are some of the limitations displayed by organic dyes. The most suitable label appropriate for optical imaging of biomolecules is one which is:

- conveniently excitable, without simultaneous excitation of the biological matrix, and detectable with conventional instrumentation;
- bright – that is, it possesses a high molar absorption coefficient at the excitation wavelength and a high fluorescence quantum yield;
- soluble in relevant buffers, cell culture media or body fluids;
- sufficiently stable under relevant conditions;
- has functional groups for site-specific labeling;
- has reported data regarding its photophysics; and
- is available in a reproducible quality.

Inorganic fluorophores termed as QDs display most of the above-mentioned characteristics; in fact, QDs gained their name from the well-known quantum confinement effect.

Today, QDs are among the most promising items in the "nanomedicine toolbox," with such nanocrystal fluorophores having several potential medical applications including nanodiagnostics, imaging, targeted drug delivery, and photodynamic therapy. QDs are basically semiconductor nanocrystals composed of a core of a semiconductor material with a smaller bandgap material, enclosed within a shell of another semiconductor that has a larger spectral band gap (see Figure 5.2) [8–12]. They were initially prepared in 1982 for use as probes to investigate surface kinetics, when it was found that the quantum yield of nanocrystals was sensitive to the concentration of surface-adsorbed species that could undergo reduction [13]. Since the number of "dangling bonds" at the core particle surface determines the fluorescence quantum yields, decay kinetics and stability, then inorganic passivation layers and/or organic capping ligands may be bound to the particle surface so as to optimize these features. Addition of the passivation shell often results in a slight red shift in absorption and emission as compared to the core QD, because of the tunneling of charge carriers into the shell. The core of the QDs are usually composed of elements from Groups II and VI (e.g., CdSe, the most common) or Groups III and V (e.g., InP) of the Periodic Table, whereas the shell is typically a

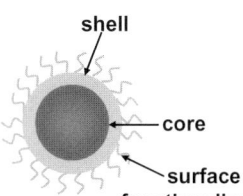

Figure 5.2 Schematic of a colloidal QD with core, shell and surface functionalization.

high-bandgap material such as ZnS [14]. A typical QD has a diameter of about 2–10 nm, and this can allow one-on-one interaction with biomolecules such as proteins, where the typical size ranges from 1 to 20 nm [9, 15]. Today, commercial materials composed of CdSe (from Sigma-Aldrich, Invitrogen, Evident, and Plasmachem), CdTe (from Plasmachem) and InP or InGaP (from Evident) are available. Thus, typical QDs are either core–shell (e.g., a CdSe core with a ZnS shell) or core-only (e.g., CdTe) structures functionalized with different coatings.

Notably, QDs are inorganic fluorophores that have a size-tunable emission, a strong light absorbance, bright fluorescence, narrow symmetric emission bands, no photobleaching, and a high photostability (the high photostability of QDs relative to dyes allows the real-time monitoring or tracking of intracellular processes over long periods of time varying from minutes to hours). The fluorescence quantum yields of properly surface-passivated QDs are in most cases high in the visible light range (400–700 nm): 0.65–0.85 for CdSe [16, 17], ≤0.6 for CdS [18] and 0.1–0.4 for InP [19, 20]; and high for the visible–near infrared (NIR) wavelength (≥700 nm) emitters CdTe and CdHgTe (0.3–0.75) [21, 22], as well as for the NIR wavelength (≥800 nm) emitters PbS (0.3–0.7) [23, 24] and PbSe (0.1–0.8) [25, 26]. In contrast, organic dyes have fluorescence quantum yields that are high in the visible light range, but moderate in the NIR wavelength range [27, 28]. Typical molar absorption coefficients are 100 000 to 1 000 000 $M^{-1} cm^{-1}$ [29, 30], whereas for dyes the molar absorption coefficients at the main (long-wavelength) absorption maximum range from 25 000 to 250 000 $M^{-1} cm^{-1}$ [31–36]. QDs retain an excellent stability of their optical properties upon conjugation to biomolecules, and can be simultaneously excited by a single light source, as their broad absorption allows free selection of the excitation wavelength and thus a straightforward separation of excitation and emission [10, 14, 37]. Due to their single excitation wavelength, several biomarkers can be probed simultaneously, thereby opening up several multiplexing potentials, including the high-throughput screening of biological samples [38]. This multiplexing potential would be particularly significant for cancer diagnostics and research, where multiple cancer biomarkers can be simultaneously detected, which would in turn help to untangle the associated complex gene expression patterns.

Quantum dots exhibit an electronic structure that is intermediate between bands and bonds, and results in a direct correlation between size and band gap energy (emitted wavelength). Notably, as the size of the QD is decreased, then the band gap energy will be increased – a property which enables QDs to be designed to emit at a wavelength of interest. The absorption and emission spectra for CdSe colloidal

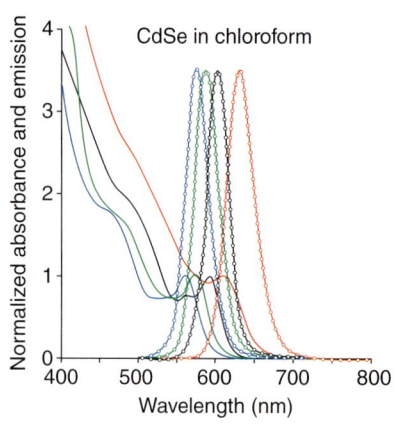

Figure 5.3 Absorption (solid lines) and emission spectra (lines with symbol) of colloidal QDs (CdSe) dispersed in chloroform. Reproduced with permission from Nann, T. (2008) Quantum dots versus organic dyes as fluorescent labels. *Nature Methods*, **5**, 763–775.

Table 5.1 Comparison between the variations in the physical properties displayed by colloidal QDs and organic dyes (in this case fluorescein is considered).

Physical properties	QDs	Organic dyes	Reference(s)
Excitation	Very broad range	Narrow excitation spectra	[39, 40]
Emission bandwidth	20–40 nm	50–100 nm	[39, 40]
Fluorescence lifetime	10–40 ns	Few nanoseconds	[14]
Photostability (upon constant illumination with a 50 mW, 488 nm laser)	Stable for over 14 h	Fluorescein photobleaches completely in less than 20 min	[14, 41]
Molar extinction coefficient	~10^5–10^6 M^{-1} cm^{-1} (for CdSe QDs)	10- to 100-fold smaller than that of CdSe QDs	[13]

Reproduced with permission from Hassan M.E. Azzazy; *From diagnostics to therapy*.

QDs of varying size are shown in Figure 5.3. The narrow emission spectrum without the long tail at red wavelengths which is characteristic of dyes reduces or eliminates spectral cross-talk in detection. Moreover, the large Stokes' shift enables fluorescence signals from the QDs to be easily separated out from the excitation wavelength of the light which is scattered. The other attractive quality of QDs is that their lifetime of emission is longer (hundreds of nanoseconds) compared to that of organic fluorophores. This allows the use of time-gated detection to suppress any autofluorescence, which has a considerably shorter lifetime. The distinct variations in the physical properties of colloidal QDs and organic dyes are listed in Table 5.1.

5.2
Protein Arrays

5.2.1
The Various Types of Protein Array

In order to analyze protein interactions with other molecules, protein microarrays require various types of molecule to be immobilized on their slide surface, to serve as "capture molecules" in the protein microarray assay. Consequently, protein microarrays can be classified into various formats, namely function protein microarrays, detection microarrays, and reverse-phase protein microarrays (RPPMs):

- **Detection (analytical) microarrays:** In protein detection microarrays, the antigens or antibodies, rather than the native proteins themselves, are usually immobilized on a solid support and are further used for the determination of protein abundances in a complex matrix, such as serum (Figure 5.4). Analytical microarrays can also be used for several purposes, for example to assay antibodies or autoimmune diseases, or to monitor protein expression on a large scale.

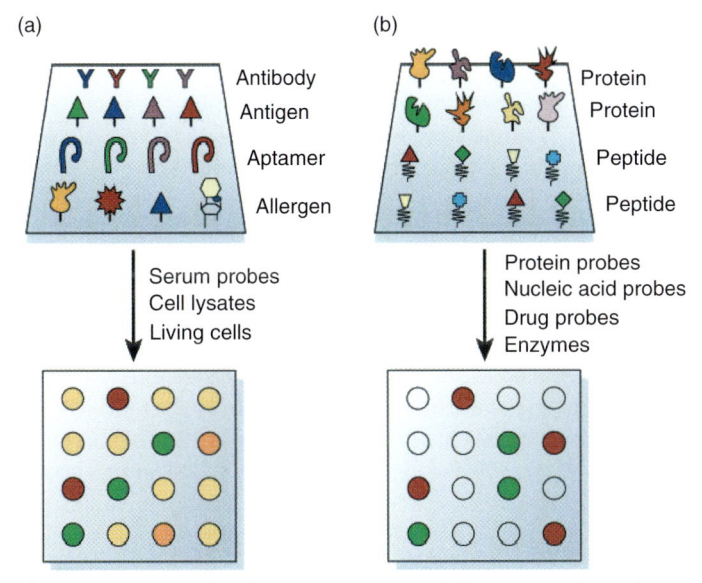

Figure 5.4 (a) An analytical protein microarray depicting different types of ligands, including antibodies, antigens, DNA or RNA aptamers, carbohydrates or small molecules, which are arrayed onto a derivatized surface. These chips are commonly used for monitoring protein expression level, protein profiling and clinical diagnostics; (b) Schematic of a functional protein microarray showing different types of proteins or peptides which, after being purified, are arrayed onto a suitable surface to form the functional protein microarrays. These types of biochip play an important role in drug and drug-target identification, and also in building biological networks. Reproduced with permission from Eric Phizicky (2003), Nature, **422**, 208.

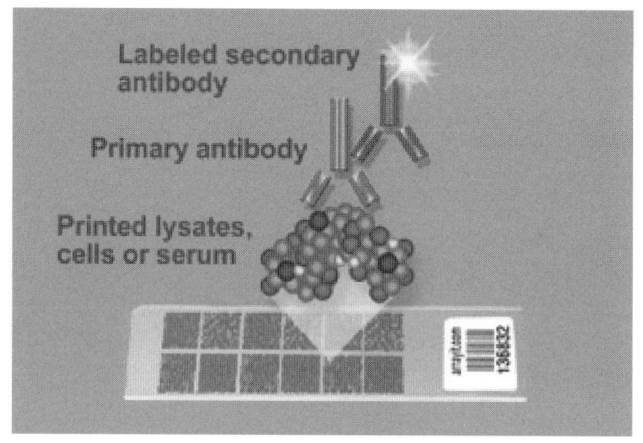

Figure 5.5 Schematic depicting a reverse-phase protein microarray (RPPM) printed on a glass slide. Reproduced with permission from Array IT International Co.

- **Functional protein microarrays:** These types of microarray consist of purified recombinant proteins or peptides, although even an entire proteome may be spotted and immobilized. Functional chips are used to discover additional information and properties concerning a particular protein. These properties include binding strength, biochemical functions and protein–protein interactions. The array is utilized for the parallel screening of a range of biochemical interactions. Various types of study can be conducted with these types of microarray, namely the effect of substrates or inhibitors on enzyme activities, protein–drug or hormone effector interactions, as well as epitope mapping studies.

- **RPPMs:** In this third category of protein arrays, tissues, cell lysates or serum samples may be spotted on the array surface and probed with one antibody per analyte for a multiplex readout. A typical RPPM is depicted schematically in Figure 5.5. RPPMs are related to analytical or detection microarrays, and are used to identify different levels of expression of the proteins present. RPPMs allow the protein to be immobilized in order to be analyzed, instead of the typical protein microarrays which immobilize the antibody probe. It was for this reason that this type of microarray was named "reverse phase." The main advantages of RPPMs over the other types include an ability to run different test samples in each individual array spot, and to require only a single antibody to probe an entire array slide.

5.2.2
Methods of Optical Detection in Protein Arrays Using Labeled Probes

These methods are distinguished as three types:

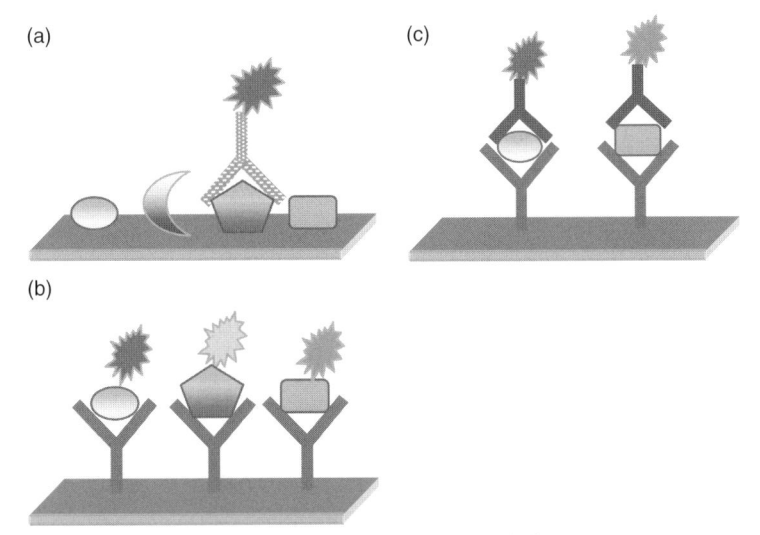

Figure 5.6 Schematics explaining: (a) the direct method of labeling; (b) the indirect method of labeling; and (c) the sandwich method of labeling for detection of protein–protein interaction.

a) **Direct method:** In this method, a mixture of different types of proteins is immobilized on a substrate, and thereafter detection is performed using labeled binding molecules such as antibodies. This type of an array is composed of several sera or cellular lysates from different patients, and contains a complex mixture of proteins. The array is first incubated with one detection protein (typically an antibody) which is labeled for optical readout of the interaction, if any. Direct labeling is mainly performed in two ways: (i) by a radioactive technique, wherein ^{125}I or ^{3}H is used; or (ii) by using organic dyes such as Cyanine and Alexa. Recently, fluorescent dyes have become the method of choice for the labeling and detection of molecules in the microarray format (Figure 5.6a).

b) **Indirect method:** This method is employed when immobilized antibodies are used as capture ligands and probed by labeled proteins. In this approach, a known capture ligand is immobilized on the surface and probed by a labeled complex mixture of proteins. The test sample may be a cellular lysate or a serum in which multiple analytes are measured simultaneously (Figure 5.6b).

c) **Sandwich method:** In this method, the initially immobilized antibody functions as a capture agent for the assayed protein, which is revealed by a recognition with a secondary labeled antibody (Figure 5.6c).

5.3
From Microarrays to Nanoarrays: Why Nanoarrays?

The concept of protein or DNA microarrays is based on the arraying of small amounts of individual proteins, wherein the individual spot size is almost 100–300 µm in size. These microarrays contain protein numbers that vary from a few to a thousand (these are termed "low-to-medium density" microarrays) or slightly larger arrays composed of 10 000 proteins ("low-to-high density" microarrays) which have also been fabricated. However, if a complex proteome such as a cell or a tissue lysate consisting of >100 000 proteins must be detected, then mega dense arrays must be fabricated, and herein lies the role of nanotechnology. Today, nanotechnology-based approaches can be employed for the fabrication of high-density arrays to generate nanosized features (nanometer range) in miniaturized arrays. The key challenge in protein nanoarrays lies in the fabrication and functionalization of these high-density arrays in an efficient manner.

> "The ability to make protein nanoarrays on a surface with well-defined feature size, shape, and spacing should increase the capabilities of researchers studying the fundamental interactions between biological structures (cells, complementary proteins, and viruses) and surfaces patterned with proteins."
>
> Chad A. Mirkin, Northwestern University in Evanston, Illinois, in *Science* Express.

In comparison to microarrays, nanoarrays can hold 10^4–10^5 more features than conventional microarrays and, as a result, the multiplexed and rapid screening of a large number of targets is possible within an individual experiment. Moreover, the total area occupied by a fixed number of targets can be dramatically reduced. Hence, the sample volumes required are very low and a smaller number of target molecules for a given analyte concentration can be detected. This can lead to significantly lower limits of detection (orders of magnitude) than could otherwise be achieved with microarray technology. Finally, nanoarrays can be used to address important fundamental questions pertaining to biomolecular recognition, since biorecognition is inherently a nanoscopic rather than a microscopic or macroscopic phenomenon.

A variety of techniques is available for the immobilization of proteins at the nanoscale. Some of the most widely used methods include soft-lithographic approaches, microcontact printing, nanoimprint lithography, electron beam lithography, focused ion-beam lithography, and dip-pen nanolithography (DPN). With the help of these techniques, patterns of different shapes and sizes can be suitably generated. Moreover, a variety of biological molecules, including DNA [42, 43], peptides [44–46], proteins [47–50], viruses [51–53] and bacteria [54], have been patterned by using either a direct-write or an indirect absorption approach. Yet,

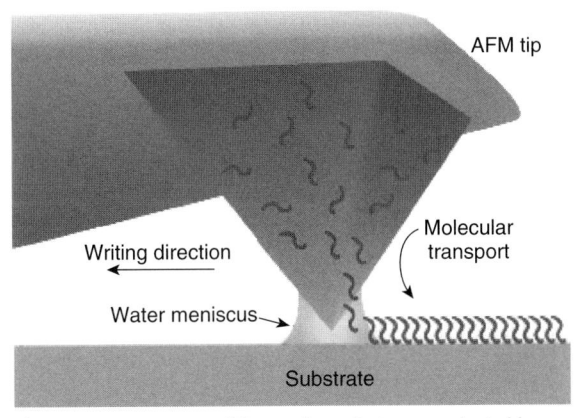

Figure 5.7 Dip-pen nanolithography technique conducted by using an atomic force microscope (AFM) tip.

there remain some persistent key issues which need to be resolved. Initially, all of the features in a nanoarray were functionalized with the same type of a probe, but in the case of protein arrays where the multiplexing of proteins is necessary, each feature must be individually addressed or functionalized with a different type of probe. This has proved to be a daunting task in the case of nanoarrays, and is one of the main problems that require attention, other than equally important issues such as the time required for printing and compatibility with the printing reagents. In this chapter, attention will be mainly focused on nanoarrays fabricated using the DPN technique, a schematic of which is shown in Figure 5.7. Some of the more important requisites for generating multicomponent system-based nanoarrays include new surface analytical tools, and the development of a complementary chemistry for the direct placement of different protein structures of interest onto a surface with nanoscale resolution, without destroying their biological activity.

5.4
Functionalization of QDs for Attachment to Proteins or Other Biomolecules

Quantum dots synthesized by traditional chemical methods are nonpolar and insoluble in aqueous solvents and, therefore, are not compatible with biological systems [55–60]. They are hydrophobic after synthesis because of the coordinating agent. Typically, QDs are synthesized in the presence of hydrophobic inorganic surfactants, such as trioctylphosphine oxide (TOPO). Here, the phosphine oxide functionality chelates the QD surface, while the long alkyl chains allow for solubility in nonpolar solvents such as hexane and toluene. Hence, a polar QD surface must be created before QDs can be used in biology.

For QD biofunctionalization, the QDs are first rendered water-dispersible, and are then bound to biomolecules (Figure 5.8). The binding can be achieved

(a)

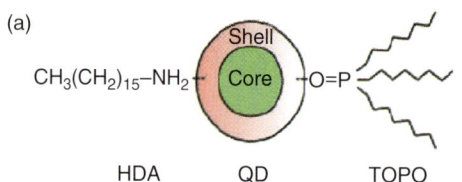

CH₃(CH₂)₁₅–NH₂ — (Shell/Core) — O=P

HDA   QD   TOPO
Hydrophobic ligands, sterically stabilized

(b) Phase transfer and ligand exchange

Electrostatic stabilization
Ligand exchange with small, charged adsorbants[48,58]

Hybrid
Bulky, (partially) charged ligands (polyelectrolytes)[50]

Steric stabilization
Intercalation with bulky uncharged molecules[53,56,57]

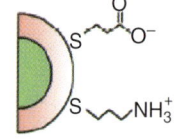

Intercalation with charged surfactants[53,54]

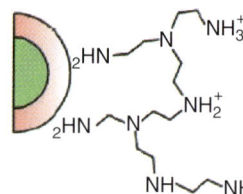

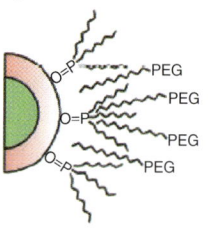

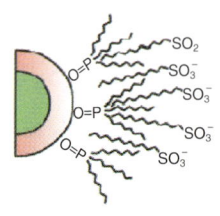

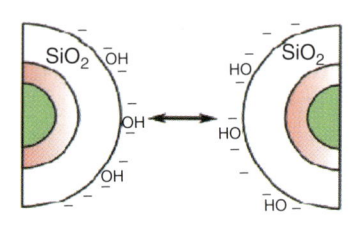

Additional inorganic shells[53]

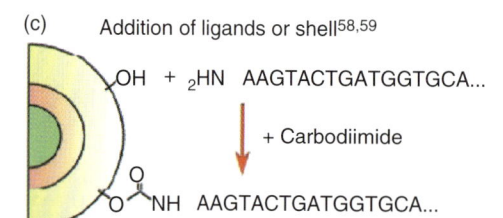

(c) Addition of ligands or shell[58,59]

OH + ₂HN AAGTACTGATGGTGCA...

+ Carbodiimide

O
‖
O–NH AAGTACTGATGGTGCA...

(d) Direct ligand exchange[55]

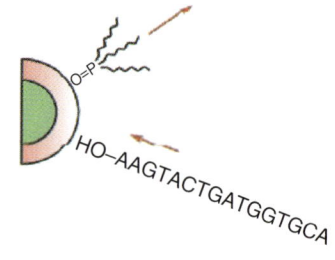

HO–AAGTACTGATGGTGCA

Figure 5.8 Overview of strategies to prepare water-dispersible QDs and QD bioconjugates. (a) QDs bearing hydrophobic ligands after preparation in organic solvent. HDA, hexadecylamine; TOPO, trioctylphosphineoxide; (b) Ligand-exchange strategies to generate water-dispersible QDs. Illustrated are electrostatic colloidal stabilization (left), electrostatic and steric stabilization (middle) and steric stabilization of colloid (right); (c) Coupling of water-dispersible QDs to biomolecules; oligonucleotides are shown here as an example; (d) Alternatively, the QDs bearing hydrophobic ligands can be subjected to direct ligand exchange. Reproduced with permission from *Nature Methods* (2008), **5** (9), 769.

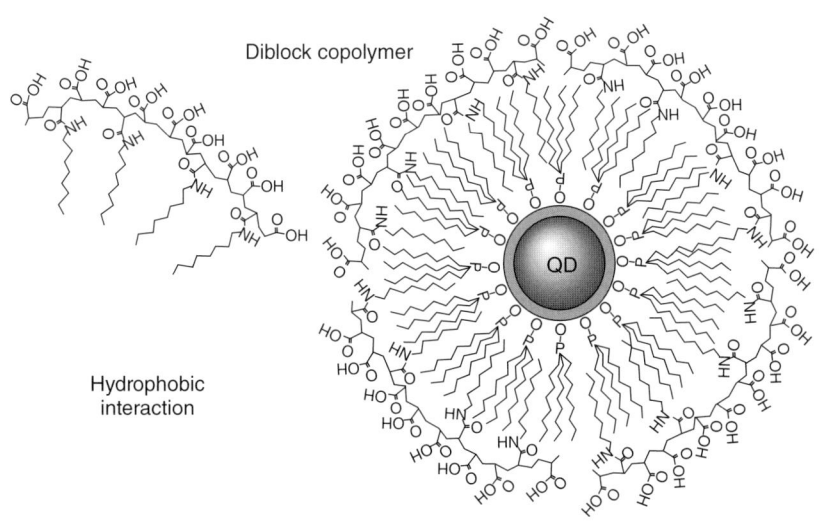

Figure 5.9 QDs coated with amphiphilic diblock copolymers for the fabrication of biocompatible QDs.

electrostatically, via biotin–avidin interactions, by covalent crosslinking (e.g., via carbodiimide-activated coupling between amine and carboxylic groups, maleinimide-catalyzed coupling between amine and sulfhydryl groups, and between aldehyde and hydrazide functions), or by binding to polyhistidine tags [61–64]. Alternatively, ligands present during the synthesis can be exchanged for biomolecules containing active groups on the surface [65]. Such ligands function very well for labeling oligonucleotides such that, during the past few years, amphiphilic di- and tri-block copolymers, typically containing polyacrylic acids, have also been developed for encapsulating QDs via spontaneous self-assembly [66].

In most designs of the amphiphilic polymers, a fraction of the carboxylic acid functionalities in the polyacrylic acid is coupled to hydrophobic alkyl chains, which stabilize the TOPO-capped QDs (Figure 5.9). The remaining carboxylic acids in the polymer provide solubility in water, and can be utilized as "chemical handles" for conjugation to primary amines in proteins through water-soluble crosslinking reagents such as 1-ethyl-3-(3-dimethylaminopropyl) carbodiimide (EDAC).

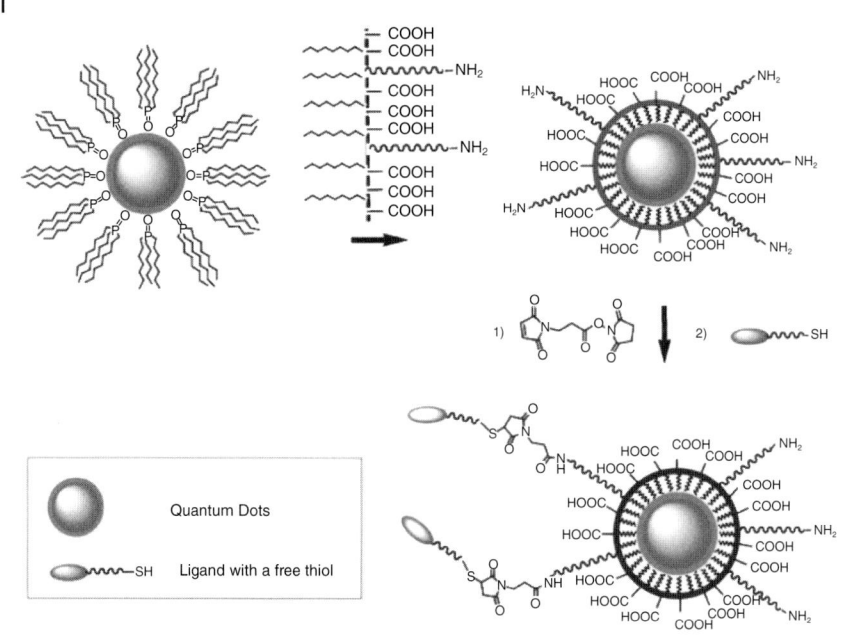

Figure 5.10 Maleimide-functionalized QDs for conjugating thiol-containing ligands. TOPO-stabilized QDs are coated with a primary amine functionalized triblock amphiphilic copolymer for producing water-soluble QDs, which facilitate further conjugation to ligands with free thiols through bifunctional crosslinkers.

Recently, free primary amines have been incorporated in an amphiphilic polymer for QD coating that can be further modified with standard N-hydroxylsuccinimyl ester-maleimide containing bifunctional crosslinkers. These maleimide-functionalized QDs can be conjugated to any macromolecule selectively through engineered cysteines or thiols on the target ligands (Figure 5.10) [67]. Currently, only a few standard protocols for labeling biomolecules with QDs are available, and the choice of suitable coupling chemistries depends on surface functionalization. It is difficult to define general principles here, because the QD surfaces are unique to a large extent, depending on the preparation procedure. Consequently, research groups using commercially available QDs must obtain sufficient knowledge of the surface functionalization of QDs before they can be used as biolabels.

The labeling of biomolecules such as peptides, proteins or oligonucleotides with a fluorophore, requires suitable functional groups for covalent binding or for the noncovalent attachment of fluorophores. The advantage of organic dyes is that the small size of such labels minimizes the possible steric hindrance, which may interfere with the functioning of the biomolecules. Organic dyes can be easily conjugated to biomolecules, as many established labeling protocols are available commercially. Although several fluorophores can be attached simultaneously to a

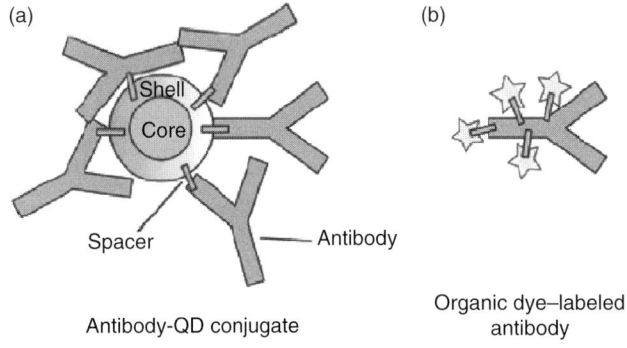

Figure 5.11 Schematics of a QD–antibody conjugate and a dye-labeled antibody, reflecting the proportions of the components. Reproduced with permission from *Nature Methods* (2008), **5** (9), 769.

single biomolecule so as to maximize the fluorescence signal, site-specificity may at the same time be a problem in the case of organic dyes. In particular, the higher density of the dye molecules can lead to fluorescence quenching, depending on the structure of the dye molecules, their charges, and their hydrophilicity [27, 34, 35].

Although, in contrast, there is a clear lack of well-defined methods for labeling biomolecules with QDs, some of the challenges encountered with organic dye biofunctionalization also apply to QDs, the main difference being that QDs do not exhibit fluorescence quenching as do high-density organic dyes. Notably, QDs may form aggregates when the surface chemistry is not perfectly optimized, and it may also be difficult to control the orientation of the biomolecules when several biomolecules are conjugated to a single QD [68] (see Figure 5.11b). This may have an adverse effect not only on the function of the biomolecule but also on the colloidal stability of the QD.

In general, CdSe/ZnS (core–shell) QDs are used for the labeling of protein microarrays, as their dangling bonds may be passivated by capping agents. QDs, after suitable surface functionalization, can be directly conjugated to the antibodies specific to the selected markers. Alternatively, streptavidin-coated QDs may be conjugated to the biotinylated detector antibodies.

5.5
Fabrication of Protein Biochips

5.5.1
Printing of QDs-Conjugated Protein Microarray Biochips

One method of fabricating QDs-labeled protein microarray samples is shown schematically in Figure 5.12. In this case, a suitably functionalized glass slide (e.g.,

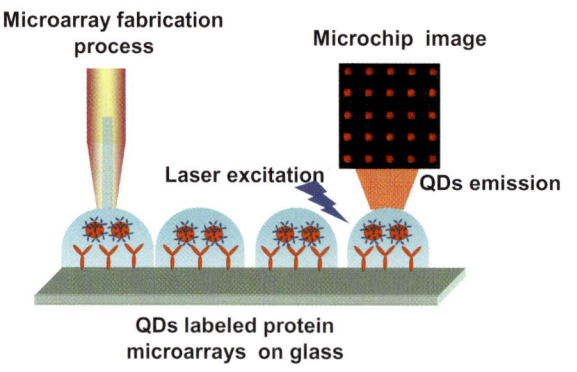

Figure 5.12 Schematic diagram depicting the fabrication process of QD-conjugated cancer protein microarrays and their readout.

aldehyde-or amine nitrocellulose-coated) [69]) can serve as a substrate for the microarray fabrication process.

In this procedure, spots of protein antigens are first arrayed onto the functionalized glass substrate after having diluted them with phosphate-buffered saline (PBS) containing glycerol. A high-precision contact printing-type robotic microarrayer is used for the microarray spotting process, with stealth pins of different sizes being used to fabricate the microarrays. The spotting process is followed by incubation of the patterned samples for an appropriate time period, at room temperature, inside a humid chamber, after which the samples are rinsed with deionized water, and finally dried with a stream of nitrogen gas. The incubation period allows coupling between the antigens and the functionalized groups present on the substrate surface. In order to block any residual free functional groups present on the antigen-patterned substrate, Gokarna et al. [70] used a solution of bovine serum albumin (BSA) in PBS, this being spread over the patterned area. The blocking step was followed by incubation of the samples at room temperature for an appropriate time, after which they were re-washed with PBS or deionized water and dried with nitrogen gas. When conducting protein–protein interactions, either the QDs-conjugated antibodies can be re-spotted over the initial arrayed area, using a microarrayer, or the QDs-antibodies solution can be spread over the antigen-patterned area and allowed to incubate. The incubation is followed by re-washing with PBS or PBS–Tween 20 (PBST) and deionized water, and final drying with nitrogen gas.

5.5.2
Nanoarray Protein Chips Using QDs

The nanoarray design must allow an efficient fabrication of high-density nanoscale arrays compatible with sensitive read-out systems. Wingren et al. [71] reported that

5.5 Fabrication of Protein Biochips | 195

Table 5.2 An overview of currently available types of protein nanoarray design.

Design features	Features (features size/array densities/ proof-of-concept applications)	Reference(s)
Planar arrays	From 10 to 350 nm features, up to Φ 2 μm-sized dots 1×10^4 dots mm^{-2} to 1×10^6 spots mm^{-2} Detection of antigen (antibody arrays)	[72–77]
Well-based arrays	6 to 8 nl reaction volumes 25-well array; enzymatic assays	[78]
Nanovial arrays	100 nl reaction volumes, 12 × 8 nanovial array Enzymatic assays	[79]
Attovial arrays	6 (Φ 200 nm) to 4000 al (Φ 5 μm) -sized vials 225 vials mm^{-2} Tentative densities of 90 000 to 225 000 vials mm^{-2} Detection of antigen (antibody arrays)	[72, 80]
Nanowire arrays	nm scale Single biosensors to 200 individually addressable devices per array Antigen (e.g., cancer biomarker or pathogen) detection (antibody arrays)	[81–85]
Random arrays	nm scale – Antigen detection (antibody arrays)	[86]
Bead arrays	900 nm-sized features (using 1 mm beads) (dependent on the size of bead applied) Enzymatic assay	[87]
Nanoparticle arrays	100 nm-sized features 300 spots per sensing surface Antigen detection (antibody arrays)	[88]
Cantilever arrays	500 μm long and 100 μm wide (common size of a single cantilever). Mainly eight cantilever silicon arrays Antigen detection (e.g., proteins or pathogens). Detection of specific protein conformations, protein–DNA interactions, and protein–ligand interactions (protein arrays)	[89–98]

the precise choice of nanoarray design would depend on an intricate combination of several factors, including the choice of nanopatterning technique, the properties of the substrate and the probes, and compatibility with the detection method. To date, proof-of concept studies have been performed for a wide range of first-generation nanoarray designs (Table 5.2), including planar arrays, well-based arrays [78], nanovial arrays [79], attovial arrays [80, 99], nanowire arrays [81–85], random arrays [86], bead arrays [87], nanoparticle arrays [88], and cantilever arrays [89–98].

A range of protein probes has been used in these studies, including enzymes, polyclonal and monoclonal antibodies, recombinant single-chain Fv (scFv) antibody fragments, as well as various model proteins such as biotin–streptavidin and integrin $\alpha_v\beta_3$–vitronectin.

The fabrication of protein nanoarrays using DPN can be achieved in a variety of ways. For example, Lee et al. [100] used nanoarrays of the anti-p24 antibody to screen for the human immunodeficiency virus-1 (HIV-1) p24 antigen in serum samples. For this, the antibody nanoarrays were fabricated using DPN-patterned 16-mercaptohexadecanoic acid (MHA) dot features as small as 60 nm (10 × 10 spot array) on a gold thin film as templates for antibody immobilization. Notably, the large spacing between features improved the ability to locate the original pattern after reaction with the biomolecules or gold nanoparticle probes. At pH 7.4, the MHA was deprotonated, such that the nanofeatures were negatively charged. In order to minimize the nonspecific binding of proteins on the inactive portions of the array, the areas surrounding the MHA-patterned features were passivated with poly(ethylene glycol) (PEG)-alkythiol (11-mercaptoundecyl-tri(ethylene glycol)), and this was followed by copious rinsing, first with ethanol and then with Nanopure water. After passivation, mouse monoclonal antibodies to the HIV-1 p24 antigen (anti-p24) were immobilized on the patterned MHA dot features by immersing the template in a solution containing the anti-p24 IgG for 1 h. Any unmodified MHA features were passivated with BSA (10% solution in 10 mM PBS) to prevent unwanted binding.

The nanoarrayed sample, which consisted of anti-p24 patterns, was next immersed in a plasma sample containing HIV-1 p24. On capturing p24, the anti-p24 features increased in height by 2.3 ± 0.6 nm (measured using atomic force microscopy; AFM). This height increase could be further amplified by sandwiching the captured p24 protein with anti-p24-functionalized gold nanoparticle probes (Figure 5.13). Most importantly, these investigations showed that nanoarray-based assays could exceed the detection limit of conventional enzyme-linked immunosorbent assays by orders of magnitude.

Gokarna et al. [70] used aldehyde-functionalized silicon substrates for the screening of prostate-specific antigen (PSA), which serves as a biomarker for prostate cancer. For this, aldehyde functional groups were formed on silicon by the self-assembled monolayer technique, whilst for the patterning of nanoarrays a special type of polymer-coated silicon cantilever was fabricated by the self-assembly technique. Patterns of PSA were created on the aldehyde-functionalized silicon substrate by using DPN, followed by a sufficient incubation time. In the next step, functionalized PEG molecules were used as a blocking layer, followed by incubation for 1 h, and washing and drying of the sample. In order to monitor the protein–protein interactions, one drop of colloidal QDs-conjugated proteins was placed over the protein-arrayed area, and allowed to incubate for a few hours. After allowing sufficient time for protein–protein interactions to be effected, the sample was washed with deionized water and dried with nitrogen gas. During this process AFM topographic imaging was carried out at several points to observe the changes in spot height before and after protein–protein interaction.

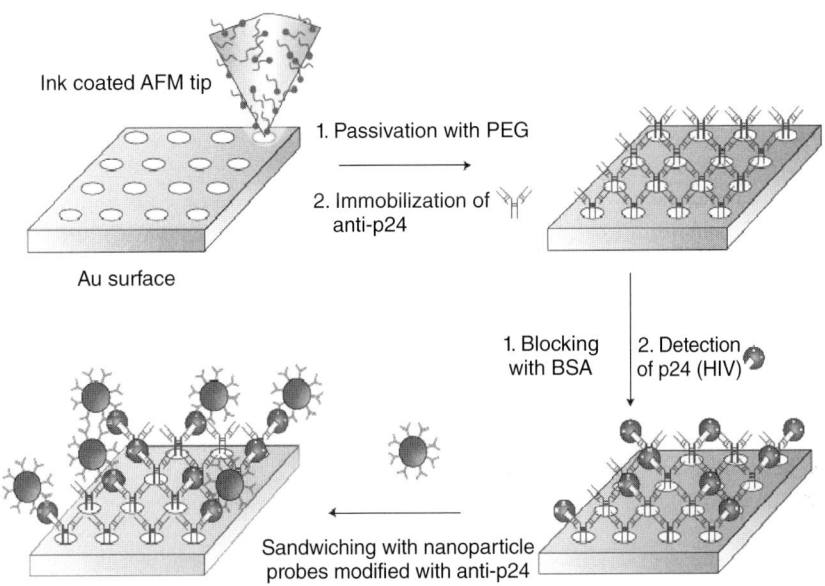

Figure 5.13 Schematic representing the detection of HIV-1 p24 antigen using a nanoarray of anti-p24 antibody. DPN-generated nanoarrays of MHA are used to immobilize anti-p24 antibodies on a Au substrate. The bare Au regions were passivated with 11-mercaptoundecyl-tri(ethylene glycol) (PEG), and nonspecific protein adsorption was minimized by blocking with bovine serum albumin (BSA). The presence of HIV-1 p24 in patient plasma is probed by measuring the height profile of the anti-p24 antibody array. The height increase signal as a result of specific antigen–antibody binding is further amplified by a nanoparticle sandwich assay.

5.6
QDs Probes in Clinical Applications

Today, interest in the detection and analysis of disease biomarkers constitutes a major research area, with proteomics and related technologies being widely implemented to optimize the detection of these biomarkers. Currently, intensive efforts are under way to extract information from proteins that can be used for the early diagnosis of diseases such as cancer. The physico-chemical malleability and high surface areas of colloidal QDs make them ideal candidates for developing biomarker harvesting platforms. Consequently, bridging together the unique features of QDs and protein microarrays may lead to the design of ultrasensitive, robust, and feasible assays for the early detection of cancer.

5.6.1
Detection of Disease (Cancer) Biomarkers Using QD Labels

Kim et al. [101] have attempted to assess the efficacy of QDs in the quantitative analysis of antibody–antigen interaction by utilizing angiogenin (ANG) and QDs

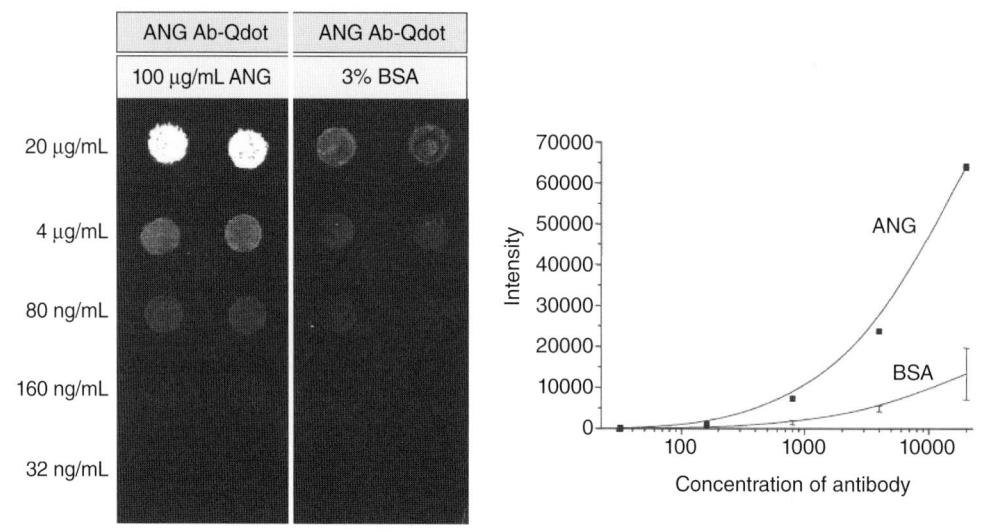

Figure 5.14 Interaction of ANG with anti-ANG antibody-QD on a ProteoChip. The image on the left-hand side is the fluorescence emission observed after the protein–protein interaction occurs between ANG and QD-labeled anti-ANG and BSA and QD-labeled anti-ANG. The spectra on the right-hand side depict the variation in fluorescence intensity as a function of the antibody concentration. The fluorescence intensity is observed to increase very rapidly in the case of ANG proteins with increasing concentration of antibody, but shows a very gradual increase in BSA. In these experiments, BSA was used as a control protein.

labeled with an anti-angiogenin (anti-ANG) antibody to monitor protein–protein interactions. In these studies, the polyclonal antibody was labeled with QD or Cy5, and the signal-to-noise ratios (SNRs) of the Cy5-labeling and the QD-labeling methods were also compared. In particular, the SNR plays a critical role in the study of quantitative biomarker analysis, expression profiling, and pathogen detection. In order to assess the applicability of QDs in a protein chip detection system, antibody–antigen interaction assays were conducted using the QD-labeling method on a ProteoChip system. Here, the ANG protein was initially immobilized in a well-on-a-chip ProteoChip base plate, an then incubated with the anti-ANG antibody labeled with QD at different concentrations ranging from $20\,\mu g\,ml^{-1}$ to $32\,ng\,ml^{-1}$ (Figure 5.14). The image analysis demonstrated a good correlation between the fluorescence intensities and QD-labeled ANG antibody concentrations (Figure 5.14). Clearly, the QD-labeled ANG antibodies had been able to bind specifically to ANG, as compared to BSA.

Recent studies relating to protein–protein interactions have also been conducted with PSA and QDs-conjugated PSA antibodies [70]. In these experiments, BSA was utilized as a control sample, and the biochips were fabricated in order to study the specificity of the protein–protein interactions in QDs-conjugated proteins. Another study target was to determine the effect of incubation time necessary for these interactions to occur. Both, PSA and BSA were arrayed on the same

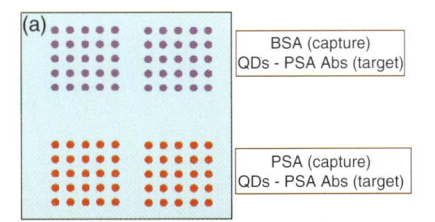

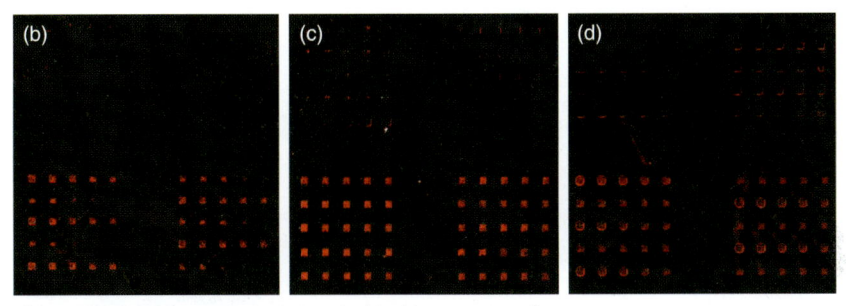

Figure 5.15 (a) Schematic depicting the biochip structure of protein–protein interaction in PSA and BSA microarrays fabricated with a variation in the incubation time and without any blocking layer; (b–d) Fluorescence images obtained by a microarray scanner depicting the emission arising from the QD-conjugated protein microarrays incubated for time periods of (b) 1 h, (c) 2 h, and (d) 5 h, respectively.

aldehyde-functionalized glass substrates to study the effect of incubation time simultaneously on both proteins. The schematic in Figure 5.15a shows the manner in which these samples were fabricated; here, the top arrays correspond to BSA antigens, while the lower arrays depict PSA interacting with labeled PSA antibodies. The incubation time of the QDs-conjugated PSA antibodies, after being arrayed onto PSA or BSA antigens, varied from 1 h to 5 h, respectively.

The emission arising from the samples incubated for periods of 1, 2, and 5 h is depicted in Figure 5.15b–d. After an incubation period of 1 h some of the microarray spots in the PSA microarrays had disappeared, although this may be attributed to an insufficient time for protein–protein interaction. An incubation time of 2 h was superior to either 1 h or 5 h. The control proteins (i.e., BSA microarrays) in all three samples showed a very weak fluorescence emission compared to the PSA microarrays, which confirmed the high specificity of interaction of QDs-conjugated PSA antibodies with PSA. Notably, in these samples the nonspecific binding effect was also observed to be quite negligible.

A novel nano- and micro-integrated protein chip that incorporates the advantages of both microfluidic networks (μFN) and QDs, and successfully demonstrates the combined advantages of using both QDs probes and μFN, has been fabricated by Yan et al. [102]. Here, the carcinoma embryonic antigen (CEA) was used as a model assay target, while a μFN was used to construct the protein chips;

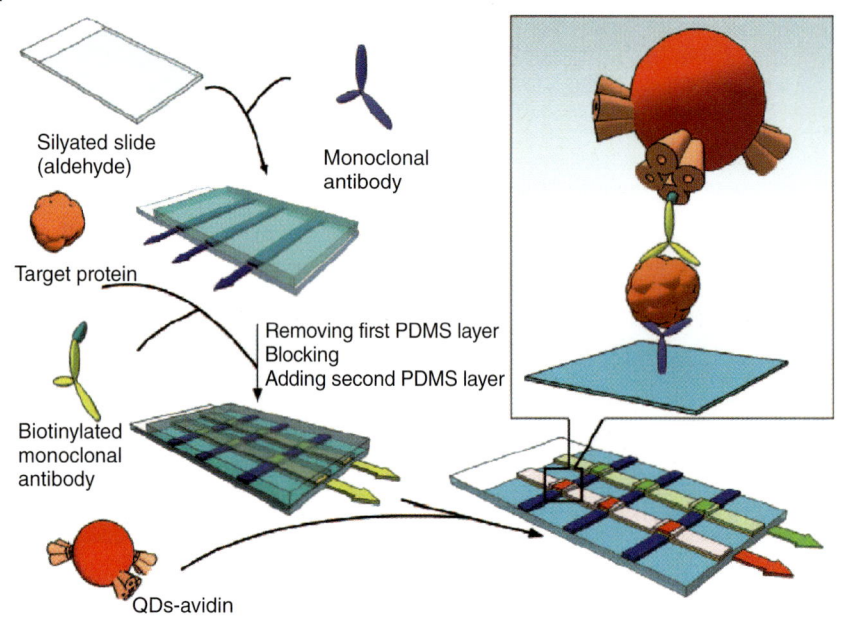

Figure 5.16 Schematic illustration of protein chips based on QDs probes and microfluidic network. Reproduced with permission from Ref. [102] (Wang, et al., (2008), Nano Res. **1**, 490).

this allowed a straightforward patterning of proteins and subsequent biomolecular recognition. More recently, microfluidic technology has emerged as a promising approach, offering convenient operation for both protein immobilization and subsequent binding assays [103–108].

In these experiments, capture antibodies (monoclonal antibody, mAbs) were initially passed through the microfluidic channels in the first polydimethylsiloxane (PDMS) layer, which resulted in their immobilization on the aldehyde-activated slide (see the scheme in Figure 5.16). The remaining area of the slide was blocked with BSA to prevent nonspecific adsorption in subsequent steps and thus reduce the background signal. After this step, the first PDMS layer was peeled off and a second PDMS layer attached to the slide, resulting in a crossed channel configuration. Target proteins (CEA solutions) were then passed through these channels, crossing the original capture mAb-bound channels, and forming the μFN. It should be noted that the target proteins were captured in the cross-section of the μFN. After binding, biotinylated detection antibodies (mAbs) were passed through the same channels, wherein they bound to the target protein. This step was followed by passing the signaling probe, namely, avidin-coated CdTe/CdS core–shell QDs, through the channels that illuminated the protein-binding assays.

This set-up was applied to the molecular diagnostics of a tumor marker (CEA) in a sandwich-based configuration, and avidin-coated QDs (λ_{em} = 579 nm) were used as signaling probes. Dose-dependent experiments were conducted over a wide concentration range of CEA covering six orders of magnitude, from

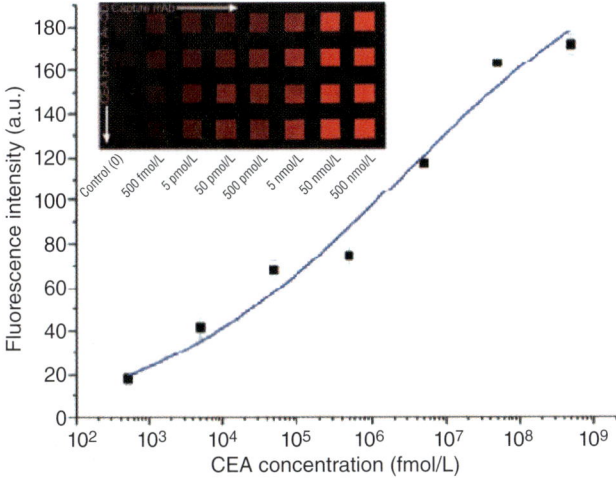

Figure 5.17 Dose–response calibration curve for CEA based on the QDs probe (λ_{em} = 579 nm) and microfluidic network chip. The standard curve was fitted by a logistic model (Equation: $y=225.18\pm29.21+[-3.43\pm12.45-225.18\pm29.21]/[1+(x/2\,529\,665.20\pm204\,3790.60)\,0.26\pm0.06]$, $R^2 = 0.99$). The background-subtracted fluorescence intensity was used as the index. Insert (upper left): fluorescent image of parallel detection of CEA samples (Concentrations: 0.500 fmol l^{-1}, 5 pmol l^{-1}, 50 pmol l^{-1}, 500 pmol l^{-1}, 5 nmol l^{-1}, 50 nmol l^{-1}, 500 nmol l^{-1}) in four parallel microfluidic channels. The arrows indicate the flow directions in the microfluidic system.

500 nmol l^{-1} to 500 fmol l^{-1}. Based on the dose–response data obtained (Figure 5.17), the presence of protein–protein interactions could be proved in these samples, and the limit of detection was identified as 500 fmol l^{-1}. In contrast, protein chips in which avidin–FITC (fluoroscein isothiocyanate) was used as the signaling probe showed a detectable signal only when the target concentration reached 500 nmol l^{-1}. Thus, protein–protein interactions were successfully observed in CEA proteins, with a high detection specificity.

5.6.2
Certain Limitations in QDs Labeled Microarrays: Nonspecific Binding Effects

Traditionally, BSA is used as a blocking agent during the fabrication of organic dye-labeled protein biochips. However, when it is used as a blocking agent in QDs-conjugated protein chips, on occasion it tends to show nonspecific interactions with the QDs, and to a high extent. Gokarna *et al.* [70] fabricated two series of protein microarray samples using colloidal QDs as labels for the purpose of optical readout. In one series, BSA was used as a blocking agent during protein chip fabrication, while the other series was fabricated in the absence of BSA. In the BSA-positive series a high degree of nonspecific interactions was observed (Figure 5.18). Any unconjugated or bare QDs also interacted with PSA or BSA as capture proteins, with such interactions mainly observed to arise due to the

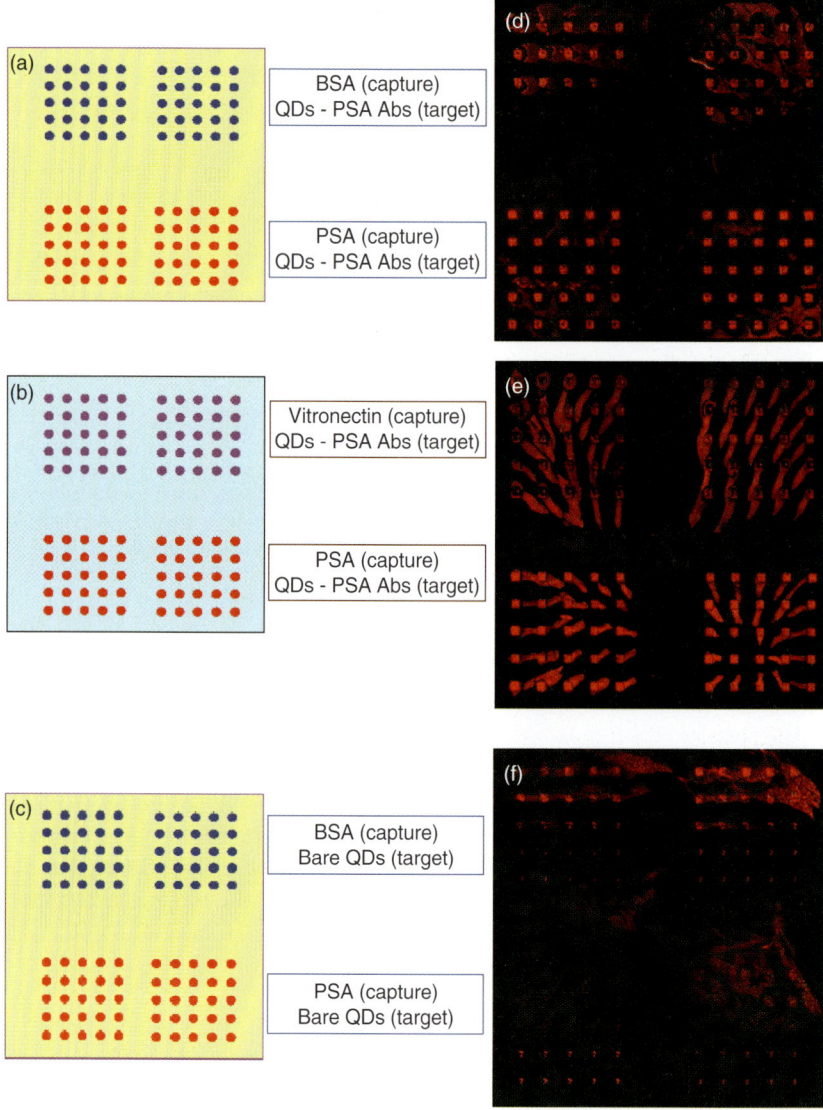

Figure 5.18 Investigation of the nonspecific binding effects (BSA used as blocking layer) in biochips wherein different control proteins and bare QDs were used as control samples. (a–c) Schematics showing various types of biochips which were fabricated. The fluorescence emission arising from the interaction of PSA with QDs-conjugated PSA antibodies, is compared with the emission due to the interaction of (d) BSA with the conjugated QDs and (e) vitronectin with the conjugated QDs; (f) Emission arising due to the nonspecific binding effects when bare (nonconjugated) QDs were used as the target material.

interaction or adherence of BSA with QDs, as had been confirmed previously [109, 110]. In contrast, those samples fabricated by totally eliminating the BSA blocking step exhibited a high specificity of protein–protein interactions (Figure 5.19).

In a study reported in *Biochip Journal* [101], Kim *et al.* explored the applications of QDs in protein microarray techniques, using angiogenin (ANG) and an anti-ANG antibody on a ProteoChip. For this, the nonspecific binding ability of the QD material was evaluated by assessing the SNR of the QDs, and comparing it with that of Cy5, a conventional fluorescent dye. The saturated signal intensity of the scanned images was 65 535. When the SNR was calculated from 65 000 to 5000, the ratio of the QD-labeled antibody was higher than that of the Cy5-labeled antibody, and was also invariable, regardless of the antibody concentration. However, the SNR of the Cy5-labeled antibody fluctuated in a concentration-dependent manner (Figure 5.20). These data indicated that the QD labeling system was significantly less nonspecific than the Cy5 labeling system, and proved that this technique could be applied to solid-phase biomarker assays on protein microarrays, as well to protein expression profiling on an antibody microarray.

5.6.3
QDs versus Organic Dyes in Protein Microarrays

Yan *et al.* [102] tested the sensitivity of their microfluidic-based protein microarray system by using different concentrations of capture antibodies (mouse IgG), with either avidin–QDs or avidin–FITC as the signaling probe. In addition, a biotinylated polyclonal antibody (goat anti-mouse IgG) was employed as a ligand which could bind to both mouse IgG and avidin. As shown in Figure 5.21, the green-colored avidin–QDs probe (λ_{em} = 554 nm) exhibited a concentration-dependent change in fluorescence intensity over the range 32 pmol l^{-1} to 3.2 μmol l^{-1}. In contrast, the organic dye avidin–FITC [111] afforded significantly lower fluorescence intensities. In the latter case, the signal could not be distinguished from the background when the target concentration was below 320 nmol l^{-1} (four orders of magnitude higher than the corresponding value for QDs).

In contrast, Kim *et al.* [101], when fabricating protein arrays in a Well-on-a-Chip, found the Cy5-labeled antibodies to show a lower limit of detection than did the QD-labeled antibodies (Figure 5.22), but attributed such variation to the size difference between QD and Cy5. Typically, the diameter of the QD–antibody was ≥20 nm, but the Cy5-labeled antibody was much smaller. The limited surface area of the Well-on-a-Chip might allow less binding of the QD-labeled antibody (due to its larger size) when compared to the Cy5-labeled counterpart, which indicated that the QD-labeled system was less sensitive than the Cy5-labeled system. However, it was stressed that the fluorescence intensity of QD was not necessarily less sensitive than that of Cy5, on a molecule-to molecule basis. It was also assumed that these results might be attributable, at least in part, to differences in the SNRs of the QD and Cy5 labeling methods.

With regards to the photostability of colloidal QDs compared to organic dyes in protein microarrays, the latter proved to be highly stable even after continuous

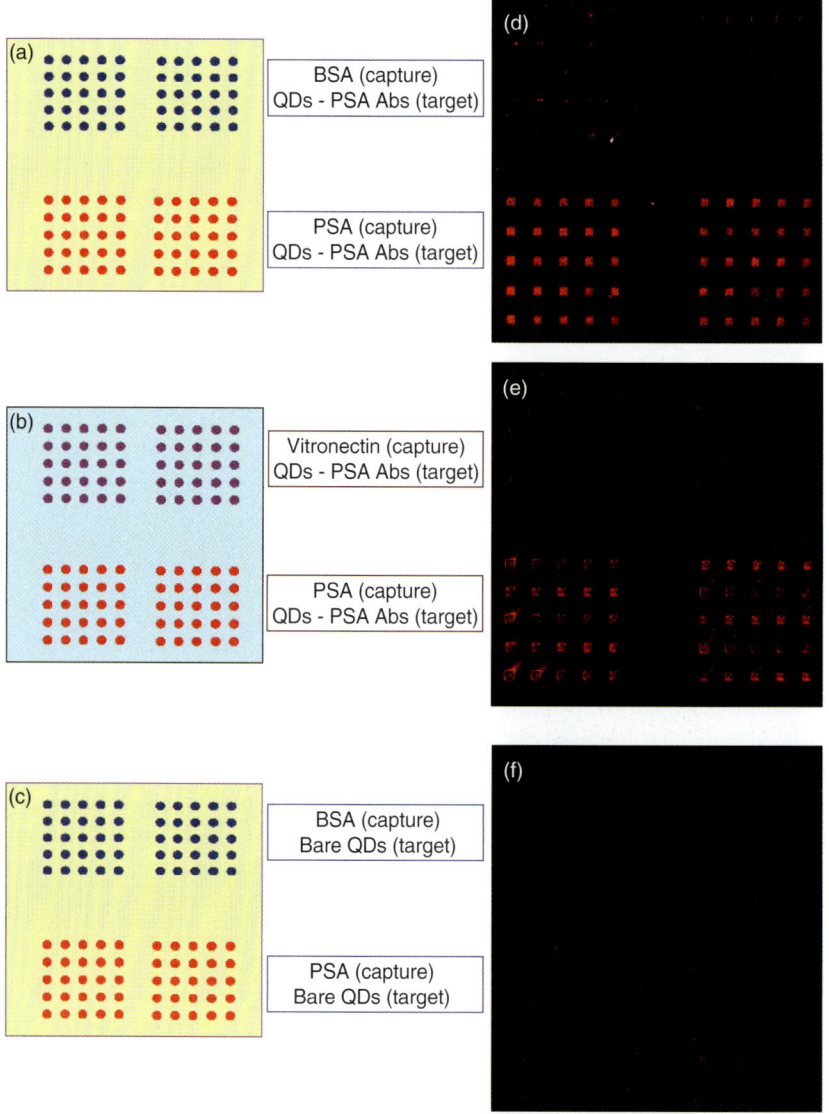

Figure 5.19 (a–c) Biochips fabricated wherein different control proteins and bare QDs were used as control samples. These biochips were fabricated without using any blocking layer. The fluorescence emission arising from the interaction of PSA with QDs-conjugated PSA antibodies, is compared with the emission from the interaction of (d) BSA with the conjugated QDs and (e) vitronectin with the conjugated QDs; (f) Total absence of emission in both PSA and BSA microarrays when bare (non-conjugated) QDs were used as targets.

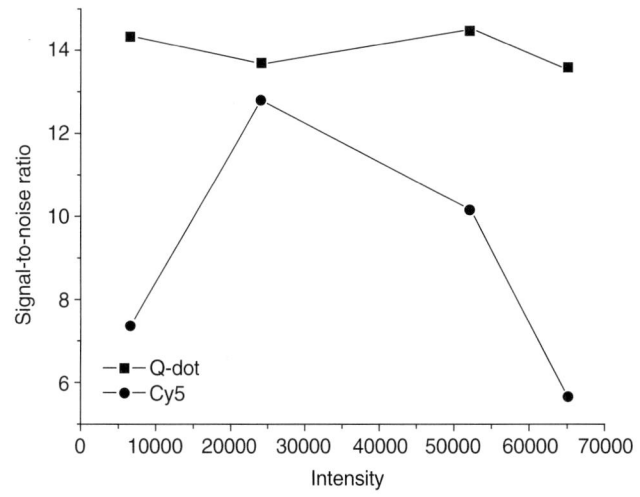

Figure 5.20 Comparison of signal-to-noise ratio (SNR) between Cy5 and QD on the ProteoChip. SNRs were calculated via the following equation: signal = V_s, noise = V_n; SNR = $20\log 10\,(V_s/V_n)$.

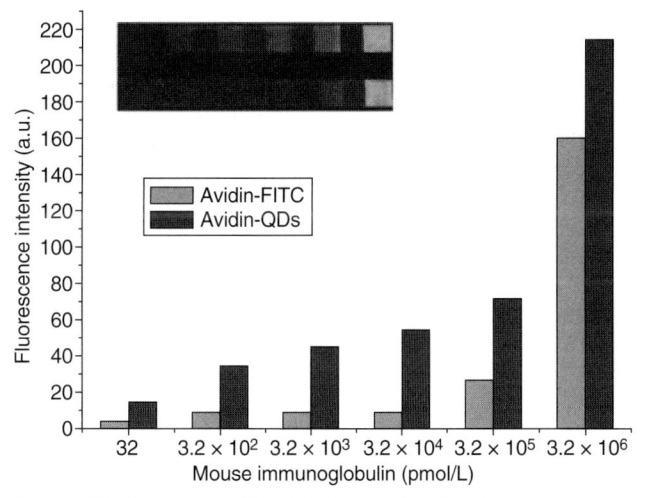

Figure 5.21 Comparison of the brightness of avidin-QDs and avidin-FITC as signal probes. The inset shows fluorescence images of the microfluidic network arrays using the two fluorescent probes with the target protein concentration range from $3.2 \times 10^{-6}\,\mathrm{mol\,l^{-1}}$ to $3.2 \times 10^{-11}\,\mathrm{mol\,l^{-1}}$, respectively (upper row = avidin-QDs; lower row = avidin-FITC).

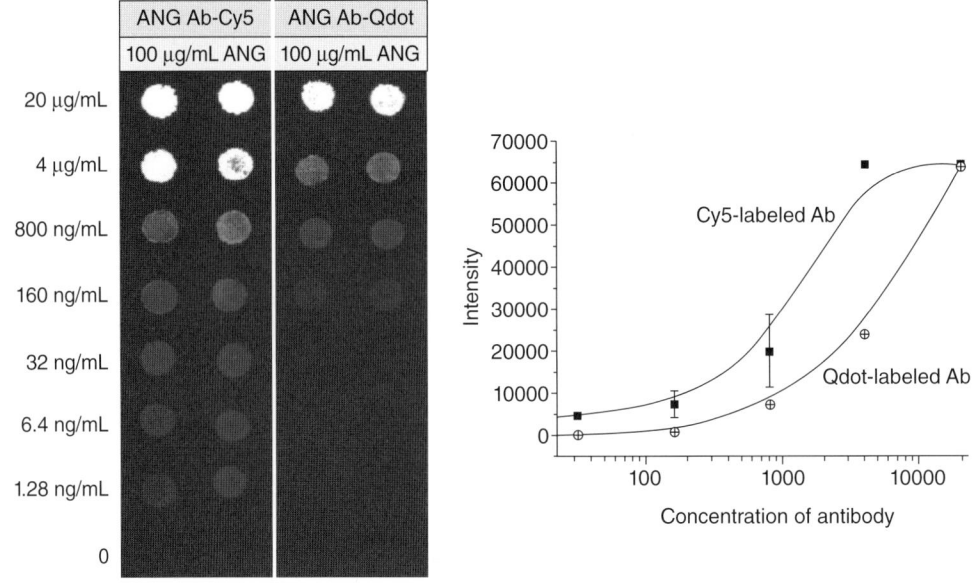

Figure 5.22 Comparison of detection efficiency between Cy5 and QD on the ProteoChip. Cy5-labeled ANG antibodies or QD-labeled ANG antibodies were used for analyses of antibody–antigen interactions on a ProteoChip immobilized with ANG.

(30 min) exposure to laser excitation, as reported by Yan et al. [102] (for further information, see Section 5.6.4).

5.6.4
Multiplexing in Protein Microarrays Using QDs

The ability of QDs to be excited by a single source makes them a unique label for multiplexed experiments, where different species of QDs can simultaneously track various biomarkers and unlike organic dyes, be excited with the same source.

It is well known that different-colored QDs can be excited with a single wavelength due to their broad excitation bands. As shown in Figure 5.23, protein chips labeled with QDs of two different colors can be effectively excited with a single laser source, implying that QDs-based protein chips allow convenient multiplex protein detection [102]. In addition, QDs also showed a high stability towards photobleaching. When Yan et al. compared the photostability of FITC (A) and two types of QD (B) and (C) (with λ_{em} of 554 nm and 579 nm, respectively) on protein chips, it was clearly proven that the fluorescent signals from the organic dye, FITC (A1 in Figure 5.23) were largely diminished after irradiation with blue light (λ = 450–490 nm) for 30 min (A2 in Figure 5.23). In contrast to the organic dyes, the signals from both the types of QDs (B1, C1 before continuous laser

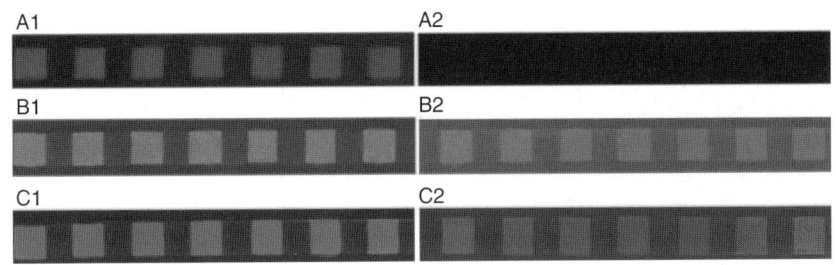

Figure 5.23 Photostability of avidin-QDs and avidin-FITC. A1: fluorescence images of microfluidic arrays using avidin-FITC probes. B1 and C1: fluorescence images of microfluidic arrays using two avidin-QDs with emission bands at 554 nm and 579 nm, respectively. A2, B2, and C2: corresponding fluorescence image after 30 min photobleaching. In all experiments, mAbs were directly immobilized on the protein chip; biotinylated secondary antibodies, avidin-QDs or avidin-FITC probes were associated on the modified chip by flowing through the channels. Reproduced with permission from Ref. [102] (Wang et al., (2008), Nano Res. 1, 490).

irradiation; B2, C2 after 30 min of irradiation) were observed to decrease by only a small degree.

Zajac et al. [112] investigated the multiplexing ability of QDs to detect six different cytokines in protein solution. (Cytokines are proteins that have the ability to stimulate or inhibit cell growth, regulate cell differentiation, induce cell chemotaxis, and modulate the expression of other cytokines.) By using this approach, cytokines such as tumor necrosis factor-alpha (TNF-α), interleukin (IL)-8, IL-6, macrophage inflammatory protein 1 beta (MIP-1β), IL-13 and IL-1β, could be detected down to picomolar concentrations, thus demonstrating the extreme sensitivity of this detection system using QDs. For this, nitrocellulose slides spotted with capture antibodies against TNF-α, IL-8, IL-6, MIP-1β, IL-13 and IL-1β were incubated with eight mixes of fourfold step cytokine dilutions in buffer, and detected with a biotinylated Ab/streptavidin–QD655 complex. Plots of data obtained through analysis of the fluorescent images showed the sensitivity of the assay to be in the femtogram range. In a control experiment, the cytokine concentration was zero; consequently, whilst fluorescence was not observed from the control spots, it was emitted from spots incubated with a cytokine. Thus, it was concluded that the QD-based microarray assay allowed the detection of cytokines with different sensitivities, varying from the $pg\,ml^{-1}$ to $ng\,ml^{-1}$ range. The assays for IL-8 and TNF-α showed the best sensitivities, with a detection limit between 10 and $100\,pg\,ml^{-1}$; by comparison, for all other cytokines the detection limit was in the $ng\,ml^{-1}$ range.

5.7
QDs-Labeled Protein Nanoarrays

Nanopatterned protein arrays offer significant advantages for sensing applications, including short diffusion times, the parallel detection of multiple targets, and the requirement for only tiny amounts of sample. The controlled assembly of proteins into bioactive nanostructures is a key challenge in nanobiotechnology, including the potential to develop protein chips with single-molecule resolution. A variety of reports have been made wherein AFM has been used for the development of high-density protein chips which would serve primarily as an ultraminiaturized bioanalysis platform for the near future [113–122].

Kang et al. used DPN to directly print integrin $α_vβ_3$ nanoarrays [50] that would provide a platform for investigating the molecular interaction between integrin $α_vβ_3$ and vitronectin cell adhesion protein. With protein recognition and binding, AFM topography measurements indicated a 30 ± 5 nm height increase. By using BSA nanoarrays as a control, it was further confirmed that the patterned integrin proteins retained their biological selectivity after surface immobilization [50].

Again using DPN, proteins have been deposited on gold, nickel oxide, and pretreated-glass substrates from chemically modified single cantilever tips. A variety of protein–surface interactions have been used in DPN, such as chemisorption [47, 77] or electrostatic attraction [48]. However, adsorption often leads to full or partial denaturation of the proteins, and therefore a loss of function. Lee et al. [123] presented an indirect approach for the fabrication of features with biologically active antibodies using templates made by the covalent attachment of protein A/G to arrays of DPN-generated 11-mercaptoundecanoyl-N-hydroxysuccinimide ester ($NHSC_{11}SH$) on gold surfaces. (Protein A/G is a genetically engineered protein that is designed to contain four Fc-binding domains from protein A and two from protein G; this allows for an enhanced binding affinity [124].) This was the first example of the use of parallel SPM to generate arrays of protein structures.

In order to generate nanoscale antibody-based arrays, highly dense DPN-generated dot arrays (23 400 dots with a 1 μm dot-to-dot spacing) of $NHSC_{11}SH$ covalently coupled to protein A/G were incubated in a solution of Alexa Fluor 594-labeled human IgG solution (Alexa Fluor 594 is an organic dye) [123]. The patterned human IgG nanoarrays were 8.0 nm in height. By performing fluorescence microscopy imaging wherein a red emission was obtained from the organic dye, it was possible to confirm the adsorption of Alexa Fluor 594-labeled human IgG immobilized on the generated protein A/G arrays.

The generality of the above-mentioned approach for protein patterning was further demonstrated by developing anti-β-galactosidase and anti-ubiquitin nanoarrays (14 000 dots with a 2 μm dot-to-dot spacing), prepared in a similar manner to that used for human IgG. The patterned anti-β-galactosidase IgG nanoarrays were approximately 6.8 nm high. The biological activity of the patterned antibodies on the protein A/G was evaluated by incubating the substrates in a solution containing either Alexa Fluor 594-labeled β-galactosidase (15 μg ml^{-1}) or Alexa Fluor 488-labeled ubiquitin in PBS for 2 h. The binding of the Alexa Fluor 594-labeled β-galactosidase molecules to anti-β-galactosidase increased the height of the patterned features from 6.8 to 9.5 nm, as observed with tapping mode AFM imaging (Figure 5.24 a,b). Uniform β-galactosidase binding could be observed from the fluorescence imaging, as shown below in Figure 5.24c. Importantly, an

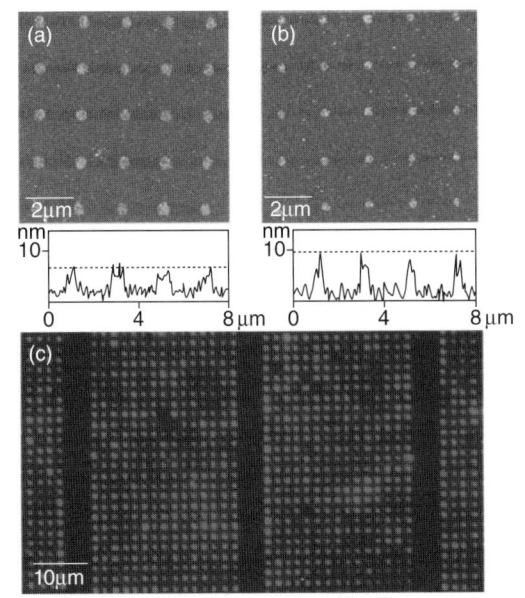

Figure 5.24 Topographical tapping mode AFM images and their corresponding height profiles of anti-β-galactosidase nanoarrays immobilized on protein A/G templates (a) before and (b) after incubation in a β-galactosidase protein solution; (c) Fluorescence microscopy image of the Alexa 594-labeled β-galactosidase complex nanoarrays: (a) topographical tapping mode AFM image and its corresponding height profile of FITC Alexa Fluor 594-labeled human IgG nanoarrays immobilized onto protein A/G templates; (b) Representative fluorescence microscopy image of Alexa Fluor 594-labeled antibody nanoarray patterns. The patterns span a distance of 1 cm.

absence of binding when the arrays were exposed to fluorophore-labeled ubiquitin protein proved that the arrays of immobilized antibodies had retained their biological activity after patterning.

Gokarna et al. [70] fabricated nanoarrays of cancer proteins using QD-conjugated antibodies of PSA by implementing the DPN technique. AFM topographic images revealed spot sizes of 500–900 nm and heights varying between 200 and 300 nm in these nanoarrays (Figure 5.25).

Whilst most of the above-described reports concerned the fabrication of *stable* proteins on a protein chip using organic dyes or QDs as labels, Tinazli et al. [124] introduced native protein nanolithography for the nanostructured assembly of *fragile* proteins or multiprotein complexes under native conditions. For this, a novel and easy-to use technology was presented for the fabrication of protein nanoarrays under fully physiological conditions, based on metal-chelating self-assembled monolayers combined with AFM-based nanolithography. The immobilized proteins were detached by a novel vibrational AFM mode (contact oscillation mode; COM), and replaced simultaneously or sequentially by different His-tagged proteins (Figure 5.26) that were selectively self-assembled from the bulk. Most importantly, during the structuring process using COM, the self-assembled

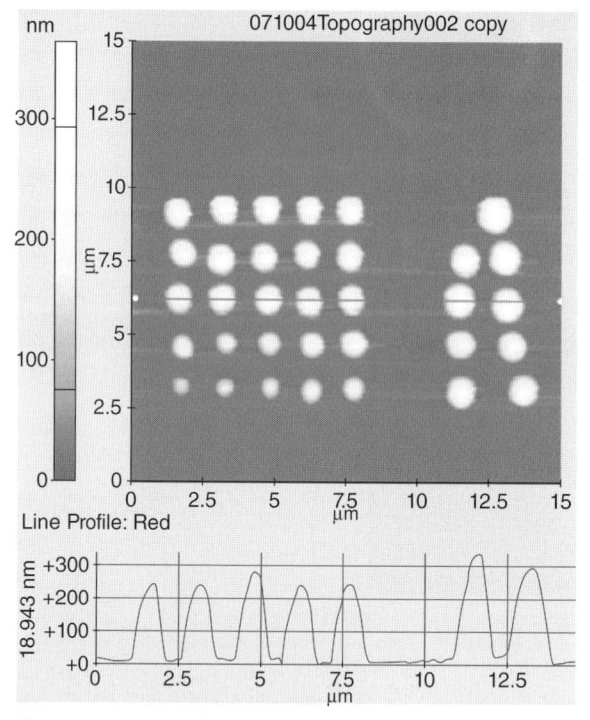

Figure 5.25 Topographical AFM images in noncontact mode of QD-conjugated protein nanoarrays fabricated on aldehyde-functionalized silicon substrates by DPN. The line profile of these spots is shown in the lower region of the image. These nanoarray spots had a diameter varying between 500 and 900 nm, and a height of 200–300 nm.

monolayer beneath the protein layer was not functionally impaired, and could be further used for repeated affinity capturing of His-tagged proteins. A high degree of specificity and a multiplexing capability of this patterning technique were also demonstrated. By implementing this type of a nanolithography technique they could perform rapid writing, reading and erasing of protein arrays in a versatile manner; indeed, rewritable protein biochips had been fabricated!

As a representative example to demonstrate the biological activity of this new nanofabricated biochip, the specific protein–protein interaction between a ligand and its receptor was monitored. For this purpose, the extracellular domain of the human type I interferon receptor (ifnar2-His10) was nanoarrayed and the binding of interferon-a2 (IFNa2) labeled with QDs was monitored using fluorescence microscopy (Figure 5.27). This medically relevant antivirus defense system had earlier been proved to be highly sensitive to immobilization procedures [125], and was well suited for proving the functional organization of fragile and dynamic protein assemblies. The ligand IFNa2 was observed to bind exclusively to the

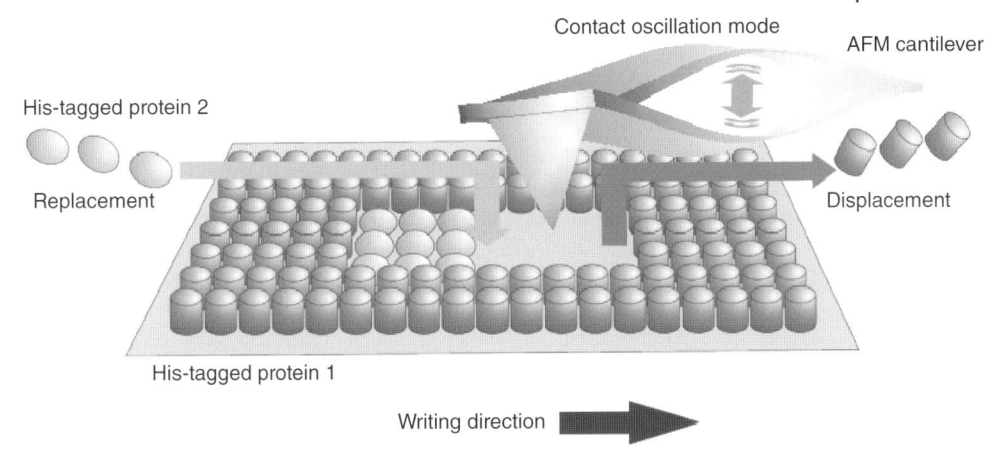

Figure 5.26 Fabrication of rewritable protein nanoarrays on self-assembled monolayers by native protein nanolithography. Uniformly oriented His-tagged proteins are removed ("displacement") with an AFM tip in COM and substituted either simultaneously or sequentially with other His-tagged proteins ("replacement"). (Nature Nanotechnology, (2007), **2**, 220).

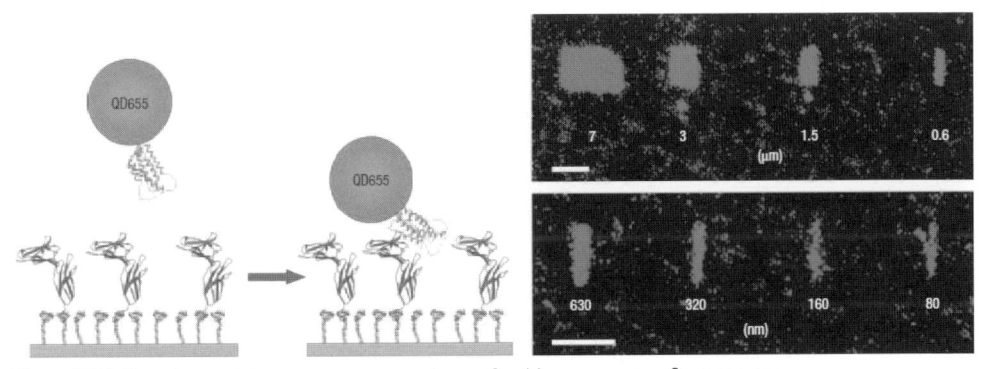

Figure 5.27 Bioactive protein nanoarrays fabricated down to 50 nm. Protein–protein interactions in micro- and nanostructured arrays: specific binding of IFNa2 S136C (red) site-specifically labeled with QD 655 through the additional cysteine residue (green) to immobilized ifnar2-His10 (blue) was probed by confocal laser scanning fluorescence microscopy. Strong fluorescence emission at 655 nm was detected only in the arrays. The few nonspecifically adsorbed QD aggregates can be easily distinguished from the prominent micro- and nanostructures. The scale bars correspond to 5 μm.

nanoarrayed receptor ifnar2, but not to the surrounding protein matrix (Figure 5.27). These results thereby demonstrated that this new AFM nanolithography-based technology might be suitable for probing protein–protein immobilization in nanoscale dimensions.

5.8
Summary and Future Perspectives

Many factors determine the success of detection techniques, including high sensitivity, resolution and detection limit; simplicity of operation; the avoidance of interference from tags; real-time monitoring; broad applicability; multiplexing and high-throughput capability; and biomedical relevance. Yet, to date no single technology has achieved this ideal status.

Although the capabilities of most of the new detection techniques have been demonstrated using very strong antibody–antigen interactions, it will be important also to conduct investigations with real-life, weak protein interactions. Nevertheless, those working in biomedical research will surely benefit from having a surfeit of methodological options available.

Acknowledgments

The studies conducted by the present authors were supported by National Research Laboratory Program (No. R0A-2005-000-10130-0) and by World Class University Program (No. R-31-2008-000-10071-0) funded by the Korean government (MEST).

References

1 Feinberg, J.G. (1961) *Nature*, **192**, 985–6.
2 Ekins, R.P. and Chu, F.W. (1991) Multianalyte microspot immunoassay–microanalytical "compact disk" of the future. *Clinical Chemistry*, **37**, 1955–67.
3 Ekins, R., Chu, F. and Micallef, J. (1989) High specific activity chemiluminescent and fluorescent markers: their potential application to high sensitivity and "multi-analyte" immunoassays. *Journal of Bioluminescence and Chemiluminescence*, **4**, 59–78.
4 MacBeath, G. and Schreiber, S.L. (2000) Printing proteins as microarrays for high-throughput function determination. *Science*, **289**, 1760–3.
5 Zhu, H., Bilgin, M., Bangham, R., Hall, D. et al. (2001) Global analysis of protein activities using proteome chips. *Science*, **293**, 2101–5.
6 Nolte, D.D. and Zhao, M. (2006) Scaling mass sensitivity of the BioCD at 0.25 pg/mm. *Proceedings of the SPIE*, **6380**, 63800J.
7 Honda, N., Lindberg, U., Andersson, P., Hoffmann, S. and Takei, H. (2005) Simultaneous multiple immunoassays in a compact disc- shaped microfluidic device based on centrifugal force. *Clinical Chemistry*, **51**, 1955–61.
8 Weiss, S. (1999) Fluorescence spectroscopy for single biomolecules. *Science*, **283**, 1676–83.
9 Jain, K.K. (2003) Nanodiagnostics: application of nanotechnology in molecular diagnostics. *Expert Review of Molecular Diagnostics*, **3**, 153–61.
10 Salata, O. (2004) Applications of nanoparticles in biology and medicine. *Journal of Nanobiotechnology*, **2**, 3.
11 Jain, K.K. (2005) Nanotechnology in clinical laboratory diagnostics. *Clinica Chimica Acta*, **358**, 37–54.
12 Michalet, X., Pinaud, F.F., Bentolila, L.A., Tsay, J.M., Doose, S., Li, J.J. et al. (2005) Quantum dots for live cells,

in vivo imaging, and diagnostics. *Science*, 307, 538–44.
13. Rossetti, R. and Brus, L. (1982) Electron-hole recombination emission as a probe of surface chemistry in aqueous CdS colloids. *Journal of Physical Chemistry*, **86**, 4470–2.
14. Alivisatos, A.P., Gu, W. and Larabell, C. (2005) Quantum dots as cellular probes. *Annual Review of Biomedical Engineering*, **7**, 55–76.
15. West, J.L. and Halas, N.J. (2003) Engineered nanomaterials for biophotonics applications: improving sensing, imaging, and therapeutics. *Annual Review of Biomedical Engineering*, **5**, 285–92.
16. Wang, X., Qu, L., Zhang, J., Peng, X. and Xiao, M. (2003) Surface-related emission in highly luminescent CdSe QDs. *Nano Letters*, **3**, 1103–6.
17. Talapin, D.V., Mekis, I., Götzinger, S., Kornowski, A., Benson, A. and Weller, H. (2004) CdSe/CdS/ZnS and CdSe/ZnSe/ZnS core-shell-shell nanocrystals. *Journal of Physical Chemistry, B*, **108**, 18826–31.
18. Spanhel, L., Haase, M., Weller, H. and Henglein, A. (1987) Photochemistry of colloidal semiconductors. 20. Surface modification and stability of strong luminescing CdS particles. *Journal of the American Chemical Society*, **109**, 5649–55.
19. Xu, S., Kumar, S. and Nann, T. (2006) Rapid synthesis of high-quality InP nanocrystals. *Journal of the American Chemical Society*, **128**, 1054–5.
20. Xu, S., Ziegler, J. and Nann, T. (2008) Synthesis of highly luminescent InP and InP/ZnS nanocrystals via one pot route. *Journal of Materials Chemistry*, **18**, 2653–6.
21. Jiang, W., Singhal, A., Zheng, J., Wang, C. and Chan, W.C. (2006) Optimizing the synthesis of red- to near-IR-emitting CdS-capped CdTe$_x$Se$_{1-x}$ alloyed quantum dots for biomedical imaging. *Chemistry of Materials*, **18**, 4845–54.
22. Shavel, A., Gaponik, N. and Eychmüller, A. (2006) Factors governing the quality of aqueous CdTe nanocrystals: calculations and experiment. *Journal of Physical Chemistry, B*, **110**, 19280–4.
23. Hinds, S., Myrskog, S., Levina, L., Koleilat, G., Yang, J., Kelley, S.O. and Sargent, E.H. (2007) NIR-Emitting colloidal quantum dots having 26% luminescence quantum yield in buffer solution. *Journal of the American Chemical Society*, **129**, 7218–19.
24. Fernee, M.J., Jensen, P. and Rubinsztein-Dunlop, H. (2006) Origin of the large homogeneous line widths obtained from strongly quantum confined PbS nanocrystals at room temperature. *Nanotechnology*, **17**, 956–62.
25. Du, H., Chen, C., Krishnan, R., Krauss, T.D., Harbold, J.M., Wise, F.W., Thomas, M.G. and Silcox, J. (2002) Optical properties of colloidal PbSe nanocrystals. *Nano Letters*, **2**, 1321–4.
26. Lifshitz, E., Brumer, M., Kigel, A., Sashchiuk, A., Bashouti, M., Sirota, M., Galun, E., Burshtein, Z., Le Quang, A.Q., Ledoux-Rak, I. and Zyss, J. (2006) Air-stable PbSe/PbS and PbSe/PbSe$_x$S$_{1-x}$ core shell nanocrystal quantum dots and their applications. *Journal of Physical Chemistry, B*, **110**, 25356–65.
27. Dähne, S., Resch-Genger, U. and Wolfbeis, O.S. (eds) (1998) *Near-Infrared Dyes for High Technology Applications*, NATO ASI Series, 3, Hightechnology, vol. **52**, Kluwer Academic Publishers, Dordrecht, The Netherlands.
28. Soper, S.A., Nutter, H.L., Keller, R.A., Davis, L.M. and Shera, E.B. (1993) The photophysical constants of several fluorescent dyes pertaining to ultrasensitive fluorescence spectroscopy. *Photochemistry and Photobiology*, **57**, 972–7.
29. Yu, W.W., Qu, L., Guo, W. and Peng, X. (2003) Experimental determination of the extinction coefficient of CdTe, CdSe and CdS nanocrystals. *Chemistry of Materials*, **15**, 2854–60.
30. Kuçur, E., Boldt, F.M., Cavaliere-Jaricot, S., Ziegler, J. and Nann, T. (2007) Quantitative analysis of the CdSe nanocrystal concentration by comparative techniques. *Analytical Chemistry*, **79**, 8987–93.
31. Sackett, D.L. and Wolff, J. (1987) Nile red as a polarity-sensitive fluorescent probe of hydrophobic protein surfaces. *Analytical Biochemistry*, **167**, 228–34.

32 Rueda, D. and Walter, N.G. (2006) Fluorescent energy transfer readout of an aptazyme based biosensor. *Methods in Molecular Biology*, **335**, 289–310.

33 Seybold, P.G., Gouterman, M. and Callis, J. (1969) Calorimetric, photometric and lifetime determinations of fluorescence yields of fluorescein dyes. *Photochemistry and Photobiology*, **9**, 229–42.

34 Mujumdar, R.B., Ernst, L.A., Mujumdar, S.R., Lewis, C.J. and Waggoner, A.S. (1993) Cyanine dye labeling agents: sulfoindocyanine succidimidyl esters. *Bioconjugate Chemistry*, **4**, 105–11.

35 Gruber, H.J., Hahn, C.D., Kada, G., Riener, C.K., Harms, G.S., Ahrer, W., Dax, T.G. and Knaus, H.G. (2000) Anomalous fluorescence enhancement of Cy3 and Cy3.5 versus anomalous fluorescence loss of Cy5 and Cy7 upon covalently linking to IgC and noncovalent binding to avidin. *Bioconjugate Chemistry*, **11**, 696–704.

36 Soper, S.A. and Mattingly, Q.L. (1994) Steady-state and picosecond laser fluorescence studies of nonradiative pathways in tricarbocyanine dyes: implications to the design of near-IR fluorochromes with high fluorescence efficiencies. *Journal of the American Chemical Society*, **116**, 3744–52.

37 Fortina, P., Kricka, L.J., Surrey, S. and Grodzinski, P. (2005) Nanobiotechnology: the promise and reality of new approaches to molecular recognition. *Trends in Biotechnology*, **23**, 168–73.

38 Smith, A.M., Dave, S., Nie, S., True, L. and Gao, X. (2006) Multicolor quantum dots for molecular diagnostics of cancer. *Expert Review of Molecular Diagnostics*, **6**, 231–44.

39 Ozkan, M. (2004) Quantum dots and other nanoparticles: what can they offer to drug discovery? *Drug Discovery Today*, **9**, 1065–71.

40 Jaiswal, J.K. and Simon, S.M. (2004) Potentials and pitfalls of fluorescent quantum dots for biological imaging. *Trends in Cell Biology*, **14**, 497–504.

41 Jaiswal, J.K., Mattoussi, H., Mauro, J.M. and Simon, S.M. (2003) Long-term multiple color imaging of live cells using quantum dot bioconjugates. *Nature Biotechnology*, **21**, 47–51.

42 Demers, L.M., Ginger, D.S., Park, S.J., Li, Z., Chung, S.W. and Mirkin, C.A. (2002) Direct patterning of modified oligonucleotides on metals and insulators by dip-pen nanolithography. *Science*, **296**, 1836–8.

43 Chung, S.W., Ginger, D.S., Morales, M.W., Zhang, Z., Chandrasekhar, V., Ratner, M.A. and Mirkin, C.A. (2005) Top-down meets bottom-up: dip-pen nanolithography and DNA-directed assembly of nanoscale electrical circuits. *Small*, **1**, 64–9.

44 Cho, Y. and Ivanisevic, A. (2005) TAT peptide immobilization on gold surfaces: a comparison study with a thiolated peptide and alkylthiols using AFM, XPS, and FT-IRRAS. *Journal of Physical Chemistry, B*, **109**, 6225–32.

45 Cho, Y. and Ivanisevic, A. (2004) SiO_x surfaces with lithographic features composed of a TAT peptide. *Journal of Physical Chemistry, B*, **108**, 15223–8.

46 Jiang, H.Z. and Stupp, S.I. (2005) Dip-pen patterning and surface assembly of peptide amphiphiles. *Langmuir*, **21**, 5242–6.

47 Lee, K.B., Lim, J.H. and Mirkin, C.A. (2003) Protein nanostructures formed via direct-write dip-pen nanolithography. *Journal of the American Chemical Society*, **125**, 5588–9.

48 Lim, J.H., Ginger, D.S., Lee, K.-B., Heo, J.S., Nam, J.-M. and Mirkin, C.A. (2003) Direct-write dip-pen nanolithography of proteins on modified silicon oxide surfaces, *Angewandte Chemie, International Edition*, **42**, 2309–12.

49 Lee, K.B., Park, S.J., Mirkin, C.A., Smith, J.C. and Mrksich, M. (2002) Protein nanoarrays generated by dip-pen nanolithography. *Science*, **295**, 1702–5.

50 Lee, M., Lee, M., Kang, D.-K, Yang, H.-K., Park, K.-H., Choe, S.Y., Kang, C.S., Chang, S.-I., Han, M.H. and Kang, I.C. (2006) Protein nanoarray on Prolinker TM surface constructed by atomic force microscopy dip-pen nanolithography for analysis of protein interaction. *Proteomics*, **6**, 1094–103.

51 Cheung, C.L., Camarero, J.A., Woods, B.W., Lin, T.W., Johnson, J.E. and De

Yoreo, J.E. (2003) Fabrication of assembled virus nanostructures on templates of chemoselective linkers formed by scanning probe nanolithography. *Journal of the American Chemical Society*, **125**, 6848–9.

52 Smith, J.C., Smith, J.C., Lee, K.B., Wang, Q., Finn, M.G., Johnson, J.E., Mrksich, M.E. and Mirkin, C.A. (2003) Nanopatterning the chemospecific immobilization of cowpea mosaic virus capsid. *Nano Letters*, **3**, 883–6.

53 Vega, R.A., Maspoch, D., Salaita, K. and Mirkin, C.A. (2005) Nanoarrays of single virus particle. *Angewandte Chemie, International Edition*, **44**, 6013–15.

54 Rozhok, S., Shen, C.K.-F., Littler, P.L.-H., Fan, Z., Liu, C., Mirkin, C.A. and Holz, R.C. (2005) Methods for fabricating microarrays of motile bacteria. *Small*, **1**, 445–51.

55 Talapin, D.V., Rogach, A.L., Haase, M. and Weller, H. (2001) *Journal of Physical Chemistry, B*, **105**, 12278–85.

56 Dabbousi, B.O., Rodriguez-Viejo, J., Mikulec, F.V., Heine, J.R., Mattoussi, H., Ober, R., Jensen, K.F. and Bawendi, M.G. (1997) (CdSe)ZnS core-shell quantum dots: synthesis and characterization of a size series of highly luminescent nanocrystallites. *Journal of Physical Chemistry, B*, **101**, 9463–75.

57 Klostranec, J.M. and Chan, W.C.W. (2006) Quantum dots in biological and biomedical research: Recent progress and present challenges. *Advanced Materials*, **18**, 1953–1964.

58 Van Sark, W.G.J.H.M., Frederix, P.L.T.M., Van den Heuvel, D.J., Gerritsen, H.C., Bol, A.A., van Lingen, J.N.J., Doncga, C.D.M. and Meijerink, A. (2001) Photooxidation and photobleaching of single CdSe/ZnS quantum dots probed by room-temperature time-resolved spectroscopy. *Journal of Physical Chemistry, B*, **105**, 8281–4.

59 Derfus, A.M., Chan, W.C.W. and Bhatia, S.N. (2004) Probing the cytotoxicity of semiconductor quantum dots. *Nano Letters*, **4**, 11–18.

60 Kirchner, C., Liedl, T., Kudera, S., Pellegrino, T., Javier, A.M., Gaub, H.E., Stolzle, S., Fertig, N. and Parak, W.J. (2005) Cytotoxicity of colloidal CdSe and CdSe/ZnS nanoparticles. *Nano Letters*, **5**, 331–8.

61 Parak, W.J., Gerion, D., Zanchet, D., Woerz, A.S., Pellegrino, T., Micheel, C., Williams, S.C., Seitz, M., Bruehl, R.E., Bryant, Z., Bustamante, C., Bertozzi, C.R. and Alivisatos, A.P. (2002) Conjugation of DNA to silanized colloidal semiconductor nanocrystalline quantum dots. *Chemistry of Materials*, **14**, 2113–19.

62 Xing, Y., Chaudry, Q., Shen, C., Kong, K.Y., Zhau, H.E., Chung, L.W., Petros, J.A., O'Regan, R.M., Yezhelyev, M.V., Simons, J.W., Wang, M.D. and Nie, S. (2007) Bioconjugated quantum dots for multiplexed and quantitative immunohistochemistry. *Nature Protocols*, **2**, 1152–65.

63 Medintz, I.L., Uyeda, H.T., Goldman, E.R. and Mattoussi, H. (2005) QD bioconjugates for imaging, labelling and sensing. *Nature Materials*, **4**, 435–46.

64 Mason, J.N., Farmer, H., Tomlinson, I.D., Schwartz, J.W., Savchenko, V., DeFelice, L.J., Rosenthal, S.J. and Blakely, R.D. (2005) Novel fluorescence-based approaches for the study of biogenic amine transporter localization, activity and regulation. *Journal of Neuroscience Methods*, **143**, 3–25.

65 Dubertret, B., Skourides, P., Norris, D.J., Noireaux, V., Brivanlou, A.H. and Libchaber, A. (2002) *In vivo* imaging of quantum dots encapsulated in phospholipid micelles. *Science*, **298**, 1759–62.

66 Gao, X.H., Cui, Y.Y., Levenson, R.M., Chung, L.W.K. and Nie, S.M. (2004) *Nature Biotechnology*, **22**, 969–76.

67 Zhou, M., Nakatani, E., Gronenberg, L.S., Tokimoto, T., Wirth, M.J., Hruby, V.J., Roberts, A., Lynch, R.M. and Ghosh, I. (2007) Peptide-labeled quantum dots for imaging GPCRs in whole cells and as single molecules. *Bioconjugate Chemistry*, **18**, 323–32.

68 Goldman, E.R., Clapp, A.R., Anderson, G.P., Uyeda, T., Mauro, J.M., Medintz, I.L. and Mattoussi, H. (2004) Multiplexed toxin analysis using four colors of quantum dot fluororeagents. *Analytical Chemistry*, **76**, 684–8.

69 Kusnezow, W., Jacob, A., Walijew, A., Diehl, F. and Hoheisel, J.D. (2003) Antibody microarrays: An evaluation of production parameters. *Proteomics*, **3**, 254–264.

70 Gokarna, A., Jin, L.-H., Hwang, J.S., Cho, Y.-H., Lim, Y.T. et al. (2008) *Proteomics*, **8**, 1809–18.

71 Wingren, C. and Borrebaeck, C.A.K. (2007) Progress in miniaturization of protein arrays – a step closer to high-density nanoarrays. *Drug Discovery Today*, **12** (19/20), 813–19.

72 Hoff, J.D., Cheng, L.-J, Meyhöfer, E., Guo, L.J. and Hunt, A.J. (2004) Nanoscale protein patterning by imprint lithography. *Nano Letters*, **4**, 853–8.

73 Zhang, G.J., Tanii, T., Kanari, Y. and Ohdomari, I. (2007) Production of nanopatterns by a combination of electron beam lithography and a self-assembled monolayer for an antibody nanoarray. *Journal of Nanoscience and Nanotechnology*, **7**, 410–17.

74 Agarwal, G., Naik, R.R. and Stone, M.O. (2003) Immobilization of histidine-tagged proteins on nickel by electrochemical dip pen nanolithography. *Journal of the American Chemical Society*, **125**, 7408–12.

75 Lee, K.B., Park, S.-J., Mirkin, C.A., Smith, J.C. and Mrksich, M. (2002) Protein nanoarrays generated by dip-pen nanolithography. *Science*, **295**, 1702–5.

76 Lee, M., Kang, D.-K, Yang, H.-K., Park, K.-H., Choe, S.Y., Kang, C.S., Chang, S.-I., Han, M.H. and Kang, I.C. (2006) Protein nanoarray on prolinker surface constructed by atomic force microscopy dip-pen nanolithography for analysis of protein interaction. *Proteomics*, **6**, 1094–103.

77 Nam, J.M., Han, S.W., Lee, K.-B, Liu, X., Ratner, M.A. and Mirkin, C.A. (2004) Bioactive protein nanoarrays on nickel oxide surfaces formed by dip-pen nanolithography. *Angewandte Chemie, International Edition*, **43**, 1246–9.

78 Dietrich, H.R., Knoll, J., Doel, L.R., van Dedem, G.W.K., Daran-Lapujade, P.A.S., Vliet, L.J., Moerman, R., Pronk, J.T. and Young, I.T. (2004) Nanoarrays: a method for performing enzymatic assays. *Analytical Chemistry*, **76**, 4112–17.

79 Angenendt, P., Nyarsik, P.L., Szaflarski, W., Glökler, J., Nierhaus, K.H., Lehrach, H., Cahill, D.J. and Lueking, A. (2004) Cell-free protein expression and functional assay in nanowell chip format. *Analytical Chemistry*, **76**, 1844–9.

80 Ghatnekar-Nilsson, S., Dexlin, L., Wingren, C., Montelius, L. and Borrebaeck, C.A. (2007) Design of atto-vial based recombinant antibody arrays combined with a planar waveguide detection system. *Proteomics*, **7**, 540–7.

81 Basu, M., Basu, M., Seggerson, S., Henshaw, J., Jiang, J., Cordona, R.A., Lefave, C., Boyle, P.J., Miller, A., Pugia, M. and Basu, S. (2004) Nano-biosensor development for bacterial detection during human kidney infection: use of glycoconjugate-specific antibody-bound gold NanoWire arrays (GNWA). *Glycoconjugate Journal*, **21**, 487–96.

82 Cui, Y., Wei, Q., Park, H. and Lieber, C.M. (2001) Nanowire nanosensors for highly sensitive and selective detection of biological and chemical species. *Science*, **293**, 1289–92.

83 Gamby, J., Abid, J.-P., Abid, M., Ansermet, J.-P. and Girault, H.H. (2006) Nanowires network for biomolecular detection using contactless impedance tomoscopy technique. *Analytical Chemistry*, **78**, 5289–95.

84 Pal, S., Alocilja, E.C. and Downes, F.P. (2007) Nanowire labeled direct-charge transfer biosensor for detecting Bacillus species. *Biosensors & Bioelectronics*, **22**, 2329–36.

85 Zheng, G., Patolsky, F., Cui, Y., Wang, W.U. and Lieber, C.M. (2005) Multiplexed electrical detection of cancer markers with nanowire sensor arrays. *Nature Biotechnology*, **23**, 1294–301.

86 Tamiya, E. et al. (2005) Nanosystems for biosensing: multianalyte immunoassay on a protein chip. *Methods in Molecular Biology*, **300**, 369–81.

87 Pammer, P., Schlapak, R., Sonnleitner, M., Ebner, A., Zhu, R., Hinterdorfer, P., Höglinger, O., Schindler, H. and Howorka, S. (2005) Nanopatterning of biomolecules with microscale beads. *ChemPhysChem*, **6**, 900–3.

88 Endo, T., Kerman, K., Nagatani, N., Hiepa, H.M., Kim, D.-K., Yonezawa, Y.,

Nakano, K. and Tamiya, E. (2006) Multiple label-free detection of antigen-antibody reaction using localized surface plasmon resonance-based core-shell structured nanoparticle layer nanochip. *Analytical Chemistry*, **78**, 6465–75.

89 Arntz, Y., Arntz, Y., Seelig, J.D., Lang, H.P., Zhang, J., Hunziker, P., Ramseyer, J.P., Meyer, E., Hegner, M. and Gerber, C.H. (2003) Label-free protein assay based on a nanomechanical cantilever array. *Nanotechnology*, **14**, 86–90.

90 Backmann, N., Zahnd, C., Huber, F., Bietsch, A., Plückthun, A., Lang, H.-P., Güntherodt, H.-J., Hegner, M. and Gerber, C. (2005) A label-free immunosensor array using single-chain antibody fragments. *Proceedings of the National Academy of Sciences of the United States of America*, **102**, 14587–92.

91 Braun, T., Backmann, N., Vögtli, M., Bietsch, A., Engel, A., Lang, H.-P., Gerber, C. and Hegner, M. (2006) Conformational change of bacteriorhodopsin quantitatively monitored by microcantilever sensors. *Biophysical Journal*, **90**, 2970–7.

92 Dutta, P., Tipple, C.A., Lavrik, N.V., Datskos, P.G., Hofstetter, H., Hofstetter, O. and Sepaniak, M.J. (2003) Enantioselective sensors based on antibody-mediated nanomechanics. *Analytical Chemistry*, **75**, 2342–8.

93 Huber, F., Hegner, M., Gerber, C., Güntherodt, H.-J. and Lang, H.P. (2006) Label free analysis of transcription factors using microcantilever arrays. *Biosensors & Bioelectronics*, **21**, 1599–605.

94 Ji, H.F., Yang, X., Zhang, J. and Thundat, T. (2004) Molecular recognition of biowarfare agents using micromechanical sensors. *Expert Review of Molecular Diagnostics*, **4**, 859–66.

95 Mukhopadhyay, R., Sumbayev, V.V., Lorentzen, M., Kjems, J., Andreasen, P.A. and Besenbacher, F. (2005) Cantilever sensor for nanomechanical detection of specific protein conformations. *Nano Letters*, **5**, 2385–8.

96 Tian, F., Hansen, K.F., Ferrell, T.L. and Thundat, T. (2005) Dynamic microcantilever sensors for discerning biomolecular interactions. *Analytical Chemistry*, **77**, 1601–6.

97 Weeks, B.L., Camarero, J., Noy, A., Miller, A.E., De Yoreo, J.J. and Stanker, L. (2003) A microcantilever-based pathogen detector. *Scanning*, **25**, 297–9.

98 Shekhawat, G., Tark, S.-H. and Dravid, V.-P. (2006) MOSFET-Embedded microcantilevers for measuring deflection in biomolecular sensors. *Science*, **311**, 1592–5.

99 Bruckbauer, A., Zhou, D., Kang, D.J., Korchev, Y.E., Abell, C. and Klenerman, D. (2004) An addressable antibody nanoarray produced on a nanostructured surface. *Journal of the American Chemical Society*, **126**, 6508–9.

100 Lee, K.-B., Kim, E.-Y., Mirkin, C.A. and Wolinsky, S.M. (2004) The use of nanoarrays for highly sensitive and selective detection of human immunodeficiency virus type 1 in plasma. *Nano Letters*, **4** (10), 1869–72.

101 Kim, E.-Y., Chang, S.-I. and Kang, I.C. (2007) Detection technology for antibody-antigen interaction on ProteoChip using quantum dot. *Biochip Journal*, **1** (2), 102–6.

102 Yan, J., Hu, M., Li, D., He, Y., Zhao, R., Jiang, X., Song, S., Wang, L. and Fan, C. (2008) A nano- and micro- integrated protein chip based on quantum dot probes and a microfluidic network. *Nano Research*, **1**, 490–6.

103 Jiang, X.Y., Xu, Q., Dertinger, S.K.W., Stroock, A.D., Fu, T. and Whitesides, G.M. (2005) A general method for patterning gradients of biomolecules on surfaces using microfluidic networks. *Analytical Chemistry*, **77**, 2338–47.

104 McDonald, J.C., Duffy, D.C., Anderson, J.R., Chiu, D.T., Wu, H., Schueller, O.J. and Whitesides, G.M. (2000) Fabrication of microfluidic systems in poly(dimethylsiloxane). *Electrophoresis*, **21**, 27–40.

105 Whitesides, G.M. (2006) The origins and the future of microfluidics. *Nature*, **442**, 368–73.

106 Psaltis, D., Quake, S.R. and Yang, C. (2006) Developing optofluidic technology through the fusion of microfluidics and optics. *Nature*, **442**, 381–6.

107 Bernard, A., Michel, B. and Delamarche, E. (2001) Micromosaic immunoassays. *Analytical Chemistry*, **73**, 8–12.

108 Wolf, M., Juncker, D., Michel, B., Hunziker, P. and Delamarche, E. (2004) Simultaneous detection of C-reactive protein and other cardiac markers in human plasma using micromosaic immunoassays and self-regulating microfluidic networks. *Biosensors & Bioelectronics*, **19**, 1193–202.

109 Hanaki, K., Momo, A., Oku, T., Komoto, A., Maenosono, S., Yamaguchi, Y. and Yamamoto, K. (2003) Semiconductor quantum dot/albumin complex is a long-life and highly photostable endosome marker. *Biochemical and Biophysical Research Communications*, **302**, 496–501.

110 Sapsford, K., Medintz, I., Golden, J., Deschamps, J.P., Uyeda, H.T. and Mattoussi, H. (2004) Surface-immobilized self assembled protein-based quantum dots nanoassemblies. *Langmuir*, **20**, 7720–8.

111 Gyorvary, E.S., O'Riordan, A., Quinn, A.J., Redmond, G., Pum, D. and Sleyt, U.B. (2003) Biomimetic nanostructure fabrication: nonlithographic lateral patterning and self-assembly of functional bacterial S-layers at silicon supports. *Nano Letters*, **3**, 315–19.

112 Zajac, A., Song, D., Qian, W. and Zhukov, T. (2007) Protein microarrays and quantum dot probes for early cancer detection. *Colloids and Surfaces B, Biointerfaces*, **58**, 309–14.

113 Unal, K., Frommer, J. and Wickramasinghe, H.K. (2006) Ultrafast molecule sorting and delivery by atomic force microscopy. *Applied Physics Letters*, **88**, 183105–7.

114 Jaschke, M. and Butt, H.-J. (1995) Deposition of organic material by the tip of a scanning force microscope. *Langmuir*, **11**, 1061–4.

115 Ginger, D.S., Zhang, H. and Mirkin, C.A. (2004) The evolution of dip-pen nanolithography. *Angewandte Chemie, International Edition*, **43**, 30–45.

116 Piner, R.D., Zhu, J., Xu, F., Hong, S. and Mirkin, C. (1997) A. "dip-pen" nanolithography. *Science*, **283**, 661–3.

117 Xu, S. and Liu, G.-Y. (1997) Nanometer-scale fabrication by simultaneous nanoshaving and molecular self-assembly. *Langmuir*, **13**, 127–9.

118 Wadu-Mesthrige, K., Xu, S., Amro, N.A. and Liu, G.Y. (1999) Fabrication and imaging of nanometer-sized protein patterns. *Langmuir*, **15**, 8580–3.

119 Wadu-Mesthrige, K., Amro, N.A., Garno, J.C., Xu, S. and Liu, G. (2001) Fabrication of nanometer-sized protein patterns using atomic force microscopy and selective immobilization. *Biophysical Journal*, **80**, 1891–9.

120 Xu, S., Miller, S., Laibinis, P.E. and Liu, G.Y. (1999) Fabrication of nanometer scale patterns within self-assembled monolayers by nanografting. *Langmuir*, **15**, 7244–51.

121 Wouters, D. and Schubert, U.S. (2004) Nanolithography and nanochemistry: probe-related patterning techniques and chemical modification for nanometer-sized devices. *Angewandte Chemie, International Edition*, **43**, 2480–95.

122 Bruckbauer, A., Zhou, D., Ying, L., Korchev, Y.E., Abell, C. and Klenerman, D. (2003) Multicomponent submicron features of biomolecules created by voltage controlled deposition from a nanopipet. *Journal of the American Chemical Society*, **125**, 9834–9.

123 Lee, S.W., Oh, B.-K., Sanedrin, R.G., Salaita, K., Fujigaya, T. and Mirkin, C.A. (2006) Biologically active protein nanoarrays generated using parallel dip-pen nanolithography. *Advanced Materials*, **18**, 1133–6.

124 Tinazli, A., Piehler, J., Beuttler, M., Guckenberger, R. and Tampe, R. (2007) Native protein nanolithography that can write, read and erase. *Nature Nanotechnology*, **2**, 220–5.

125 Piehler, J. and Schreiber, G. (2001) Fast transient cytokine-receptor interactions monitored in real time by reflectometric interference spectroscopy. *Analytical Biochemistry*, **289**, 173–86.

126 Lynch, M., Mosher, C., Huff, J., Nettikadan, S., Johnson, J. and Henderson, E. (2003) *Proteomics*, **4**, 1695–702.

6
Imaging and Tracking of Viruses Using Quantum Dots
Kye-Il Joo, April Tai and Pin Wang

6.1
Introduction

Although the quantum dot (QD) was initially developed for the semiconductor field, its novel properties have enabled it to be engineered for biological applications. Traditionally, organic fluorophores are used to detect and track biomolecules, such as antibodies, peptides, and viruses. However, problems with metabolic degradation, or photobleaching, have limited their applicability for the long-term imaging of biological processes. The use of QDs can potentially mitigate these concerns, as well as allow for the development of novel detection techniques. QDs are resistant to metabolic degradation and possess a wide absorption spectrum and a narrow emission spectrum. In this chapter, the methods of forming QD–biomolecule hybrids, and the use of QDs in virus tracking and detection, will be examined.

First, the optical properties of QDs and the different methods of creating functionalized QDs for the synthesis of nanoparticle–biomolecule hybrids will be discussed. For this, two studies linking cowpea mosaic virus (CPMV) to QDs are detailed: one where CPMV-QD networks were produced, and another where immobilized CPMV were decorated with QDs. A method is also described of incorporating QDs into the capsid of brome mosaic virus (BMV) by utilizing the unique reaction of the BMV capsid to pH change. Next, different virus-tracking techniques are reviewed, notably the traditional methods using organic fluorophores, and the limitations of these methods are outlined. The advantages of using QDs are also detailed, and their use in tracking both enveloped and nonenveloped viruses is discussed. The use of QDs to detect viruses is then discussed, notably of respiratory syncytial virus (RSV), which is responsible for many serious respiratory illnesses in young children. These studies, in which QDs are used to both label and detect viruses, in combination with confocal microscopy, confirm the rapid and efficient nature of these techniques. Finally, the use of pH-sensitive QDs as biosensors is discussed, in which a complex system linking ATPase, pH-sensitive QDs conjugated with viral antibody, and chromatophores is employed in the rapid and simultaneous detection of different virus types.

Nanomaterials for the Life Sciences Vol.6: Semiconductor Nanomaterials.
Edited by Challa S. S. R. Kumar
Copyright © 2010 WILEY-VCH Verlag GmbH & Co. KGaA, Weinheim
ISBN: 978-3-527-32166-7

6.2
Quantum Dot–Biomolecule Hybrids

6.2.1
Optical Properties of QDs

Quantum dots are tiny, light-emitting crystals with sizes in the nanometer scale. Their small size causes a confinement of the electron-hole pairs in the crystal and, as a result, they display discrete energy levels [1]. These energy levels can absorb and emit wavelengths from ultraviolet to infrared, depending on the size of the QD [2]. As QDs have several advantages over conventional organic dyes, they are emerging as a fluorescent probe for biological imaging and medical diagnostics. Organic dyes generally encounter problems in fluorescing continuously for long periods of time, due to their metabolic degradation. It is also difficult to use them for multicolor applications because, although they can only be excited by light of a narrow wavelength, they emit light with a broad spectrum, and this often causes an overlapping of the signals for different dyes. By contrast, QDs are highly resistant to metabolic degradation and exhibit a high quantum yield – the ratio of the amount of light emitted from a sample to the amount of light that it absorbs [3, 4]. QDs also have molar extinction coefficients which are 10- to 100-fold larger than those of most organic dyes [5, 6]. These properties allow QDs to fluoresce more brightly and for longer periods of time than conventional organic dyes. Another advantage that QDs have an over organic dyes relates to their optical spectra (Figure 6.1). In addition to their narrow absorption and broad emission spectra, organic dyes also experience "red-tail," a phenomenon which is caused by the asymmetric shape of their emission spectra and the consequent leakage of signal towards the red end of the spectrum [7–9]. Quantum dots, on the other hand, have broad absorption spectra but having narrow and symmetric emission spectra. Thus, a single wavelength is sufficient to excite multicolor QDs of different sizes, and this results in discrete emission wavelengths with minimal signal overlap.

6.2.2
Functionalized QDs Conjugated with Biomolecules

Before QDs can be used for biological applications, such as probing live cells, they must first be conjugated with biological molecules. This must be carried out with great care in order to avoid disrupting the functions of the cell or of the biological molecules. Currently, QD conjugation with biological molecules has been successfully achieved in several ways, including electrostatic interaction, mercapto (–SH) exchange, and covalent linkage.

With the *electrostatic exchange* approach, the negatively charged QD interacts electrostatically with the positively charged domain of the engineered protein of interest. The resultant protein–QD conjugates were very stable and, somewhat remarkably, displayed a fluorescence yield that was even higher than that of non-conjugated QDs [10].

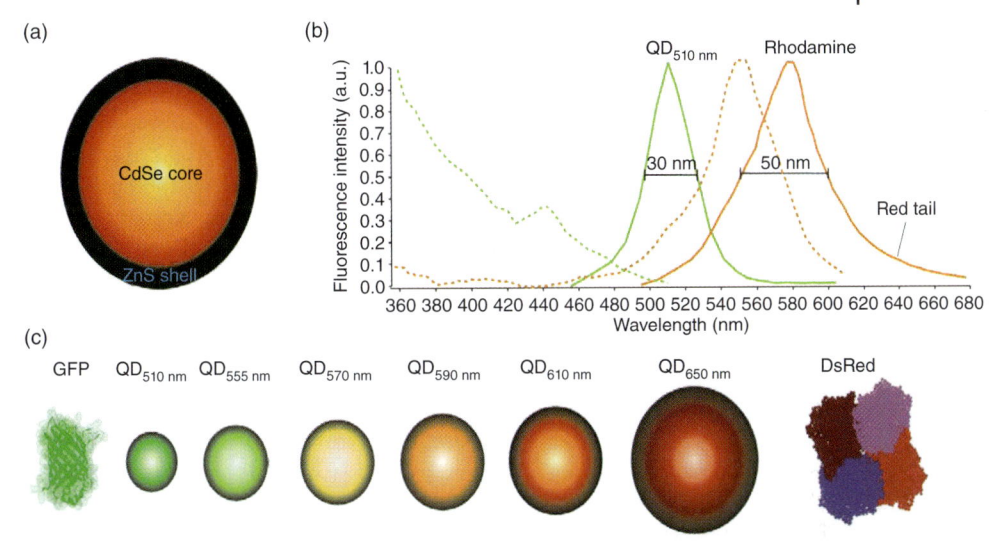

Figure 6.1 Properties of bioconjugatable quantum dots (QDs). (a) QDs are inorganic fluorophores and consist of a cadmium selenide (CdSe) core with several layers of a thick zinc sulfide (ZnS) shell to improve quantum yield and photostability; (b) The excitation spectrum (broken lines) of a QD (green) is very broad, whereas that of an organic dye (rhodamine, orange) is narrow. The emission spectrum (unbroken lines) is narrower for a QD (green) than for organic dyes (rhodamine, orange). Values indicate the full spectral width at half-maximum intensity (FWHM value); (c) The emission of the QDs can be tuned by controlling the size of the CdSe core: an increase in the size of the core shifts the emission to the red end of the spectrum. The combined size of the core and the shell of QDs emitting in the visible region of spectra are in the size range of commonly used fluorescent proteins such as green (GFP) and red (DsRed) fluorescent proteins. From Ref. [7].

A *mercapto exchange* process can also be used to conjugate biological molecules containing thiol groups, which serve as an anchor to the ZnS shell of QDs via ligand exchange [11–16]. However, as the bond between Zn and thiol is dynamic and weak, the biological molecules can easily detach and the QDs will precipitate out of solution. Biological molecules can also be covalently linked to functional groups on the QD surfaces [2, 8, 9, 17–20]; this is achieved by crosslinking certain functional groups (e.g., –COOH, –SH, or –NH$_2$) on the QD surface to corresponding reaction groups on the biological molecules. The resultant QD–biological molecule structures are much more stably linked than those produced by the mercapto exchange process. An example of a crosslinker is 1-ethyl-3-(3-dimethylaminopropyl)carbodiimide (EDC), which is able to covalently link –NH$_2$ and –COOH groups. Another crosslinker, 4-(*N*-maleimidomethyl)-cyclohexanecarboxylic acid *N*-hydroxysuccinimide ester (SMCC) can join –SH and –NH$_2$ groups [21]. Avidin/streptavidin-conjugated QDs have also been employed to tag biotinylated proteins of interest via the affinity of avidin/streptavidin for biotin [22, 23].

6.2.3
Formation of Virus–QD Networks

Hybrid nanoparticle–biomaterial networks have a variety of useful properties that can be utilized for biosensing devices, diagnostic agents, and novel drug development [1, 24–26]. Viruses are ideal models for nanoassembly due to their nanoscale dimensions and relatively well-characterized surface properties. Recently, hybrid virus networks between the plant cowpea mosaic virus (CPMV) and QDs have been successfully constructed through the use of carbodiimide coupling chemistry [27]. The two-step coupling procedure is shown in Figure 6.2. For this, water-soluble ZnS-capped CdSe QDs were first prepared using 2-mercaptoacetic acid (MAA) via ligand exchange, after which the carboxylates on the QDs were stably activated by EDC and sulfo-NHS (N-hydroxysulfosuccinimide). When CPMV was incubated with the activated QDs, virus–QD networks were produced via covalent amide bond linkages forming between the carboxylate groups on the QDs and the primary amines of the virus capsid protein. The heterostructures were linked by covalent amide bonds, as confirmed by Fourier transform infrared (FTIR) spectroscopy, while the large-scale network assembly was visualized using scanning electron microscopy (SEM), and visualization of QD incorporation into the microscale networks was monitored using fluorescence microscopy. Taken together, these results showed that viruses and QDs could be covalently linked with EDC chemistry, and that enhancements and modifications could be applied in order to improve the conjugation of the system. The techniques used allowed for great versatility in the size and shape of the networks designed. The advantages of such a system include novel large-scale material design, enhanced biological applications for targeting, and more densely arrayed nanoelectronic devices. Future studies include quantifying the porosity and size of the network by varying the concentrations of carboxyl and amine sources. Extending this method to other protein systems might allow for a larger range of material designs and applications.

6.2.4
Decoration of Discretely Immobilized Virus with QDs

In another study, two methods of functionalizing CPMV with QDs were initiated [28], whereby metal affinity coordination was used to immobilize CPMV on a substrate functionalized with NeutrAvidin. This was achieved by using a nickel-activated heterobifunctional biotin-nitrilotriacetic acid (NTA) moiety to link the histidines on the CPMV to the NeutrAvidin. Two different linking chemistries were then used to functionalize the CPMV with hydrophilic CdSe/ZnS core–shell QDs.

In the first method, biotin-X-NTA was exposed to NeutrAvidin-functionalized glass slides through the biotin–NeutrAvidin interaction (Figure 6.3a). The Ni-NTA end of the bound biotin-X-NTA became bound to the histidine residues on the CPMV particles. Then, the immobilized CPMV was further exposed to biotin-

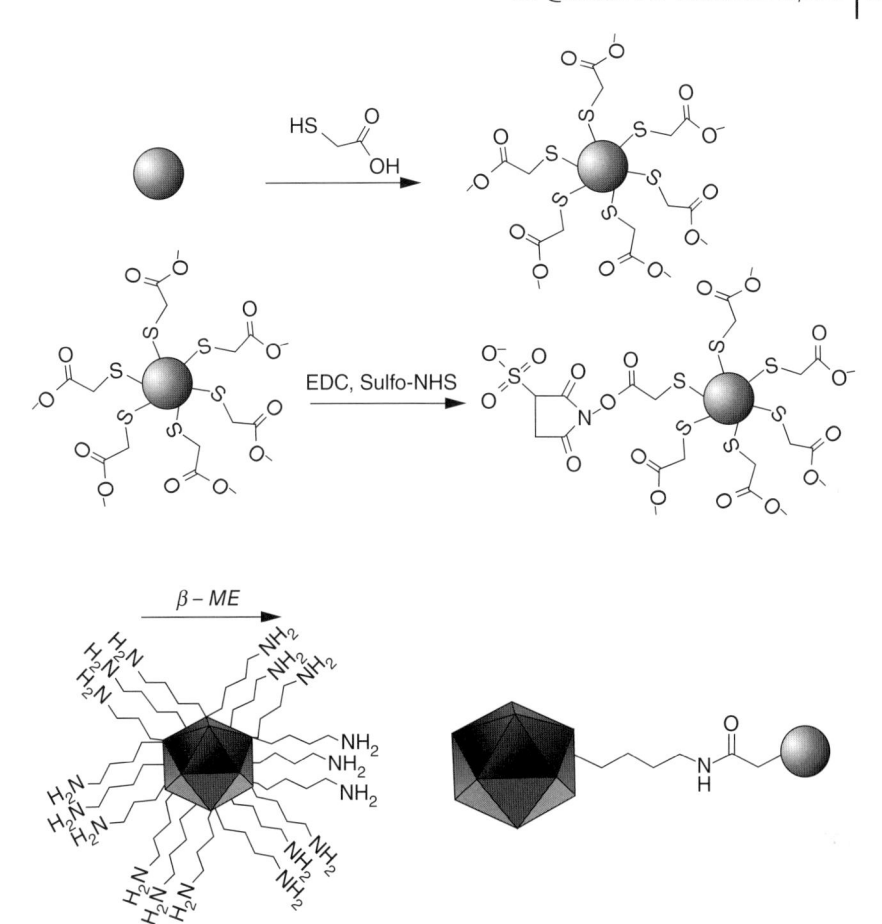

Figure 6.2 Chemical synthesis of virus networks with inorganic nanoparticles. QDs are chemically modified with MAA by thiol-zinc attachment to decorate the surface with carboxyl groups. The addition of EDC activates the carboxylic group, and subsequent addition of sulfo-NHS stabilizes the O-acylisourea intermediate long enough to react with primary amine nucleophiles (virus lysines) depicted below to produce amide linkages. 2-Mercaptoethanol chelates excess EDC to avoid virus polymerization prior to virus addition. After 4 h of coupling, hydroxylamine was added to deactivate any remaining active carboxyl esters to prevent nonspecific aggregation. Modified after Ref. [27].

X-NTA, which served as a bridge, where one end bound again through Ni-NTA with the remaining histidines on the surface of virions, and the other end bound to dihydrolipoic acid (DHLA)-capped and avidin-conjugated QDs. In the second method, lysine residues on the CPMV viral surface were functionalized with NHS-ester-LC (long-chain)-biotin (Figure 6.3b), after which the mutants were bound to the NeutrAvidin-functionalized glass slide and exposed to maltose-binding protein (MBP)-QD-avidin conjugates. Both methods were able to successfully bind QDs

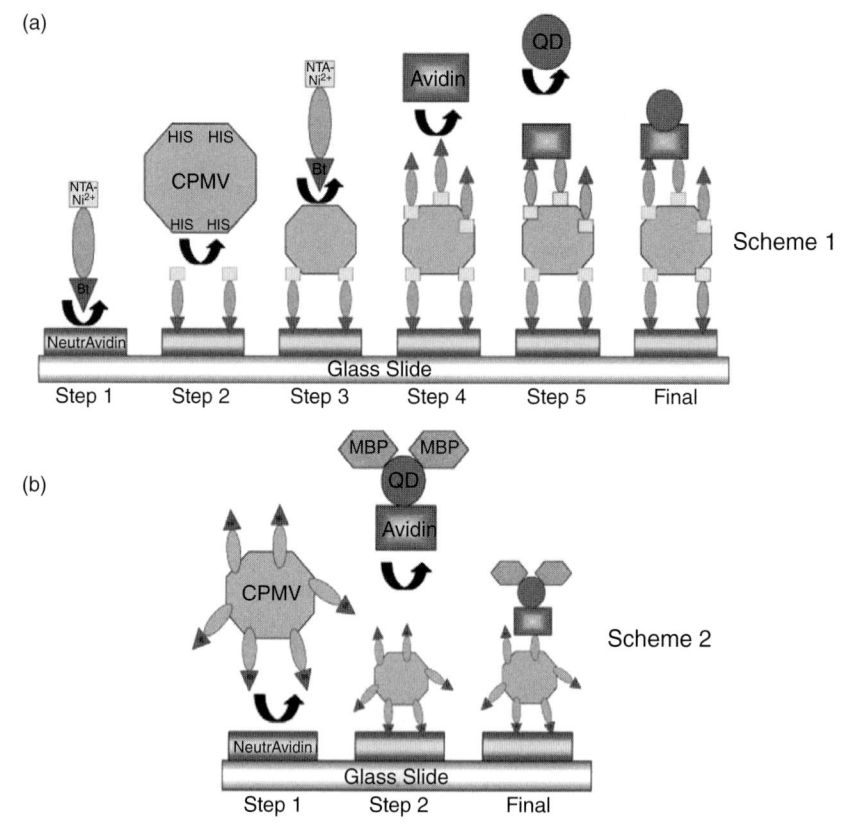

Figure 6.3 Schemes for patterning CPMV virus on a substrate and decoration with QDs. (a) Scheme 1. Glass slides were functionalized with NeutrAvidin and the surface exposed to biotin-X-NTA. Biotin binds NeutrAvidin leaving the Ni-NTA function available to coordinate to HIS_6 on CPMV particles. The upper exposed surface of immobilized CPMV is then exposed to biotin-X-NTA, then DHLA-capped and avidin-conjugated QDs; (b) Scheme 2. CPMV mutants labeled with NHS-ester-LC-biotin (LC-Bt) were first immobilized on the NeutrAvidin functionalized glass slide. The biotinylated CPMV mutants were then exposed to MBP-QD-avidin conjugates. Modified after Ref. [28].

to the immobilized CPMV. The resultant structures could then be characterized and the QD structures could either act as a scaffold for other proteins, or be used to construct new nanocomposite materials with possible biomedical or bioelectronic applications.

6.2.5
QD Encapsulation in Viral Capsids

One strategy for incorporating QDs into viral particles is to encapsulate them in the viral capsid by partial or total replacement of the viral nucleic acid [29]. In one study, the BMV—a small virus that features a capsid composed of 180 identical proteins – was used for this purpose [30]. These proteins form pentameric and hexameric subunits on the viral capsid that are stabilized by weak interactions [31]. As a result of three decades of research with BMV, several tools have been developed to study the replication, protein–RNA interactions, RNA-dependent RNA replication, and capsid assembly and disassembly of the virus [31–33]. In BMV, the capsid subunits are tightly associated and positively charged, while the nucleic acid is negatively charged. This results in a high degree of ionic interactions between the two, and it is this property which allows the foreign core encapsulation of the negatively charged nanoparticles in the viral capsid. In these studies, the BMV virions were first disassembled, and then reassembled in the presence of the nanoparticles; the resultant BMV particles were then purified. This capsid dissociation and reassembly was enabled by a unique property of the BVM capsid. In the presence of low- to moderate-ionic-strength buffers of pH < 5.0, the capsid remains stable, but if the pH is increased from 5.0 to 7.0 then a profound structural transition occurs, whereby Mg^{2+} stabilizes the interaction of the capsid subunits and a reversible expansion occurs, without dissociation. At pH ~7.5 and an ionic strength greater than 0.5 M, however, the capsid dissociates and the viral RNA precipitates. Yet, when a low pH and ionic strength is re-established, then reassociation will occur.

Different types of QDs have been incorporated into the BMV capsid and subsequently investigated, including those coated with lipid-micelles, DHLA, and single-stranded DNA (ssDNA) which contains the base sequence of the RNA motif required for BMV assembly in cells. When the QDs had been functionalized with the different coatings, and their hydrodynamic radii determined using dynamic light scattering (DLS) (Figure 6.4a), the results indicated that there was no significant aggregation in all cases. Recently, it has been shown that at a ratio of 1:180 poly(ethylene glycol) (PEG)-coated cores to BMV capsid units, the PEG-coated QDs produced the regularly shaped particles expected for BMV. Also, virus-like particle (VLP) assembly was optimal when the stoichiometry of the PEG-coated core was at two or three equivalents of the capsid (Figure 6.4b–d).

One concern with the QD labeling of BMV is that capsid reassembly with QDs remains somewhat unstable in storage, as determined by DLS and transmission electron microscopy (TEM); moreover, the degradation was exacerbated by exposure to visible light [29]. Although PEG-coated QD cores remain stable under laser exposure, VLPs with these cores will precipitate out of solution. Consequently, excitation of the QDs may have an effect on the protein shell in this set-up. Overall, however, these studies have provided a general self-assembly protocol for the encapsulation of QDs into viral capsids.

6.3
Quantum Dots for Single Virus Tracking in Living Cells

One key requirement for viral infection is an efficient transport of the genetic material from the surface of the host cell to its nucleus. During viral entry, the virus interacts with various cellular structures and takes advantage of their

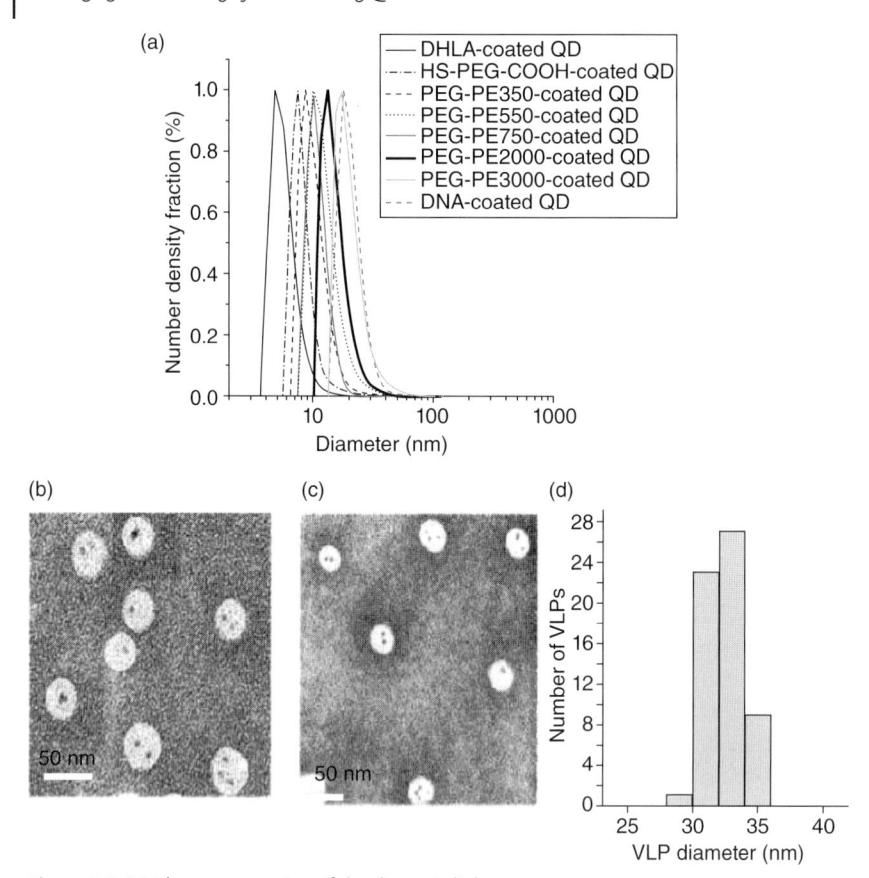

Figure 6.4 (a) The concentration of the dynamic light scattering distributions of functionalized QDs in aqueous solution; (b,c) TEM images of VLPs containing HS-PEG-COOH-coated QDs. The equivalent ratios of QD to CP are 2:1 (b) and 3:1 (c), respectively; (d) Size histogram of VLPs corresponding to (c). Modified after Ref. [29].

environments to optimize delivery of the viral genome to the nucleus and promote efficient viral replication [34]. In order to understand virus trafficking, it is necessary in turn also to understand not only the interactions that occur between the virus and the cell (including the cellular structures), but also the route of viral infection [35]. An improved understanding of these interactions may provide crucial insights into preventing virus-triggered diseases, as well as enhancing the efficacy of virus-mediated gene delivery.

Many routes may be taken in viral infection, and these vary by virus type. In general, viruses first attach to the receptors on the cell surface, such as proteins, carbohydrates, or lipids, and then deliver their viral genome into the cellular cytoplasm [36–40]. Viruses can use either an *endocytic* or *nonendocytic* route to enter the cells (Figure 6.5). Many viruses enter cells via the endocytic pathway, which

(a) Endocytic route: clathrin-mediated endocytosis and penetration

(b) Non-endocytic route: fusion at the cell surface

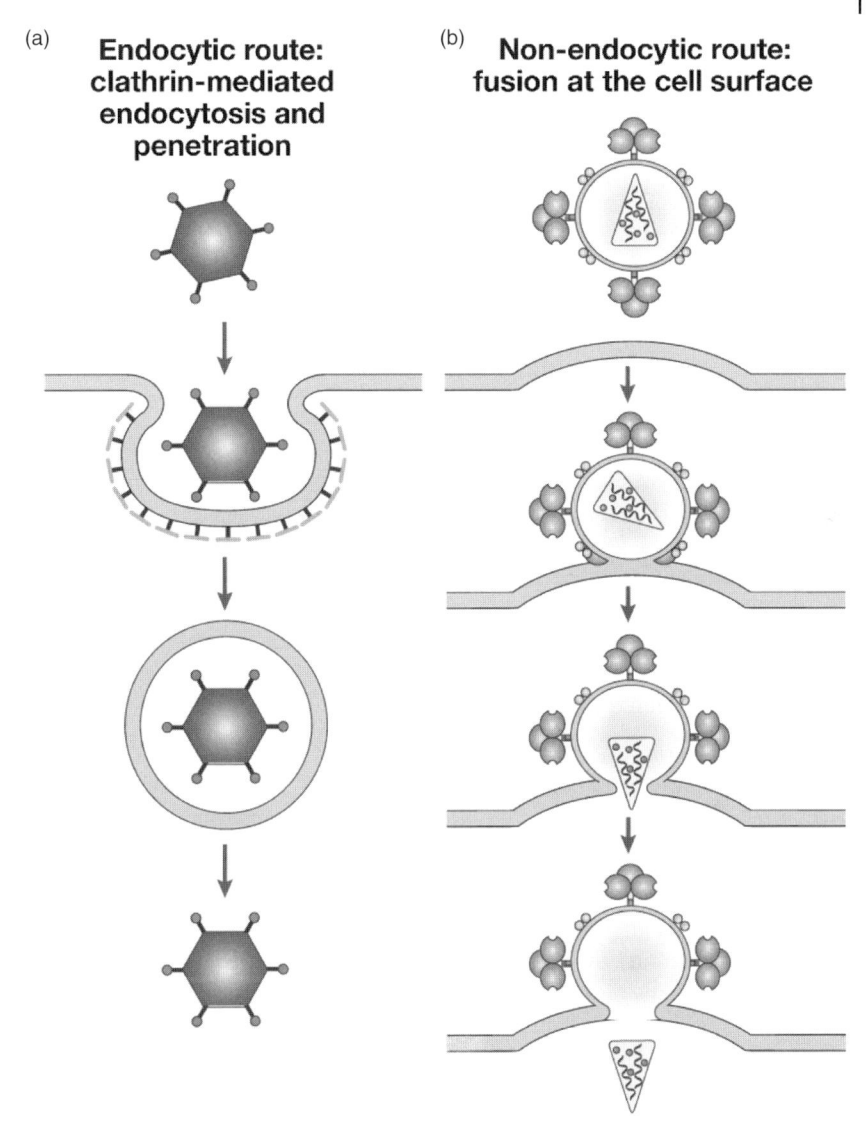

Figure 6.5 The two main virus entry pathways. (a) Endocytic route; (b) Nonendocytic route. Modified after Ref. [36].

may involve either clathrin- or caveolin-dependent pathways or, alternatively, a clathrin- and caveolin-independent pathway [34, 41–46]. Those viruses that use the endocytic route undergo a conformational change of the viral entry proteins or target cell receptors, triggered by low endosomal pH, followed by endosomal fusion for enveloped viruses, or by endosomal escape for nonenveloped viruses [47–49]. Viruses that use the nonendocytic route enter cells by directly crossing the plasma membrane of the cell (i.e., cell membrane fusion) at neutral pH. Some examples of viruses that use this route include the human immunodeficiency virus type-1 (HIV-1) and the herpes simplex virus 1 (HSV-1) [36]. The cell membrane components (e.g., lipid raft) [50–52] and the cytoskeletons, such as microtubules and actin-filaments [38, 53, 54], have also been suggested to have essential roles in viral entry. The ability to track individual viruses enables the possible elucidation of previously unknown routes of entry. Additionally, the critical steps involved in the penetration of viruses into cells, and in their dissemination, might also reveal novel therapeutic opportunities for controlling virus pandemics and pathogenesis [34, 35].

6.3.1
Conventional Labeling for Single-Virus Tracking in Live Cells

Traditionally, in order to track single viruses in live cells both the virus and the relevant cellular structures must be fluorescently labeled; fluorescence microscopy can then be used to detect and then track the labeled particles. One key problem here is that the virus and appropriate cellular structures must be labeled fluorescently in such a way that they can be easily visualized, but not to a level where the functions of the cell and virus could be affected. It is for this reason that antibodies to viral proteins cannot be used for virus labeling, as they block the function of the viral proteins after binding. Currently, two methods are used to incorporate a fluorescent label onto the cell or virus:

- *Fluorescent proteins* may be used by encoding them in the viral genes, such that they are incorporated with the viral structures [55–57]. As several copies are required per virion, capsid or tegument proteins are normally selected; consequently, this method can only be used with certain types of virus.

- A *chemical label* can be used which can be attached either covalently or noncovalently to the proteins. For example, a virus has many $-NH_2$ groups on its surface to which amino-reactive dyes can be attached. Although concern has been expressed that the nonspecific modification of lysine residues on the viral protein may disrupt viral functions and infectivity, this labeling method has been relatively well-characterized for labeling the capsid proteins of nonenveloped viruses [58, 59].

Lipophilic dyes (e.g., 1,1′-dioctadecyl-3,3,3′,3′-tetramethylindodicarbocyanine; DiD) can be incorporated spontaneously into the outer membrane of enveloped viruses, and can be used to label the viral envelope membrane [55, 60, 61]. Although this membrane-labeling method is widely used for studies of virus-target membrane fusion mechanisms, the virus cannot be tracked after the membrane fusion process has occurred. Consequently, these dyes have limited applications for imaging.

Following their fluorescent labeling, the viruses can be tracked in live cells by detection with fluorescence microscopy, for which three principal set-ups are available [35]:

- **Epi-fluorescence microscopy:** This has the simplest set-up and features low signal loss, a large imaging depth, and rapid wide-field detection. However, the autofluorescence inherent in cells results in a high background noise with this method, which makes it unfavorable for situations where a small number of fluorescent molecules must be detected.

- **Confocal fluorescence microscopy:** This can be either in the form of the laser scanning confocal microscope (for typical use) or the spinning-disc confocal microscope, when the imaging of rapid dynamics in live cells is required. Confocal microscopes have much lower background noise and are able to produce three-dimensional images, but they lose much of the fluorescence signal during the imaging process.

- **Total internal reflection fluorescence (TIRF) microscopy:** This allows wide-field imaging with less signal loss than confocal microscopy, and less background noise than epi-fluorescence microscopy. Although the limited imaging depth provides a superior resolution for cell-surface imaging [62], the main problem with this method is its imaging depth. Typically, TIRF microscopy is able to image only at a depth of 100–200 nm, although events occurring near the cell surface, such as viral entry, can be clearly detected.

6.3.2
Problems Encountered in Single-Virus Tracking

Today, several difficulties may be faced in the fluorescence labeling and imaging of viral particles. First, the small size of viral particles limits the number of fluorescent probes that can be attached without causing a self-quenching effect; the attachment of too many dye molecules may also affect the infectivity of the virus. For example, if the adeno-associated virus (AAV) has more than a few dye molecules attached to it, it will become noninfectious [63]. Another problem lies in the size of the fluorescent proteins, with large and bulky molecules possibly leading to the disruption of viral structures and a marked loss of viral infectivity [64, 65]. The major problem when viruses are labeled with either dyes or fluorescent proteins is that the fluorophores may be destroyed photochemically in an excitation-induced phenomenon called *photobleaching*. This limits the long-term tracking of single viruses in living cells, which is necessary when monitoring the dynamic interactions between the virus and cellular structures. The small size of the virus also necessitates longer exposure times for signal detection, which also leads to photobleaching.

The use of QDs can potentially mitigate many of these obstacles in single-virus tracking, as they exhibit remarkable photostability and brightness while maintaining a broad absorption spectrum and narrow emission spectrum. The inorganic nature of QDs provides them with an improved resistance to metabolic degradation, or photobleaching, over the organic fluorophores that, in turn, enables longer tracking of viral particles in living cells.

6.3.3
Use of QDs for Labeling Enveloped Viruses

Enveloped viruses are enclosed by a lipid membrane, which is gathered from the host cell membrane during the budding process, or when the virus moves through the cell's membrane system such as the endoplasmic reticulum or the Golgi apparatus. Examples of enveloped viruses include HIV, influenza virus, severe acute respiratory syndrome (SARS) virus, and hepatitis C virus. As conditions that damage membranes will also damage the envelopes of these viruses, they are usually more fragile than nonenveloped viruses. Hence, more gentle and nondisruptive labeling methods are needed.

One method of labeling enveloped virus with QDs is through a site-specific scheme (Figure 6.6) [66]. The strategy was to first incorporate a 15-amino acid

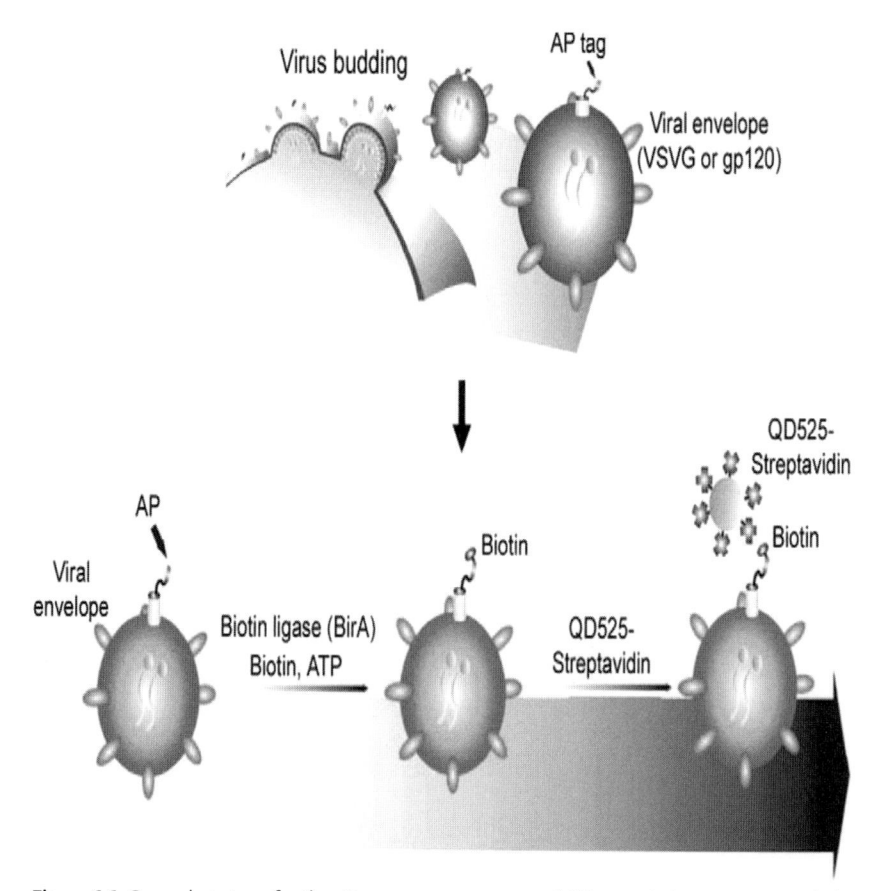

Figure 6.6 General strategy for the site-specific labeling of enveloped viruses with QDs. The biotin acceptor peptide (AP) tag is incorporated on the surface of viruses while they are budding from virus producing cells expressing AP tag on the cell surface. The concentrated AP-tagged viruses resuspended in PBS/MgCl$_2$ were biotinylated by adding biotin ligase (BirA), ATP, and biotin into virus. Streptavidin-conjugated QDs are then added to bind the biotinylated surface of viruses. Modified after Ref. [66].

biotin acceptor peptide (AP) tag onto the surface of a virion [67, 68], after which biotin ligase (BirA) was used to specifically modify the AP-tag so as to introduce the biotin moiety to the viral surface. Due to the tight interaction between biotin and streptavidin ($K_d = 10^{-13}$ M) [69], the further addition of streptavidin-conjugated QDs allowed the site-specific labeling of viral particles with photostable and fluorescent QDs. The colocalization assay with an antibody specific for HIV capsid protein (p24) confirmed that the AP-tag could be incorporated onto the surface of lentivirus for QD labeling (Figure 6.7a). Additionally, this approach has been shown to have no effect on virus infectivity for both the lentivirus and gamma-retrovirus (Figure 6.7b,c).

Good photostability of the labeling fluorophores is desirable for the continuous tracking of individual viruses, since a high-magnification objective must be used in order to detect these tiny viral particles (20–100 nm), generating a high excitation light intensity in the focal plane of the objective [70]. Also, with the development of new imaging techniques requiring the rapid and continuous excitation of fluorophores for z-stack image acquisition (3-D reconstruction) and time-lapse imaging [71], greater photostability is necessary for detailed trafficking studies. Compared to traditionally labeled fluorophores, these QD-labeled viruses were shown to exhibit excellent photostability against photobleaching (Figure 6.8) [66].

As vesicular stomatitis virus (VSV) is known to enter the cell via endocytosis [72], QD-labeled VSV glycoprotein-pseudotyped lentiviruses were used to test the intracellular tracking of viruses in live cells (Figure 6.9). For this, fluorescent protein-tagged Rab5 was overexpressed in target cells to label early endosomes, then QD-labeled viruses were incubated with the cells for various time periods before confocal imaging. Based on the results obtained, it was determined that the viruses could be effectively tracked in their movement through early endosomes with the use of fluorescence labeling provided by QDs.

The use of QD-labeled viruses was studied further by applying a site-specific scheme to the HIV virus. As HIV is known to bind to CD4 and CCR5, cells expressing CD4 were incubated with the QD-labeled viruses and cell-virus binding was observed through confocal imaging, confirming the validity of this method to label HIV with QDs (Figure 6.10a). Furthermore, studies of the interaction between HIV and DC-SIGN, a dendritic cell marker, revealed that HIV enters the DC-SIGN-expressing cell through clathrin-dependent endocytosis (Figure 6.10b–d). In theory, this site-specific QD-labeling method could be applied to any enveloped virus in order to image trafficking mechanisms through different cell types.

6.3.4
Use of QDs for Labeling Nonenveloped Viruses

Nonenveloped viruses are small but stable particles that lack a lipid bilayer membrane. As they can form crystals that diffract to good resolutions, the structures of a relatively large number of representatives from different virus families have been determined using X-ray crystallography at high resolution [36]. However, despite these viruses being relatively small and simple in structure, their

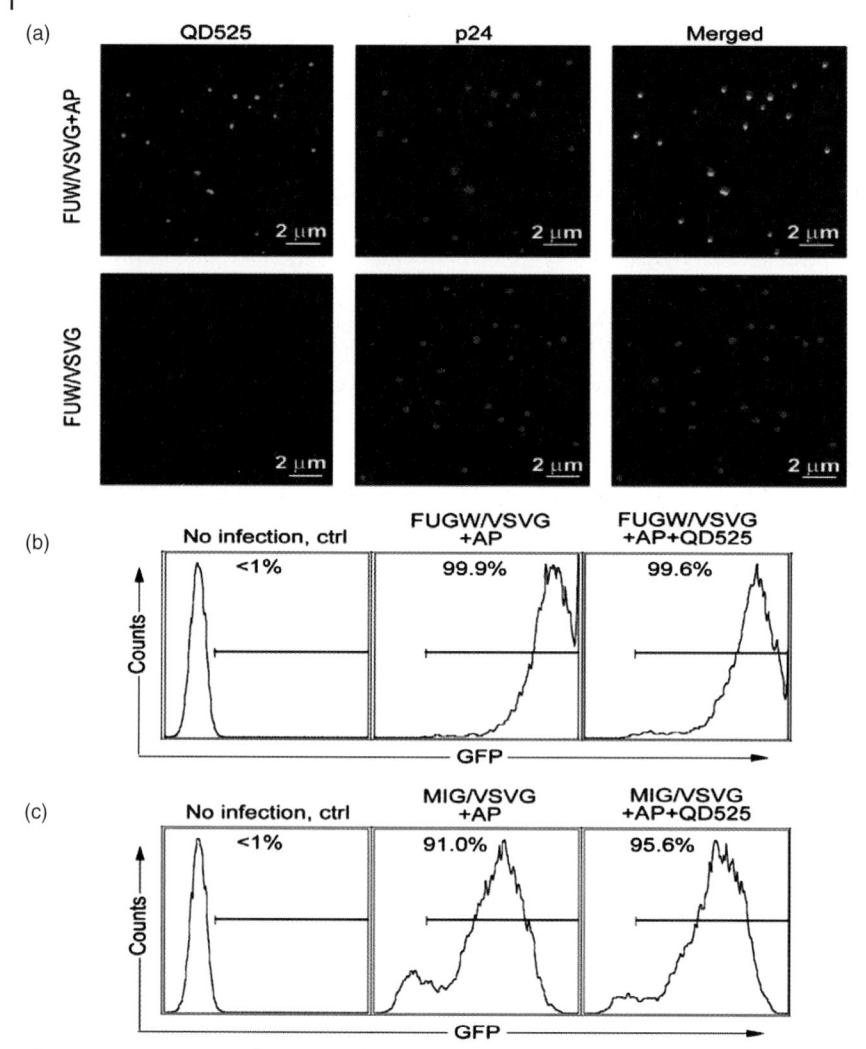

Figure 6.7 Incorporation of AP tag on the surface of lentivirus for QD525 labeling. (a) VSVG-pseudotyped lentiviruses produced by cotransfection either with AP-TM (FUW/VSVG + AP) or without AP-TM (FUW/VSVG) were incubated with BirA, ATP, and biotin, followed by labeling viruses with streptavidin-QD525. The viruses (green) were overlaid upon poly-lysine coated coverslips for 60 min at 37 °C. The coverslips were fixed, permeabilized, and immunostained with antibodies specific to HIV capsid protein p24 (red). Overlapping green and red signals appears as yellow in a merged image. Scale bars = 2 μm; (b) AP-tag bearing VSVG-pseudotyped lentiviruses encoding a GFP reporter gene were produced and biotinylated by adding BirA, ATP, and biotin. The half of biotinylated viruses were further labeled with streptavidin-QD525 (FUGW/VSVG + AP + QD525), and anther half was used without QD labeling as the control (FUGW/VSVG + AP). 293T cells (2×10^5) were spin-infected with QD-labeled or unlabeled viruses. The resulting GFP expression was analyzed by FACS; (c) Similarly, 293T cells were spin-infected with QD-labeled gamma-retrovirus (MIG/VSVG + AP + QD) or unlabeled gamma-retrovirus (MIG/VSVG + AP). The resulting GFP expression was analyzed by fluorescence-activated cell sorting (FACS). From Ref. [66].

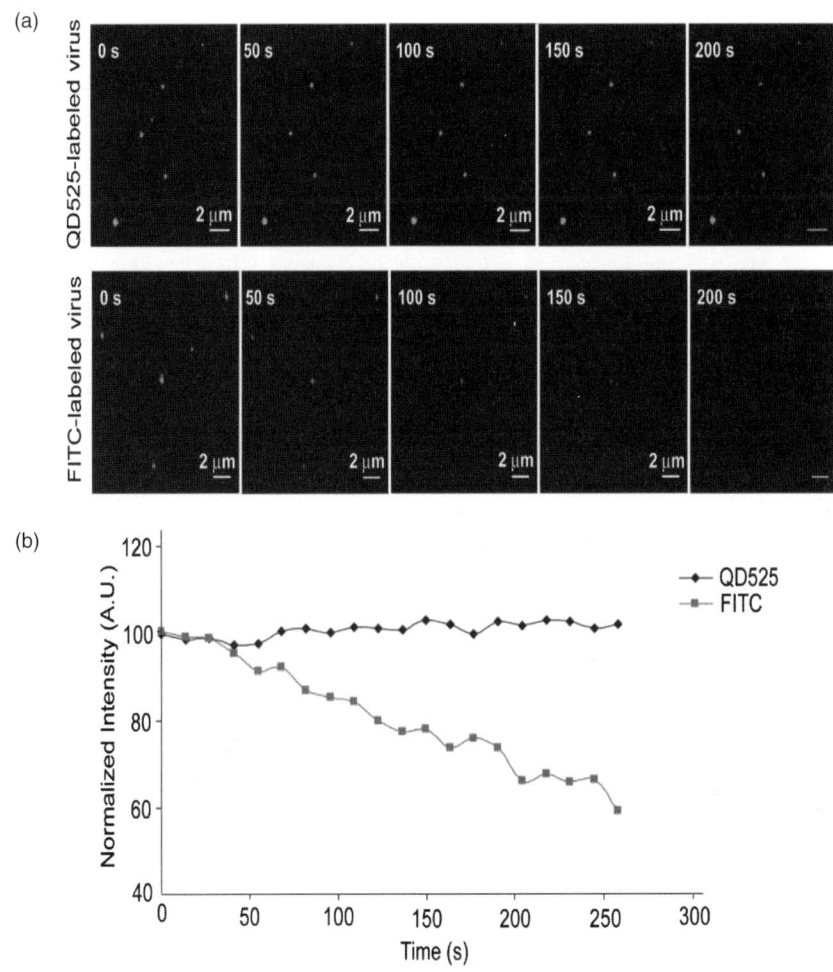

Figure 6.8 Photostability comparison between QD-labeled or FITC-labeled viral particles. (a) Biotinylated VSVG-pseudotyped lentiviruses were labeled with streptavidin-QD525 or streptavidin-FITC and overlaid upon poly-lysine-coated coverslips. The specimens were continuously illuminated by argon laser at 488 nm over 3 min. Images were captured at ~10 s intervals. Scale bars represent 2 μm; (b) Kinetics of the fluorescence intensity of QD-labeled or FITC-labeled viral particles. The fluorescent intensity of viral particles was measured using the software package for the Zeiss LSM 510. From Ref. [66].

membrane penetration and uncoating mechanisms are poorly understood [73]. The membrane penetration process requires a significant reorganization of the viral nucleoprotein complex compared to enveloped viruses [36, 73]. Some examples of nonenveloped viruses include the poliovirus, rotavirus, parvovirus, including the AAV, and the adenovirus. Due to the robust nature of nonenveloped viruses, the capsids of purified viruses can be labeled by the covalent attachment of dyes, without any significant loss of virus infectivity [58, 59].

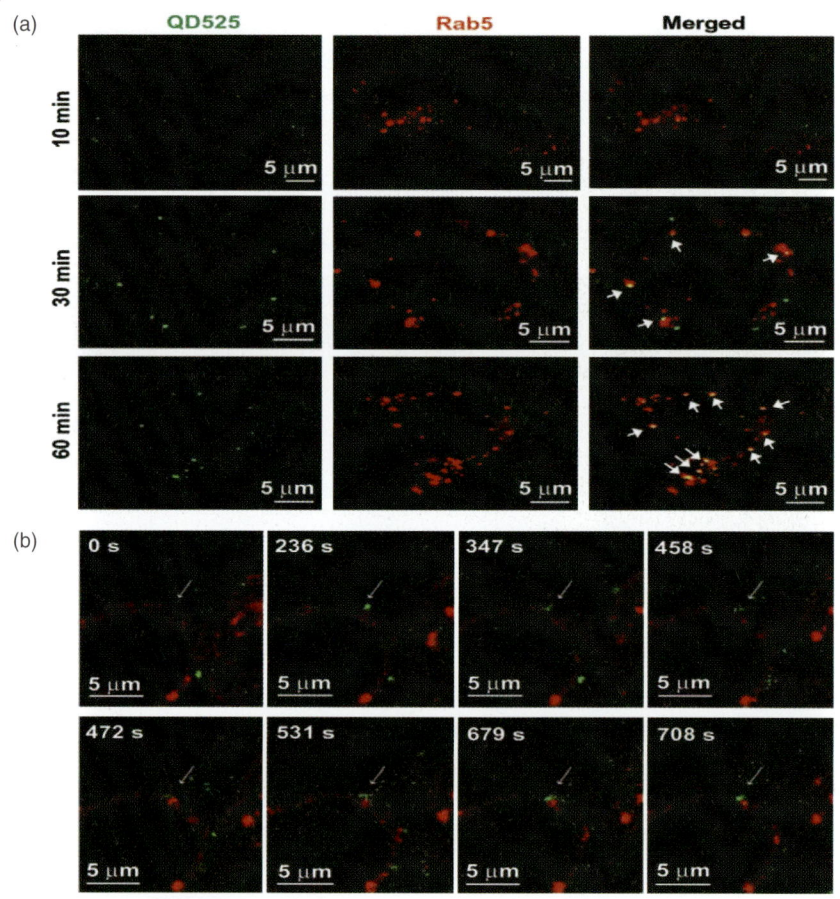

Figure 6.9 The trafficking of QD-labeled viral particles through endosomes. (a) 293T cells transiently transfected with DsRed-Rab5 (red) were seeded on the glass-bottomed dish at 48 h post-transfection. QD-labeled VSVG-pseudotyped lentiviruses (green) were then incubated with the cells for 30 min at 4 °C to synchronize infection. The cells were shifted to 37 °C for various time periods (10, 30, 60 min) and then fixed; (b) Real-time monitoring of QD-labeled viral transport to endosomes. Rab5 (red) expression 293T cells were incubated with QD-labeled VSVG-pseudotyped lentiviruses (green) for 30 min at 4 °C and shifted to 37 °C for 10 min to initiate virus internalization. Confocal time-lapse images were then recorded. The arrows indicate the internalized viral particle. Scale bars = 5 μm. From Ref. [66].

AAV, in particular, has attracted considerable interest because it shows great promise for use in human gene therapy [74]. As the AAV virion is only about 20 nm in diameter, the number of dye molecules that can be attached to a single virus without causing self-quenching or affecting viral infectivity is very limited [63]. However, the much greater brightness of QDs compared to conventional fluorophores means that the viruses can be detected with a much lower loading of the labeling molecules. It is believed that AAV virions are transported into the

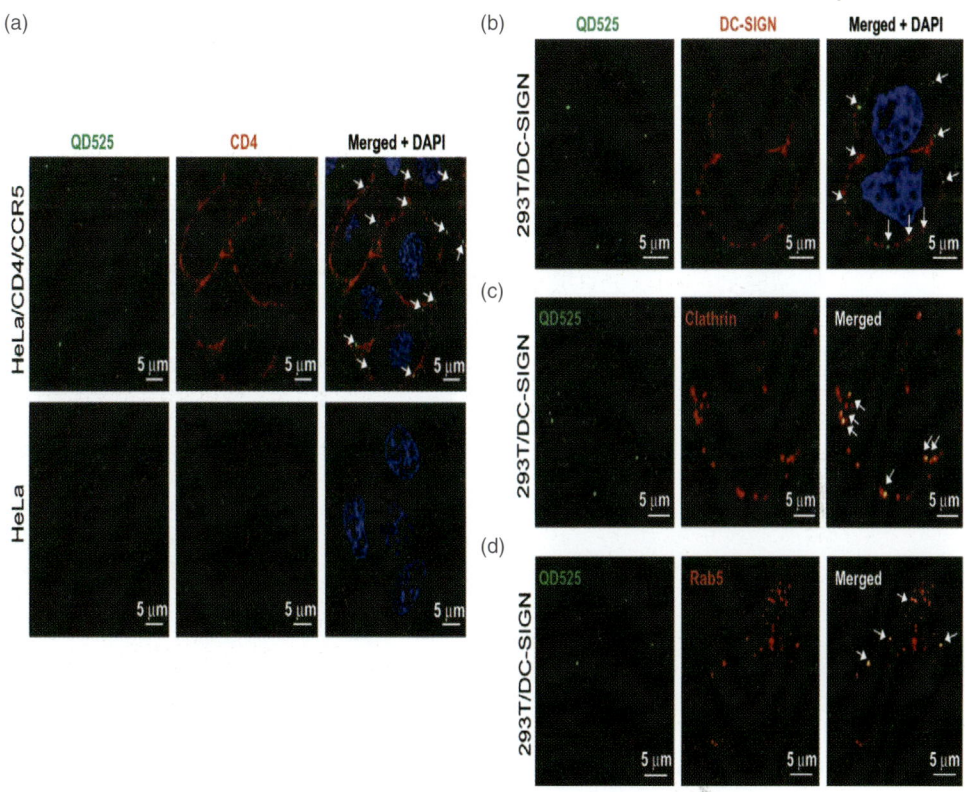

Figure 6.10 Intracellular trafficking of QD-labeled HIV. (a) Visualization of HIV viral particles on the surface of HeLa cells expressing viral receptor CD4 and coreceptor CCR5. HeLa cells or HeLa/CD4/CCR5 cells were seeded on the glass-bottomed dish overnight. The cells were incubated with QD-labeled HIV viral particles (green) for 30 min at 4 °C, fixed, and then immunostained with anti-human CD4 antibodies (red) and counterstained with DAPI (blue). The arrows indicate QD-labeled HIV viral particles bound to the cell surface; (b) 293T/DC-SIGN cells were seeded and incubated with QD-labeled HIV viral particles (green) for 30 min at 4 °C. The cells were fixed and immunostained anti-human DC-SIGN antibodies (red) and counterstained with DAPI (blue). The arrows denote HIV viral particles bound to DC-SIGN; (c) Involvement of clathrin-dependent pathway of HIV viral entry. 293T/DC-SIGN cells were transiently transfected with DsRed-clathrin and seeded on the glass bottom dish at 48 h post-transfection. The cells were incubated with QD-labeled HIV viral particles for 30 min at 4 °C to synchronize infection, sifted to 37 °C for 10 min, and then fixed. The arrows denote HIV viral particles that were colocalized with clathrin; (d) HIV viral transport to endosomes. 293T/DC-SIGN cells transiently transfected with DsRed-Rab5 were seeded on the glass-bottomed dish. The cells were incubated with QD-labeled HIV viral particles for 30 min at 4 °C, shifted to 37 °C for 30 min, and then fixed. The arrows indicate HIV viral particles that were colocalized with early endosomes. Scale bars = 5 μm. From Ref. [66].

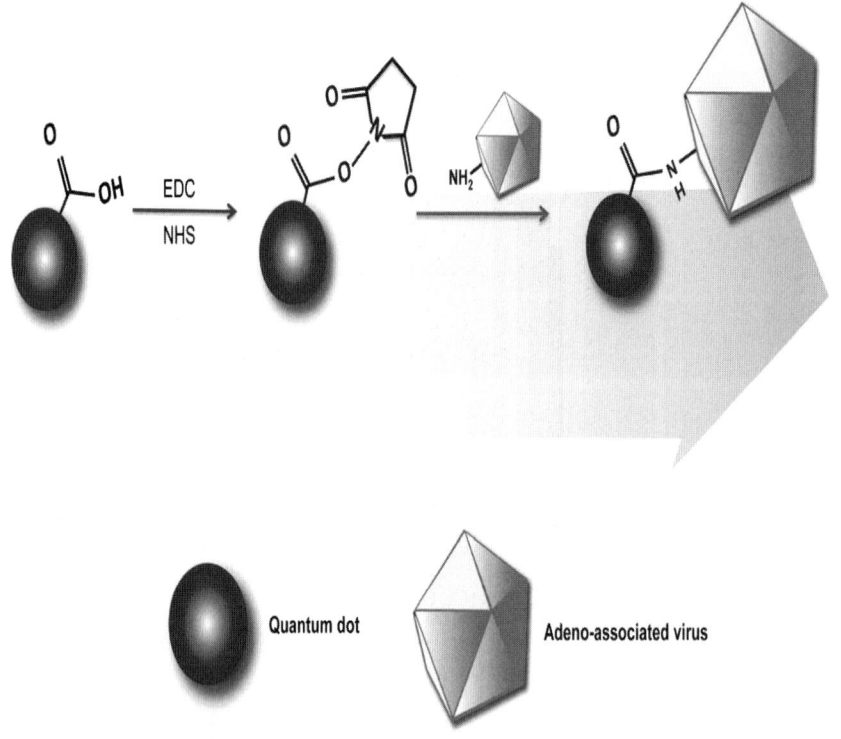

Figure 6.11 Covalent attachment of QDs dots on adeno-associated virus serotype 2 (AAV2). The QDs-AAV2 networks are generated by an amide-bond formation between the carboxylic source of QDs and primary amines from lysine residues on the AAV capsid, via carbodiimide chemistry.

nucleus before viral uncoating occurs [49, 63]; thus, the viral capsid can be targeted for labeling by QDs to track viral particle movements in the cell.

The general strategy for linking adeno-associated virus serotype 2 (AAV2) with QDs through a coupling reaction is illustrated in Figure 6.11. The QD-AAV2 networks were produced by covalent amide bonds formed by carbodiimide chemistry between carboxylic moieties on the QDs and the primary amines from the lysine residues on the virus capsid protein. The carboxyl groups on the QDs were incubated with EDC and NHS to modify the carboxyl group to an amine-reactive NHS ester. Excess EDC and NHS were then removed using a gel filtration column, and the primary amine source, AAV2, was added for a coupling reaction. Conformation that the QDs dots were indeed coupled to AAV2 was made by overlaying the particle solutions onto coverslips and immunostaining the adhered viral particles with an antibody specific for intact AAV2. The fluorescence signal of the QDs on the viral surface was readily detected, and almost all of the QD signals were colocalized with the signal generated by anti-AAV2 antibody staining (Figure 6.12a). The results of many studies have suggested that AAV2 enters the cell

(a)
Merged | anti-AAV2 | QD705

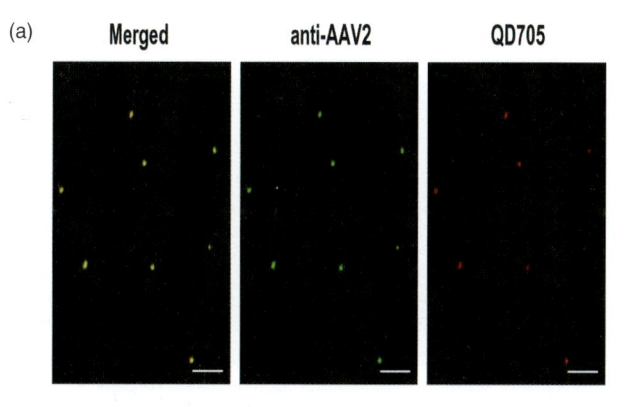

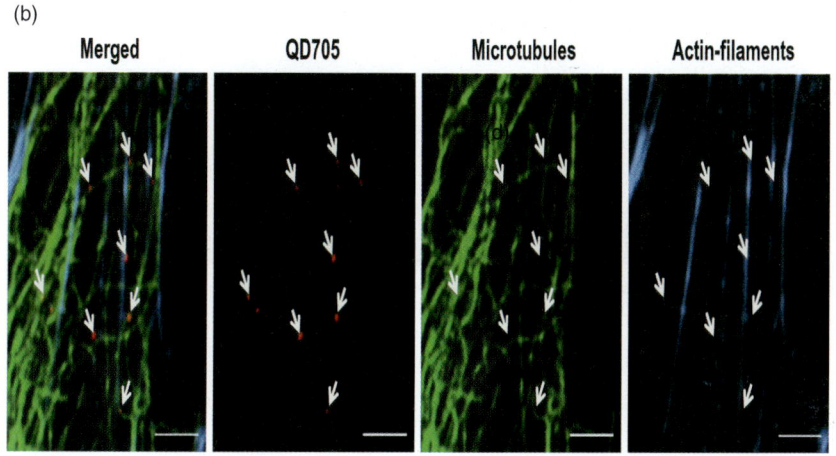

(b)
Merged | QD705 | Microtubules | Actin-filaments

Figure 6.12 Intracellular tracking of QD705-labeled adeno-associated virus 2. (a) Colocalization of QD705-labeled AAV2 with anti-AAV2 antibody. QD705-labeled AAV2 (red) were overlaid upon poly-lysine-coated coverslips for 60 min at 37 °C. The coverslips were fixed and immunostained with an antibody specific to AAV capsid protein (green). Overlapping green and red signals appear as yellow in a merged image. Scale bars = 2 μm; (b) Cytoskeletons-mediated transport of AAV2. QD705-labeled AAV2 (red) were incubated with HeLa cells for 30 min at 37 °C. The cells were then fixed, permeabilized, and stained for microtubules (green) and actin-filaments (blue) with the monoclonal antibody to α-tubulin and rhodamine-conjugated phalloidin, respectively. The arrows indicate viral particles on either microtubules or actin-filaments. Scale bars = 5 μm.

through both clathrin- and dynamin-dependent endocytosis [75, 76]. It has also been suggested, based on the results of drug-inhibition studies, that transport of the virus to the nucleus is dependent on microtubules and/or actin-filaments [77]. The direct visualization of QD-AAV2 virions in living cells suggested that viral transport involved both microtubules and actin-filaments (Figure 6.12b).

Live-cell imaging technology with fluorescent microscopy (e.g., spinning disk confocal microscopy) has allowed tracking of the dynamic interactions between

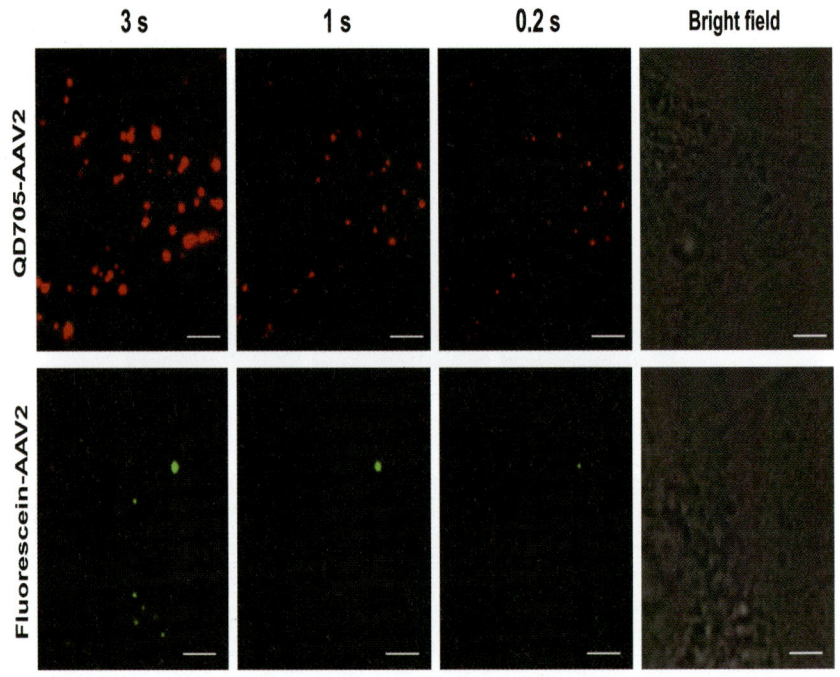

Figure 6.13 Detection sensitivity comparison between QD705-labeled or FITC-labeled viral particles at different exposure times. AAV2 were labeled with either QD705 or an amine-reactive dye (fluorescein-NHS) and incubated with HeLa cells for 30 min at 37 °C. The cells were then fixed and illuminated by an argon laser at 491 nm at different exposure times. Scale bars = 5 μm.

viruses and cellular structures. In multicolor real-time imaging studies, it is always desirable to acquire an image with exposure times as short as possible, to monitor the dynamics of virus trafficking in living cells in greater detail. However, the limited number of fluorescent dyes that can be attached to a single virus without affecting viral infectivity sets an upper limit to the exposure time needed to detect the fluorescent signal emitted from a tiny virus. The use of QDs can, at least potentially, offer advantages for fast, multicolor time-lapse imaging due to their remarkable brightness compared to conventional fluorophores. When the sensitivity of the QD-AAV2 particles was compared to that of fluorophore-labeled AAV2, the QD signal was seen to be much brighter and detectable after a much shorter exposure time (Figure 6.13). Thus, QD labeling not only allows the detection of surface proteins that are present in smaller numbers, but also permits the observation of the dynamics of virus trafficking during live cell imaging in more detail.

6.4
Quantum Dots for the Sensitive Detection of Virus and Infection

6.4.1
Monitoring the Progression of Viral Infection with Fluorescent QD Probes

Respiratory syncytial virus (RSV) is an enveloped RNA paramyxovirus that is primarily responsible for the lower-respiratory tract infections typically seen in infants and children aged under 5 years [78, 79]. The virus is enclosed in a lipid membrane envelope incorporated with fusion and attachment proteins (Figure 6.14a). Inside, the matrix proteins support the structure and surround the nucleocapsid core that contains the viral genome, nucleoprotein, phosphoprotein, and polymerase proteins [80, 81]. The attachment proteins help the virus bind to the cell surface, and the fusion protein is then responsible for mediating fusion of the viral and cell membranes to deliver the nucleocapsid into the cellular cytoplasm [81]. When the viral nucleocapsid has entered the cell, transcription of the viral genes occurs and both mRNA and viral proteins are produced. When the full-length genomic RNA has been replicated, the RNAs and viral proteins are packaged into new virions that bud out from the surface of the host cell; this results in fully functional virions capable of infecting other cells. An interesting phenomenon that occurs with these viruses is that the fusion proteins on the surface of the infected cells may cause fusion with neighboring cells, producing giant multinucleated cells termed *syncytia* (Figure 6.14b). As the fusion and attachment proteins are displayed on the infected cell surface, they may serve as good antigen markers for RSV detection. Also, as the virus replicates, the number of these proteins is amplified as a function of time; consequently, QDs can be used not only to detect the presence of RSV infection but also to monitor its progression [81]. The QDs were attached through antibodies and protein molecules to the fusion and attachment proteins on the viral surface. By using this method, it proved possible to link 605 nm and 525 nm streptavidin-QDs to the fusion protein and attachment protein, respectively.

For the detection of RSV infection, confocal microscopy was used four days after the initial viral infection to examine the presence of the fusion and attachment proteins. These were observed mostly on the surface of the infected cells, as evidenced by the colocalization of the signals from the 525 and 605 nm channels (Figure 6.14c). This was consistent with the pathophysiology of RSV infection determined previously [80, 82]. Interestingly, although the fusion and attachment proteins were colocalized in the images, orthogonal slices of the same images showed that the proteins were actually segregated in some areas at opposite surfaces of the cell.

After determining that QDs could be used successfully to label the RSV fusion and attachment proteins, the development of viral infection was monitored through the expression of those proteins over time. Images were taken of the cells at various time points after infection, and amounts of both fusion and attachment

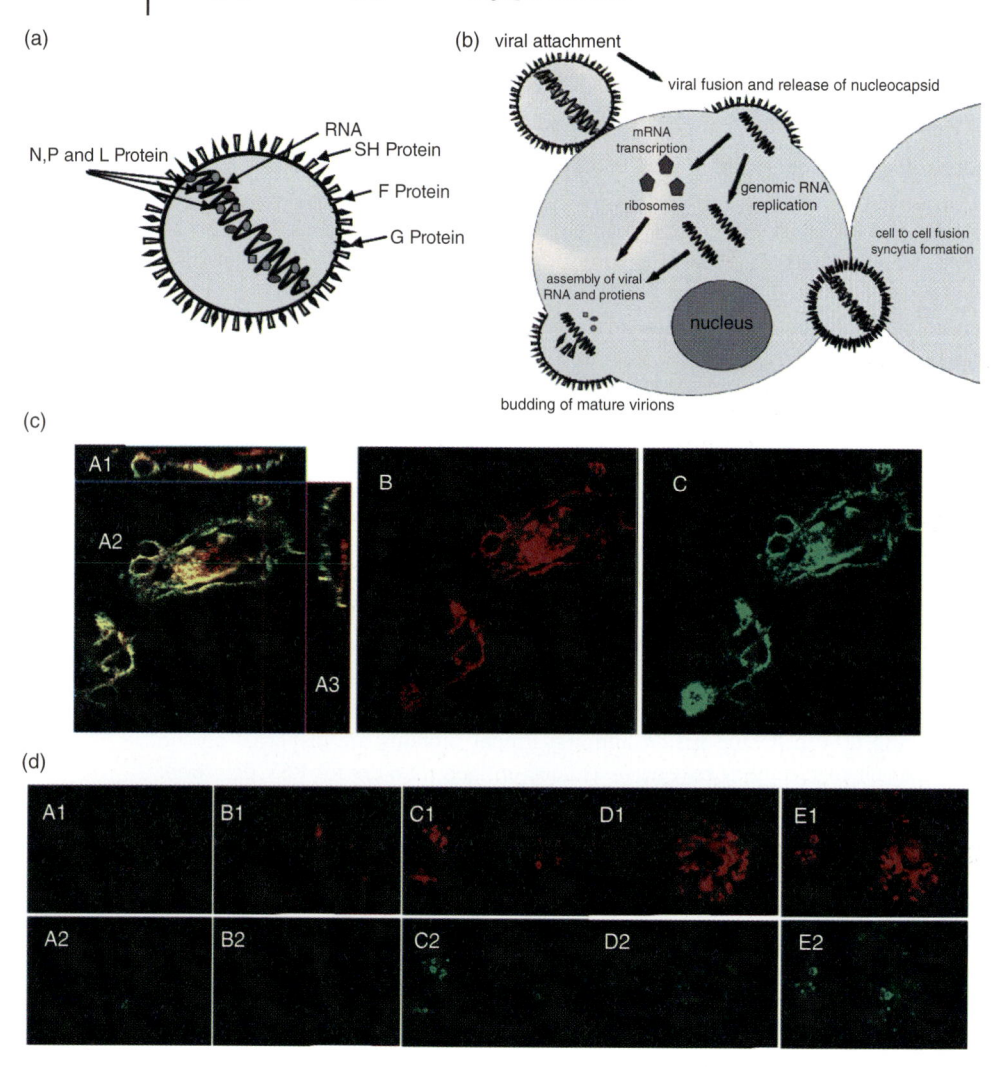

240 | *6 Imaging and Tracking of Viruses Using Quantum Dots*

(a)

N,P and L Protein — RNA — SH Protein — F Protein — G Protein

(b) viral attachment → viral fusion and release of nucleocapsid

mRNA transcription
ribosomes
genomic RNA replication
assembly of viral RNA and protiens
nucleus
cell to cell fusion syncytia formation
budding of mature virions

(c) A1, A2, A3, B, C

(d) A1, B1, C1, D1, E1, A2, B2, C2, D2, E2

Figure 6.14 The structure and life cycle of respiratory syncytial virus. (a) The structure shows the location of the fusion (F) and attachment (G) proteins at the surface of the virion particle in the lipid bilayer envelope. Antibodies to these two surface glycoproteins were used to label RSV-infected HEp-2 cells; (b) The replication cycle of RSV. The RSV virion attaches to the host cell via the G protein. The F protein then mediates fusion of the viral membrane into the host cell membrane. The RSV virion nucleocapsid is released into the host cell cytoplasm. The host cell ribosomes are used to translate the RSV mRNAs transcribed by the viral polymerase to produce viral proteins; the full-length RSV genomic RNA is replicated at later stages of infection. These newly synthesized viral proteins and genomic RNA are assembled and a new virion buds from the plasma membrane at the surface of the infected cell. Additionally, F protein on the infected cell and the newly budding virions may fuse with neighboring cells to form syncytia; (c) Confocal microscopic image showing colocalization of the F and G proteins in RSV-infected HEp-2 cell monolayer cultures at 4 days post infection. The composite image (left) shows orthogonal slices XZ (A1) and YZ (A3), suggesting that the F and G proteins are located predominantly on the surface of the infected cells. In addition to the colocalization, the images show segregated areas of F and G protein at opposing surfaces. The images in panel (B) (605 nm channel) and panel (C) (525 nm channel) are of the same syncytium; (d) Fluorescent images of the F protein labeled with 605 nm streptavidin QDs at 1 h (A1), 24 h (B1), 48 h (C1), 72 h (D1), and 96 h (E1) after infection. The G protein (A2, B2, C2, D2, and E2) was subsequently labeled using 525 nm streptavidin QDs on the same infected monolayer; images are shown for the same time intervals. Modified after Ref. [81].

proteins were seen to increase over time (Figure 6.14d). Both proteins were detectable at the earliest time point, which was at 1 h after infection. To determine the difference between QD labeling and organic fluorophore labeling, the fusion protein was labeled with fluorescein isothiocyanate (FITC), a common organic fluorophore, and imaged over the same time period. The FITC-conjugated antibodies were able to detect fusion proteins at time points after 24 h; however, exposures 46-fold greater than those of the QD-labeled cells were required, and the organic fluorophores suffered from near-complete photodegradation within 15 min [81].

Subsequently, a system was designed that was capable of utilizing multiple probes for simultaneous determination of the spatial distribution of different viral proteins during the various stages of infection. This system may be applied for a quick and early detection of viral infection using QDs, on the basis of their advantages over conventional fluorophores. It can also provide a novel method to study several aspects of viral infection, such as the spatial features of infected cells or the intracellular movements of viral proteins.

6.4.2
The Use of Two-Color QDs for the Real-Time Detection of Viral Particles and Viral Protein Expression

Another method used to detect viral particles employs dual-color QDs capable of simultaneous excitation with a single light source [83, 84]. For this, a microcapillary flow cytometry system was set up with a fixed-point confocal microscope for the rapid and accurate detection of RSV. At the time, the currently available methods of detection were limited by sensitivity and involved organic dyes or fluorescent proteins, which lacked the brightness and stability to study single molecules. However, by using dual-color QDs capable of simultaneous excitation with a single light source, a time-correlation procedure could be used to detect single viral particles [84].

Specifically, a microcapillary flow system was coupled with a fixed-point confocal microscope to detect photon emission from the QDs in the probe volume. The photons were then spectrally separated and analyzed for their time of detection. In order to detect the presence of RSV, the QDs were linked to monoclonal antibodies of either the anti-RSV fusion protein or the anti-RSV attachment protein, and incubated with RSV. The photons released by excitation of the nanoparticles were detected by the confocal probe in real time. The detected signals from the two different nanoparticle types should only coincide if the fusion and attachment proteins were present on the same particle [84]. Also, the photon emission intensity should be a quantitative representation of the relative amounts of protein expression on the viral surface [84]. A particle with a greater amount of surface protein expression should bind more of the corresponding QD, resulting in a greater photon emission and a higher intensity in the photon count spectrum. Using this method, the level of RSV fusion protein expression was compared for different RSV strains [84].

As a result, a method was demonstrated which overcame the difficulties of labeling viruses with fluorescent dyes that required excitation with two different wavelengths. Dual-color QDs were utilized in the detection system for the rapid and simultaneous, qualitative and quantitative detection of different viral proteins. This information could subsequently be used to determine the presence, and possibly also the type, of RSV in fluid samples.

6.4.3
Viral Detection with pH-Sensitive QDs in a Biological Motor

Multiplexed detection systems are extremely useful in the simultaneous detection of different biological targets. Indeed, a system has been created which detects different viruses with multicolored QDs by embedding the QDs into polymeric microbeads and using ATPase as a biological motor (Figure 6.15a) [85]. In this procedure, the chromatophores (pigment-containing cells) were first labeled with pH-sensitive CdTe QDs that were either green or orange. ATPase was then linked to a herpes virus antibody and added to the orange QD chromatophores, and also linked to a H9 Flu virus antibody for the green QD chromatophores. When ADP was added to start the reaction, the ATPase caused protons to be pumped out of the chromatophores. Then, when the viruses were added and the corresponding viruses bound to the ATPase-QD-chromatophore sensors, the activity of the ATPase was enhanced such that more protons were pumped out of the chromatophores; this led to a change in the pH-sensitive fluorescence intensity of the QDs. By measuring the rate of fluorescent intensity change, the activity of ATPase could

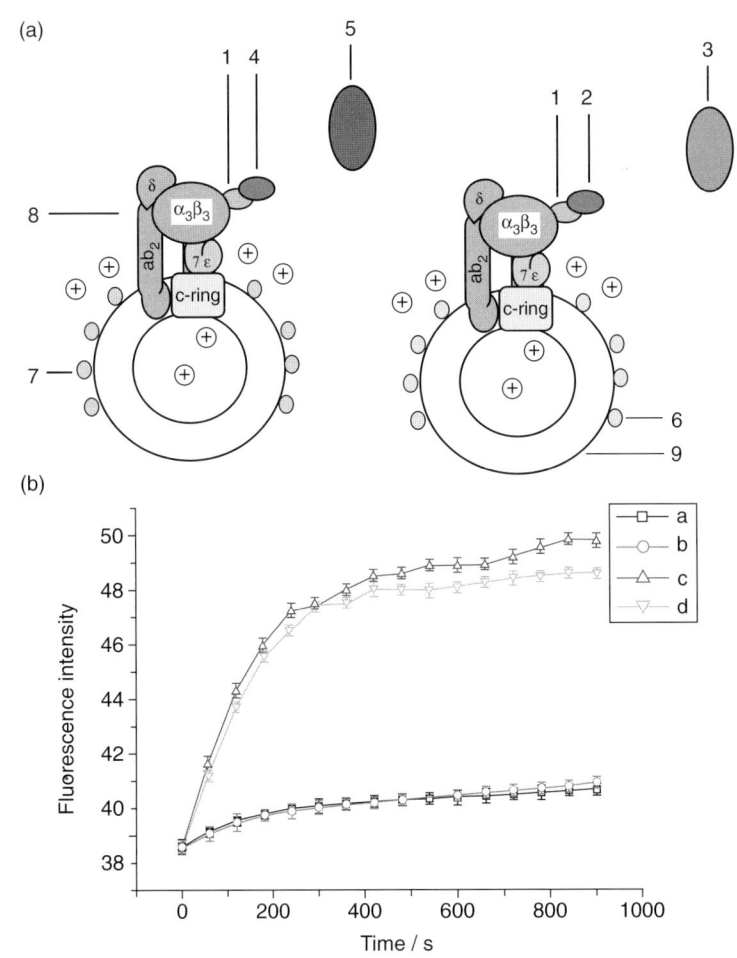

Figure 6.15 (a) The basic design of QD biosensors based on F0F1-ATPase: (1) antibody of β-subunit; (2) the antibody of MHV68; (3) MHV68; (4) the antibody of H9 avian influenza virus; (5) H9 avian influenza virus; (6) CdTe QDs with emission wavelength at 585 nm; (7) CdTe QDs with emission wavelength at 535 nm; (8) F0F1-ATPase within chromatophores; (9) chromatophores; (b) Changes of fluorescence intensity of QD biosensors with and without viruses. Curve a: changes of fluorescence intensity of orange QD biosensors without MHV68 when ADP is added to start the reaction. Curve b: changes of fluorescence intensity of green QD biosensors without H9 avian influenza virus when ADP is added to start the reaction. Curve c: changes of fluorescence intensity of orange QD biosensors with capturing MHV68 when ADP is added to start the reaction. Curve d: changes of fluorescence intensity of green QD biosensors with capturing H9 avian influenza virus when ADP is added to start the reaction. From Ref. [85].

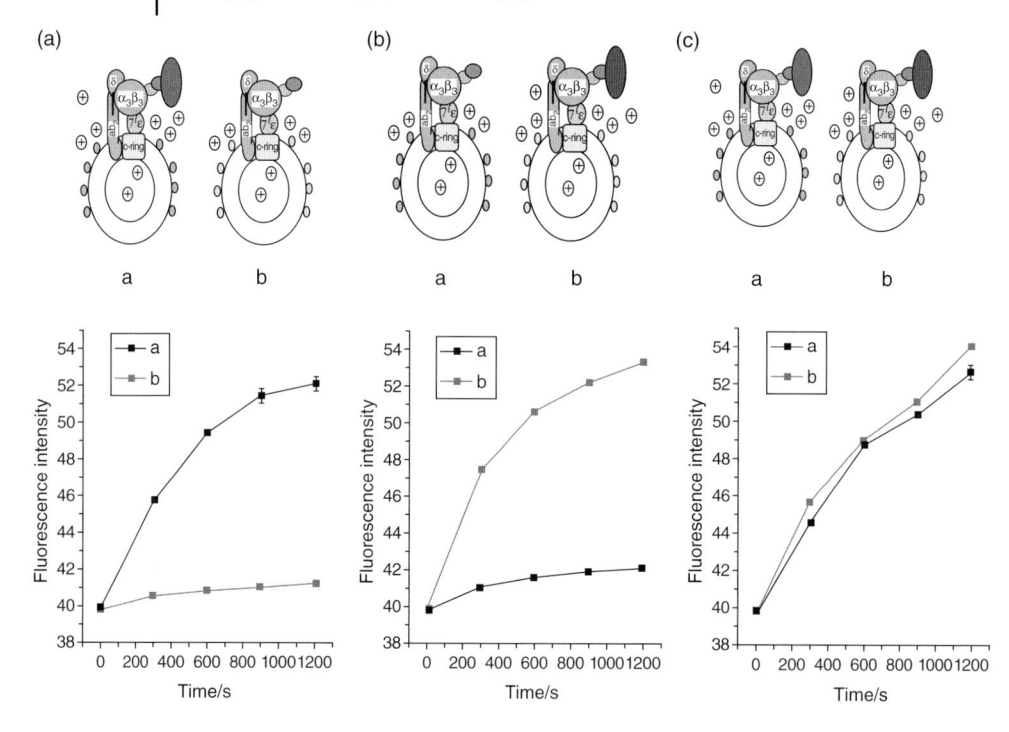

Figure 6.16 Detection of viruses by the QDs-biosensors. (A) H9 avian influenza virus detected by the QD biosensors: (a) QD biosensors (emission at 535 nm) loaded by antibodies (antibody of β-subunit and second antibody of H9 avian influenza virus) with capture of H9 avian influenza virus; (b) QD biosensors (emission at 585 nm) loaded by antibodies (antibody of β-subunit and second antibody of MHV68). Schematic illustrations are shown in insets a and b; (B) MHV68 detected by the QDs-biosensors: (a) QD biosensors (emission at 535 nm) loaded by antibodies (antibody of β-subunit and antibody of H9 avian influenza virus); (b) QD biosensors (emission at 585 nm) loaded by antibodies (antibody of β-subunit with antibody of MHV68) with capture of MHV68. Schematic illustrations are shown in insets a and b; (C) H9 avian influenza virus and MHV68 detected by the QD biosensors simultaneously and independently: (a) QD biosensors (emission at 535 nm) loaded by antibodies (antibody of β-subunit and antibody of H9 avian influenza virus) with capture of H9 avian influenza virus; (b) QD biosensors (emission at 585 nm) loaded by antibodies (antibody of β-subunit and antibody of MHV68) with capture of MHV68. Schematic illustrations are shown in insets a and b. From Ref. [85].

be determined and consequently, the presence of viral particles detected (Figure 6.15b). This procedure was demonstrated successfully with both herpes virus and the H9 Flu virus.

The applicability of this system for the simultaneous sensing of different viral types was then tested by mixing the green and orange QD biosensors and applying different viruses (Figure 6.16). Both, herpes virus and H9 Flu virus were added to the QD mixture, individually and simultaneously, and were detected successfully. Even with a population of mixed green and orange biosensors, the corresponding viruses were detected with no interference by the nonrelevant biosensor. Thus, this system could be used as a cheap, rapid, convenient, and effective method to detect several types of viruses simultaneously with a mixture of pH-sensitive CdTe QDs. It is likely that, with further engineering, the QD biosensors could also be used to detect other biomolecules, and applied for the optical encoding of other quantitative and qualitative detection systems that require simultaneous and independent multicolor sensing.

6.5
Conclusions

In recent years, QDs have been shown to be versatile and effective for different fluorescence studies, including virus detection and tracking. Notably, QDs possess several clear advantages over conventional organic dyes, including a greatly improved photostability against metabolic degradation, or photobleaching, which is of great concern in confocal microscopy, and a higher emission intensity. QDs also have the ability to achieve excitation over a wider range of wavelengths while possessing a narrower emission spectrum, which allows for the simultaneous, dual-color excitation of fluorescent particles. These unique qualities have prompted the successful incorporation of QDs in novel systems for the determination of viral structures, the tracking of virus trafficking, and virus detection.

The fact that QDs remain more bulky than conventional organic fluorophores may pose certain difficulties for their use in the labeling of internal viral components, without disrupting viral functionality. Nonetheless, studies are under way to reduce the size of QDs so as to improve their applicability for biological detection processes. Difficulties have also been encountered in purifying conjugated QDs, such as functionalized or antibody-bound, from nonconjugated QDs through conventional purification methods. Currently, molecular weight cut-off (MWCO) is typically used to purify molecules by filtering them to sort by size. However, as QDs are large and similar in size to proteins, they cannot be distinguished and sorted from a mixed population with such a set-up. Although, to date, size-exclusion chromatography has been the most promising and widely used method for purifying conjugated QDs, further studies are required to determine a quicker and more convenient method.

Acknowledgments

The financial support of the National Institute of Health and the Keck foundation is acknowledged. The authors thank Chi-Lin Lee for his help in the preparation of figures.

References

1 Michalet, X., Pinaud, F.F., Bentolila, L.A., Tsay, J.M., Doose, S., Li, J.J., Sundaresan, G., Wu, A.M., Gambhir, S.S. and Weiss, S. (2005) Quantum dots for live cells, in vivo imaging, and diagnostics. *Science*, **307**, 538–44.

2 Bruchez, M. Jr, Moronne, M., Gin, P., Weiss, S. and Alivisatos, A.P. (1998) Semiconductor nanocrystals as fluorescent biological labels. *Science*, **281**, 2013–16.

3 Dabbousi, B.O., Rodriguez-Viejo, J., Mikulec, F.V., Heine, J.R., Mattoussi, H., Ober, R., Jensen, K.F. and Bawendi, M.G. (1997) (CdSe)ZnS core – shell quantum dots: synthesis and characterization of a size series of highly luminescent nanocrystallites. *Journal of Physical Chemistry*, **101**, 9463–75.

4 Hines, M.A. and Guyot-Sionnest, P. (1996) Synthesis and characterization of strongly luminescing ZnS-Capped CdSe nanocrystals. *Journal of Physical Chemistry*, **100**, 468–71.

5 Dubertret, B., Skourides, P., Norris, D.J., Noireaux, V., Brivanlou, A.H. and Libchaber, A. (2002) In vivo imaging of quantum dots encapsulated in phospholipid micelles. *Science*, **298**, 1759–62.

6 Ballou, B., Lagerholm, B.C., Ernst, L.A., Bruchez, M.P. and Waggoner, A.S. (2004) Noninvasive imaging of quantum dots in mice. *Bioconjugate Chemistry*, **15**, 79–86.

7 Jaiswal, J.K. and Simon, S.M. (2004) Potentials and pitfalls of fluorescent quantum dots for biological imaging. *Trends in Cell Biology*, **14**, 497–504.

8 Gerion, D., Pinaud, F., Williams, S.C., Parak, W.J., Zanchet, D., Weiss, S. and Alivisatos, A.P. (2001) Synthesis and properties of biocompatible water-soluble silica-coated CdSe/ZnS semiconductor quantum dots. *Journal of Physical Chemistry, B*, **105**, 8861–71.

9 Alivisatos, A.P., Gu, W. and Larabell, C. (2005) Quantum dots as cellular probes. *Annual Reviews of Biomedical Engineering*, **7**, 55–76.

10 Mattoussi, H., Mauro, J.M., Goldman, E.R., Anderson, G.P., Sundar, V.C., Mikulec, F.V. and Bawendi, M.G. (2000) Self-assembly of CdSe-ZnS quantum dot bioconjugates using an engineered recombinant protein. *Journal of the American Chemical Society*, **122**, 12142–50.

11 Mitchell, G.P., Mirkin, C.A. and Letsinger, R.L. (1999) Programmed assembly of DNA functionalized quantum dots. *Journal of the American Chemical Society*, **121**, 8122–3.

12 Willard, D.M., Carillo, L.L., Jung, J. and van Orden, A. (2001) Cdse-ZnS quantum dots as resonance energy transfer donors in a model protein-protein binding assay. *Nano Letters*, **1**, 469–74.

13 Zhang, C., Ma, H., Nie, S., Ding, Y., Jin, L. and Chen, D. (2000) Quantum dot-labeled trichosanthin. *Analyst*, **125**, 1029–31.

14 Akerman, M.E., Chan, W.C., Laakkonen, P., Bhatia, S.N. and Ruoslahti, E. (2002) Nanocrystal targeting in vivo. *Proceedings of the National Academy of Sciences of the United States of America*, **99**, 12617–21.

15 Rosenthal, S.J., Tomlinson, I., Adkins, E.M., Schroeter, S., Adams, S., Swafford, L., Mcbride, J., Wang, Y., Defelice, L.J. and Blakely, R.D. (2002) Targeting cell surface receptors with ligand-conjugated nanocrystals. *Journal of the American Chemical Society*, **124**, 4586–94.

16 Winter, J., Liu, T., Korgel, B. and Schmidt, C. (2001) Recognition molecule directed interfacing between semiconductor quantum dots and nerve cells. *Advanced Materials*, **13**, 1673–7.

17 Chan, W.C., Maxwell, D.J., Gao, X., Bailey, R.E., Han, M. and Nie, S. (2002) Luminescent quantum dots for multiplexed biological detection and imaging. *Current Opinion in Biotechnology*, **13**, 40–6.

18 Chan, W.C. and Nie, S. (1998) Quantum dot bioconjugates for ultrasensitive nonisotopic detection. *Science*, **281**, 2016–18.

19 Pathak, S., Choi, S., Arnheim, N. and Thompson, M. (2001) Hydroxylated quantum dots as luminescent probes for in situ hybridization. *Journal of the American Chemical Society*, **123**, 4103–4.

20 Gerion, D., Parak, W.J., Williams, S.C., Zanchet, D., Micheel, C.M. and Alivisatos, A.P. (2002) Sorting fluorescent nanocrystals with DNA. *Journal of the American Chemical Society*, **124**, 7070–4.

21 Gao, X.H., Yang, L.L., Petros, J.A., Marshal, F.F., Simons, J.W. and Nie, S.M. (2005) In vivo molecular and cellular imaging with quantum dots. *Current Opinion in Biotechnology*, **16**, 63–72.

22 Howarth, M., Takao, K., Hayashi, Y. and Ting, A.Y. (2005) Targeting quantum dots to surface proteins in living cells with biotin ligase. *Proceedings of the National Academy of Sciences of the United States of America*, **102**, 7583–8.

23 Cambi, A., Lidke, D.S., Arndt-Jovin, D.J., Figdor, C.G. and Jovin, T.M. (2007) Ligand-conjugated quantum dots monitor antigen uptake and processing by dendritic cells. *Nano Letters*, **7**, 970–7.

24 Sukhanova, A., Devy, M., Venteo, L., Kaplan, H., Artemyev, M., Oleinikov, V., Klinov, D., Pluot, M., Cohen, J.H.M. and Nabiev, I. (2004) Biocompatible fluorescent nanocrystals for immunolabeling of membrane proteins and cells. *Analytical Biochemistry*, **324**, 60–7.

25 Zanchet, D., Micheel, C.M., Parak, W.J., Gerion, D. and Alivisatos, A.P. (2001) Electrophoretic isolation of discrete Au nanocrystal/DNA conjugates. *Nano Letters*, **1**, 32–5.

26 Gao, X.H., Chan, W.C.W. and Nie, S.M. (2002) Quantum-dot nanocrystals for ultrasensitive biological labeling and multicolor optical encoding. *Journal of Biomedical Optics*, **7**, 532–7.

27 Portney, N.G., Singh, K., Chaudhary, S., Destito, G., Schneemann, A., Manchester, M. and Ozkan, M. (2005) Organic and inorganic nanoparticle hybrids. *Langmuir*, **21**, 2098–103.

28 Medintz, I.L., Sapsford, K.E., Konnert, J.H., Chatterji, A., Lin, T.W., Johnson, J.E. and Mattoussi, H. (2005) Decoration of discretely immobilized cowpea mosaic virus with luminescent quantum dots. *Langmuir*, **21**, 5501–10.

29 Dixit, S.K., Goicochea, N.L., Daniel, M.C., Murali, A., Bronstein, L., De, M., Stein, B., Rotello, V.M., Kao, C.C. and Dragnea, B. (2006) Quantum dot encapsulation in viral capsids. *Nano Letters*, **6**, 1993–9.

30 Lucas, R.W., Larson, S.B. and McPherson, A. (2002) The crystallographic structure of brome mosaic virus. *Journal of Molecular Biology*, **317**, 95–108.

31 Bancroft, J.B., Hills, G.J. and Markham, R. (1967) A study of self-assembly process in a small spherical virus – formation of organized structures from protein subunits in vitro. *Virology*, **31**, 354–79.

32 Dragnea, B., Chen, C., Kwak, E.S., Stein, B. and Kao, C.C. (2003) Gold nanoparticles as spectroscopic enhancers for in vitro studies on single viruses. *Journal of the American Chemical Society*, **125**, 6374–5.

33 Ahlquist, P., Allison, R., Dejong, W., Janda, M., Kroner, P., Pacha, R. and Traynor, P. (1990) *Viral Genes and Plant Pathogenesis* (eds T.P. Pirone and J.G. Shaw), Springer-Verlag, New York, pp. 144–55.

34 Marsh, M. and Helenius, A. (2006) Virus entry: open sesame. *Cell*, **124**, 729–40.

35 Brandenburg, B. and Zhuang, X.W. (2007) Virus trafficking – learning from single-virus tracking. *Nature Reviews. Microbiology*, **5**, 197–208.

36 Dimitrov, D.S. (2004) Virus entry: molecular mechanisms and biomedical applications. *Nature Reviews. Microbiology*, **2**, 109–22.

37 Gruenberg, L. (2001) The endocytic pathway: a mosaic of domains. *Nature Reviews. Molecular Cell Biology*, **2**, 721–30.

38 Anderson, J.L. and Hope, T.J. (2005) Intracellular trafficking of retroviral vectors: obstacles and advances. *Gene Therapy*, **12**, 1667–78.

39 Klasse, P.J., Bron, R. and Marsh, M. (1998) Mechanisms of enveloped virus entry into animal cells. *Advanced Drug Delivery Reviews*, **34**, 65–91.

40 Pelkmans, L. and Helenius, A. (2003) Insider information: what viruses tell us about endocytosis. *Current Opinion in Cell Biology*, **15**, 414–22.

41 Sieczkarski, S.B. and Whittaker, G.R. (2002) Dissecting virus entry via endocytosis. *Journal of General Virology*, **83**, 1535–45.

42 Nichols, B.J. and Lippincott-Schwartz, J. (2001) Endocytosis without clathrin coats. *Trends in Cell Biology*, **11**, 406–12.

43 Brodsky, F.M., Chen, C.Y., Knuehl, C., Towler, M.C. and Wakeham, D.E. (2001) Biological basket weaving: formation and function of clathrin-coated vesicles. *Annual Review of Cell and Developmental Biology*, **17**, 517–68.

44 Kirchhausen, T. (2000) Clathrin. *Annual Review of Biochemistry*, **69**, 699–727.

45 Nabi, I.R. and Le, P.U. (2003) Caveolae/raft-dependent endocytosis. *Journal of Cell Biology*, **161**, 673–7.

46 De Tulleo, L. and Kirchhausen, T. (1998) The clathrin endocytic pathway in viral infection. *The EMBO Journal*, **17**, 4585–93.

47 Harrison, S.C. (2008) Viral membrane fusion. *Nature Structural & Molecular Biology*, **15**, 690–8.

48 Kielian, M. and Rey, F.A. (2006) Virus membrane-fusion proteins: more than one way to make a hairpin. *Nature Reviews. Microbiology*, **4**, 67–76.

49 Ding, W., Zhang, L., Yan, Z. and Engelhardt, J.F. (2005) Intracellular trafficking of adeno-associated viral vectors. *Gene Therapy*, **12**, 873–80.

50 Manes, S., Del Real, G. and Martinez, A.C. (2003) Pathogens: raft hijackers. *Nature Reviews. Immunology*, **3**, 557–68.

51 Nayak, D.P. and Barman, S. (2002) Role of lipid rafts in virus assembly and budding. *Advances in Virus Research*, **58**, 1–28.

52 Rawat, S.S., Viard, M., Gallo, S.A., Rein, A., Blumenthal, R. and Puri, A. (2003) Modulation of entry of enveloped viruses by cholesterol and sphingolipids (Review). *Molecular Membrane Biology*, **20**, 243–54.

53 Apodaca, G. (2001) Endocytic traffic in polarized epithelial cells: role of the actin and microtubule cytoskeleton. *Traffic*, **2**, 149–59.

54 Mallik, R. and Gross, S.P. (2004) Molecular motors: strategies to get along. *Current Biology*, **14**, R971–82.

55 Joo, K.I. and Wang, P. (2008) Visualization of targeted transduction by engineered lentiviral vectors. *Gene Therapy*, **15**, 1384–96.

56 McDonald, D., Vodicka, M.A., Lucero, G., Svitkina, T.M., Borisy, G.G., Emerman, M. and Hope, T.J. (2002) Visualization of the intracellular behavior of HIV in living cells. *Journal of Cell Biology*, **159**, 441–52.

57 Melikyan, G.B., Barnard, R.J.O., Abrahamyan, L.G., Mothes, W. and Young, J.A.T. (2005) Imaging individual retroviral fusion events: from hemifusion to pore formation and growth. *Proceedings of the National Academy of Sciences of the United States of America*, **102**, 8728–33.

58 Bartlett, J.S., Wilcher, R. and Samulski, R.J. (2000) Infectious entry pathway of adeno-associated virus and adeno-associated virus vectors. *Journal of Virology*, **74**, 2777–85.

59 Greber, U.F., Suomalainen, M., Stidwill, R.P., Boucke, K., Ebersold, M.W. and Helenius, A. (1997) The role of the nuclear pore complex in adenovirus DNA entry. *The EMBO Journal*, **16**, 5998–6007.

60 Lakadamyali, M., Rust, M.J., Babcock, H.P. and Zhuang, X.W. (2003) Visualizing infection of individual influenza viruses. *Proceedings of the National Academy of Sciences of the United States of America*, **100**, 9280–5.

61 Sakai, T., Ohuchi, M., Imai, M., Mizuno, T., Kawasaki, K., Kuroda, K. and Yamashina, S. (2006) Dual wavelength imaging allows analysis of membrane fusion of influenza virus inside cells. *Journal of Virology*, **80**, 2013–18.

62 Jaiswal, J.K. and Simon, S.M. (2007) Imaging single events at the cell membrane. *Nature Chemical Biology*, **3**, 92–8.

63 Seisenberger, G., Ried, M.U., Endress, T., Buning, H., Hallek, M. and Brauchle, C. (2001) Real-time single-molecule imaging of the infection pathway of an adeno-associated virus. *Science*, **294**, 1929–32.

64 Muller, B., Daecke, J., Fackler, O.T., Dittmar, M.T., Zentgraf, H. and Krausslich, H.G. (2004) Construction and characterization of a fluorescently labeled infectious human immunodeficiency virus type 1 derivative. *Journal of Virology*, **78**, 10803–13.

65 Engelman, A., Englund, G., Orenstein, J.M., Martin, M.A. and Craigie, R. (1995) Multiple effects of mutations in human-immunodeficiency-virus type-1 integrase on viral replication. *Journal of Virology*, **69**, 2729–36.

66 Joo, K.I., Lei, Y.N., Lee, C.L., Lo, J., Xie, J.S., Hamm-Alvarez, S.F. and Wang, P. (2008) Site-specific labeling of enveloped viruses with quantum dots for single virus tracking. *ACS Nano*, **2**, 1553–62.

67 Beckett, D., Kovaleva, E. and Schatz, P.J. (1999) A minimal peptide substrate in biotin holoenzyme synthetase-catalyzed biotinylation. *Protein Science*, **8**, 921–9.

68 Howarth, M., Takao, K., Hayashi, Y. and Ting, A.Y. (2005) Targeting quantum dots to surface proteins in living cells with biotin ligase. *Proceedings of the National Academy of Sciences of the United States of America*, **102**, 7583–8.

69 Piran, U. and Riordan, W.J. (1990) Dissociation rate-constant of the biotin-streptavidin complex. *Journal of Immunological Methods*, **133**, 141–3.

70 Wu, X.Y., Liu, H.J., Liu, J.Q., Haley, K.N., Treadway, J.A., Larson, J.P., Ge, N.F., Peale, F. and Bruchez, M.P. (2003) Immunofluorescent labeling of cancer marker Her2 and other cellular targets with semiconductor quantum dots. *Nature Biotechnology*, **21**, 41–6.

71 Levi, V., Ruan, Q.Q. and Gratton, E. (2005) 3-D particle tracking in a two-photon microscope: application to the study of molecular dynamics in cells. *Biophysical Journal*, **88**, 2919–28.

72 Sun, X.J., Yau, V.K., Briggs, B.J. and Whittaker, G.R. (2005) Role of clathrin-mediated endocytosis during vesicular stomatitis virus entry into host cells. *Virology*, **338**, 53–60.

73 Hogle, J.M. (2002) Poliovirus cell entry: common structural themes in viral cell entry pathways. *Annual Reviews of Microbiology*, **56**, 677–702.

74 Mueller, C. and Flotte, T.R. (2008) Clinical gene therapy using recombinant adeno-associated virus vectors. *Gene Therapy*, **15**, 858–63.

75 Parker, J.S.L., Murphy, W.J., Wang, D., O'Brien, S.J. and Parrish, C.R. (2001) Canine and feline parvoviruses can use human or feline transferrin receptors to bind, enter, and infect cells. *Journal of Virology*, **75**, 3896–902.

76 Parker, J.S.L. and Parrish, C.R. (2000) Cellular uptake and infection by canine parvovirus involves rapid dynamin-regulated clathrin-mediated endocytosis, followed by slower intracellular trafficking. *Journal of Virology*, **74**, 1919–30.

77 Sanlioglu, S., Benson, P.K., Yang, J.S., Atkinson, E.M., Reynolds, T. and Engelhardt, J.F. (2000) Endocytosis and nuclear trafficking of adeno-associated virus type 2 are controlled by Rac1 and phosphatidylinositol-3 kinase activation. *Journal of Virology*, **74**, 9184–96.

78 Davidson, T. (1999) Respiratory syncytial virus, in *Gale Encyclopedia of Medicine* (eds D. Olendorf, C. Jeryan, K. Boydon and M.K. Fyke), Detroit, Michigan, pp. 2478–80.

79 Falsey, A.R. and Walsh, E.E. (2000) Respiratory syncytial virus infection in adults. *Clinical Microbiology Reviews*, **13**, 371–84.

80 Collins, P.L., Chanock, R.M. and Murphy, B.R. (2001) Respiratory syncytial virus, in *Fields' Virology* (eds B.N. Fields, D.M. Knipe, P.M. Howley and D.E. Griffin), Lippincott Williams & Wilkins, Philadelphia, pp. 1443–85.

81 Bentzen, E.L., House, F., Utley, T.J., Crowe, J.E. and Wright, D.W. (2005) Progression of respiratory syncytial virus infection monitored by fluorescent quantum dot probes. *Nano Letters*, **5**, 591–5.

82 Bachi, T. (1988) Direct observation of the budding and fusion of an enveloped virus by video microscopy of viable cells. *Journal of Cell Biology*, **107**, 1689–95.

83 Agrawal, A., Zhang, C.Y., Byassee, T., Tripp, R.A. and Nie, S.M. (2006) Counting single native biomolecules and intact viruses with color-coded nanoparticles. *Analytical Chemistry*, **78**, 1061–70.

84 Agrawal, A., Tripp, R.A., Anderson, L.J. and Nie, S.M. (2005) Real-time detection of virus particles and viral protein expression with two-color nanoparticle probes. *Journal of Virology*, **79**, 8625–8.

85 Deng, Z.T., Zhang, Y., Yue, J.C., Tang, F.Q. and Wei, Q. (2007) Green and orange CdTe quantum dots as effective pH-sensitive fluorescent probes for dual simultaneous and independent detection of viruses. *Journal of Physical Chemistry, B*, **111**, 12024–31.

7
Nanomaterials for Radiation Therapy
Ke Sheng and Wensha Yang

7.1
Introduction

The first question in radiation therapy is how to delineate the geometry – the location and shape – of the tumor. This question is answered by imaging, that is, by creating a contour map encompassing the tumor and a margin for uncertainties in microscopic extension, patient motion and set-up errors. Imaging techniques make use of the fact that tumors have different physical and chemical properties than normal tissue. Computed tomography (CT) is the most commonly used imaging method for tumor delineation, and takes advantage of the difference in electron density between tumors and normal tissue. When this difference is not sufficient, contrast agents with high photoelectric cross-sections are used to enhance the ability so as to visualize the tumor. Magnetic resonance imaging (MRI) can distinguish tissues with different proton densities but, more frequently, the distinct proton relaxation time of the tumor is exploited to delineate the tumor. More subtle differences in the chemical composition of malignant tumors can also be detected with MRI spectroscopy, as when the concentration of certain organic molecules creates a "biochemical fingerprint." In addition, tumors differ from normal tissue in terms of their physiological behaviors, including their metabolic activities, proliferation rates, gene expression patterns and oxygen levels. The combined used of targeting moieties and contrast agents can thus reveal both the physiological and the physical properties of the tumor, allowing a high precision in delineating the tumor boundaries. Nanomaterials are ideal targeting moieties because they are highly customizable in both functional surface and size, and thus can be designed to infiltrate specific biological barriers such as the cell membrane. In addition, their unique physical properties, such as density, luminescence, and magnetism, make them extremely useful for CT, optical imaging, and MRI. These applications have contributed to new methods of target delineation in radiation therapy.

There is a second, equally important question in radiation therapy, namely, how to deliver a sufficient radiation dose to the tumor without excessively irradiating

Nanomaterials for the Life Sciences Vol.6: Semiconductor Nanomaterials.
Edited by Challa S. S. R. Kumar
Copyright © 2010 WILEY-VCH Verlag GmbH & Co. KGaA, Weinheim
ISBN: 978-3-527-32166-7

the surrounding normal tissue. Although the working mechanisms of nanomaterials for therapeutic and imaging purposes are often parallel and overlapping, the direct utilization of nanomaterials to increase the effectiveness of radiation therapy is relatively new. Most therapeutic applications of nanomaterials are still in the infant stage of their development, with few human trials having been conducted. One major difference between imaging and therapeutic applications is that the dose used in therapy is at least two orders of magnitude higher than that in an imaging procedure. For example, it is considered acceptable in positron emission tomography (PET) that the brain and bladder receive moderate doses of radioactivity during imaging, but this would be hazardous in a therapeutic procedure due to the much higher radiation dose. This difference, and other unique problems, have made the introduction of nanomaterials in radiation therapy particularly challenging.

This chapter is dedicated to the therapeutic applications of nanoparticles, and is structured to contain a brief introduction to radiation therapy and its limitations, followed by discussions of physical radiosensitizers and the use of nanoparticle semiconductors as energy mediators during the combination of radiation therapy with photodynamic therapy. The following sections are on nanobrachytherapy, the distinguishing characteristic of which is that the radioactive nanoparticles are targeted specifically to tumor cells, and radioprotection, which focuses on decreasing the dosage delivered to healthy tissue. Next is described the use of nanoparticles as semiconductors in radiation dosimetry. Finally, the outlook for future research into nanomaterials for radiation therapy is assessed.

7.2
A Brief Introduction to Radiation Therapy, and its Limitations

In radiation therapy, photons or charged particles with various energy levels are projected into cancer patients, such that radiation energy is delivered to the lesions that need to be treated. The spectrum of electromagnetic (EM) radiation is shown in Figure 7.1. Depending on the energy of a single particle, this spectrum can be divided into ionizing radiation and nonionizing radiation:

- Ionizing radiation can act *indirectly*, where single particles with energy higher than the first ionization potential of a water molecule (12.6 eV) are used to excite or ionize water molecules, generating a cascade of free radicals that damage cellular DNA and ultimately cause cell death if the damage is not repaired. Alternatively, ionizing radiation can act *directly* on biological molecules such as DNA, but this requires an energy of 75 eV per particle on average, based on empirical data. It can be ascertained from Figure 7.1, that the ionization potential of 12.6 eV corresponds to photons with wavelengths shorter than the ultraviolet (UV) band.

- In nonionizing radiation, the energy carried by a single particle is not high enough to generate ionization events, but collectively, the irradiated tissue is

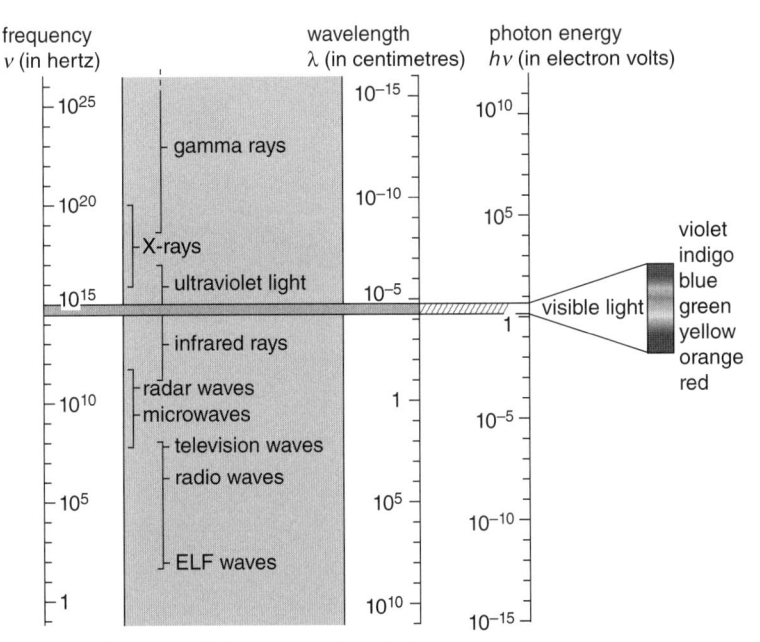

Figure 7.1 Spectrum of electromagnetic radiation.

heated by a large number of particles to a threshold temperature, thereby producing irreversible damage. The achieved temperature depends on the duration of heating; for example, conventional hyperthermia treatment using temperatures of 42–45 °C requires hours of exposure to kill tumor cells, but ablative treatments can achieve temperatures of 60–100 °C quite rapidly (in fact, a short exposure time is essential to avoid heat dissipation by the blood circulation). Increases in tissue temperature to this level for 1 s result in immediate protein denaturation and coagulative necrosis.

In either type of radiation, the treatment of tumors that are not on the surface of the patient's skin requires that the radiation enters and penetrates a segment of normal tissue, resulting in an entrance dose. For more penetrating high-energy particles, a percentage of the particles must also exit the patient, resulting in an exit dose. The amount of damage to the lesions relative to that in the normal tissue is referred to as the *therapeutic ratio*, a quantity which is dependent not only on the physical dose of radiation but also on the biological response of the target cells. In modern ionizing radiation therapy, the therapeutic ratio is optimized by redistributing the entrance and exit doses to more radioresistant normal tissue and minimizing the dosage outside the tumor volume.

State-of-the-art radiation planning and delivery can mold the shape of the dose to fit the shape of the tumor and spare most important organs at risk. However, as demonstrated in Figure 7.2, even with Helical TomoTherapy [1, 2] – a recently designed CT-based radiation therapy machine that delivers modulated 6 MV X-rays

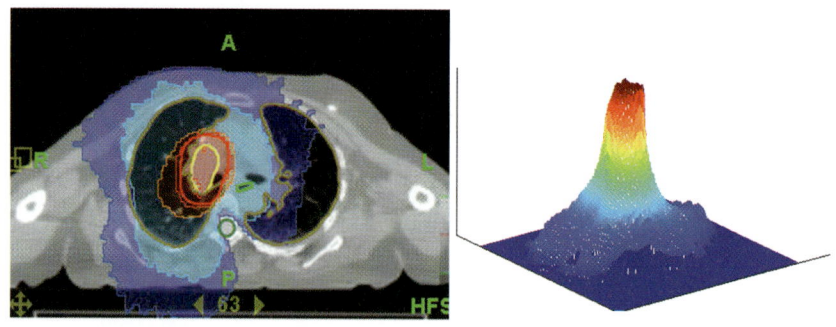

Figure 7.2 Dose distribution of a lung cancer treatment plan generated by Helical TomoTherapy. (a) Two-dimensional colorwash superimposed on an axial CT slide; (b) Three-dimensional surface plot. A "warmer" color indicates a higher dose.

from all 360 degrees around the patient – there is still substantial spillage from the entrance and exit doses, which are combined and defined as the *integral dose*. The integral dose as a physical limitation cannot be reduced by modulating radiation and optimizing beam incidence angles.

Although tumor cells are generally slightly more sensitive to radiation than normal tissue due to their higher mitotic activity and inferior ability to repair DNA damage, the dose spillage from radiation therapy can cause normal tissue damage and treatment side effects which, in rare cases, may lead to patient death. In addition, radiation injury to normal tissue is cumulative, rendering repetitive radiation therapy a last-resort option for patients with recurrent tumors at the same location. Therefore, it is essential to deliver sufficient doses of radiation therapy at the first treatment, whilst at the same time avoiding intolerable radiation injury from spillover.

To improve the therapeutic ratio, drugs have been investigated that will either enhance the radiosensitivity of tumor cells, or protect normal tissues from radiation damage. Chemical radiosensitizers [3] were developed to increase the tumor cells' sensitivity to radiation via many different biological pathways. For example, electron affinity chemicals were used to react with DNA free radicals and reduce hypoxia-associated radioresistance [4, 5]. As tirapazamin is more toxic in a hypoxic environment, it was used to treat these more radioresistant tumor cells [6]. Pyrimidines substituted with bromine or iodine have been incorporated into DNA and enhance free radical damage [7], and drugs involved in DNA repair were evaluated, but with mixed outcomes [8, 9]. Proteins involved in cell signaling, such as the Ras family, are attractive targets linked to radioresistance [10, 11]. The suppression of radioprotective thiols was also investigated as a pathway for radiosensitization [12]. Although these applications have shown promise in one or more areas, they are generally toxic to normal tissues, with uncertainties in the mechanism, and sometimes rely on a tumor cell target subject to change. It was concluded that the clinical gains from these chemical radiosensitizers has been insignificant [3, 13].

This less than satisfactory outcome can be largely attributed to the complex and often unpredictable mechanisms of chemical radiosensitizers, and to the fact that the activity of the sensitizers is not linearly controlled by radiation. As a more straightforward approach, physical sensitizers have been utilized to absorb radiation energy at the site of the tumor using materials that have a higher absorption coefficient than tissue, and this has led to the first important role of nanomaterials as energy mediator. The role of nanoparticles in radioprotection will be introduced briefly later in the chapter.

7.3
Physical Radiosensitizers

7.3.1
Enhanced Radiation Therapy Using High Z but Non-Nanoscaled Materials

X-rays interact with matter and lose energy by one or more of the following mechanisms: the photoelectric effect; coherence; Compton scatter; and pair production. At different energy ranges, the cross-section of each interaction varies, as shown in Figure 7.3 At medium to high energy (100 kV–10 MV), X-rays lose energy primarily through Compton scatter, the cross-section of which is proportional to the electron density of the medium. Therefore, the relative absorption coefficient for an electron-dense material such as gold (density 19.3 g cm^{-3}) is approximately

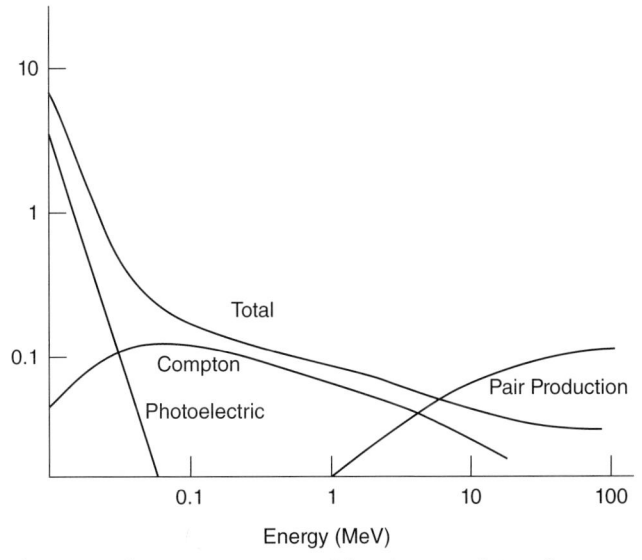

Figure 7.3 Schematic presentation of the relative weighting of X-ray–matter interaction cross-sections at different photon energies.

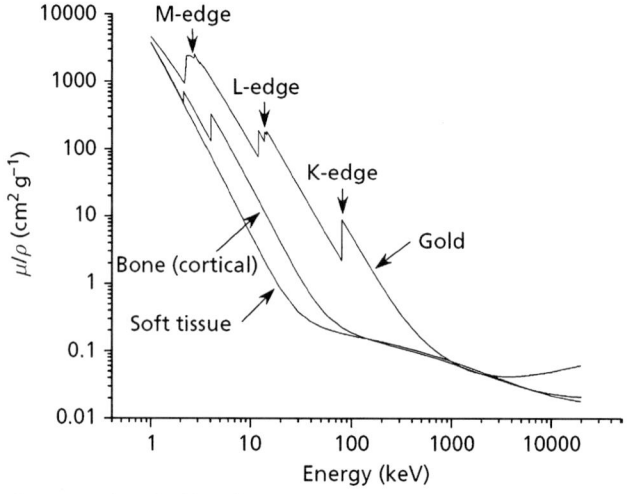

Figure 7.4 Plot of gold, soft tissue, and bone attenuation versus X-ray energy. For gold, absorption edges occur at: K (80.7 keV), L (L1 14.4 keV; L2 13.7 keV; L3 11.9 keV), and M (2.2–3.4 keV); the K-edge is where the incident energy is just enough to eject an electron from the inner K shell [15].

20-fold that of normal human tissue. Lower-energy X-rays react with high-atomic number (high Z) materials more efficiently by the photoelectric effect, which allows for a higher absorption of energy proportional to the cube of the atomic number [14]. A typical absorption coefficient plot of tissue, bone and gold is shown in Figure 7.4. At the K-edge of gold, the relative absorption coefficient of the gold is approximately 1217 ($79^3/7.4^3$) times that of normal tissue with an average atomic number of 7.4. Therefore, high-Z materials can be used to intercept more X-ray energy to improve the therapeutic ratio of radiation therapy. Although coherent interaction also has a higher cross-section at lower energy and with high-Z material, this type of interaction does not involve energy transfer from the incident particle to the tissue, and therefore does not contribute to therapy. In pair production, a paired positron and electron is produced by converting kinetic energy to resting mass. At higher energy, pair production becomes more important. The cross-section is proportional to Z^2, but the advantage of a high-Z material is not significant until the photon energy exceeds 10 MV, where the photoelectric effect becomes negligible.

It is also important to clarify the beam properties associated with X-rays generated by X-ray tubes or an accelerator. A nominal energy of 100 kVp means that the electrons are accelerated to 100 keV before hitting the target, where they are then decelerated and bremsstrahlung X-ray photons are generated. These bremsstrahlung photons are no longer monoenergetic, but rather have widely distributed spectra (polychromatic) with 100 kVp as the maximum (peak) energy. The actual output of the X-ray unit is normally filtered before leaving the X-ray head so as to "harden" the X-rays for more penetrating beams. The average energy of the X-rays from a modern megavoltage clinical accelerator is approximately one-third of the peak energy. Therefore, with a kilovoltage X-ray machine operating at 100 kVp, there will still be large numbers of photons at a much lower energy than the nominal value, and these interact with matter by the photoelectric effect.

The large cross-section of kV X-rays with high-Z materials has been exploited in various cell and animal apparatus. Iodine (Z = 53) was the first element to be tested *in vitro* [16, 17], and showed the effect of radiosensitization. When iododeoxyuridine (IUdR) is integrated into cellular DNA, a threefold increase in tumor cell killing was observed with radiation [18]. Likewise, an improved tumor growth delay was observed [16, 19] with iodine contrast medium and 100 kVp X-rays. Improved canine survival (53%) was also observed with the same medium and orthovoltage (140 kVp) X-rays delivered by a CT scanner [20] modified by the insertion of a collimator to narrow the beams so as to encompass the tumor only and spare the normal tissue surrounding it. (This concept was similar to Helical TomoTherapy [1, 2], but with kV X-rays replaced by MV X-ray.) This apparatus was then tested in a Phase I clinical trial of brain tumors in humans, whereby the method was shown to be safe and effective, albeit in only a few patients [21]. One difficulty associated with iodine as a radiosensitizer is that the highest absorbance energy for iodine, the K-edge, is close that of bone and tissues. In order to further improve the specificity, X-rays at 50 keV with narrow bandwidth were desired; consequently, synchrotrons capable of generating near-monochromatic X-rays were used in combination with iodine to deliver stereotactic radiosurgery to mice [22, 23].

Compared with iodine, gold (Au, Z = 79) has a higher and more desired K-edge at 80.7 keV, which is farther away from that of the bones and tissues. When gold was first applied in the form of gold foil. it showed an ability to enhance cell killing by approximately 100-fold [24]. Subsequently, micron-sized gold particles injected directly into tumors before radiation and also reduced the cells' viability [25]. Although, during these pioneer studies, it became clear that the enhancement of radiation therapy by a locally distributed high-Z material was feasible, it was clear that these methods had several limitations. In order to achieve high dose-enhancement effects, the percentage of thymine that must be replaced by iodouracil was prohibitively high *in vivo* [18]. Synchrotrons producing monochromatic X-rays are not widely available, and the low energy with the K-edge of iodine prohibited it from being applied to larger animals or humans due to a limited X-ray penetration. With gold, it is difficult to apply bulk materials such as foil and to achieve a uniform dose enhancement, as the range of dose enhancement with the foil is on the order of 50 μm [24]. In addition, gold microspheres were unable to infiltrate densely packed tumor cells [25]. Most importantly, neither iodine contrast media nor micron-sized gold particles could be easily modified for biological targeting.

7.3.2
Enhanced Radiation Therapy by Gold Nanoparticles

Although the high-Z nanoparticles used simply to increase radiation absorption are not semiconductors, the energy transfer pathway is highly relevant to the semiconductor nanoparticle–photosensitizer pair used in combined radiation–photodynamic therapy (see Section 7.3). Therefore, to provide a complete picture of the physical radiosensitizer, it is essential to include details of these metal nanoparticles at this point. For the same reason, metal oxide nanoparticles, as radioprotectors, are briefly introduced in Section 7.5.

To overcome the difficulties in enhancing radiation therapy with gold materials, gold nanoparticles have emerged as an attractive solution. Their *in vitro* efficacy was demonstrated by Kong *et al.* [26], who compared cell survival after treatment with kV radiation only or kV radiation in the presence of gold nanoparticles. Cells containing gold nanoparticles survived significantly less, and the dark toxicity of the gold nanoparticles was negligible. Although the degree of radiosensitization is arguably the same as micron-sized gold particles for a given concentration, gold nanoparticles are more versatile and biocompatible. It has been shown that gold particles less than 2 nm (and without surface modification) can effectively evade the immune system and liver retention. They can also exploit the leaky nature of the tumor vascular structure to achieve tumor : liver concentration ratios of 1.6 [27]. In a study conducted by Hainfeld *et al.* [27], when 0.01 ml g^{-1} of 1.9 ± 0.1 nm gold nanoparticles was injected into the tail vein of mice, the gold nanoparticles appeared in the xenograft tumor shortly after injection (the contrast from X-ray images of the gold nanoparticles is shown in Figure 7.5). These mice were subsequently irradiated and compared to mice without gold nanoparticles. In the radiation-only group, the long-term survival (one year) was 20%, whereas in groups irradiated with lower (135 mg Au kg^{-1}) and higher (270 mg Au kg^{-1}) loads of gold nanoparticles the long-term survival was 50% and 86%, respectively. This study was the first proof of principle demonstrating the ability of high-Z nanoparticles to effectively enhance radiation therapy without any significant side effects. However, the ability to modify gold nanoparticles is where the true therapeutic value emerges.

Based on size selection alone, the gold nanoparticle concentrations were slightly higher in the tumors than in the liver; however, the tumor concentrations were

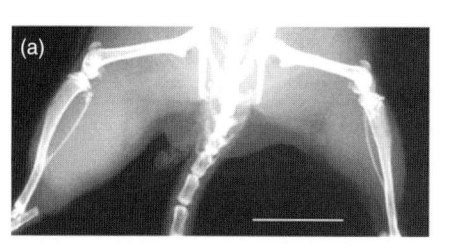

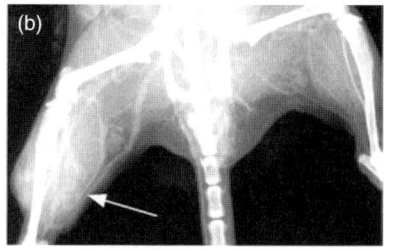

Figure 7.5 Radiographs of mouse hind legs before and after gold nanoparticle injection. (a) Before injection; (b) At 2 min after intravenous gold injection (2.7 g Au kg^{-1}). Significant contrast (white) from the gold is seen in the leg with the tumor (arrow) compared to the normal contralateral leg. Six-second exposures at 22 kVp and 40 mA. Scale bar = 1 cm. Reproduced from Ref. [27], with permission.

still lower than those in the kidney and blood, and similar to those of other tissues. although, initially, this lack of biospecificity limited the effect of gold on the therapeutic ratio, the achievement of tumor specificity was shown to require surface modification, which would allow the gold nanoparticles to bind to molecules such as polyethylene glycol (PEG), carboxyl or amino groups, thiol-derivatized drugs, DNA, lipids, and carbohydrates [28, 29]. Surface modification improved the biocompatibility of gold nanoparticles and allowed an advantage to be taken of the higher demand for glucose and the overexpression of certain receptors in rapidly metabolizing tumor cells. For example, Kong et al. showed that gold nanoparticles, when coated by glucose, would be selectively internalized by breast cancer cells, and that the selectivity could be further tuned by modifying the surface charge [26]. Li et al. demonstrated a fourfold higher tumor cell uptake of gold nanoparticles functionalized by transferrin compared to normal cells [30]. A similarly increased uptake was observed in prostate cancer cells, although the enhancement in cell killing was not proportional to the gold nanoparticle loading, which in turn indicated the existence of a saturation mechanism with regards to the effectiveness of the therapy [31]. In addition to nonspecific coating molecules, gold nanoparticles have also been conjugated with peptides [32, 33] and antibodies [34] for more specific tumor cell targeting.

In vivo applications of these surface modifications and improved tumor control have yet to be reported. This indicates the difficulties encountered with *in vivo* targeting, namely, that the uptake of the nanoparticles is not determined simply by the affinity between the particles and the tumor cells, as it would be *in vitro*; rather, the particle might be intercepted by the immune system before entering the tumor. Due to a lack of lymphatic drainage, tumors may possess a higher intracellular pressure, which in turn makes it more difficult for nanoparticles to penetrate into tumor core, especially when the lesion is not well vascularized. In addition, the very high loading required for radiosensitization (0.5–5%) may saturate the cell uptake, with or without targeting moieties. Moreover, most cancer patients today are treated with megavoltage X-rays that penetrate more deeply into the tissue, for skin sparing and a higher dose conformality. The efficacy of gold nanoparticles under these conditions would be adversely affected due to the lack of a photoelectric effect.

7.3.3
Dose Enhancement of Physical Radiosensitizers

In order to quantify the enhancement in the physical dose, a quantity referred to as the dose enhancement factor (DEF) is commonly used to describe the efficacy of a radiosensitizer. A DEF of 2 indicates that, for 1 Gy of radiation deposited to a normal tissue, 2 Gy will be deposited within the high-Z material. A DEF of 1.2 or higher is considered to be the threshold for determining whether a radiosensitizing agent is effective.

By using reported elemental attenuation and integration of the absorption over the energy spectrum of the radiation source, the theoretical DEFs for elements with atomic numbers ranging from 25 to 90 can be calculated for a range of concentrations and varying radiation energies [35]. For a tissue concentration of 0.5%, very little enhancement (DEF < 1.05) was obtained using ^{60}Co (1.25 MeV), ^{192}Ir (317 keV), ^{198}Au (412 keV), ^{137}Cs (661 keV), and 6-, 18- and 25-MV X-rays for all elements. Unlike X-rays, γ-rays from these isotopes do not have any low-energy photon components that react with the high-Z material by the photoelectric effect. With high percentages of low-energy photons, a DEF of ~1.65 was achieved with 80–140-kVp X-rays and gold nanoparticles at a single concentration. Lower-energy brachytherapy sources, such as ^{103}Pd (peak 362 keV) and ^{125}I (35 keV), resulted in DEFs of 1.60 and 1.67. However, low-dose-rate brachytherapy takes several days to several weeks to finish, which poses the question of whether the high-Z material can be retained in the tumor for such a long period.

A Monte Carlo simulation was also conducted to calculate the dose enhancement for 140-kVp X-rays, 4- and 6-MV photon beams, and ^{192}Ir gamma rays [36]. For a gold nanoparticle concentration of 7 mg g^{-1} tumor, the dose enhancement for 140-kVp X-rays was twofold, but for the MV radiation the enhancement was only between 1% and 7%. This essentially ruled out its application in state-of-the-art external radiation therapy using megavoltage X-rays. For ^{192}Ir, the dose enhancement within the tumor region ranged from 5% to 31%, which was higher than in other theoretical studies using attenuation tables. Although this number was not as significant, as for the 140-kVp X-ray, it has more realistic clinical implications as ^{192}Ir is commonly used in both high-dose-rate and low-dose-rate brachytherapy. The study also assumed a rather high biodistribution of gold nanoparticles; neither was it clear how to achieve such concentrations without simultaneously sensitizing the normal tissue. Auger electrons [37] produced by the interaction between X-rays and the gold nanoparticles were neglected in this simulation. Although Auger electrons constitute a small percentage of the energy deposition, they are high linear energy transfer (LET) particles that may cause more double-stranded DNA (dsDNA) breakage than would the physical radiation dose [38]. The main drawback to taking Auger electrons into consideration is their extremely short range (ca. 10 nm), since Auger electrons – and thus the gold nanoparticles that emit them – must be within the cell nucleus and close to the DNA in order to cause effective cell killing.

In addition to low-energy X-rays, both high- and low-energy electron beams were investigated for their potential in combination with gold nanoparticles. The stopping power of intermediate-energy electrons is proportional to the atomic number of matter, and therefore should result in a weaker enhancement compared to low-energy X-rays [39]. Nevertheless, a 10-fold enhancement in tumor cell killing was observed with 60 keV electrons, as well as enhanced apoptosis with 6 MeV electron beams.

7.3.4
Radiation Therapy Enhancement Using Nonionizing Radiation

Gold nanoparticles react not only with ionizing radiation but also with other energy sources, because of their high absorption coefficient and black color

[40] – features which are especially outstanding when a nanoshell structure is engineered. Collective oscillations of the metal shell may happen occur when it is irradiated with light, and the resonance frequency is tunable by adjusting the thickness of the shell relative to the core. In order to optimize tissue penetration, near-infrared (NIR) light is the most desirable, and silica-gold nanoshells can be readily tuned to resonate with this frequency. In a study conducted by Hirsch *et al.*, human breast carcinoma cells incubated with nanoshells *in vitro* were found to have undergone photothermally induced mortality on exposure to NIR light [41]. In this case, solid tumors treated with gold nanoshells were heated up by an average of 37.4 °C within 4–6 min, while the control tissue achieved a temperature change of less than one-quarter of that.

It was also recently discovered that the photoabsorption coefficient of a nanoparticle was dependent not only on the structure of the nanoshell but also on its concentration. By using a finite-difference time-domain analysis, Liu *et al.* showed that the closer the nanoparticles were, the higher was the absorption achieved with the gold nanoshells [42]. Whilst the theoretical development is still in its early stages, the thermometry of gold nanomaterial-facilitated hyperthermia will rely on empirical data and the real-time monitoring of temperature by modalities such as ultrasound [43].

7.4
Radiation Therapy in Combination with Photodynamic Therapy Using Semiconductor Nanoparticles as the Energy Mediator

7.4.1
Photodynamic Therapy

There are several similarities between radiation therapy and photodynamic therapy (PDT). For example, both obtain energy from photons (albeit with very different wavelengths), and both cause damage to tumor cells by way of secondary molecules such as free radical species or singlet oxygen molecules generated from the incident photons. Yet, PDT differs from radiation therapy in many other ways. Typically, in PDT a separate drug is required, referred to as the *photosensitizer*, which is activated by light and enters an excited state. Through intersystem crossing, the excited state becomes a metastable triplet state that can exist for a few microseconds. The triplet state reacts with molecules in the environment and releases energy via type I and type II mechanisms. In the type I reaction, whether through hydrogen-atom abstraction or electron transfer, free radical species such as the superoxide radical anion are generated. In contrast, in the type II reaction, which is considered to be the primary reaction in PDT, the triplet-state photosensitizer reacts directly with the ground-state triplet molecular $^3O_2$ to generate the excited singlet $^1O_2$, which is highly reactive and toxic to cell membranes, lysosomes, and mitochondria.

Whilst PDT is used to treat many types of cancer, including skin, head and neck, esophagus and bladder, it also has nononcologic applications. When applicable,

PDT is effective, potent, and shows few severe side effects. However, one major limitation of PDT is that the light required for activation has a shallow penetration. For example, the wavelength of the activating light for the only FDA-approved photosensitizer, Photofrin, is 620 nm; this has an attenuation coefficient of approximately $1\,mm^{-1}$ in tissue, and thus an effective treatment depth of 5 mm. In order to treat deep-seated solid tumors, optic fibers must be inserted into the patient, either through orifices or through specific incisions. In addition, the light dosimetry is difficult to calculate, due to uncertainties from scatter, reflective light, and the distribution of oxyhemoglobin, which is a strong absorber of red light. Recently, new classes of photosensitizers, such as the phthalocyanines (Pcs), have been developed to utilize longer wavelengths for activation. With activation in the NIR band, the treatment depth can be increased, in practical terms, from less than 1 cm to several centimeters [44].

Photosensitizers have also been tested for use as radiosensitizers [45–50], with moderate radiosensitization having been observed in several aggressive mouse and human cell lines, both *in vitro* and *in vivo*. The mechanism is not completely understood since, in theory, photosensitizers such as the porphyrins used in these studies have a narrow absorption spectra and cannot be excited by X-rays directly to generate singlet oxygen. Nonetheless, a plausible theory was proposed and tested by Luksiene *et al.* [45], that ligands of the peripheral benzodiazepine receptors (PBR), which is overexpressed in aggressive tumor cells, might diminish the cell growth. Dicarboxylic porphyrins are the ligands for such receptors, and in these experiments the primary effect of the photosensitizers was anti-proliferation rather than to cause apoptosis, which is a more common effect of singlet oxygen. Thus, the working mechanism of photosensitizers as radiosensitizers more closely resembles that of the aforementioned chemical radiosensitizers. The use of more penetrating X-rays, however, would require the presence of an energy mediator.

7.4.2
Semiconductor Nanoparticles as the Energy Mediator for Photodynamic Therapy

Nanoparticles were first used as a delivery vehicle to facilitate PDT. Most photosensitizers, including porphyrins and phthalocyanines (Pcs), are not highly water-soluble and tend to aggregate in tissue; such aggregation would then lead to an impairment of the efficiency of the photochemical activities. Gold nanoparticles coated with Zn-Pc were synthesized as a more efficient hydrophilic PDT delivery system, while biodegradable liposome nanoparticles were also used to facilitate the transportation of photosensitizer molecules to the tumor sites [51]. In addition to facilitating the localization of PDT, nanoparticles were also used to improve the energy transfer efficiency. Photosensitizers have low extinction coefficients (attenuation coefficient) that can be improved by conjugating them to nanoparticles with high extinction coefficients. In this respect, Samia *et al.* first showed that CdSe quantum dots (QDs) could be used to mediate energy from UVA light to a PDT agent via a Förster resonance energy transfer (FRET) mechanism. QDs can be used to excite conjugated Pcs [52]; indeed, Tsay *et al.* [53] reported a three- to fourfold higher singlet oxygen yield from the QD/photosensitizer conjugate compared to the photosensitizer alone.

In order to utilize the more penetrating NIR light, much effort was expended to convert NIR light to visible photons that could be used to excite the conjugated photosensitizer. Currently, two mechanisms have been recognized as capable of converting the energy from two photons with a lower energy (longer wavelength) into one photon with a higher energy (shorter wavelength). So-called *simultaneous two-photon absorption* requires a single nonlinear optical conversion with a combined energy sufficient to induce transition from the ground state to an excited electronic state. The second conversion relies on sequential discrete absorption and luminescence steps, where at least two metastable energy states are involved, the first state serving as a temporary excitation reservoir. The energy in the reservoir is later combined with the second photon, such that a second higher excitation state can be reached. A higher energy photon can then be emitted from this state. In the first mechanism, a virtual intermediate state is involved from the quantum mechanical view, and the two excitation photons must be coherent. This situation is only achievable by a laser source with an extremely fine temporal resolution (10^{-15} s). In the second mechanism, because of the intermediate metastable state, the demand for resolution from the excitation source is much lower, but that for efficiency is much higher.

The opposite of upconversion is a more intuitive and more efficient process that converts photons with higher energy to the visible range for photosensitizer excitation. This concept was first explored by Samia et al., using QDs as energy mediators to more efficiently excite conjugated photosensitizers with UV light [52]. Although UV light is less penetrative than visible light, this energy pathway is important in the combination of radiation therapy and PDT. However, before discussing the role of QDs and other fluorescent nanoparticles in radiation therapy, it might be worth examining the physical characteristics of QDs and similar photoluminescent nanoparticles.

7.4.3
Quantum Dots

For a bulk semiconductive material, the band gap is determined by the chemical composition of the material, not by its size. However, when the size of the semiconductor material is reduced to the Bohr radius of 1–5 nm, the quantum confinement effect emerges and the nanoscaled semiconductive materials, also named QDs, behave like a single atom with discrete energy states. The energy levels can be solved by the Schrödinger equation. Assuming spherical symmetry, the energy levels of a QD can be expressed by:

$$E_g(QD) = E_{g,0} + \frac{h^2 \alpha_{n,l}}{8\pi^2 m_{eh} R^2}$$

where $\alpha_{n,l}$ are the energy states (n = 1,2,3 ... , l = s,p,d ...) similar to a single atom, h is Planck's constant, $E_{g,0}$ is the band gap of the bulk material, R is the radius of the dot, and $m_{eh} = m_e \cdot m_h$ is the effective mass of an electron-hole pair (exciton). It is clear that the energy level is proportional to the inverse square of the radius. Therefore, by fine-tuning the size of the particle, a full spectrum of visible light can be obtained. Quantum confinement was first described in theory [54–56] and investigated experimentally as the QD by Alivosato and Bawendi [57, 58]. In addition to the high quantum yield, QDs also have a highly modifiable surface to attain biocompatibility and molecular targeting, which makes them attractive agents for molecular imaging. Because QDs can be excited by a broad range of photon energy, they are ideal energy mediators for the combination of radiation with photodynamic therapy.

One major obstacle to the biological application of QDs is their potential toxicity. Typically, QDs have a cadmium core, which is normally encapsulated by a bioinert shell structure composed of ZnS. However, in biological applications, it not uncommon to further coat the QDs with a layer of PEG to improve their water solubility and biocompatibility. Although the core of QDs is between 2–5 nm in diameter, when combined with the shell, PEG layer and additional biological functionalization, the hydrodynamic diameter of QDs may approach 25 nm. Although particles of this size cannot be cleared by the kidneys, the long-term circulation of QDs in the body may cause the shell structure to be broken down, allowing Cd^{2+} ions to leak into the cytoplasm and cause cytotoxic effects. Official conclusions on QD toxicity have not been easily drawn [59], largely because such toxicity can be affected by many environmental and intrinsic variables, including particle size, charge, concentration, outer coating bioactivity (capping material and functional groups), and stability towards oxidation, photolysis, and mechanical force, all of which can affect QD toxicity [60]. Due to such variation, and also to the fact that the quality of QDs varies significantly across manufacturers, it may be difficult to compare the results of toxicology studies. In any case, these uncertainties surrounding the toxicity of QDs have been a major roadblock for their further human application, and consequently efforts to develop other photoluminescent nanoparticles have been undertaken to circumvent this long-term obstacle. Notably, ZnS nanoparticles doped with Mn^{2+} [61] or Eu^{2+} [62] have been fabricated, both of which were photoluminescent and magnetic, making dual optical and MRI applications possible. In addition to ZnS semiconductive nanoparticles, silicon nanoparticles [63], carbon dots [64–66] and SiC-based QDs [67] have each been synthesized. All of these have similar photoluminescent properties and, reportedly, low or negligible toxicities. Although most current biological applications using photoluminescent nanoparticles continue to use the CdSe QD due to its more mature surface modification and availability, toxicity concerns over released Cd^{2+} ions are likely to be clarified or circumvented in the near future.

7.4.4
Photoluminescent Nanoparticles in Radiation Therapy

With its obvious photoluminescence ability and wide absorption spectrum, it is natural to consider the QD as a potential candidate to transfer its energy to chemi-

7.4 Radiation Therapy in Combination with Photodynamic Therapy | 265

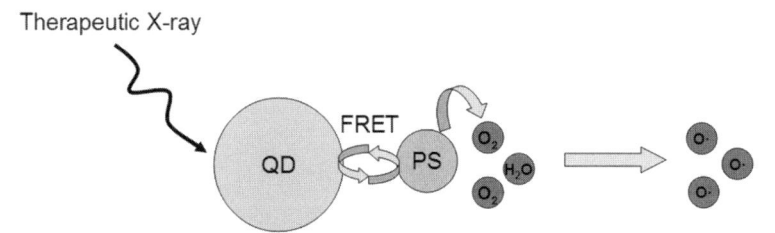

Figure 7.6 Proposed mechanism of photoactivation resulting in oxidative radicals and cellular toxicity. FRET = Förster resonance energy transfer; PS = photosensitizer; QD = quantum dot.

cally bonded photosensitizers. In order to utilize more penetrating X-ray photons as the excitation source, the innovative idea of combining radiation therapy and PDT to excite nanoparticles in the deep tissue was first proposed by Chen et al. [68]. Thus, it was demonstrated, in a proof-of-principle study, that excitation by kV X-rays would induce fluorescence and phosphorescence in nanoparticles. Although, strict speaking, the particles used were not QDs, the concept was readily adapted by Yang et al. [69, 70], who combined QDs and MV radiation therapy for more clinically relevant applications in which linear excitation with an X-ray dose was also demonstrated.

The energy transfer pathway depicted in Figure 7.6 shows that the QD is first excited by therapeutic X-rays, and its energy is then transferred to the chemically conjugated photosensitizer by FRET. The excited photosensitizer then releases energy by type I and type II reactions to generate free radicals or singlet oxygen. The conjugation chemistry is shown in Figure 7.7. The carboxylic acid group on the Photofrin is activated by 1-ethyl-3-(3-dimethylaminopropyl)-carbodiimide (EDC), and then reacts with the amine group on the QD to form a covalent bond. Figure 7.8 shows that the fluorescence emission of QDs is quenched in the conjugate. Instead, energy from the QD is transferred to the Photofrin, which emits characteristic photons with the wavelength of 630 nm.

FRET is an important pathway for the transfer of energy without fluorescence through the dipole–dipole coupling of two conjugated molecules [71]. The FRET efficiency (η) can be calculated by the separation distance between the QD (donor) and the photosensitizer molecule (acceptor) using [71, 72]:

$$\eta = \frac{1}{1+\left(\dfrac{r}{R_o}\right)^6} \tag{7.1}$$

where r is the actual separation distance and R_o is the Förster distance. It is clear that the energy transfer efficiency is 0.5 when the actual separation is the same as the Förster distance, which can be calculated from [73]

$$R_0 = (BQ_D I)^{1/6} \tag{7.2}$$

Conjugation Chemistry

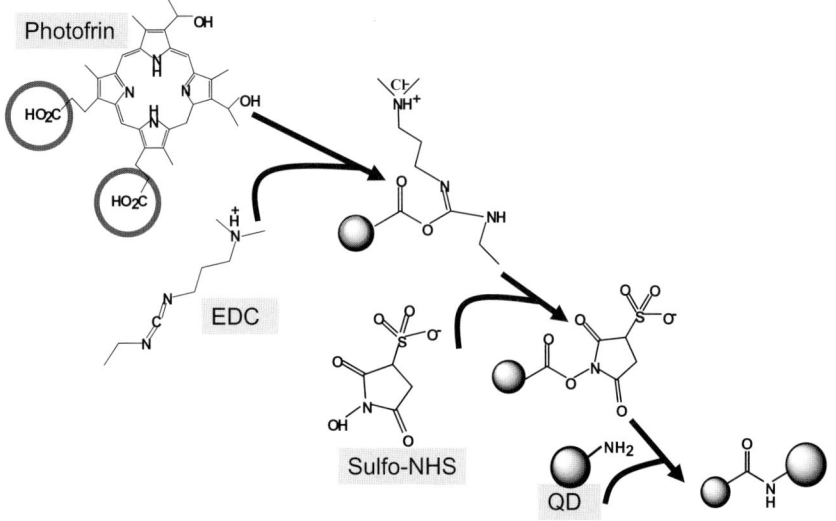

Figure 7.7 Conjugation of Photofrin to the amine-terminated quantum dots by EDC chemistry.

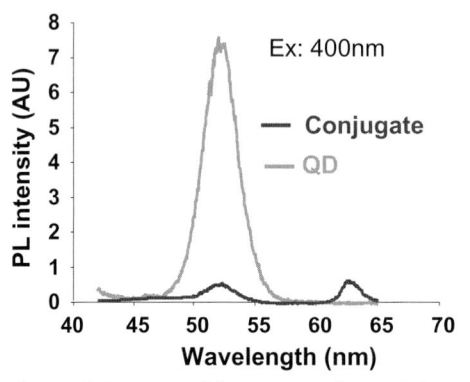

Figure 7.8 Emission of the quantum dots and the QD/Photofrin conjugate excited by 400 nm UV light. The QD emission is quenched in the conjugate, but a Photofrin emission at 630 nm is observed.

where Q_D is the quantum yield of the donor, I is the spectral overlap between the QD and Photofrin, and B is a constant that can be expressed as

$$B = \frac{[9000 \ln(10)] k_p^2}{128 \pi^5 n_D^4 N_A} \quad (7.3)$$

In this equation, n_D is the refractive index of the medium, k_p^2 is the orientation factor, which varies from 0 (when the dipoles of the donor and the acceptor are perpendicular) to 4 (when they are parallel), and N_A is Avogadro's constant. Since Equation 7.1 was traditionally used to describe the energy transfer between two molecules with smaller size than a semiconductor nanocrystal, it was not initially clear whether it could be applied to the QD/small organic molecule conjugate, because the size of a PEG-coated QD with a core shell structure is several orders of magnitude larger than the conjugated molecule. Pons et al. [74] conducted an elegant experiment using self-assembling CdSe/ZnS core–shell QDs decorated with a series of Cy3-labeled beta-strand peptides of increasing length. The bridging peptides were rigid and had fixed lengths, as confirmed by electron microscopy. Using this system, the FRET efficiency as a function of the separation distance was determined and a very good agreement between Equation 7.1 and the experimental measurement was observed.

Because of the large gradient introduced by the sixth order term, the uncertainties in estimating r, and the possibility of multiple conjugation points between the QD and photosensitizer molecules, the efficiency is more commonly determined by experimental quenching. QDs have limited channels through which to release their energy upon excitation. In this case, the energy must be released by either photon emission, FRET, or other channels such as singlet oxygen emission through the triplet state of the QD. It was demonstrated that the last pathway constitutes less than 5% of the total energy release [52], leaving the first two competing against each other for the remaining 95%. Therefore, the efficiency can be expressed as [74, 75]:

$$\eta = 1 - \frac{I_{conj}}{I_{QD}} \quad (7.4)$$

which compares the emission of photons of the QD simply mixed with the acceptor molecule without conjugation (I_{QD}) to the emission of the QD when conjugated (I_{conj}). The percentage of energy transferred to the acceptor by FRET can thus be determined. Using energy quenching, high FRET efficiencies between QDs and conjugated Photofrin were reported by several groups, ranging between 58% [76] and 77% [52].

Another factor that can affect the FRET efficiency is the number of acceptors conjugated to the donor. Because QDs have large surfaces that are usually functionalized with multiple binding sites, more than one acceptor can be attached to a single donor. It was reported that n multiply bound acceptors can increase the FRET efficiency according to the following equation [77],

$$\eta = \frac{nR_0^6}{nR_0^6 + r^6} \quad (7.5)$$

Yang et al. [70] showed that the FRET efficiency increased with the number of Photofrins conjugated to the surface of the QD, which was terminated by multiple amine groups (Figure 7.9). When the number of Photofrins per QD in the conjugation chemistry was increased to 20, the FRET efficiency approached 100% [69], following the curve shown in Figure 7.10.

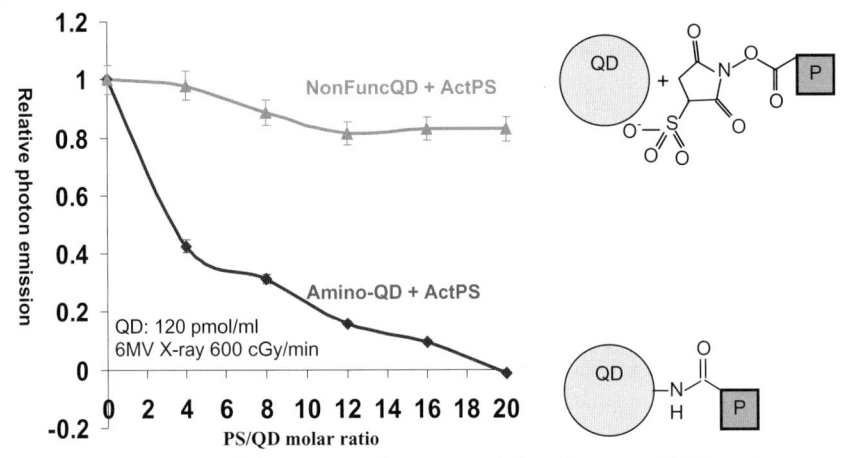

Figure 7.9 Quantitation of the quenching by conjugated Photofrin versus PS/QD molar ratio.

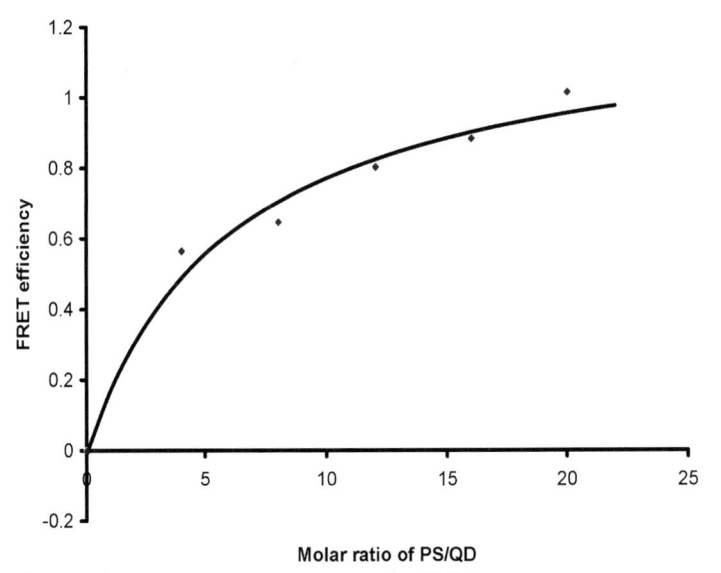

Figure 7.10 Förster resonance energy transfer (FRET) efficiency determined by the quenching experiment in Figure 7.9. The trend line follows Equation 7.5, and shows that the efficiency approaches 100% as more Photofrins are conjugated to the QD surface.

It was also pointed out [77] that Equation 7.4 is valid only when the number of acceptors bound to a QD is uniform; in practice, the heterogeneity in conjugate valence – that is, the acceptor to donor ratio – can vary significantly. In such cases, the FRET efficiency for multiple acceptors bound to the QD is more precisely described by a Poisson distribution as follows:

$$\eta(N) = \sum_{n}^{N} \frac{e^{-N}N^m}{n!}\eta(n) \tag{7.6}$$

where N is the nominal valence and n is the actual number of acceptors bound to a donor.

With the FRET efficiency determined experimentally, it is possible to estimate the singlet oxygen produced from a given amount of radiation. The number of singlet oxygens produced in a cell was estimated by Morgan et al., based on LaF$_3$ luminescent nanoparticles [78], using the following formula:

$$N_{1o_2} = 3.2 DM\nu\Phi_{1o_2} \tag{7.7}$$

where D is the radiation dose in Gy, M is the absorption of the nanoparticle cores relative to that of tissue and is heavily dependent on incident X-ray energy, ν is the concentration of the nanoparticle, and Φ_{1o_2} is the energy transfer efficiency. The conversion factor of 3.2 comes from the fact that 1 Gy of radiation deposits 3.2 MeV to a cell with an estimated volume of 0.52 pl. A wide range of nanoparticle cell loading values between 0.1% and 5% was reported [78]. The lower Niedre limit [79] of 5.6×10^7 was used as the number of singlet oxygen molecules per cell for effective tumor cell killing. The FRET efficiency was assumed to be 0.75, and the quantum yields of the LaF$_3$ nanoparticle and the photosensitizer were 0.5 and 0.89, respectively. It was also assumed to generate 3.9×10^5 photons per 1 MeV of radiation using the excitation wavelength of 480 nm. The resultant singlet oxygen production as a function of radiation dose is plotted in Figure 7.11. Because of the photoelectric effect at lower energy and higher atomic numbers of the LaF$_3$, low-energy X-rays generate significantly higher numbers of singlet oxygen molecules. For MV X-rays used in state-of-the-art intensity-modulated radiation therapy, the

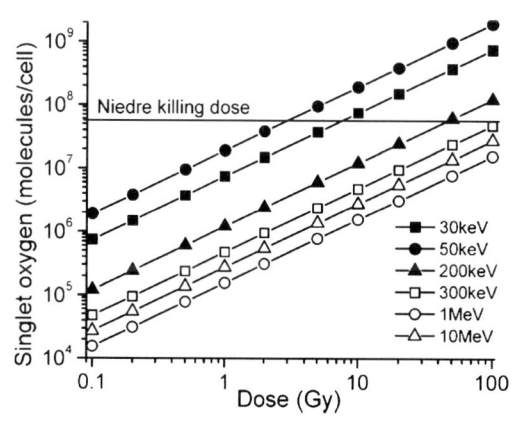

Figure 7.11 Theoretical singlet oxygen production as a function of radiation dose from various energy sources. Reproduced from Ref. [77], with permission.

required dose is close to 100 Gy, which itself can likely eliminate the tumor cells, without the use of PDT. Therefore, the role of the nanoparticle conjugate is different in different X-ray energy ranges. In the low-energy range, PDT can be the dominant mechanism for tumor cell killing, but for high-energy X-rays the role of the nanoparticle conjugate is as a radiation sensitizer or enhancer.

It is important to realize that this theoretical calculation is limited by a number of factors. The Niedre killing limit was derived based on a singlet oxygen measurement that has very low detection efficiency. A cell loading of 5% is likely to be an overestimate, particularly for *in vivo* experiments. The energy deposition to nanoparticles from radiation may have underestimated the contribution from scatter photons and electrons from surrounding molecules, such as the shell and PEG coating of the nanoparticles. The radiation dose is defined as the total energy deposited in a finite volume, and it may be different than the dose received by a microscopic particle with atomic number significantly higher than the surrounding material. Therefore, Figure 7.11 can only be used as an order of magnitude estimate; the actual biological effect will need to be quantified experimentally using *in vivo* and *in vitro* assays.

An alternative way to estimate the physical energy transfer is to compare the energy deposition from X-ray therapy and conventional PDT as follows, using CdSe QDs as the energy mediator.

Assuming that Compton scattering is the dominant effect for 6 MV radiation to interact with the media, the energy transferred to Photofrin per mass can be expressed as:

$$Ec = \eta D \rho_m M \qquad (7.8)$$

where D is the radiation dose, ρ_m is the molar concentration of the conjugates, η is the FRET efficiency, and M is the molar mass of the conjugates. η is assumed to be 0.76 based on the quenching data reported by Yang et al. [70].

The molar mass of QDs (CdSe core and ZnS shell, PEG coating, amine terminated; Evident, Troy, NY, USA) is distributed between $1 \times 10^5 \, g\,mol^{-1}$ and $3 \times 10^5 \, g\,mol^{-1}$. The molar mass of Photofrin is $600 \, g\,mol^{-1}$ (both values are provided by the manufacturer). Because the excitation efficiency of the Photofrin by 520 nm light is approximately threefold greater [80–84] than the excitation efficiency by the 630 nm light used clinically, the energy transferred to Photofrin per gram of tissue (assuming a tissue density of $1 \, g\,ml^{-1}$) at depth d in this conventional PDT is

$$Ep = \frac{1}{3} \psi e^{-(d.\kappa)} \frac{\rho t}{0.1 \, cm} \qquad (7.9)$$

where ψ is the photon energy density, converted to energy per mass by the 0.1 cm in the denominator, κ is the attenuation coefficient $\sim 1 \, mm^{-1}$ [85] and ρ is the clinically achieved tissue concentration $\sim 10^{-6}$ [86]. t is the percentage energy deposited to 1 mm of tissue, as follows:

$$t = \frac{de^{-(a \times \kappa)}}{da}\bigg|_{a=1mm} = 0.37 \tag{7.10}$$

Ec can be derived by Equation 7.8:

$$\begin{aligned} Ec &= \eta D \rho_m M \\ &= 0.76 \times 50\ \text{Gy} \times 24\ \text{pmol g}^{-1} \times 2 \times 10^5\ \text{g mol}^{-1} \\ &= 1.8 \times 10^{-7}\ \text{J g}^{-1} \end{aligned} \tag{7.11}$$

As for the conventional PDT, the energy deposited at 0.5 cm depth, based on Equation 7.9, is

$$\begin{aligned} Ep &= \frac{1}{3}\psi e^{-(d.\kappa)} \frac{\rho}{0.1\ \text{cm}} \times 0.37 \\ &= 0.33 \times 80 \times e^{-5} \times 10^{-6} \times 10 \times 0.37 \\ &= 6.6 \times 10^{-7}\ \text{J g}^{-1} \end{aligned} \tag{7.12}$$

In Equation 7.11, 50 Gy was used as the typical clinical radiotherapy dose; as a result, Ec was 27% of Ep. It is important to note that the energy deposition here is calculated to the Photofrin *only*. If converted to the tissue dose, the energy deposition would be 10^6 greater in the calculation for traditional PDT, resulting in 0.66 J g^{-1}, which agrees well with previous estimates of 0.3 to 1 J g^{-1} [87, 88]. This theoretical calculation therefore demonstrates that the energy transferred to Photofrin using standard-dose X-rays is comparable to the low-dose end of a conventional PDT procedure. The result is consistent in its order of magnitude with the calculation reported by Morgan et al. [78], confirming that the role of the conjugate in MV radiation is in the category of radiation sensitizer.

Biological verification of the efficacy of the conjugate is relatively scarce at the time of writing of this chapter. The only biological testing published to date has demonstrated that with the conjugate, radiation therapy to a lung carcinoma cell line was sensitized significantly [70], as shown by Figure 7.12. QDs alone, however, did not sensitize the tumor cells at all, confirming that the amount of singlet oxygen production by QD alone when excited by X-rays is insufficient [52] to cause any biological effects. It is also verified that Photofrin alone is insufficient to sensitize this particular tumor cell line without QDs as the energy mediator.

While progress has been made in the synthesis of material for and characterization of QDs, further *in vitro* and *in vivo* testing is clearly needed to determine the potential of this method.

7.5 Nanobrachytherapy

Radiation therapy has been used to treat almost all types of cancer, with varying degrees of success. Currently, there are two limiting factors in radiation therapy:

Dose Response of H460

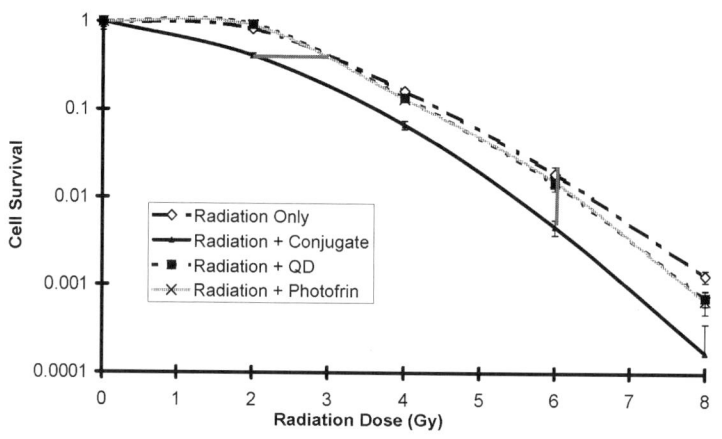

Figure 7.12 The survival fraction of cells receiving radiation or radiation and drugs [37].

- When the position and shape of the solid tumor is known, but there is not a good way to deliver a sufficiently high dose to the tumor without severe side effects.

- When the exact location of the tumor is unknown, or the tumor cells are highly fragmented, as is the case for microscopic tumors that present in 60% of cancer patients [89].

Although positive lymph nodes are commonly included in the planning target volume (PTV) of the radiation treatment, it is impossible to treat all of these micrometastases without a lethal total body irradiation. The presence of micrometastases correlates to a higher chance of recurrence and a poorer prognosis [90–92].

To overcome the first limitation in some applicable cases, one invasive means of delivering radiation more conformally to the tumor is by a technique referred to as *brachytherapy* (Latin: *brachy* = short or contact; this contrasts with external-beam radiotherapy, where the radiation source is much further from the patient). Brachytherapy is a subcategory of radiation therapy in which one or more radioactive seeds are physically inserted into the patient to make contact with the tumor for a given amount of time, depending on the prescription dose and source activity. Because of the geometrical proximity and the utilization of less-penetrating lower-energy γ-rays or β-particles in brachytherapy, the radiation dose is generally more conformal to the tumor [93]. Brachytherapy has gained wide application in the treatment of prostate, cervical, and other type of cancers [94].

Besides the problems introduced by the invasive nature of brachytherapy, unfortunately it cannot be used to treat tumors without knowing the exact location of the tumors, or if there are diffusive micrometastases instead of a solid tumor. The most common method of detecting these micrometastases is *immunocytochemistry*, using antibodies that bind preferentially to tumor cells. Interestingly, this method is very similar to *radioimmunology*, where the cells are radiolabeled by antibodies.

It is intuitive to apply these technical advances to a "smart" radiation therapy method that can use, instead of macroscopic brachytherapy sources, millions (or even billions) of small radiation sources that can target the tumor cells and deliver the therapeutic dose. In that way, both the geometric conformality can be improved and the micrometastases with uncertain locations can be tracked and killed. This is basically the concept of radioimmunotherapy or molecular radiation therapy.

Research into radioimmunotherapy began in the 1950s, when ^{131}I-labeled polyclonal antibodies were used to treat metastatic melanoma [95]. Today, two radioimmunotherapy drugs – Bexxar (^{131}I-tositumomab) and Zevalin (^{90}Y-ibritumomab tiuxetan) – have been approved to treat non-Hodgkin's lymphoma. These target different regions (i.e., epitopes) on the B-cell-associated CD20 antigen, and treat this particular type of disease very effectively. Radioimmunotherapy has also been tested on various other types of metastatic and solid tumors, using antibodies and peptides to target the tumor cells. Radionuclides, such as the α-particle emitters ^{213}Bi and ^{225}Ac, conjugated to HuM195 (a humanized antibody to CD33) are currently being (or have been) tested in clinical trials. These applications are reviewed by DeNardo *et al.* [96] and Sgouros *et al.* [97] among many other authors.

Nanoparticles were recently introduced in radioimmunotherapy as functional carriers for both the radioactivity and the antibodies. There are several distinctly different nanoscaled carriers, including liposomes, nanoparticles and dendrimers, to deliver therapeutic isotopes to the tumor cells. In order to differentiate this method from traditional radioimmunotherapy that does not use nanodevices, the term "nanobrachytherapy" was coined [98].

In general, radionuclides used for therapeutic purposes emit radiation with very short ranges so as to reduce collateral damage to surrounding normal cells. Hence, Auger electron-, α-particle- and β-particle-emitters are commonly selected, with ranges from several nanometers to 10 mm [99]. Compared to β-particles, Auger electrons and α-particles have 10- to 100-fold higher linear energy transfer (LET) coefficients that result in more irreparable DNA double-strand breakage. A list of commonly used therapeutic radionuclides is provided in Table 7.1. The selection of a radionuclide depends on the application, and how close it be maneuvered to the tumor.

The role of nanostructures in the delivery of radionuclides is to make the drug more stable before reaching the tumor cells, to allow evasion of the reticuloendothelial system (RES), if the liver or spleen are not the target, and to promote binding to the tumor cells for as long as possible so as to increase the therapeutic ratio. Depending on the composition, nanoparticles often add new properties, such as magnetism, to the drug. To facilitate these functions, the following carriers have been used.

7.5.1
Liposomes

A liposome is a lipid bilayer structure, usually with a hydrophilic outer layer and a hydrophobic inner layer, that encloses active agents. Liposomes are

Table 7.1 Characteristics of common radionuclides used for radioimmunotherapy.

Radionuclide	Emission type	Half-life	E_{max} (keV)	Range	Production
^{186}Re	β, γ (9.4%)	89.2 h	1069	5 mm	^{185}Re(n, γ), ^{186}Re
^{166}Ho	β, γ (6.7%)	26.9 h	1853	10.2 mm	^{165}Ho(n, γ), ^{166}Ho
^{188}Re	β, γ (15.1%)	17.0 h	2120	11 mm	^{188}W/^{188}Re-generator
^{89}Sr	β	52.7 days	1463	3 mm	^{88}Sr(n, γ), ^{89}Sr
^{32}P	β	14.3 days	1710	8.7 mm	^{32}S(n,p), ^{32}P
^{90}Y	β	64.1 h	2280	12 mm	^{90}Sr/^{90}Y-generator
^{225}Ac	α	10 days	5830, 5972, 5790, 5732	40–80 μm	^{225}Ra-generator
^{211}At	α	7.2 h	5870	60–80 μm	Accelerator
^{213}Bi	α	45.7 min	5869	50–80 μm	^{225}Ra-generator

self-assembling nanostructures that have been used to enclose imaging radionuclides in a large number of studies, but their therapeutic applications have been relatively few and less mature due to the more stringent requirements relating to the specificity and the higher radiation doses involved in radiotherapy. *In vitro* feasibility tests were first conducted using radioactive oxodichloroethoxy-bis-triphenylphosphine ^{186}Re and ^{188}Re in liposomes, but the *in vitro* stability was not satisfactory [100]. Further *in vivo* testing was performed with high-energy β-particle emitters [101] to deliver radiation to avascular prostate carcinoma spheroids. Because of the longer penetration of the high-energy β-particles, radiation can be delivered to the core of a tumor, even if it is not well perfused, and carefully selected surface properties can improve tissue penetration. A cholesterol-stabilized small unilamellar vesicle (SUV)-dimyristoyl-phosphatidylcholine (DMPC) showed an increased penetration into the avascular tumor and improved dose homogeneity. The preliminary testing of α-particle-emitters was also conducted, focusing primarily on the stability of the liposome structure. It was shown that for hydrophilic α-particle emitters, the retention rate can be as high as 88%, mainly because the hydrophobic inner layer of the liposome can prevent loss of the radionuclide [102]. Whilst further *in vivo* testing is clearly required needed to validate these techniques before they can be used on human patients, the preliminary results have shown much promise.

7.5.2
Nanoparticles

Although the material is different, the purpose of nanoparticle-facilitated radionuclide delivery is the same as for liposomes. However, in addition to facilitating drug delivery, nanoparticles may add functionality to the radionuclide delivery system. For example, ^{188}Re-radiolabeled magnetic nanoparticles were synthesized to deliver radiation to a superficial lesion, guided by an external magnetic field [103]. For this, amino-functionalized superparamagnetic iron oxide nanoparticles (SPIOs) were synthesized and conjugated with Hepama-1, a humanized monoclonal antibody directed against liver cancer, to prepare immunomagnetic nanoparticles (IMN) [104]. The ^{188}Re-IMN demonstrated an ability to kill SMMC-7721 liver cancer cells efficiently *in vitro*, while the SPIOs were investigated for their ability to introduce negative MRI contrast at very low concentrations [105]. Therefore, theranostic (therapeutic + diagnostic) functions may be performed with a single naonocomposite.

^{90}Y is another popular radionuclide that has been used to target the process of angiogenesis in endothelial cells of rapidly growing tumors. For example, a significant tumor growth delay was observed in mice treated with ^{90}Y-labeled nanoparticles that had been lined with a small integrin antagonist (IA) (IA-NP-90Y), or to a monoclonal antibody against murine VEGF receptor 2 (Flk-1), compared to a group of mice treated with the ^{90}Y-labeled nanoparticle alone [106].

As a further demonstration of the versatility of nanoparticles, the magnetism introduced by the nanoparticle may be utilized to kill the cells with heat generated from an external alternating magnetic field (AMF). For example, ^{111}In-mAb-conjugated iron oxide nanoparticles (^{111}In-bioprobes) have been shown to be capable of targeting tumors *in vivo* and subsequently inducing tumor cell necrosis in response to an externally applied AMF [107].

7.5.3
Dendrimers

Dendrimers are highly branched macromolecules with a central core molecule and tree-root-shaped arms that grow outwards (Figure 7.13). Dendrimers can be grown from a central core, a process known as the *divergent method* [108], or by Fréchet's *convergent method*, in which the dendrimer is synthesized from the periphery inwards [109]. Dendrimers are generally characterized by their terminal branch point, called the *generation*, such that a G5 dendrimer refers to a polymer with four generations of branch points from the central core. The synthesis of dendrimers can be precisely controlled, resulting in a monodispersed distribution in the size and chemistry with highly configurable termination groups, while their biological properties can be tuned by altering the polymer size, charge, and composition. As a consequence, dendrimers have become an ideal delivery vehicle candidate to adapt to different biological goals such as lipid bilayer interactions, cytotoxicity, internalization, blood plasma retention time, biodistribution, and filtration [110].

Dendrimers have undergone testing for both imaging and therapeutic applications. Although, in imaging applications, dendrimers are characterized by a low delivery efficiency, it is acceptable to have less than 1% of the radioactivity delivered to the tumor, even with antibody for targeting. Therapeutic applications, in

276 | *7 Nanomaterials for Radiation Therapy*

Figure 7.13 Structures of dendrimers used for delivery of cancer therapies. (1) PAMAM; (2) Melamine-based dendrimer; (3) Dendrimer based on 2,2-bis(hydroxymethyl) propionic acid; (4) PPI; (5) Dendrimer based on glycerol and succinic acid with a PEG core; (6) Dendrimer based on 5-aminolaevulinic acid.

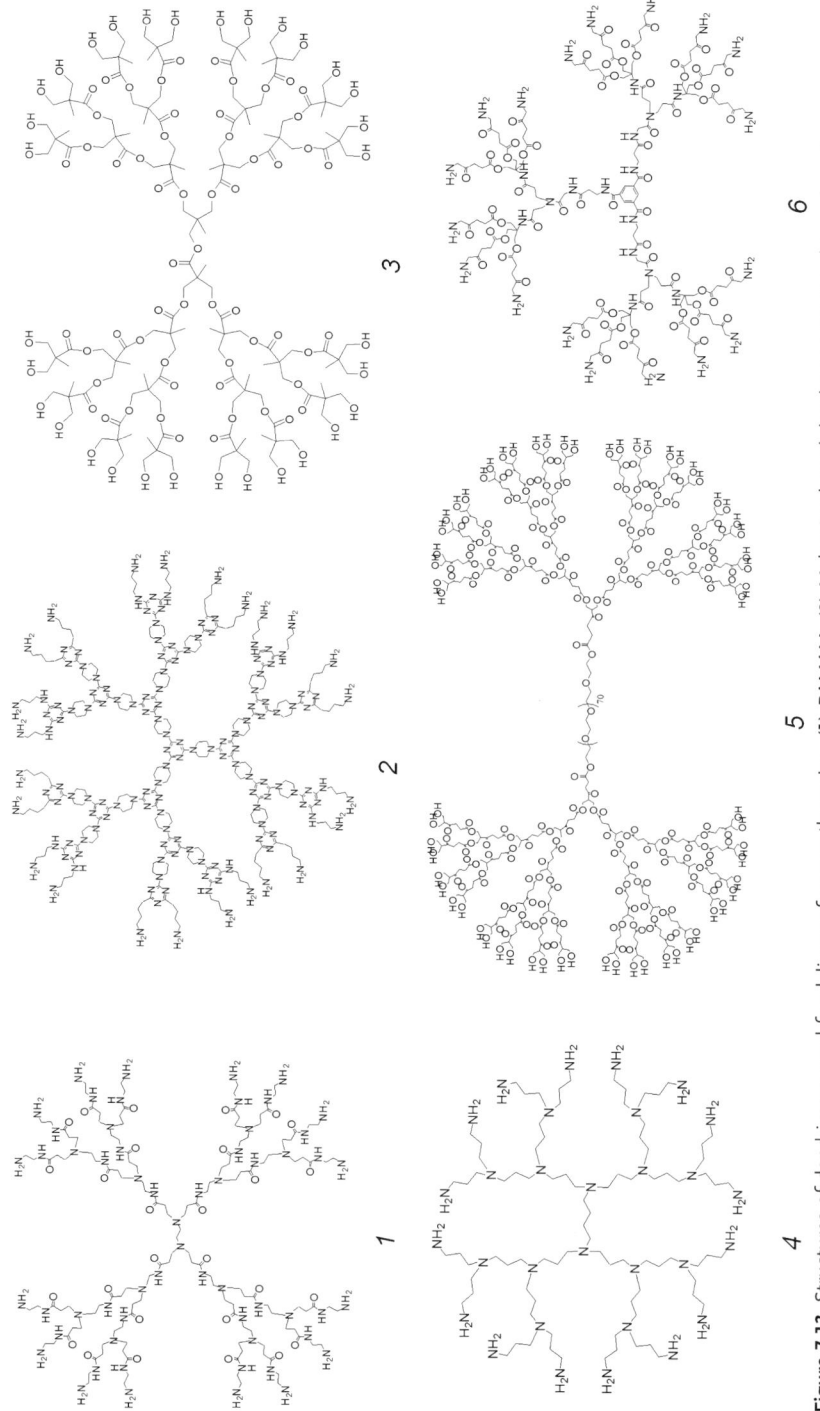

contrast, require higher delivery rates, although certain features of the tumor cells make the delivery of nanoparticles difficult. First, tumors are usually poorly vascularized, with regions that are not well perfused. Second, in the absence of lymphatic fluid in the tumor, it is common for the intratumor pressure to be higher than that of the surrounding tissues. Most importantly, tumor cells are highly mutating and may inactivate the antigen used for targeting. Whilst these limitations have kept the therapeutic applications of dendrimers to a minimum, a few studies have reported on the use of radioactively labeled dendrimers for *in vivo* tumor delivery.

Among the few *in vivo* applications of dendrimers, Khan *et al.* used an intratumoral injection for a higher tumor dose. For this, a radioactive gold-dendrimer composite nanodevices (CNDs) was fabricated in distinct sizes for targeted radiopharmaceutical dose delivery [98]. The system was based on the poly (amidoamine) (PAMAM) dendrimer nanoparticles, and poly $^{197}Au_0$ was transformed through a simultaneous γ-ray-induced polymerization process and bombardment by neutrons to generate radioactive poly ^{198}Au. A mouse melanoma tumor model was used to test whether the poly $^{198}Au_0$ CNDs could deliver a therapeutic dose. A single intratumoral injection of poly $^{198}Au_0$ CNDs (diameter 22 nm) in phosphate-buffered saline to deliver a dose of 74 μCi, resulted after eight days in a statistically significant 45% reduction in tumor volume, when compared to untreated groups and those injected with the nonradioactive nanodevice. The positively charged dendrimers were also found to be better at evading the RES system and thus to remain in the tissues for longer times. Dong *et al.* [111] used ^{32}P-nanocolloids (^{32}P is a β⁻ emitter with an energy of 1.709 MeV and a half-life of 14.28 days) with sizes distributed around 47.5 nm to treat micrometastases in the lymph nodes of rabbits. Following treatment the number of metastatic lymph nodes was seen to decrease significantly, although the nanocolloid was not functionalized to target the tumor cells in specific fashion.

Dendrimers have also been studied for their ability to carry radionuclides to tumors in nanobrachytherapy. To test such targeting ability, a murine monoclonal IgG1 was attached to ^{111}In- or ^{153}Gd-labeled G4 dendrimers, such that a markedly high specific activity was obtained with a minimal loss of immunoreactivity although, unfortunately, high concentrations of radiolabeled dendrimer were detected in the liver and spleen. A high RES retention was also observed with a dendrimer-based boron capture therapy [112]. It appears that, although antibody- or epidermal growth factor-labeled dendrimers have the ability to target and infiltrate the tumor cell membrane [113], RES retention remains problematic for *in vivo* applications.

7.6
Nanoparticles as Radioprotectors

Approaching from a different direction, the therapeutic ratio can be improved by protecting normal tissues against radiation damage. Radiation-induced injury to cells is caused primarily by the free radicals generated by excitation and ionization events during the interaction of radiation with the tissue; consequently, free radicals have been the primary target of research in radiation protection.

Amifostine is the only approved treatment for radioprotection in head-and-neck cancer patients [114]. Amifostine releases active thiol metabolites that scavenge superoxide radicals generated from ionizing radiation, although its common side effects include hypocalcemia, diarrhea, nausea, vomiting, sneezing, somnolence, and hiccoughs. More serious side effects include hypotension (in 62% of patients), erythema multiforme, Stevens–Johnson syndrome, toxic epidermal necrolysis, immune hypersensitivity syndrome, erythroderma, anaphylaxis, and loss of consciousness. These side effects have prevented the wider application of amifostine in radiation therapy.

Recent nanotechnology-based molecular engineering advancements have produced new classes of molecules, such as cerium oxide (CeO_2) (Figure 7.14), which were developed as potent free radical scavengers [115]. CeO_2 is a rare earth oxide material from the lanthanide series of the Periodic Table. Cerium nanoparticles are highly efficient redox reagents, and are used in various applications such as ultraviolet absorbents, oxygen sensors, and automotive catalytic converters. All of these applications are based on the ability of CeO_2 to reduce oxidation species in a catalytic manner [115]. The cerium atom can exist in either the +3 (fully reduced) or +4 (fully oxidized) state. In its oxidative form, CeO_2 also exhibits oxygen vacancies, or defects, in the lattice structure, through loss of oxygen and/or its electrons, alternating between CeO_2 and CeO_{2-x} during redox reactions. The change in cerium valence during a redox event subsequently alters the structure of the oxide lattice, possibly creating additional oxygen vacancies by lattice expansion [115]. This electron translation within the lattice provides reducing power for free radical

Figure 7.14 Transmission electron microscopy (TEM) image of cerium oxide nanoparticles.

scavenging. After the scavenging event, the original lattice structure may be regenerated by releasing H_2O, while the cerium atom returns to the +3 state. It was also reported that the redox efficiency of CeO_2 was inversely proportional to the CeO_2 nanoparticle size [115]. By using nanosized (10–20 nm) CeO_2 particles, both high antioxidant efficiency and cell penetration can be achieved.

CeO_2 nanoparticles may offer radiation protection to cells. It has been reported that CeO_2 particles can protect 90% of normal cells against 10 Gy of radiation, with minimal tumor cell protection [116]. The mechanism behind the different protections is not entirely clear, but it may arise from the differential uptake of the particles, though this has not been proven. Another plausible mechanism was proposed by Cohen-Jonathan et al. [117], who hypothesized that tumor cells expose more bases of the chromatin structure as targets for free-radical attack. The greater number of vulnerable sites in tumor cells renders the radiation protection by CeO_2 more difficult.

As CeO_2 can reduce large numbers of free radicals more rapidly than amifostine, it is suitable for use as a radioprotector during standard radiation therapy. As the burst of free radical generation that occurs during radiation is completed within milliseconds of treatment, the longer retention time of the CeO_2 nanoparticle means that fewer infusions will be required during radiation treatment. In a toxicology study conducted by Rzigalinski et al. [115, 118], CeO_2 did not exhibit any toxicity in neuronal and macrophage cell lines as long as the particle size was less than 20 nm [118]. In the only reported in vivo study in mammals, CeO_2 resulted in a more than twofold longer survival of mice as compared to the amifostine-treated group after total body irradiation [119], which usually results in multiple organ toxicity. The results of these initial studies have suggested that the CeO_2 nanoparticle is an attractive candidate for clinical development as a novel radioprotector. Application in thoracic radiotherapy seems a particularly logical choice, given the high sensitivity of normal lung tissue to radiation, which limits curative treatment for many patients.

Carboxyfullerene (C60) is another nanomaterial with potential to scavenge free radicals [120–122]. Generated by ionizing radiation, carboxyfullerene has been described as a free radical "sponge" that can add multiple radicals to a single nanoparticle. Carboxyfullerene has been shown to prevent hydrogen peroxide- and cumene hydroperoxide-elicited changes in the rat hippocampus, and also to provide an effective protection of human keratinocytes against UVb radiation [123]. C3, a regioisomer of water-soluble carboxyfullerene has also been tested for radioprotective function. A protection factor (defined as the ratio of survival with and without C3) of up to 2.38 was demonstrated in normal hematopoietic progenitor cells, but much less protection was observed in mouse and human tumor cell lines [124]. An anti-oxidative stress action of C3 was also observed after C3 treatment of $Sod2^{-/-}$ mice, which lack expression of mitochondrial manganese superoxide dismutase; their life span was increased by 300% [125]. Further experiments conducted by Yin et al. showed that reactive oxygen species (ROS), superoxide radical anions, singlet oxygen and hydroxyl radicals can be effectively inhibited by the fullerenes. This report also revealed that the radical-scavenging ability is affected by surface chemistry-induced differences in electron affinity and physical properties, such as the degree of aggregation [126].

An important aspect of effective radioprotection is that it should not protect tumor cells, which would render the drug useless. Amifostine is effective because it and its metabolites are present in healthy cells at 100-fold greater concentrations than in tumor cells. Although the aforementioned studies on CeO_2 and carboxyfullerene have shown the preferential protection of normal cells, the mechanisms are not as clear, and further studies must be conducted to resolve this problem.

7.7
Radiation Dosimeters Using Semiconductor Nanomaterials

To calibrate a radiation delivery device or ensure accurate dose delivery in radiation therapy, the radiation dose must be measured with a dosimeter that converts radiation energy to signals which can be stored and detected. Conventional dosimeters, such as ion chambers, diodes, thermoluminescent dosimeters (TLDs) and metal-oxide semiconductor field effect transistors (MOSFETs), are used to convert the radiation to electrical signals, photoemission or a shift in the threshold voltage. The choice of dosimeter depends on the requirements of the application, the desired accuracy, convenience and accessibility, and it can be difficult to optimize all of these factors at once. For example, an ion chamber paired with an electrometer can be used to measure radiation dosage accurately and reproducibly. This set-up is also minimally affected by energy over a wide range of megavoltage photon and electron measurements. However, ion chambers require a high voltage to operate, are cumbersome, and are too large for *in vivo* measurements. TLDs can be made tissue-equivalent and very small so that they are suitable for *in vivo* measurements, but the calibration and readout process is tedious. Semiconductor diodes are simple to operate, but they are wired, not tissue equivalent, and dependent on the radiation energy and angles. MOSFETs, depending on the model, are limited by the number of readings before the shift in threshold voltage is saturated irreversibly. These dosimeters do not provide two-dimensional (2-D) or three-dimensional (3-D) dose distribution unless hundreds of them are built into a bulky array. Radiograph film is widely available for 2-D dose measurements, but it does not provide an absolute readout and is cumbersome to process. 3-D gel dosimeters such as BANG polymer gel [127] polymerize upon radiation, and can be used for 3-D dose measurements, but they are expensive to use. In addition, the polymerization is irreversible and requires a dedicated optical scanner to read. In addition to challenges from conventional dosimetry, in nanodosimetry, the energy deposited on a scale comparable to the size of a DNA molecule must be measured using a device that is in the same scale [128–130]. Novel dosimeters are clearly needed to answer these challenges, and nanotechnology may provide some of the answers for future radiation dosimetry.

7.7 Radiation Dosimeters Using Semiconductor Nanomaterials

The principles employed by conventional dosimetry can be adapted to create nanodosimeters for radiation measurements. For example, carbon nanotubes (CNTs) are effective in converting photon energy to electronic signals; such a conversion is referred to as *optoelectronics* [131]. When a CNT is irradiated by visible light and a bias voltage is applied to it, an electrical current is generated and part of the energy from the light can be converted to electricity. The efficiency of energy conversion may exceed 10%, and this has led to the use of CNTs in solar panels and photodiodes. The same capacity may be applied to radiation dosimetry, where X-ray energy can be converted to electricity. More interestingly, short-circuit photocurrents can be theoretically generated [132] without applying a bias voltage to a nanotube, making wireless operation of the nanodosimeter possible.

When a TLD is irradiated, electrons in the crystal's atoms jump to higher energy states, where they stay trapped due to impurities in the crystal. After irradiation, the TLD is heated to release the electrons from the trapped state, during which process the electrons return to the ground state and visible photons are released. As the number of trapped electrons, and thereafter the amount of photon emission, is proportional to the radiation dose within a linear response range, the radiation dose can be determined by measuring the photon emission using a photomultiplier tube. Nanometer-sized phosphorescence (afterglow) crystals such as LiF:Mg and MgF_2:Eu can be manufactured for this purpose [133, 134]. Compared with macrosized TLDs, nano-TLDs are more stable, with high luminescent intensity, low photobleaching, and a large Stokes' shift [135]. The dynamic range of TLDs was improved with the nanometer-scaled phosphors. Sahare *et al.* synthesized a $K_3Na(SO_4)_2$:Eu nanocrystalline powder that was fourfold more sensitive than a LiF:Mg,Ti (TLD-100) phosphor, and had a near-linear response up to a radiation dose of 70 000 Gy [135].

MOSFETs can also be made on the nanometer scale. A schematic presentation of a typical MOSFET is shown in Figure 7.15. When a positive voltage is applied

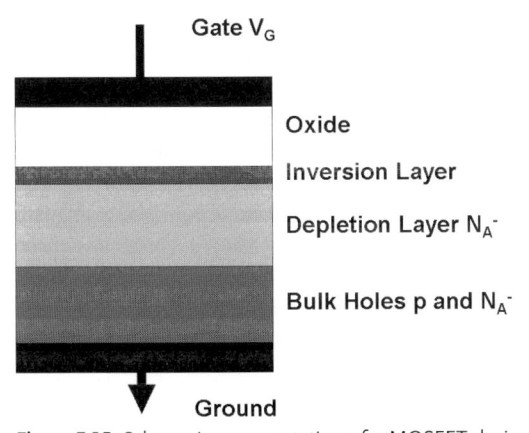

Figure 7.15 Schematic representation of a MOSFET device.

to the gate, the electric field causes the holes to be repelled from the interface, creating a depletion region containing negatively charged acceptor ions. A further increase in the gate voltage eventually causes a sufficient number of electrons to appear at the inversion layer, such that the MOSFET becomes a conductor. The voltage that turns on the MOSFET is referred as the *threshold voltage*. When a MOSFET device is irradiated, trapped charges are built up in the oxide layer, the number of interface traps increases, and the number of bulk oxide traps increases. With the excitation from radiation, electron-hole pairs are generated, whereupon electrons quickly move out of the gate electrode while holes move slowly in towards the Si/SiO_2 interface where they become trapped, causing a negative threshold voltage shift that can be measured. The voltage shift is proportional to the amount of radiation received [136, 137]. Very small MOSFETs suitable for *in vivo* measurements can be manufactured. Compared to ion chambers, they do not need a high voltage to operate, and can be used in the wireless mode. Compared to diodes, MOSFETs are less dependent on the radiation energy and angle. The readout of a MOSFET is instantaneous and much less elaborate than that of a TLD. As opposed to the traditional bulk MOSFET material, field effect transistors (FETs) based on single-wall carbon nanotubes (SWNT) have been fabricated for molecular, chemical, and biological sensing [138, 139]. By using the same platform, Tang et al. [140] synthesized SWNT-FETs on a SiO_2/Si substrate using patterned chemical vapor deposition (CVD) on top of W/Pt electrodes. The structure of the SWCN-FET was similar to the conventional MOSFET, except that the p-type Si was replaced by the nanotubes. The SWNT-FET was shown to be stable against doses up to 1 Gy, and to be about two orders of magnitude more sensitive than a conventional MOSFET.

7.8
Conclusions

Although the application of nanomaterials in radiation therapy is relatively new, the early results have been encouraging. As carriers, nanomaterials have been utilized in nanobrachytherapy to bring radioactive nuclides into close contact with tumor cells. As energy mediators, high-atomic-number nanomaterials targeted to tumor cells are used to deposit more radiation energy. In a more sophisticated form, semiconductor nanomaterials were employed as an energy reservoir that absorbs a wide range of X-rays and converts them to visible light with specific wavelength tuned to the absorption peak of the photosensitizer, which in turn generates cytotoxic singlet oxygen molecules for enhanced tumor cell killing. This application, compared to the simple energy sink with high-Z materials, may activate new biological pathways for tumor cell death, because PDT has very different cellular targets than radiation. Many semiconductor materials have a high redox ability that enables them to be free radical scavengers for radioprotection. The nanoengineering of these semiconductor materials has led to the production of a new generation of dosimeters that are more nimble, sensitive, and convenient to use. On the other hand, the fact that nanomaterials are not routinely used in patient treatment indicates that this field is still in the early stages of research and development. To realize the full potential of nanotechnology in radiation therapy, technological breakthroughs are needed in the following areas:

1. Effectiveness
2. Tolerable toxicity
3. Water-solubility and biocompatibility
4. Predictable and modifiable pharmacokinetics
5. Long serum half-life for solid tumor uptake
6. Standardized targeting moieties for different tumors
7. Reproducibility and cost-effectiveness in manufacturing
8. Biological clearance after the function is performed

Many of these problems are longstanding with regards to the biological application of nanomaterials. Although remarkable progress has been made, there are at present no well-established general protocols, with success usually being achieved on a case-by-case basis. For example, toxicity is a less-prominent concern with gold nanoparticles compared to CdSe QDs, while RES retention is helpful when the liver or spleen is the target organ, but is not desirable for other applications. It is difficult–if not impossible–to fabricate a nanodevice capable of satisfying all of these requirements, and it is important therefore to prioritize these properties for specific applications. In the case of radiotherapy, long-term toxicity and biological clearance may be arguably less important than effectiveness, particularly for patients with terminal-stage cancer.

Last, but not least, it is clear from the many reports made to date that research into nanomedicine for radiation therapy is highly segmented and specialized. Whilst an individual group may achieve technical breakthroughs in one or more areas, the lack of a broader interdisciplinary collaboration may hamper the development and clinical application of these highly promising nanomaterials. The involvement of more clinicians, including radiation oncologists and physicists, will be essential as investigations are continued to create nanodevices with novel properties.

References

1 Mackie, T.R. (2006) History of tomotherapy. *Physics in Medicine and Biology*, **51** (13), R427–53.
2 Mackie, T.R. *et al.* (1993) Tomotherapy: a new concept for the delivery of dynamic conformal radiotherapy. *Medical Physics*, **20** (6), 1709–19.
3 Wardman, P. (2007) Chemical radiosensitizers for use in radiotherapy. *Clinical Oncology (Royal College of Radiologists, Great Britain)*, **19** (6), 397–417.
4 Adams, G.E. (1973) Chemical radiosensitization of hypoxic cells. *British Medical Bulletin*, **29** (1), 48–53.
5 Fowler, J.F., Adams, G.E. and Denekamp, J. (1976) Radiosensitizers of hypoxic cells in solid tumors. *Cancer Treatment Reviews*, **3** (4), 227–56.

6 Brown, J.M. and Wilson, W.R. (2004) Exploiting tumour hypoxia in cancer treatment. *Nature Reviews. Cancer*, **4** (6), 437–47.

7 Poggi, M.M., Coleman, C.N. and Mitchell, J.B. (2001) Sensitizers and protectors of radiation and chemotherapy. *Current Problems in Cancer*, **25** (6), 334–411.

8 Eberhardt, W., Pottgen, C. and Stuschke, M. (2006) Chemoradiation paradigm for the treatment of lung cancer. *Nature Clinical Practice. Oncology*, **3** (4), 188–99.

9 Hao, D. et al. (2006) Platinum-based concurrent chemoradiotherapy for tumors of the head and neck and the esophagus. *Seminars in Radiation Oncology*, **16** (1), 10–19.

10 Chinnaiyan, P., Allen, G.W. and Harari, P.M. (2006) Radiation and new molecular agents, part II: targeting HDAC, HSP90, IGF-1R, PI3K, and Ras. *Seminars in Radiation Oncology*, **16** (1), 59–64.

11 Choudhury, A., Cuddihy, A. and Bristow, R.G. (2006) Radiation and new molecular agents part I: targeting ATM-ATR checkpoints, DNA repair, and the proteasome. *Seminars in Radiation Oncology*, **16** (1), 51–8.

12 Minchinton, A.I. et al. (1984) Glutathione depletion in tissues after administration of buthionine sulphoximine. *International Journal of Radiation Oncology, Biology, Physics*, **10** (8), 1261–4.

13 Tannock, I.F. (1996) Treatment of cancer with radiation and drugs. *Journal of Clinical Oncology*, **14** (12), 3156–74.

14 Attix, F. (1986) *Introduction to Radiological Physics and Radiation Dosimetry*, Wiley-Interscience.

15 Tang, X.W. et al. (2005) Measurement of ionizing radiation using carbon nanotube field effect transistor. *Physics in Medicine and Biology*, **50** (3), N23–31.

16 Santos Mello, R. et al. (1983) Radiation dose enhancement in tumors with iodine. *Medical Physics*, **10** (1), 75–8.

17 Matsudaira, H., Ueno, A.M. and Furuno, I. (1980) Iodine contrast medium sensitizes cultured mammalian cells to X rays but not to gamma rays. *Radiation Research*, **84** (1), 144–8.

18 Nath, R., Bongiorni, P. and Rockwell, S. (1990) Iododeoxyuridine radiosensitization by low- and high-energy photons for brachytherapy dose rates. *Radiation Research*, **124** (3), 249–58.

19 Iwamoto, K.S. et al. (1987) Radiation dose enhancement therapy with iodine in rabbit VX-2 brain tumors. *Radiotherapy and Oncology*, **8** (2), 161–70.

20 Norman, A. et al. (1997) X-ray phototherapy for canine brain masses. *Radiation Oncology Investigations*, **5** (1), 8–14.

21 Rose, J.H. et al. (1999) First radiotherapy of human metastatic brain tumors delivered by a computerized tomography scanner (CTRx). *International Journal of Radiation Oncology, Biology, Physics*, **45** (5), 1127–32.

22 Rousseau, J. et al. (2007) Convection-enhanced delivery of an iodine tracer into rat brain for synchrotron stereotactic radiotherapy. *International Journal of Radiation Oncology, Biology, Physics*, **68** (3), 943–51.

23 Adam, J.F. et al. (2005) Enhanced delivery of iodine for synchrotron stereotactic radiotherapy by means of intracarotid injection and blood-brain barrier disruption: quantitative iodine biodistribution studies and associated dosimetry. *International Journal of Radiation Oncology, Biology, Physics*, **61** (4), 1173–82.

24 Regulla, D.F., Hieber, L.B. and Seidenbusch, M. (1998) Physical and biological interface dose effects in tissue due to X-ray-induced release of secondary radiation from metallic gold surfaces. *Radiation Research*, **150** (1), 92–100.

25 Herold, D.M. et al. (2000) Gold microspheres: a selective technique for producing biologically effective dose enhancement. *International Journal of Radiation Biology*, **76** (10), 1357–64.

26 Kong, T. et al. (2008) Enhancement of radiation cytotoxicity in breast-cancer cells by localized attachment of gold nanoparticles. *Small*, **4** (9), 1537–43.

27 Hainfeld, J.F., Slatkin, D.N. and Smilowitz, H.M. (2004) The use of gold nanoparticles to enhance radiotherapy in

mice. *Physics in Medicine and Biology*, **49** (18), N309–15.
28 Zhou, J. et al. (2009) Functionalized gold nanoparticles: synthesis, structure and colloid stability. *Journal of Colloid and Interface Science*, **331** (2), 251–62.
29 Rosi, N.L. and Mirkin, C.A. (2005) Nanostructures in biodiagnostics. *Chemical Reviews*, **105** (4), 1547–62.
30 Li, J.L. et al. (2009) In vitro cancer cell imaging and therapy using transferrin-conjugated gold nanoparticles. *Cancer Letters*, **274** (2), 319–26.
31 Zhang, X.J. et al. (2008) Enhanced radiation sensitivity in prostate cancer by gold-nanoparticles. *Clinical and Investigative Medicine*, **31** (3), E160–7.
32 Porta, F. et al. (2007) Gold nanoparticles capped by peptides. *Materials Science and Engineering B-Solid State Materials for Advanced Technology*, **140** (3), 187–94.
33 Surujpaul, P.P. et al. (2008) Gold nanoparticles conjugated to [Tyr(3)] Octreotide peptide. *Biophysical Chemistry*, **138** (3), 83–90.
34 Pissuwan, D. et al. (2007) Gold nanosphere-antibody conjugates for hyperthermal therapeutic applications. *Gold Bulletin*, **40** (2), 121–9.
35 Roeske, J.C. et al. (2007) Characterization of the theoretical radiation dose enhancement from nanoparticles. *Technology in Cancer Research & Treatment*, **6** (5), 395–401.
36 Cho, S.H. (2005) Estimation of tumour dose enhancement due to gold nanoparticles during typical radiation treatments: a preliminary Monte Carlo study. *Physics in Medicine and Biology*, **50** (15), N163–73.
37 Hainfeld, J.F. et al. (2008) Radiotherapy enhancement with gold nanoparticles. *Journal of Pharmacy and Pharmacology*, **60** (8), 977–85.
38 Buchegger, F. et al. (2006) Auger radiation targeted into DNA: a therapy perspective. *European Journal of Nuclear Medicine and Molecular Imaging*, **33** (11), 1352–63.
39 Sugiyama, H. (1985) Stopping power formula for intermediate energy electrons. *Physics in Medicine and Biology*, **30** (4), 331–5.
40 Sonvico, F. et al. (2005) Metallic colloid nanotechnology, applications in diagnosis and therapeutics. *Current Pharmaceutical Design*, **11** (16), 2091–105.
41 Hirsch, L.R. et al. (2003) Nanoshell-mediated near-infrared thermal therapy of tumors under magnetic resonance guidance. *Proceedings of the National Academy of Sciences of the United States of America*, **100** (23), 13549–54.
42 Liu, C.H., Mi, C.C. and Li, B.Q. (2008) Energy absorption of gold nanoshells in hyperthermia therapy. *IEEE Transactions on Nanobioscience*, **7** (3), 206–14.
43 Shah, J. et al. (2008) Ultrasound imaging to monitor photothermal therapy – feasibility study. *Optics Express*, **16** (6), 3776–85.
44 Moan, J. and Anholt, H. (1990) Phthalocyanine fluorescence in tumors during PDT. *Photochemistry and Photobiology*, **51** (3), 379–81.
45 Luksiene, Z., Juzenas, P. and Moan, J. (2006) Radiosensitization of tumours by porphyrins. *Cancer Letters*, **235** (1), 40–7.
46 Luksiene, Z. et al. (2006) Mechanism of radiosensitization by porphyrins. *Journal of Environmental Pathology, Toxicology and Oncology*, **25** (1–2), 293–306.
47 Schaffer, M. et al. (2005) The Application of Photofrin II as a sensitizing agent for ionizing radiation – a new approach in tumor therapy? *Current Medicinal Chemistry*, **12** (10), 1209–15.
48 Schaffer, M. et al. (2002) Application of Photofrin II as a specific radiosensitising agent in patients with bladder cancer – a report of two cases. *Photochemical & Photobiological Sciences*, **1** (9), 686–9.
49 Kulka, U. et al. (2003) Photofrin as a radiosensitizer in an in vitro cell survival assay. *Biochemical and Biophysical Research Communications*, **311** (1), 98–103.
50 Schaffer, M. et al. (2003) Porphyrins as radiosensitizing agents for solid neoplasms. *Current Pharmaceutical Design*, **9** (25), 2024–35.
51 Konan, Y.N., Gurny, R. and Allemann, E. (2002) State of the art in the delivery of photosensitizers for photodynamic therapy. *Journal of Photochemistry and Photobiology. B, Biology*, **66** (2), 89–106.

52 Samia, A.C.S., Chen, X.B. and Burda, C. (2003) Semiconductor quantum dots for photodynamic therapy. *Journal of the American Chemical Society*, **125** (51), 15736–7.

53 Tsay, J.M. et al. (2007) Singlet oxygen production by peptide-coated quantum dot-photosensitizer conjugates. *Journal of the American Chemical Society*, **129** (21), 6865–71.

54 Stucky, G.D. and Mac Dougall, J.E. (1990) Quantum confinement and host/guest chemistry: probing a new dimension. *Science*, **247** (4943), 669–78.

55 Bryant, G.W. (1988) Excitons in quantum boxes: correlation effects and quantum confinement. *Physical Review. B, Condensed Matter*, **37** (15), 8763–72.

56 Norris, D.J. (1994) Electronic structure in semiconductor nanocrystals, in *Semiconductor and Metal Nanocrystals: Synthesis and Electronic and Optical Properties* (ed. V.I. Klimov), Marcel Dekker, New York, pp. 65–102.

57 Alivisatos, A.P. (1996) Perspectives on the physical chemistry of semiconductor nanocrystals. *Journal of Physical Chemistry*, **100** (31), 13226–39.

58 Bawendi, M.G., Steigerwald, M.L. and Brus, L.E. (1990) The quantum-mechanics of larger semiconductor clusters (quantum dots). *Annual Review of Physical Chemistry*, **41**, 477–96.

59 Hardman, R. (2006) A toxicologic review of quantum dots: toxicity depends on physicochemical and environmental factors. *Environmental Health Perspectives*, **114** (2), 165–72.

60 Bouldin, J.L. et al. (2008) Aqueous toxicity and food chain transfer of quantum dots in freshwater algae and *Ceriodaphnia dubia*. *Environmental Toxicology and Chemistry*, **27** (9), 1958–63.

61 Chen, W., Joly, A.G. and Zhang, J.Z. (2001) Up-conversion luminescence of Mn2+ in ZnS : Mn2+ nanoparticles. *Physical Review B*, **6404** (4), 041202-1–041202-4.

62 Chen, W. et al. (2000) Energy structure and fluorescence of Eu2+ in ZnS: Eu nanoparticles. *Physical Review B*, **61** (16), 11021–4.

63 He, Y. et al. (2009) Photo and pH stable, highly-luminescent silicon nanospheres and their bioconjugates for immunofluorescent cell imaging. *Journal of the American Chemical Society*, **131** (12), 4434–8.

64 Cao, L. et al. (2007) Carbon dots for multiphoton bioimaging. *Journal of the American Chemical Society*, **129** (37), 11318–19.

65 Sun, Y.P. et al. (2006) Quantum-sized carbon dots for bright and colorful photoluminescence. *Journal of the American Chemical Society*, **128** (24), 7756–7.

66 Buitelaar, M.R. et al. (2002) Multiwall carbon nanotubes as quantum dots. *Physical Review Letters*, **88** (15), 156801.

67 Botsoa, J. et al. (2008) Application of 3C-SiC quantum dots for living cell imaging. *Applied Physics Letters*, **92** (17), 173902-1–173902-3.

68 Chen, W. and Zhang, J. (2006) Using nanoparticles to enable simultaneous radiation and photodynamic therapies for cancer treatment. *Journal of Nanoscience and Nanotechnology*, **6** (4), 1159–66.

69 Yang, W. et al. (2007) *Novel FRET-Based Radiosensitization Using Quantum Dot-Photosensitizer Conjugates Signals, Systems and Computers*, ACSSC, pp. 1861–5.

70 Yang, W. et al. (2008) Semiconductor nanoparticles as energy mediators for photosensitizer-enhanced radiotherapy. *International Journal of Radiation Oncology, Biology, Physics*, **72** (3), 633–5.

71 Andrews, D.L. and Unified, A. (1989) Theory of radiative and radiationless molecular-energy transfer. *Chemical Physics*, **135** (2), 195–201.

72 Lakowicz, J.R. (1999) *Principles of Fluorescence Spectroscopy*, 2nd edn, Plenum Publishing Corporation.

73 Lakowicz, J. (2006) *Principles of Fluorescence Spectroscopy*, Springer, New York.

74 Pons, T. et al. (2007) On the quenching of semiconductor quantum dot photoluminescence by proximal gold nanoparticles. *Nano Letters*, **7** (10), 3157–64.

75 Biju, V. et al. (2006) Quenching of photoluminescence in conjugates of quantum dots and single-walled carbon

nanotube. *Journal of Physical Chemistry, B*, **110** (51), 26068–74.

76 Idowu, M., Chen, J.Y. and Nyokong, T. (2008) Photoinduced energy transfer between water-soluble CdTe quantum dots and aluminium tetrasulfonated phthalocyanine. *New Journal of Chemistry*, **32**, 290–6.

77 Sapsford, K.E. et al. (2007) Kinetics of metal-affinity driven self-assembly between proteins or peptides and CdSe-ZnS quantum dots. *Journal of Physical Chemistry, C*, **111** (31), 11528–38.

78 Morgan, N.Y. et al. (2009) Nanoscintillator conjugates as photodynamic therapy-based radiosensitizers: calculation of required physical parameters. *Radiation Research*, **171** (2), 236–44.

79 Hainfeld, J.F. et al. (2008) Radiotherapy enhancement with gold nanoparticles. *Journal of Pharmacy and Pharmacology*, **60** (8), 977–85.

80 Karotki, A. et al. (2006) Simultaneous two-photon excitation of photofrin in relation to photodynamic therapy. *Photochemistry and Photobiology*, **82** (2), 443–52.

81 Dougherty, T.J. et al. (1998) Photodynamic therapy. *Journal of the National Cancer Institute*, **90**, 889–905.

82 Macdonald, I. and Dougherty, T. (2001) Basic principles of photodynamic therapy. *Journal of Porphyrins and Phthalocyanines*, **5**, 105.

83 Allison, R.R. et al. (2004) Photosensitizer in clinical PDT. *Photodiagnosis and Photodynamic Therapy*, **1**, 27.

84 Bonnett, R. (1995) Photosensitizers of the porphyrin and phthalocynanine series for photodynamic therapy. *Chemical Society Reviews*, **24**, 19.

85 Whitehurst, C., Pantelides, M.L., Moore, J.V., King, T.A. and Blacklock, N.J. (1990) *In vivo* to post-mortem change in tissue penetration of red light. *Lasers in Medical Science*, **5** (4), 395–8.

86 Hahn, S.M. et al. (2006) Photofrin uptake in the tumor and normal tissues of patients receiving intraperitoneal photodynamic therapy. *Clinical Cancer Research*, **12** (18), 5464–70.

87 Farrell, T.J. et al. (1998) Comparison of the *in vivo* photodynamic threshold dose for photofrin, mono- and tetrasulfonated aluminum phthalocyanine using a rat liver model. *Photochemistry and Photobiology*, **68** (3), 394–9.

88 Lilge, L. et al. (1996) The sensitivity of normal brain and intracranially implanted VX2 tumour to interstitial photodynamic therapy. *British Journal of Cancer*, **73** (3), 332–43.

89 Ji, R.C. (2006) Lymphatic endothelial cells, tumor lymphangiogenesis and metastasis: new insights into intratumoral and peritumoral lymphatics. *Cancer Metastasis Reviews*, **25** (4), 677–94.

90 Kuijt, G.P. et al. (2005) The prognostic significance of axillary lymph-node micrometastases in breast cancer patients. *European Journal of Surgical Oncology*, **31** (5), 500–5.

91 Fehm, T. et al. (2008) Micrometastatic spread in breast cancer: detection, molecular characterization and clinical relevance. *Breast Cancer Research*, **10** (Suppl. 1), S1.

92 Riethdorf, S., Wikman, H. and Pantel, K. (2008) Review: biological relevance of disseminated tumor cells in cancer patients. *International Journal of Cancer*, **123** (9), 1991–2006.

93 Eng, T.Y., Luh, J.Y. and Thomas, C.R. Jr (2005) The efficacy of conventional external beam, three-dimensional conformal, intensity-modulated, particle beam radiation, and brachytherapy for localized prostate cancer. *Current Urology Reports*, **6** (3), 194–209.

94 Khan, F. (2003) *The Physics of Radiation Therapy*, 3rd edn, Lippincott Williams & Wilkins.

95 Beierwaltes, W. (1974) Radioiodine-labeled compounds previously or currently used for tumor localization. *Proceedings of an Advisory Group Meeting on Tumor Localization with Radioactive Agents, (Agency IAE)*, pp. 47–56.

96 DeNardo, S.J. and Denardo, G.L. (2006) Targeted radionuclide therapy for solid tumors: an overview. *International Journal of Radiation Oncology, Biology, Physics*, **66** (2 Suppl.), S89–95.

97 Sgouros, G. (2008) Update: molecular radiotherapy: survey and current status. *Cancer Biotherapy & Radiopharmaceuticals*, **23** (5), 531–40.

98 Khan, M.K. et al. (2008) Fabrication of {198Au0} radioactive composite nanodevices and their use for nanobrachytherapy. *Nanomedicine*, **4** (1), 57–69.

99 Sofou, S. (2008) Radionuclide carriers for targeting of cancer. *International Journal of Nanomedicine*, **3** (2), 181–99.

100 Hafeli, U. et al. (1991) A lipophilic complex with Re-186/Re-188 incorporated in liposomes suitable for radiotherapy. *Nuclear Medicine and Biology*, **18** (5), 449–54.

101 Emfietzoglou, D., Kostarelos, K. and Sgouros, G. (2001) An analytic dosimetry study for the use of radionuclide-liposome conjugates in internal radiotherapy. *Journal of Nuclear Medicine*, **42** (3), 499–504.

102 Henriksen, G. et al. (2004) Sterically stabilized liposomes as a carrier for alpha-emitting radium and actinium radionuclides. *Nuclear Medicine and Biology*, **31** (4), 441–9.

103 Hafeli, U.O. et al. (2001) Stability of biodegradable radioactive rhenium (Re-186 and Re-188) microspheres after neutron-activation. *Applied Radiation and Isotopes*, **54** (6), 869–79.

104 Liang, S. et al. (2007) Surface modified superparamagnetic iron oxide nanoparticles: as a new carrier for bio-magnetically targeted therapy. *Journal of Materials Science. Materials in Medicine*, **18** (12), 2297–302.

105 Bonnemain, B. (1998) Superparamagnetic agents in magnetic resonance imaging: physicochemical characteristics and clinical applications. A review. *Journal of Drug Targeting*, **6** (3), 167–74.

106 Li, L. et al. (2004) A novel antiangiogenesis therapy using an integrin antagonist or anti-Flk-1 antibody coated 90Y-labeled nanoparticles. *International Journal of Radiation Oncology, Biology, Physics*, **58** (4), 1215–27.

107 DeNardo, S.J. et al. (2005) Development of tumor targeting bioprobes ((111)In-chimeric L6 monoclonal antibody nanoparticles) for alternating magnetic field cancer therapy. *Clinical Cancer Research*, **11** (19 Pt 2), 7087s–92s.

108 Tomalia, D.A. et al. (1986) Dendritic macromolecules – synthesis of starburst dendrimers. *Macromolecules*, **19** (9), 2466–8.

109 Hawker, C.J. and Frechet, J.M.J. (1990) Preparation of polymers with controlled molecular architecture – a new convergent approach to dendritic macromolecules. *Journal of the American Chemical Society*, **112** (21), 7638–47.

110 Wolinsky, J.B. and Grinstaff, M.W. (2008) Therapeutic and diagnostic applications of dendrimers for cancer treatment. *Advanced Drug Delivery Reviews*, **60** (9), 1037–55.

111 Dong, S. et al. (2008) Efficacy and safety of (32)P-nanocolloid for treatment of distant lymph node metastasis in VX2 tumor-bearing rabbits. *Annals of Nuclear Medicine*, **22** (10), 849–58.

112 Barth, R.F. et al. (1994) Boronated starburst dendrimer-monoclonal antibody immunoconjugates: evaluation as a potential delivery system for neutron capture therapy. *Bioconjugate Chemistry*, **5** (1), 58–66.

113 Capala, J. et al. (1996) Boronated epidermal growth factor as a potential targeting agent for boron neutron capture therapy of brain tumors. *Bioconjugate Chemistry*, **7** (1), 7–15.

114 Brizel, D.M. et al. (2000) Phase III randomized trial of amifostine as a radioprotector in head and neck cancer. *Journal of Clinical Oncology*, **18** (19), 3339–45.

115 Rzigalinski, B.A. et al. (2006) Radical nanomedicine. *Nanomedicine*, **1** (4), 399–412.

116 Tarnuzzer, R.W. et al. (2005) Vacancy engineered ceria nanostructures for protection from radiation-induced cellular damage. *Nano Letters*, **5** (12), 2573–7.

117 Jonathan, E.C., Bernhard, E.J. and McKenna, W.G. (1999) How does radiation kill cells? *Current Opinion in Chemical Biology*, **3** (1), 77–83.

118 Rzigalinski, B.A. (2005) Nanoparticles and cell longevity. *Technology in Cancer Research & Treatment*, **4** (6), 651–9.

119 Colon, J. et al. (2008) Selective radioprotection of normal tissues with

cerium oxide nanoparticles. *International Journal of Radiation Oncology, Biology, Physics*, **72** (1), S700–1.
120 Dugan, L.L. *et al.* (1996) Buckminsterfullerenol free radical scavengers reduce excitotoxic and apoptotic death of cultured cortical neurons. *Neurobiology of Disease*, **3** (2), 129–35.
121 Krusic, P.J. *et al.* (1991) Radical reactions of C60. *Science*, **254** (5035), 1183–5.
122 Monti, D. *et al.* (2000) C60 carboxyfullerene exerts a protective activity against oxidative stress-induced apoptosis in human peripheral blood mononuclear cells. *Biochemical and Biophysical Research Communications*, **277** (3), 711–17.
123 Fumelli, C. *et al.* (2000) Carboxyfullerenes protect human keratinocytes from ultraviolet-B-induced apoptosis. *Journal of Investigative Dermatology*, **115** (5), 835–41.
124 Lin, H.S. *et al.* (2001) Fullerenes as a new class of radioprotectors. *International Journal of Radiation Biology*, **77** (2), 235–9.
125 Ali, S.S. *et al.* (2004) A biologically effective fullerene (C-60) derivative with superoxide dismutase mimetic properties. *Free Radical Biology and Medicine*, **37** (8), 1191–202.
126 Yin, J.J. *et al.* (2009) The scavenging of reactive oxygen species and the potential for cell protection by functionalized fullerene materials. *Biomaterials*, **30** (4), 611–21.
127 Scheib, S., Schenkel, Y. and Gianolini, S. (2003) Absolute dosimetric verification of 3d dose distributions in radiosurgery using BANG gel. *Medical Physics*, **30** (6), 1522–.
128 Schulte, R.W. *et al.* (2008) Nanodosimetry-based quality factors for radiation protection in space. *Zeitschrift Für Medizinische Physik*, **18** (4), 286–96.
129 De Nardo, L. *et al.* (2002) Track nanodosimetry of an alpha particle. *Radiation Protection Dosimetry*, **99** (1–4), 355–8.
130 De Nardo, L. *et al.* (2002) A detector for track-nanodosimetry. *Nuclear Instruments & Methods in Physics Research Section A – Accelerators Spectrometers Detectors and Associated Equipment*, **484** (1–3), 312–26.
131 Stewart, D.A. and Leonard, F. (2005) Energy conversion efficiency in nanotube optoelectronics. *Nano Letters*, **5** (2), 219–22.
132 Stewart, D.A. and Leonard, F. (2004) Photocurrents in nanotube junctions. *Physical Review Letters*, **93** (10), 107401.
133 Chen, W. *et al.* (2008) Dose dependent X-ray luminescence in MgF_2: Eu^{2+}, Mn^{2+} phosphors. *Journal of Applied Physics*, **103** (11), 113103-1–113103-5.
134 Salah, N. *et al.* (2006) TL and PL studies on $CaSO_4$: Dy nanoparticles. *Radiation Measurements*, **41** (1), 40–7.
135 Yi, G.S. *et al.* (2001) Bionic synthesis of ZnS: Mn nanocrystals and their optical properties. *Journal of Materials Chemistry*, **11** (12), 2928–9.
136 Gladstone, D.J. *et al.* (1994) A miniature MOSFET radiation dosimeter probe. *Medical Physics*, **21** (11), 1721–8.
137 Soubra, M., Cygler, J. and Mackay, G. (1994) Evaluation of a dual bias dual metal-oxide-silicon semiconductor field-effect transistor detector as radiation dosimeter. *Medical Physics*, **21** (4), 567–72.
138 Kong, J. *et al.* (2000) Nanotube molecular wires as chemical sensors. *Science*, **287** (5453), 622–5.
139 Pengfei, Q.F. *et al.* (2003) Toward large arrays of multiplex functionalized carbon nanotube sensors for highly sensitive and selective molecular detection. *Nano Letters*, **3** (3), 347–51.
140 Sahare, P.D. *et al.* (2007) $K_3Na(SO_4)(2)$: Eu nanoparticles for high dose of ionizing radiation. *Journal of Physics D – Applied Physics*, **40** (3), 759–64.

8
Prospects of Semiconductor Quantum Dots for Imaging and Photodynamic Therapy of Cancer

Vasudevanpillai Biju, Sathish Mundayoor, Abdulaziz Anas and Mitsuru Ishikawa

8.1
Introduction

In recent years, *in vivo* fluorescence imaging and photodynamic therapy (PDT) have been investigated widely in clinical trials for the efficient detection and cure of cancer and other diseases. Although noninvasive imaging techniques such as ultrasound, X-radiography, magnetic resonance imaging (MRI), computed tomography (CT) and positron emission tomography (PET) are now well established, *in vivo* fluorescence imaging and PDT remain major challenges due mainly to the poor photostability of fluorescent dye photosensitizing drugs, less penetration depth for visible light in tissues, and the limited availability of near-infrared (NIR) dyes and photosensitizers.

Photodynamic therapy is the clinical process by which different cancers are treated by applying a photosensitizing drug, followed by exposure to light. The principle underlying PDT is that a photoexcited photosensitizer transfers energy or electrons to proximal oxygen and other molecules, which results in the formation of reactive oxygen intermediates (ROIs) that react immediately with, and cause damage to, vital biomolecules in cells; the consequence of this is an induction of cell death through necrosis, apoptosis, or autophagy. Photostable and NIR-absorbing fluorescent dyes and photosensitizers, the *in vivo* targeted delivery of molecules, and the efficient generation of ROIs by photosensitizers are essential for the successful *in vivo* fluorescence imaging and PDT of cancer. The main advantage of PDT over chemotherapy and radiation therapy is its ability to photodegrade cancer cells in selective fashion.

Although PS and PDT have been recognized since the early 1900s, remarkable advancements in PDT were made possible only recently when Boyle and coworkers showed that hematoporphyrin derivatives could produce a complete cure of breast cancer in mice [1]. Subsequently, with the introduction of purified photosensitizers such as porphyrins, phthalocyanines, and chlorine and levulinic acid derivatives, PDT has become a popular clinical trial for treating peripheral cancers. These purified photosensitizer molecules are referred to as second-generation photosensitizers, whilst the mixture of unidentified porphyrins used for PDT during its early days was known as first-generation photosensitizers. Photofrin was the first photosensitizer to show remarkable cancer-curing properties in humans. For example, superficial bladder cancer showed a curative effect after the nonspecific administration of photofrin, followed by transurethral illumination of the whole bladder with red light [2] although, unfortunately, the treatment resulted also in the damage of normal cells and severe side effects. With the advancement of targeted drug delivery, image-guided PDT, laser light sources, and the introduction of second-generation photosensitizers, PDT became a widely accepted clinical approach for the treatment of skin cancers, Barrett's esophagus, bronchial cancers, head and neck cancer, lung cancer, and bladder cancer.

Recently, PDT has also emerged as a promising clinical cure for a variety of cancers that are accessible for photoactivation, either directly or endoscopically. PDT has also been used extensively to kill cancer cells remaining after the surgical removal of tumors, with survival rates for post-surgical PDT being high. The main advantage of PDT over chemotherapy and radiation therapy is that the cancer cells can be killed selectively, using a combination of drug targeting and local illumination with nonionizing radiation, such that the normal cells in the body are left intact. In addition, PDT is cost-effective, fast, and also has higher cure rates anvd provides a greater chance of the regeneration of tissues and immunity.

8.2
Basic Principles and Challenges in PDT

The basis of PDT is that cancer cells can be killed selectively by either systemic or targeted delivery of a photosensitizer in the affected area/tissue, followed by its photoactivation using a suitable light source. The basic principle underlying PDT is that a photoactivated photosensitizer creates ROIs such as singlet oxygen ($^1O_2$), superoxide anion ($^-O_2$), hydroxyl radical ($^\cdot OH$), and hydrogen peroxide (H_2O_2) through energy and electron transfer with proximal oxygen, water and biomolecules. The ROIs react immediately with biomolecules in cells, causing damage to the cell organelles and inducing cell death.

The electronic and molecular processes involved in the photoactivation of a photosensitizer and the production of ROIs are shown in Figure 8.1. When a photosensitizer in the electronic ground state (S^0) is illuminated using a light source with wavelength within the absorption spectrum of the photosensitizer, it will be excited to the singlet excited state (S^1). Generally, the singlet excited state lifetimes for photosensitizers are only a few nanoseconds, which is too brief for collisions with molecular oxygen to occur and for energy to be transferred from the photosensitizer to oxygen. A photosensitizer in the S^1 state immediately relaxes to the S^0 state, either directly or through the triplet state (T^1). Direct (S^1-S^0) radiative relaxation produces fluorescence, and nonradiative relaxation produces thermal energy. Most photosensitizers in the S^1 state undergo an electronic spin-inversion (also known as inter-system crossing or internal conversion) to the T^1 state, followed by radiative and nonradiative relaxation. Radiative relaxation from the T^1

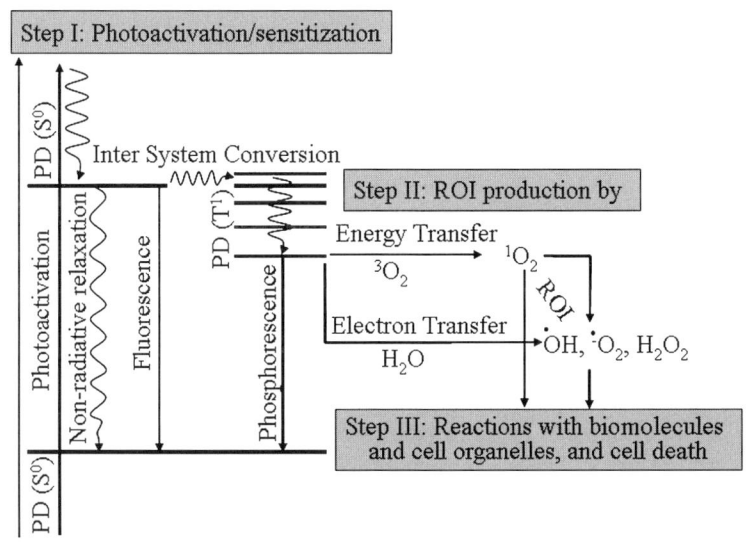

Figure 8.1 Photoactivation and relaxation processes of a photosensitizer and generation of ROI by a photoactivated photosensitizer. Reprinted with permission from V. Biju et al., (2009) Bioconjugated quantum dots for cancer research: present status, prospects and remaining issues. Biotechnology Advances, **28**, 0000 [74].

state produces phosphorescence. The triplet state lifetime ranges from several hundred nanoseconds to milliseconds – which is much longer than the singlet state lifetime – and provides sufficient time for energy and electron transfer reactions and the initiation of the ROI cycle. Thus, a photosensitizer in the T^1 state will either relax to the S^0 state by transferring its excess energy to molecular oxygen (triplet oxygen, $^3O_2$), or it will undergo chemical transformations through electron-transfer reactions with water and other molecules in the biological environment. Energy transfer from a photosensitizer to $^3O_2$ takes place by triplet–triplet annihilation to generate excited molecular oxygen (also called singlet oxygen; $^1O_2$) that subsequently reacts with water and produces other ROIs such as superoxide anion ($^-O_2$), hydroxyl radical ($^{\cdot}OH$), and hydrogen peroxide (H_2O_2). The term singlet oxygen derives from the spin-inversion of one of the electrons in the degenerate π_x^*/π_y^* states in oxygen. The energy required for this spin inversion, at 94.3 kJ mol^{-1}, is small enough for any photosensitizer to produce $^1O_2$. On the other hand, the electron-transfer reactions of photosensitizers in the T^1 state with oxygen and water directly produce ROIs such as $^-O_2$, $^{\cdot}OH$, and H_2O_2. The ROIs trigger a series of photochemical reactions on biomolecules and subcellular organelles in their proximity, which usually is within 100 nm. In PDT, the main targets of the ROIs are amino acids such as cysteine, histidine, methionine, tryptophan and tyrosine in the cell membrane, the endoplasmic reticulum, mitochondrion, lysosome, and Golgi bodies. The photochemical biodegradation of these vital molecules and organelles results in cell death through apoptosis, necrosis, or autophagy. In addition to the direct killing of cancer cells, photochemical degradation of the tumor vasculature represents a promising approach for the control of cancer.

It is necessary to optimize multiple parameters for effective and successful PDT, including selection of the photosensitizer, targeting the cancer, and selecting a light source suitable for a particular photosensitizer. In the early days of PDT, the clinical trials were carried out using first-generation photosensitizers, but this resulted in severe side effects due to poor targeting and an inferior sensitivity. However, PDT became more effective and popular with the introduction of second-generation photosensitizers, among which porphyne derivatives or porphyrins showed the most promise. More recently, however, nanoparticles have attracted attraction in PDT as either photosensitizers or as carriers of photosensitizers. The details of conventional photosensitizers, and the advantages and applications of various photosensitizers based on metallic, ceramic and polymer nanoparticles, in addition to targeted PDT using nanoparticles, are included in Chapter 2 in C. Kumar (2006) Nanomaterials for Cancer Therapy. In this chapter, attention is focused on the potential applications of semiconductor quantum dots (QDs) as both imaging probes for cancer and photosensitizers for PDT. Consequently, discussions will include the advantages of QDs for PDT over conventional photosensitizers, the synthesis of visible and NIR QDs, the optical properties of QDs relevant to PDT, the preparation of biocompatible QDs for targeting cancer, the *in vitro* and *in vivo* targeting and imaging of cancer cells with bioconjugated QDs, ROI production by QDs and QD–PS conjugates, and recent advances and prospects of QDs in PDT of cancer.

8.3
Advantages of Quantum Dots for PDT

Quantum dots are nanoparticles in which electrons and holes are three-dimensionally confined (quantum-confined) within the exciton Bohr radius of the material [3]. Such quantum confinement provides unique optical properties to the QDs, such as size-tunable and sharp band-edge emission. In addition to these unique properties, and compared to conventional photosensitizers, QDs have several advantages for the imaging and PDT of cancer, including a large surface area, photostability, bright emission, unbiased photoactivation at any wavelength below the band edge absorption, and a large two-photon absorption cross-section. Photoactivated QDs also produce ROIs, although with poor efficiency (<5%), but higher efficiencies may be achieved by interfacing QDs with conventional photosensitizers. A combination of a QD and a photosensitizer may be advantageous for the imaging and PDT of cancer, due to the exceptional photostability of the QDs and higher efficiency of ROI production by the photosensitizer. In a QD–photosensitizer (QD–PS) conjugate, the PS is indirectly excited by Förster resonance energy transfer (FRET) from photoactivated QDs. Such QD-PS systems have shown promise for prolonged PDT based on the QDs' photostability and the higher ROI production by the PS. The large surface area of QDs is promising not only for increasing the effective concentration of PS by tethering multiple energy acceptors, but also for targeted PDT by conjugating anticancer antibodies. An unusual photostability of QDs would offer a prolonged activation of the PS during PDT, while the photostability and bright fluorescence of QDs have shown promise for imaging the tumor milieu during and after PDT. The broad absorption bands of QDs allow unrestricted photoactivation; notably, QDs can be photoactivated with various NIR light sources due to a combined effect of broad-band absorption and large two-photon absorption cross-section. The main merit of NIR light over visible light is that the penetration depth in tissues is larger for the former. A direct excitation of NIR QDs or two-photon NIR excitation of visible QDs also avoids background due to tissue autofluorescence. In short, cancer-targeting QD–PS conjugates with direct/two-photon NIR absorption represent the best combination for *in vivo* imaging and PDT.

8.4
Synthesis of Quantum Dots

Biocompatible QDs with both visible and NIR absorption, and strong and stable fluorescence, are the primary requirements for the imaging and PDT of cancer. The most established method for synthesizing high-quality QDs is a colloidal synthesis of hydrophobic-capped QDs in an organic phase, and consequently the conversion of hydrophobic-capped QDs into water-soluble QDs has been necessary for preparing biocompatible QDs. Among the various QDs, core-only CdS, CdSe, CdTe, and InP QDs and core–shell QDs with single, mixed or multiple shells from ZnS, CdS and CdSe have been widely applied for targeting cancer cells both *in vitro* and *in vivo*. CdSe QDs, with or without shells excited by visible or NIR (two-photon) light, are the most widely applied QDs for *in vitro* and *in vivo* imaging of cancer cells. The most attractive property of CdSe QD is that its emission color can be tuned throughout the visible region [3, 4]. However, compared to visible excitation, NIR excitation offers a better penetration depth in tissues; thus, NIR QDs represent better candidates for the imaging and PDT of cancer.

8.4.1
Synthesis of Visible QDs

8.4.1.1   Synthesis of Cadmium-Based QDs
Although the existence of semiconductor nanoparticles has been recognized since the early 1980s, the fundamental science and applications of QDs was greatly advanced when Bawendi and coworkers synthesized CdS, CdSe, and CdTe QDs by using high-temperature colloidal reactions [5]. The pyrolysis of organometallic precursors of cadmium and S/Se/Te provides hydrophobically capped CdS, CdSe, and CdTe QDs. For example, CdSe QDs can be synthesized by reacting dimethyl cadmium (CdMe$_2$) [13.35 mmol in 25 ml trioctylphosphine (TOP)] with TOP-selenide (TOPSe) (10 mmol in 15 ml TOP) at 230–300 °C in the presence of trioctylphosphine oxide (TOPO). CdTe QDs may be synthesized in a similar fashion

by replacing TOPSe with TOPTe, whilst CdS QDs can be prepared by reacting hexamethyldisilathiane (HMDT) with $CdMe_2$. These reactions provided a series of QDs, with the individual sizes being separated using size-selective centrifugation from a mixture of 1-butanol and methanol. A modification of this method was later devised by Alivisatos and coworkers [6], the key being that, with Ostwald ripening, a gradual growth of larger QDs at a cost of gradual dissolution of smaller ones could be controlled by separating the nanocrystal growth process from the spontaneous nucleation process. The main advantage of this modified synthesis was that size-selected QDs could be prepared simply by selecting an injection temperature and a different growth temperature.

In the above methods, the use of volatile and pyrophoric $CdMe_2$ makes the synthesis of QDs less safe at high temperatures; hence, alternative cadmium precursors have been introduced for the synthesis of colloidal QDs. Examples of these "safer and greener" precursors include cadmium perchlorate, cadmium oxide, cadmium oleate, and cadmium acetate [7]. A typical example of an alternative synthesis is the nucleation of CdSe QDs by injecting TOPSe into a suspension of cadmium acetate in either TOPO or a mixture of TOPO–phosphonic acid maintained at 250–360 °C. After nucleation, the QD nanocrystals were grown at 200–320 °C. Recently, it has become possible to synthesize size-selected CdSe QDs with green fluorescence at a relatively low temperature (≤75 °C) by employing a heterogeneous reaction between cadmium acetate and TOPSe in a mixture of TOP and TOPO [8].

Among the cadmium-based QDs, core-only CdS and CdSe QDs are of less practical biological applications due to the near-UV fluorescence and low fluorescence quantum efficiency for CdS, and the poor physical and photostability of both CdS and CdSe. In contrast, CdSe QDs with ZnS shells have been widely applied to the bioimaging of cancer cells and the tumor milieu, both *in vitro* and *in vivo*. The most attractive properties of CdSe QDs for biological applications are the high fluorescence quantum yield and size-tunable fluorescence distributed throughout the visible spectrum.

8.4.1.2 Synthesis of InP QDs

Compared to QDs based on heavy metals, InP QDs have shown great promise for biological applications, due mainly to their reduced toxicity. The best method to prepare InP QDs is in the organic phase [9] where, typically, InP QDs are prepared by injecting tris-trimethylsilyl phosphine (0.58 mmol) into a hot (281 °C) solution of indium(III) myristate (1.014 mmol) in 40 ml 1-ocatadecane. After injection, an additional amount (~40 ml) of 1-octadecane is added to the reaction mixture, and the hydrophobic-capped InP QDs are grown at 180 °C for 2 h. In this reaction it is necessary that surface capping and hydrophilic molecules are introduced onto the surface of the as-prepared InP QDs.

8.4.2
Synthesis of NIR QDs

NIR absorption and fluorescence are attractive for *in vivo* imaging and PDT, as NIR excitation offers a better penetration depth in tissues than for visible light. Autofluorescence from the tissues can also be avoided or minimized under NIR excitation. Other promising applications of NIR QDs include the image-guided presurgical and surgical oncology of various cancers, such as gastrointestinal tumors, metastases of spontaneous melanoma, breast cancer, and non-small-cell lung cancer. Examples of NIR QDs include core-only CdTe, PbS, PbSe, PbTe, HgTe and InP QDs, and type II core–shell QDs such as CdSe/ZnS, CdSe/CdS, CdSe/ZnSe, CdTe/HgTe, CdS/HgS/CdS, CdTe/CdSe, CdSe/$Zn_{0.2}Cd_{0.8}S$, CdSe/CdS/ZnS, PbSe/PbS, HgTe/CdS, InAs/InP, and InAs/CdSe QDs. Among these QDs, CdTe, InP, CdSe/ZnS, CdSe/ZnCdS (mixed sulfide shell), CdTe/CdSe and CdSe/CdS/ZnS are attractive for the *in vivo* cancer imaging and PDT of cancer. The Hg- and Pb-based QDs are less attractive due to their potential toxicity, but have shown promise for electro-optical and optoelectronic devices.

8.4.2.1 Synthesis of Core-Only NIR QDs

The most attractive core-only NIR QDs for biological applications are CdTe QDs. Although the absorption and fluorescence bands of Pb- and Hg-based QDs are clearly within the NIR region, these QDs have not shown promise for biological applications for reasons of toxicity. When synthesizing CdTe QDs, the hydrophobically capped material is first prepared by reacting $CdMe_2$ with TOPTe in a of TOP and TOPO as solvent (see Section 8.4.1) [5]. Biocompatible thioglycolic acid (TGA) -capped CdTe QDs may be prepared directly an aqueous solution at 100 °C and ~11.5 pH, by reaction between freshly prepared H_2Te gas and a degassed solution of cadmium perchlorate hexahydrate (2.35 mmol) in water (125 ml) in the presence of TGA (5.7 mmol) [10]. H_2Te gas is prepared by adding aluminum telluride (0.46 mmol) to sulfuric acid (30 ml, 0.5 M) under a N_2 atmosphere. The size and optical properties of the CdTe QDs can be controlled by adjusting the reaction time.

8.4.2.2 Synthesis of Core–Shell QDs

The advantages of core–shell QDs (which are also known as type II QDS) over core-only QDs are many. For example, the inorganic and polymer shells will protect the core against chemical degradation, while shells from higher band-gap materials will preserve or even improve the fluorescence quantum efficiencies of the core by lowering the surface defects; the shells also offer new surfaces for chemical reactions and bioconjugation. Although many types of core–shell QDs have been developed, the classical example is the CdSe/ZnS QD. Notably, inorganic shells offer a higher tolerance to QDs than do organic shells; for example, core–shell QDs based on CdS, CdSe, CdTe, PbSe, HgTe and InAs cores and ZnS, ZnSe, CdS, CdSe, ZnSe, HgTe, HgS, InP and mixed metal sulfide shells show better fluorescence quantum yields, unusual photostability and physical stability, and NIR absorption and fluorescence characteristics. However, as stated above, QDs based on Pb and Hg cores or shells are unattractive for the imaging and PDT of cancer for reasons of toxicity. Thus, CdSe/ZnS, CdSe/$Zn_{0.2}Cd_{0.8}S$ and CdTe/CdSe core–shell QDs have been selected for the *in vivo* imaging and PDT of cancer.

The epitaxial growth of ZnS shells on a CdSe core is perhaps the best example of shell preparation [4], and modified versions of ZnS shell preparation have been applied for the synthesis of various core–shell QDs. Typically, ZnS shells are grown on CdSe QDs with core diameters ranging from 23 to 55 Å, as follows. First, monodispersed CdSe QDs are prepared as described in Section 8.4.1 [5], with TOPO being used as the coordinating solvent for preparing the ZnS shell. Initially, 5 g TOPO is heated to 190 °C under vacuum for several hours, cooled to 60 °C, and 0.5 ml TOP added. In parallel, a solution of CdSe QDs (0.4 mmol) in hexane is prepared and transferred into the TOP/TOPO mixture. After the transfer, the hexane is pumped off. The ideal precursors for preparing ZnS shell are diethylzinc and HMDT, with the amounts of Zn and S precursors being determined based on the average diameter of the core and the desired thickness of the shell. An equimolar mixture of diethyl zinc and HMDT dissolved in TOP is added drop-wise into the CdSe suspension, maintained at 140–220 °C. The growth of ZnS shell is quenched by adding 1-butanol, and the core–shell QDs are purified by their precipitation from a mixture of 1-butanol and methanol.

A similar procedure can be applied to the synthesis of CdSe/CdS core–shell QDs [11]. The CdSe QDs are prepared by first adding the desired amount of $CdMe_2$ to a solution of tributylphosphine selenide (TBPSe), prepared by dissolving Se powder in TBP, and the cadmium : selenium molar ratio is set at 1.0 : 0.7 or 1.0 : 0.9. For a typical synthesis, 12 g TOPO is first transferred into a flask and heated to 360 °C in an argon atmosphere. The stock solution of Cd and Se precursors is quickly injected into the TOPO, after which the temperature of the reaction mixture is lowered to 300 °C and the reaction continued until the desired size of CdSe QDs is obtained, based on the absorption and fluorescence spectra. The CdSe QDs are then purified by precipitation with methanol, followed by centrifugation. The precursors for the preparation of CdS shell are HMDT (100 µl) and $CdMe_2$ (0.033 g) dissolved in TBP (3.81 g). For shell growth, TOPO-capped CdSe nanocrystals (2–13 mg) are transferred into a three-necked flask, degassed, and 15 ml anhydrous pyridine added. This solution is refluxed overnight under an argon atmosphere, the temperature is lowered to 100 °C, and 0.5–2 ml diluted CdS precursor solution is then added dropwise. The heating is stopped when the desired shell-thickness is obtained, based on the absorption spectrum, and the CdSe/CdS QDs are then precipitated by adding dodecylamine. The formation of CdSe/CdS core–shell QDs is associated with a considerable red-shift in the absorption and fluorescence bands towards the NIR region. Thus, CdSe/CdS QDs with suitable core diameters and shell thicknesses are suitable for the NIR imaging and PDT of cancer.

8.4.2.3 Synthesis of CdTe/CdSe QDs

Core–shell CdTe/CdSe QDs show stable fluorescence in the NIR region, and are prepared by growing CdSe shells on colloidal CdTe core QDs [12]. $CdMe_2$ and bis(trimethylsilyl) selenide are ideal precursors for preparing the CdSe shell. Initially, CdTe core QDs are prepared by the pyrolysis of $CdMe_2$ and TOPTe (see Section 8.4.1), after which the as-prepared CdTe QDs (400 mg) are centrifuged into a solid mass and resuspended in TOPO (20 g). This mixture is dried under vacuum at 160 °C for 2 h, and the temperature then lowered to 130 °C. Finally, a precursor solution consisting of equimolar $CdMe_2$ and bis (trimethylsilyl) selenide in TOP (10 ml) is added dropwise into the hot solution of CdTe. After such addition, the reaction mixture is stirred at 130–300 °C until a desired thickness for CdSe shell is obtained.

8.5
Optical Properties of Quantum Dots

Unique optical properties, such as size-tunable absorption and photoluminescence (hereafter termed fluorescence) bands in the visible and NIR regions, an uncompromised photostability, broad absorption and narrow fluorescence bands and large two-photon absorption coefficients, are the most important properties of QDs for the imaging and PDT of cancer. These properties originate from the strong confinements of charge carriers within the exciton Bohr radius of the core material, and the large surface-to-volume ratios of QDs.

8.5.1
Absorption and Fluorescence Properties

Semiconductor QDs show broad absorption spectra with sharp band-edge transitions and narrow and symmetrical fluorescence spectra. Both, the absorption and fluorescence spectra vary with the size of the core QD and the core material. For example, the size-dependent absorption and fluorescence spectra and fluorescence colors of CdSe QDs shelled with ZnS are shown in Figure 8.2 [1, 4, 13]. The band-edge absorption band of CdSe QDs can be shifted from ~450 nm for a ~2 nm-diameter QD to ~650 nm for a ~6 nm-diameter QD. Similarly, the fluorescence color can be tuned from purple for a ~2 nm-diameter QD to deep red for a ~6 nm-diameter QD. The absorption and emission characteristics of QDs are discussed using an energy level diagram for CdSe (Figure 8.3). Here, the 4p orbitals of selenium constitute the highest occupied states, while the 5s orbitals of cadmium constitute the lowest unoccupied states. The broad absorption spectra of QDs originate from a distribution of electronic transitions such as s-s, p-p and d-d. The quantum confinement effect also provides discrete energy states in the band edge (Figure 8.3). Size-dependent and sharp emission spectra of QDs originate from carrier recombination at the band edge. Red-shifted fluorescence, if observed, should derive from carrier recombination in deep-trap states.

Analogous to the size-dependent absorption and fluorescence spectra for a given core QD, the absorption and fluorescence spectra can be tuned from the near-UV to the NIR region by changing the core material. Thus, QDs with suitable absorption and fluorescence bands can be selected for bioimaging and PDT, based on the size when there is no option for changing the material, and on the material when there is no option for changing the size. The broad absorption bands of QDs

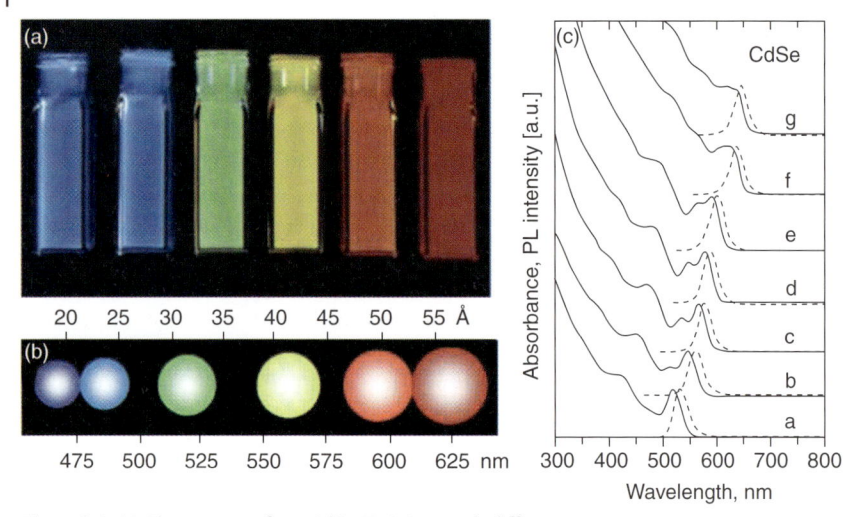

Figure 8.2 (a) Fluorescence from CdSe/ZnS QDs with different size under UV excitation; (b) Representation of size of CdSe QDs versus fluorescence color/wavelength; (c) Absorption (solid lines) and fluorescence (broken lines) spectra of CdSe QDs with different size. Reprinted with permission from Refs [4] (a) and [13] (c), copyright (1997, 2001) American Chemical Society. Part (b) was reprinted from Ref. [3] with kind permission from Springer Science and Business Media.

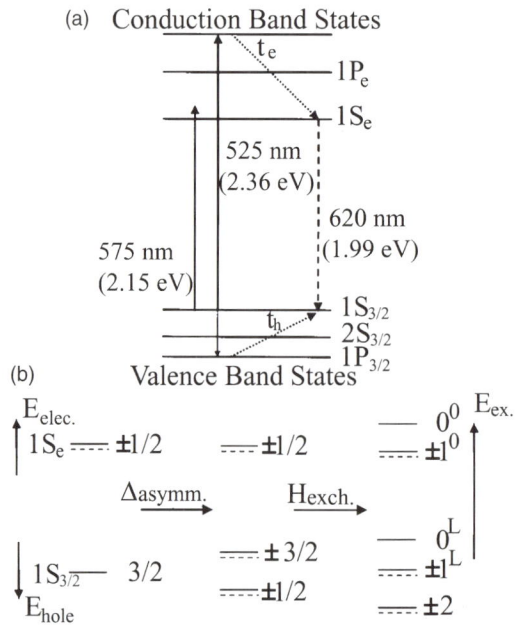

Figure 8.3 Energy states in CdSe QDs. (a) Overall picture; (b) Band edge states. Reprinted with permission from Refs [14] (a) and [15] (b).

offer many advantages for the imaging and PDT of cancer. For example, the broad absorption bands allow QDs to be photoactivated at any wavelength below the band edge absorption, to separate the excitation wavelength clearly from the fluorescence band, and to excite multiple QDs with a single wavelength. Additional merits of QDs include their huge molar extinction coefficient and sharp fluorescence bands; the latter characteristic allows the multiplexed tagging of cells and tissues, and the production of well-resolved multicolor images. In contrast, specific light sources are required to excite conventional organic dyes and photosensitizers because of their sharp absorption bands. Multiple tagging and multicolor imaging of cells and tissues using organic dyes are also limited, due to the overlapping of relatively broad fluorescence bands.

8.5.2
Photostability of QDs

The photostability of QDs is perhaps the most attractive property for their use in PDT and long-term bioimaging. The exceptional photostability of QDs means that cells labeled with QDs can be imaged continuously for long periods of time, without photobleaching, whereas conventional dyes and photosensitizers would photobleach much more quickly. The photostability of CdSe/ZnS QDs and Alexafluor488 dye are compared in Figure 8.4 [16]; here, fluorescence from organelles labeled with QDs (red) can be seen to remain intact under a 100 W mercury lamp for 3 min whereas, under the same conditions, the fluorescence from organelles labeled with Alexafluor488 (green), a relatively stable dye, disappeared within 1 min, due to photobleaching. Indeed, the fluorescence quantum efficiency of the as-synthesized CdSe QDs was seen to increase under photoactivation (Figure 8.5) [17]. This increase in fluorescence quantum efficiency was considered to be due to a photoinduced passivation of the surface defect states. The exceptional photostability of QDs is an important and attractive property for highly sensitive detection, for prolonged *in vitro* and *in vivo* imaging, and for continuous and repeated PDT with a single dose of QD, and even for monitoring the status of PDT. Unfortunately, core-only QDs are not photostable in the aqueous phase, due mainly to photo-oxidation and photo-etching of the surface atoms. The photo-etching of cadmium-based QDs is undesirable for PDT, as it might lead to toxic effects of the dissolved metal ions. On the other hand, core–shell QDs with protective inorganic and polymer shells offer biocompatibility, minimal or no toxicity, and both photo- and chemical stability.

8.5.3
Two-Photon Absorption by QDs

Two-photon absorption is a nonlinear process in which QDs or organic molecules are excited from the ground electronic state to an excited electronic state by simultaneously absorbing two low-energy photons with the same or different energy. Here, the sum energy of the two photons should be the energy difference between

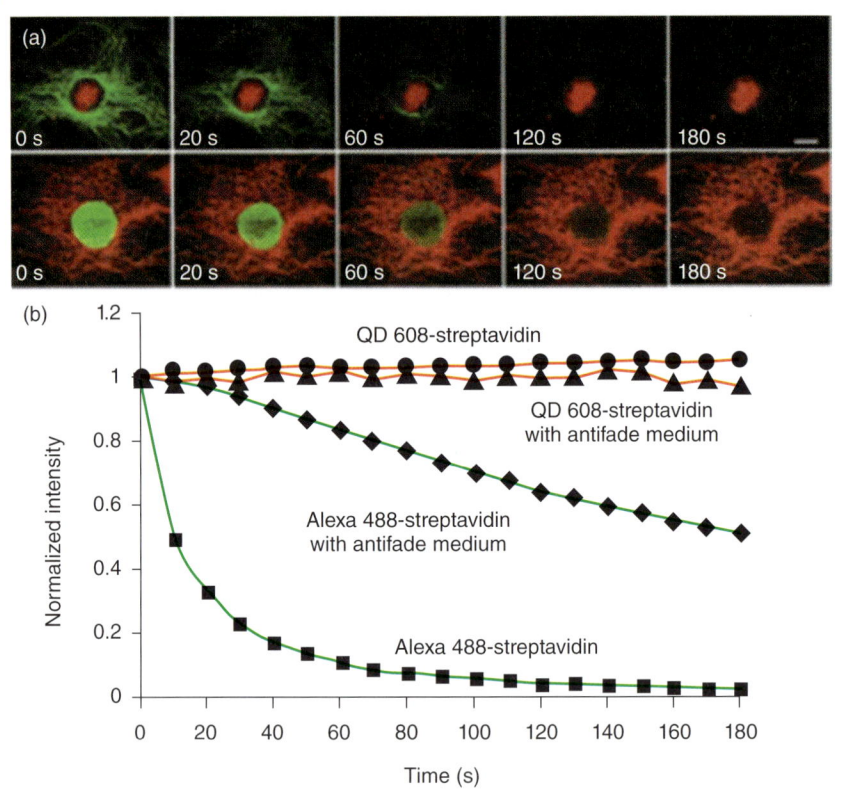

Figure 8.4 (a) Fluorescence images of 3T3 cells acquired at different times under photoactivation with a 100 W mercury lamp (filter 485 ± 20 nm). Top row: nuclear antigen stained with QD–streptavidin conjugate (red) and microtubules with Alexafluor488 (green). Bottom row: nuclear antigen stained with Alexafluor488 (green) and microtubules with QD–streptavidin conjugate (red); (b) Time-dependent fluorescence intensity of QDs and Alexafluor488 photoactivated under different conditions. Reprinted with permission from Ref. Macmillan Publishers Ltd: Nature Biotechnology, Ref. [16], copyright (2003).

the ground and excited states of QDs. The significance of two or multiphoton absorption by QDs is that visible QDs can be excited with NIR light; a deeper penetration into the tissues by NIR light is necessary for efficient *in vivo* imaging and PDT of cancer. The *in vivo* excitation of QDs at wavelengths below 700 nm is undesirable because of poor tissue penetration and bright tissue autofluorescence. Also, *in vivo* excitation of QDs at wavelengths beyond 1200 nm results in the generation of heat due to the vibrational excitation of water and other biomolecules. Thus, QDs with absorption bands in the 700–1200 nm range are necessary for *in vivo* applications and, indeed, such NIR QDs suitable for *in vivo* applications are few in number; examples include the core-only CdTe QDs and core–shell CdTe/CdSe and CdSe/CdS/ZnS QDs. Other NIR QDs based on Pb and Hg are less attractive for *in vivo* applications for reasons of toxicity. Thus, two-photon or

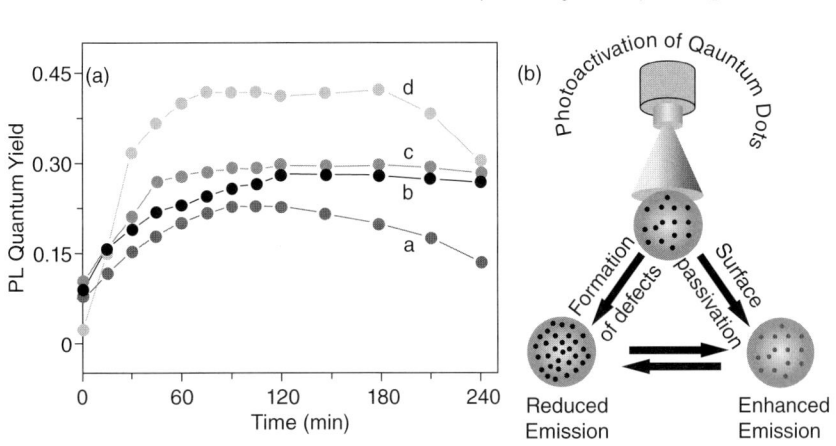

Figure 8.5 (a) Fluorescence intensity of CdSe QD solutions in chloroform under photoactivation at 400 nm: (curve a) without any polymer, (curve b) in the presence of poly (dimethylsiloxane), (curve c) in the presence of polybutadiene, and (curve d) in the presence of polyvinyl pyrrolidone; (b) Schematic representation of the surface passivation and the formation of surface defects on CdSe QDs. Reprinted with permission from Ref. [17]. Copyright (2007) American Chemical Society.

multiphoton NIR excitation is valuable when using CdSe and CdSe/ZnS QDs for the imaging and PDT of cancer. The broad absorption bands of the visible QDs are advantageous when randomly selecting the NIR wavelengths for two-photon excitation. The two-photon excitation wavelength is determined based on the energy of a QD; for example, Figure 8.3 shows the energy levels in CdSe QDs, where the energy states are defined by applying a particle in a sphere model approximation to the bulk Wannier Hamiltonian. The lowest hole-state for CdSe is denoted as $1S_{3/2}$, and the next hole-states as $2S_{3/2}$, $1P_{3/2}$, and so on. The electron states or emitting states are defined by the total angular momentum $F = L \pm 1/2$, where $\Delta L = 0, \pm 2$ and $\Delta F = 0, \pm 1$ are the selection rules for one-photon excitation, and $\Delta L = \pm 1, \pm 3$ and $\Delta F = 0, \pm 1, \pm 2$ for two-photon excitation [18]. The large two-photon absorption cross-section (10^3–10^4 Goeppert–Mayer [GM] units) for QDs [19] compared to organic dyes and PS (10^1–10^2 GM units) is promising for the two-photon excitation of CdSe/ZnS and other visible QDs for *in vivo* imaging and PDT.

8.6
Preparation of Biocompatible Quantum Dots

In general, QDs with high fluorescence quantum yields and physico-chemical stability are prepared in the organic phase. For this, the surfaces of the QDs are covered with hydrophobic molecules, which makes it difficult for them to be dispersed in water. But, for biological applications it is essential that the QDs have a hydrophilic surface coating and reactive functional groups. Although the direct synthesis of CdSe and CdTe QDs in the aqueous phase has recently become possible, the synthesis of high-quality QDs under such conditions remains a major challenge. Thus, the conversion of hydrophobic-capped core and core–shell QDs into their water-dispersible counterparts is necessary for biological applications. The coating or conjugation of hydrophilic and amphiphilic molecules onto the surfaces of core and core–shell QDs represents an attractive route for preparing biocompatible QDs. Typical examples are the exchange of alkylphosphines and alkylphosphine oxides on the surfaces of as-synthesized CdSe, CdTe, InP, CdSe/ZnS and CdTe/CdSe QDs with TGA, hydrophilic dendrimers, silica-shells, amphiphilic polymers, proteins, and sugars [20]. Poly(ethylene glycol) (PEG) shells are also useful for avoiding the nonspecific attachment of QDs with biomolecules *in vitro* and *in vivo*.

Two important methods for preparing biocompatible QDs are to exchange TOPO and TOP on ZnS-shelled QDs with either a monodentate mercaptoacetic acid [21] or a bidentate dihydrolipoic acid (DHLA) [22]. The formation of a disulfide bond with the ZnS shell by the thiol group is the key to these preparations. Once dispersed in water, additional functional groups such as avidin, biotin, peptides, antibodies, and oligonucleotides will be introduced onto the surface for intracellular delivery, *in vitro* and *in vivo* imaging, and therapeutic interventions [3]. However, irrespective of which reaction is occurring on the surface of the QDs, care must be taken to preserve their fluorescence.

In general, antibodies and peptides have shown much promise as carriers of QDs to the cells. In this case, the crosslinking chemistry of carbodiimide, maleimide and succinimide can be used to attach antibodies and peptides to QDs functionalized with carboxylic acids, thiols and primary amines, respectively. An additional, and very versatile, method for preparing antibody- and peptide-conjugated QDs is that of avidin–biotin crosslinking. Some general methods for preparing antibody- and peptide-conjugated QDs are shown in Figure 8.6, while examples of bioconjugated QDs, and their targets within the cells, are shown in Figure 8.7.

8.7
Nontargeted Intracellular Delivery of QDs

The *in vivo*-targeted delivery of QDs and QD–PS conjugates into cancer cells is an essential requirement for effective PDT, because singlet oxygen produced by QDs and other photosensitizers exists for less than 3.5 μs and diffuses only 0.01– 0.02 μm during this period. Cancer targeting with QDs can be aimed either at overexpressed receptors on cancer cell membranes, or at intracellular organelles through receptor-mediated endocytosis. A variety of methods are available for delivering QDs inside cells, and these are exploited for both the *in vitro* and *in vivo* imaging of cells and tissues, for the selective detection of cancer cells and the tumor milieu, and for PDT. Based on the mode of intracellular delivery, these methods are classified broadly as either physical or biochemical. Physical tech-

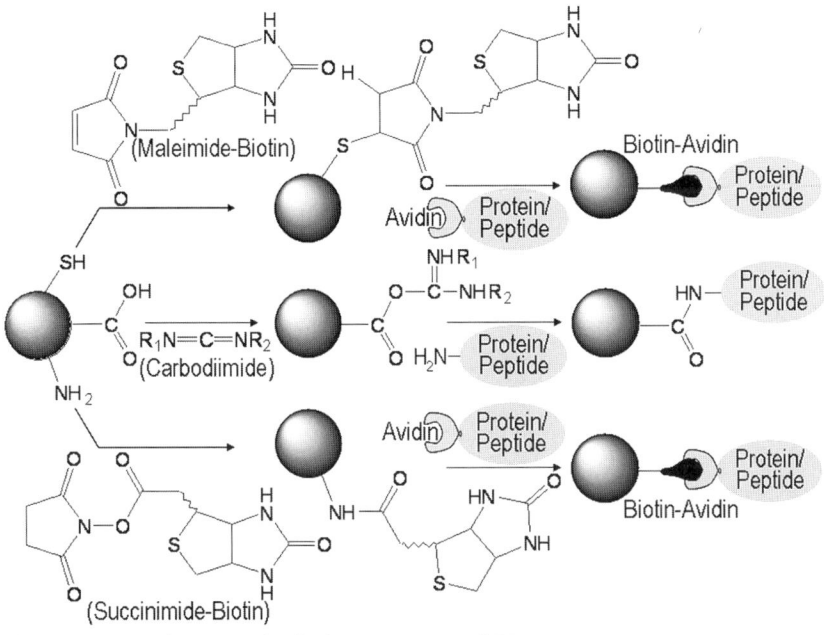

Figure 8.6 General reactions for the bioconjugations of QDs.
Reprinted with permission from Ref. [3].

niques include electroporation and microinjection, whereas biochemical methods include passive uptake, cell penetration, and receptor-mediated endocytosis. For example, QDs bearing polymers, lipids, alkane thiols, oligonucleotides and nonspecific peptides may be delivered into both normal and cancer cells, and will be either uniformly distributed or aggregated inside the cytoplasm. In contrast, QDs conjugated with antibodies, secondary antibodies, or with certain peptides will target specific organelles/proteins in the cancer cells. Among these techniques, nonspecific approaches such as physical delivery and passive intake have not shown much promise for the imaging and PDT of cancer.

8.7.1
Physical Methods of Intracellular Delivery

Both, electroporation and microinjection are useful for the transformation and transfection of recombinant genetic materials into prokaryotic and eukaryotic cells.

In *electroporation*, a short-duration (from microseconds to milliseconds) electric pulse is applied in the extracellular medium; this pulse causes the trans-membrane voltage to rise from ~0.5 to 1.0 V and, along with thermal fluctuations, creates a transient and heterogeneous population of pores on the cell wall. For electroporation, the target cells and QDs are mixed in the cell culture medium and loaded in a plastic cuvette fitted with aluminum electrodes [38]. A high voltage is then applied to the medium, using a hand-held or bench-top electroporator. The

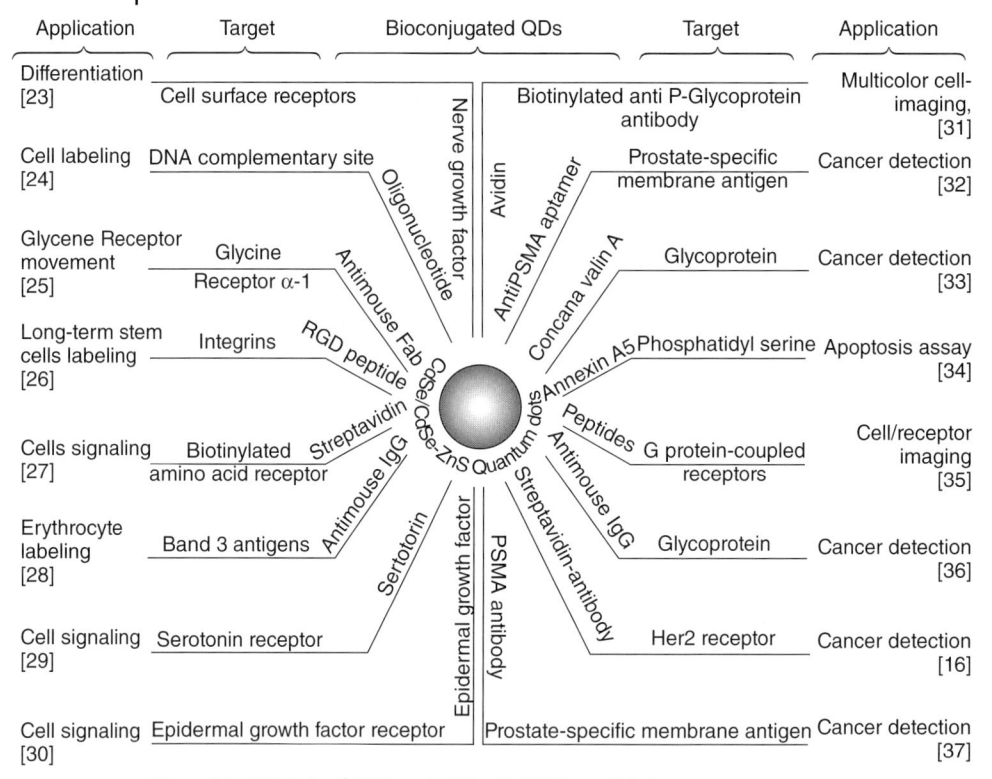

Figure 8.7 Biolabels of CdSe and CdSe–ZnS QDs and their targets in cells. Reprinted with permission from Ref. [3].

Application	Target	Bioconjugated QDs	Target	Application
Differentiation [23]	Cell surface receptors	Nerve growth factor	Biotinylated anti P-Glycoprotein antibody	Multicolor cell-imaging, [31]
Cell labeling [24]	DNA complementary site	Avidin / Oligonucleotide	Prostate-specific membrane antigen	Cancer detection [32]
Glycene Receptor movement [25]	Glycine Receptor α-1	Antimouse Fab / AntiPSMA aptamer	Glycoprotein	Cancer detection [33]
Long-term stem cells labeling [26]	Integrins	RGD peptide / Concana valin A	Annexin A5 / Phosphatidyl serine	Apoptosis assay [34]
Cells signaling [27]	Biotinylated amino acid receptor	Streptavidin / CdSe / CdSe–ZnS Quantum dots / Peptides	G protein-coupled receptors	Cell/receptor imaging [35]
Erythrocyte labeling [28]	Band 3 antigens	Antimouse IgG / Sertotorin / Antimouse IgG / Streptavidin-antibody	Glycoprotein	Cancer detection [36]
Cell signaling [29]	Serotonin receptor	Epidermal growth factor / PSMA antibody	Her2 receptor	Cancer detection [16]
Cell signaling [30]	Epidermal growth factor receptor		Prostate-specific membrane antigen	Cancer detection [37]

QDs, along with some extracellular fluid, are internalized through the pores that have formed on the cell wall. However, the internalized QDs tend to aggregate inside the cytosol. The efficiency of electroporation depends on various factors, including the time integral of the electric pulse, the ionic composition of the medium, and the type of the cell.

Microinjection is the process by which foreign molecules or materials are injected into the cytoplasm or cell nucleus through a microneedle. The microinjection apparatus consists of a specialized optical microscope and a digitally controlled microneedle of approximately 0.5 to 5 µm diameter. The microneedle penetrates the cell membrane and/or the nuclear membrane, and the contents are injected by applying either a pneumatic pressure or an electric pulse. A typical example is the microinjection of QD–DNA conjugates in a *Xenopus* embryo [38], where the microinjected QDs are distributed homogeneously inside the embryo.

In general, electroporation and microinjection serve only as vectors for the intracellular delivery of QDs, and the conjugation of specific biomarkers on the surface of QDs is required for targeting the intracellular organelles. For example, microinjected QDs will target the nucleus and mitochondria when conjugated with a 23-mer nuclear localization peptide and a 28-mer mitochondrial localization peptide [38]. Despite their simple approach, physical methods have practical limitations for *in vivo* applications, because electroporation can be applied only for cells dispersed in a medium, while microinjection can be applied only on single-cell basis.

8.7.2
Biochemical Techniques

8.7.2.1 Nonspecific Intracellular Delivery of QDs

An electrostatic interaction between the hydrophilic-capped QDs and the plasma membrane and subsequent nonspecific endocytosis is applicable for labeling a variety of cells. QDs conjugated with thioglycolic acid (QD-COOH), cysteamine (QD-NH$_2$), thioglycerol (QD-OH), phalloidin, or dihydrolipoic acid (QD-DHLA) are internalized by nonspecific endocytosis. Such intracellular delivery is limited by the size and zeta potential of the QD conjugates and the type of cell used. QDs capped with carboxylic acids, thioglycerol or dihydrolipoic acid are negatively charged, whereas cysteamine-capped QDs are positively charged. For example, QDs conjugated with a combination of a carboxylic acid and transferrin [21], or DHLA are efficiently delivered into HeLa cells [31]. In general, QDs delivered by these methods mostly aggregate in the cytosol due to endosomal arrest, although this situation can be resolved if the QDs are conjugated with an endosome-disrupting polymer, such as PEG-grafted, hyperbranched polyethylenimine (PEG-g-PEI) [39]. In this case, the PEI creates an osmotic imbalance inside the endosome by sequestering protons and their counterions via the "proton-sponge effect," causing the endosome to swell and rupture.

8.7.2.2 Cell-Penetrating Peptide-Mediated Delivery

Cell-penetrating peptides (CPPs) are positively charged peptides that are able to transport various therapeutic macromolecules, including oligonucleotides, proteins, DNA, and nanoparticles, across the plasma membrane. CPPs are mainly derived from natural proteins, but they may also be synthesized. In general, electrostatic interactions between CPP and the plasma membrane, followed by macropinocytsis or direct cell penetration, account for the intracellular delivery of CPPs. The concept of CPP originated from the HIV-1 transcriptional transactivation (Tat) protein, the Tat peptide segment (residues 48–60) derived from this protein having since emerged as a novel CPP for the intracellular delivery of nanoparticles and macromolecules. The intracellular delivery of Tat peptides takes place via a charge-based attachment to the cell membrane, followed by macropinocytosis, whereby multiple arginine units provide a net positive charge to Tat. The same mechanism allows the efficient intracellular delivery of noncovalent mixtures and supramolecular nanostructures composed of QDs and Tat, for example in HeLa [40], HEK293T/17 [41] and COS-1 [41] cells. QDs delivered by Tat eventually accumulate in the microtubule organizing center [40].

Other examples of CPPs employed for the intracellular delivery of QDs include Pep-I [42] and allatostatin [42, 43]. Pep-I is a synthetic peptide consisting of 21 amino acids, which acts as a hydrophobic linker between the plasma membrane

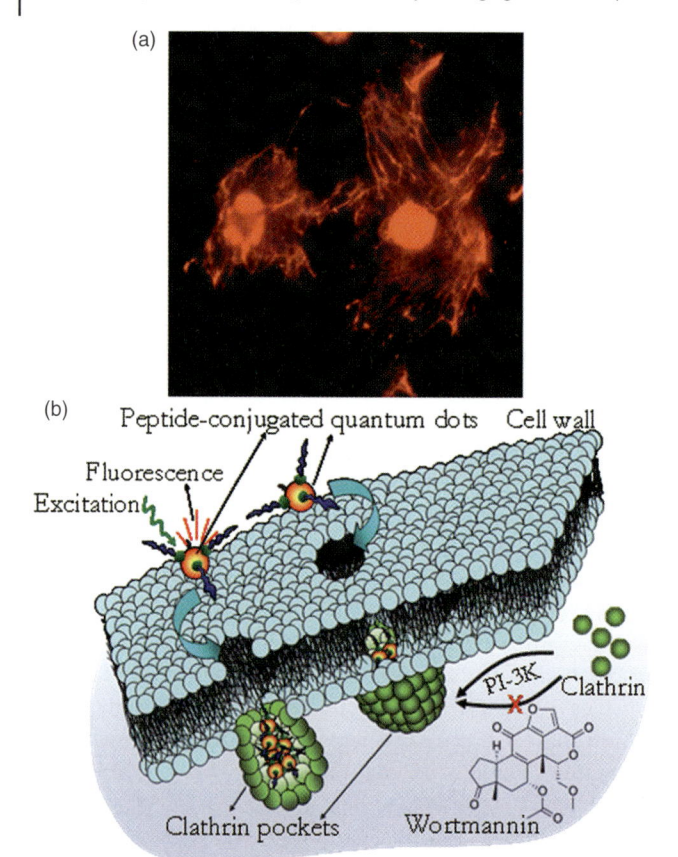

Figure 8.8 (a) Fluorescence image of 3T3 cells incubated with QD–allatostatin conjugates. Here, the QDs are delivered inside the cells by clathrin-mediated endocytosis; (b) Schematic presentation of clathrin-mediated endocytosis of QD–peptide conjugates. Reprinted with permission from Refs [43] (a) and [44] (b). Copyright (2007, 2009), American Chemical Society.

and a protein bound to QDs. This method can be tested by delivering CdSe/ZnS QDs into Jurkat cells, with the intracellular delivery of QDs conjugated to polyarginines and poly-L-lysines following essentially the same pathway as for QD-Tat conjugates, due to the positive charge of the peptides.

Allatostatin, a neuropeptide derived from insects and crustaceans, is recognized as one of the most promising peptides for delivering QDs into living animal and human cells [43]. Typical fluorescence images of 3T3 cells internalized with QD-allatostatin conjugates are shown in Figure 8.8a. Interestingly, QD-allatostatin conjugates bind preferentially with the microtubules and are eventually transported into the nucleus. The intracellular delivery of QD-allatostatin conjugates occurs via multiple pathways such as clathrin-mediated endocytosis, charge-based cell-penetration, and receptor-mediated endocytosis, among which clathrin-mediated endocytosis is the most important [44]. A schematic representation of the clathrin-mediated endocytosis of QD-allatostatin conjugates into a cell is shown in Figure 8.8b; the process is characterized by the inhibition of phosphoinositide-3OH kinase (PI-3K), an essential enzyme for the formation of clathrin-coated vesicles, with wortmannin (a steroidal furanoid that irreversibly inhibits PI-3K by blocking its ATP-binding pocket).

8.7.2.3 Chitosan- and Liposome-Mediated Delivery

The aggregation of QDs inside the cytosol, as observed with many of the above-mentioned methods, represents a major limitation for *in vivo* imaging and PDT applications. However, aggregation can be minimized if the QDs are conjugated with chitosan and lipids. *Chitosan*, a copolymer of glucosamine and N-acetylglucosamine, is a safe and efficient nonviral vector for the delivery of genetic materials inside the cytosol. Chitosan has two advantages: (i) its amino group can be conjugated with carboxylic acid-functionalized QDs by utilizing carbodiimide chemistry; and (ii) it avoids QD agglomeration by inducing electrostatic repulsion. Thus, QDs encapsulated in a chitosan matrix would be useful for intracellular delivery and imaging [45].

Liposome is another class of prototype nanoparticle system that has been fully optimized for *in vivo* use as a drug carrier in humans, with QDs that have either been directly conjugated with, or encapsulated in, liposomes (termed "lipodots") having shown promise for intracellular deliveries. Phospholipid micelles [46], polymer-based immunoliposomes [47], and cationic Lipofectamine 2000 [38] are all typically used to prepare lipodots, but among these Lipofectamine 2000 seems to provide the most efficient delivery of QDs into HeLa cells [38]. The main advantage of liposomes over chitosan and peptides is that both hydrophobic and hydrophilic QDs can be incorporated, through either oil-in-water- or water-in-oil-type micellar assemblies. The hydrocarbon chains of liposomes may also interact with the hydrophobic plasma membrane, thus facilitating the nonspecific endocytosis of lipodots.

8.8
Targeting Cancer Cells with QDs

Although the physical and biochemical methods described above are valuable for the nonspecific delivery of QDs into living, fixed, normal and cancer cells, in order to conduct the imaging and PDT of cancer it is necessary to target the cancer cells and tumor milieu in a specific manner. The targeting of cells with QDs is made possible by attaching specific proteins, antibodies and peptides onto the surfaces of the QDs; the latter may then be conjugated, for example, with cholera toxin B (CTB) and phallotoxin (also called phalloidin) to produce organelle-specific QDs. The QD–CTB conjugate is able to recognize ganglioside receptors on mammalian cells [48], its uptake being initiated by caveolae-mediated endocytosis and retrograde transport through the Golgi apparatus into the endoplasmic reticulum. In similar fashion, phallotoxin will bind specifically to actin filaments, such that the QD–phallotoxin conjugate can be used to image the actin network and analyze any changes in the actin filaments within the cell [49]. Apart from these two proteins, however, the targeted intracellular delivery of QDs is focused essentially on the detection, imaging, and therapy of cancer.

The targeting of cancer cells with specific probes represents a major challenge not only for the imaging of cancers but also for conducting effective and targeted PDT. The targeted delivery of QDs depends on the type of cancer cell and the surface functionality of QDs. For example, specific receptors which are overexpressed in many cancers can serve as the target of anticancer drugs and imaging probes. Thus, QDs conjugated with peptides, antibodies and ligands, which selectively recognize overexpressed receptors in cancers, are widely applied for the *in vitro* and *in vivo* imaging of cancer cells, tumors, and tumor vasculature.

8.8.1
In Vitro Targeting of Cancer Cells with QDs

During recent years, one of the most exciting developments in the application of QDs has been the *in vitro* imaging of cancer cells. Shortly after the introduction of biocompatible QDs in 1998, their use for the imaging of cancer cells was investigated by conjugating them with cancer-specific ligands/antibodies/peptides. The result of these studies was a system whereby various types of human cancer cell, whether derived from prostate cancer, breast cancer, pancreatic cancer, B16F10 melanoma cells, glioblastoma, or cancers of the bone marrow and tongue, could be detected and imaged.

8.8.1.1 Targeting Cancer Cells with QD–Antibody Conjugates

QDs conjugated with primary or secondary antibodies detect specific receptors that are overexpressed in cancer cells. For example, QDs conjugated with an antibody (Ab) for prostate-specific membrane antigen (PSMA) will bind specifically with human prostate cancer cells (C4-2). The QD–PSMA-Ab conjugate will bind efficiently with PSMA-positive C4-2 cells, but not PSMA-negative PC-3 cells [37]. Likewise, QDs without PSMA-Ab will not bind to either C4-2 or PC-3 cells. Similarly, QDs conjugated with immunoglobulin G (IgG) will bind specifically with Her2 receptors [16], a cancer marker which is overexpressed in many breast cancers, on human breast cancer cells, and in mouse mammary tumor sections. In this case, the human breast cancer cells (SK-BR-3) and mouse mammary tumor tissues were effectively labeled with QDs by incubating the cells with Herceptin (anti-Her2 antibody), followed by QD–IgG conjugates. The targeting of breast cancer cells (e.g., KPL-4) with QDs was also seen to be possible by conjugating Herceptin directly onto the surface of the QDs. An recent advanced application of this method has been the targeted delivery of QDs and anticancer drugs by using multifunctional immunoliposomes (Figure 8.9a) [47]. In this case, immunoliposomes conjugated with CdSe/ZnS QDs and with the anti-Her2 antibody on the surface and encompassed with doxorubicin, an anticancer drug, efficiently tar-

8.8 Targeting Cancer Cells with QDs | 311

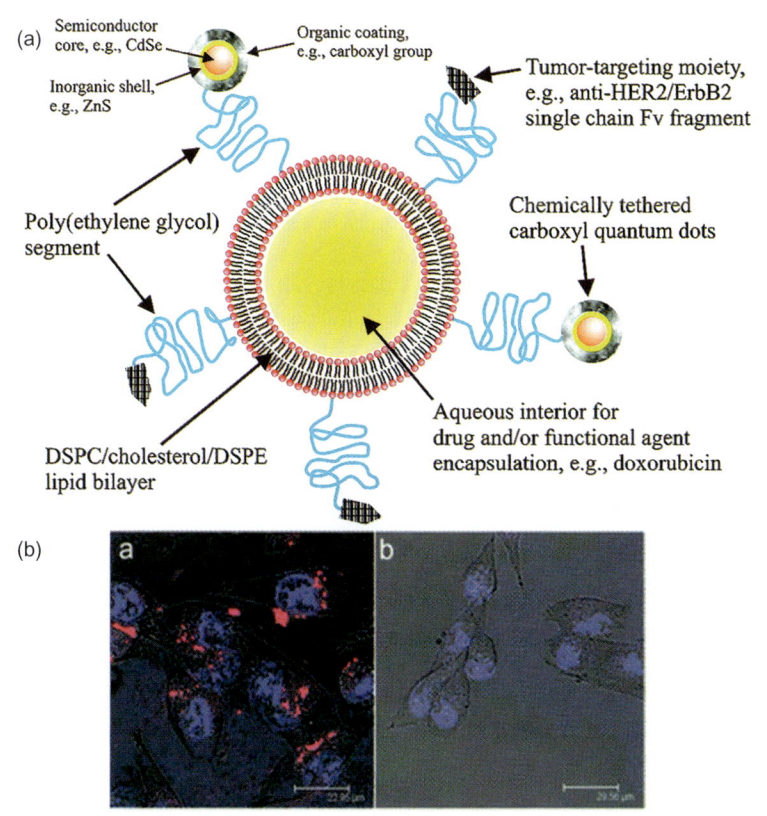

Figure 8.9 (a) Schematic presentation of an immunoliposome internalized with doxorubicin and conjugated with QDs and anti-Her2 antibody; (b) Fluorescence image of human pancreatic cancer cells incubated with InP QD–anti-Claudin-4 antibody conjugate (a) and InP QD without antibody (b). Reprinted with permission from Refs [47] (a) and [51] (b). Copyright (2008, 2009) American Chemical Society.

geted Her2-overexpressing breast cancer cells such as SK-BR-3 and MCF-7/Her2. The advantages of this immunoliposome-based targeting over the targeting of cancer cells with QD–Ab conjugates were three fold: (i) the anti-Her2 antibody targeted breast cancer cells; (ii) the QDs served as imaging probes; and (iii) the liposomes delivered anticancer drugs inside the cancer cells. Yet another candidate for the targeting of breast cancer cells has been that of QDs conjugated to an antibody to anti-type 1 insulin-like growth factor receptor (IGF1R) [50]. The QD–anti-IGFR1 conjugate can detect upregulated IGF1R levels in MCF-7 breast cancer cells, such that pancreatic cancer cells can be selectively labeled and imaged using QDs conjugated with anti-Claudin-4 antibody and anti-prostate stem cell antigen (anti-PSCA) [51]. The principle underlying this selective labeling is that the membrane proteins Claudin-4 and PSCA are each overexpressed in both primary and metastatic pancreatic cancer cells. A fluorescence image of human pancreatic cancer cells (MiaPaCa) treated with QDs (InP/ZnS)-anti-Claudin-4 conjugates is shown in Figure 8.9b; notably, those QDs without the antibody present were not internalized.

The intracellular distribution pattern of *mortalin* protein represents another approach to distinguishing normal cells from cancer cells. Although mortalin is distributed uniformly in the cytoplasm of normal human cells, in transformed cells (e.g., human cancer cells) it is found only in the perinuclear region. QD–streptavidin conjugates were able to distinguish between normal human fetal fibroblasts (WI-38) and osteogenic sarcoma (U2OS) cells, both of which had been immunostained with an anti-mortalin antibody, followed by a biotinylated secondary antibody [52]. Although the labeling pattern was able to distinguish the cancer cells from normal cells, the selective labeling of cancer cells could not be achieved using this method.

8.8.1.2 Targeting Cancer Cells with QD–Ligand Conjugates

Ligands based on peptides, proteins and aptamers which specifically recognize certain receptors that are overexpressed in cancer cells, have shown promise for the targeting of cancer cells. Examples of promising cancer-specific ligands include arginine-glycine-aspartic acid (RGD peptide), folic acid, epidermal growth factor (EGF), and transferrin. The RGD peptide targets the $\alpha_v\beta_3$ integrin, which is overexpressed in human breast cancer cells (MDA-MB-435) and, more importantly, in human glioblastoma cells (U87MG). Thus, MDA-MB-435 and U87MG cells may be selectively labeled with QD–RGD peptide conjugates [53]. The RGD peptide is also able to distinguish the human breast cancer cell MCF-7, in which the $\alpha_v\beta_3$ integrin is not upregulated, from MDA-MB-435 and U87MG cells.

The *folate receptor* represents another potential target for the imaging and PDT of human nasopharyngeal epidermal carcinoma cells (KB cells). For example, InP QDs conjugated with folic acid were shown to bind selectively with KB cells in which folate receptors were overexpressed [54]. In contrast, human lung carcinoma cells (e.g., A549) in which the folate receptor was not upregulated could not be detected with QD–folic acid conjugates.

Aptamers represent another candidate for the selective delivery of QDs into prostate cancer cells. Typically, QDs conjugated with an A10 RNA aptamer selectively labeled PSMA-positive LNCaP prostate cells, but not PC3 prostate adenocarcinoma cells which were PSMA-negative [55]. The efficiency for targeting LNCaP prostate cells with the QD–aptamer conjugates was essentially comparable to that of QD–anti-PSMA conjugates [37].

Yet another target in cancer cells is the G-protein-coupled epidermal growth factor receptor (EGFR), which is overexpressed in many cancer cells (e.g., Chinese hamster ovary and A431 cells); as a consequence, QD–EGF conjugates may be efficiently delivered into these cells [30]. Since all of these receptors, as signaling proteins, are important for the growth, proliferation, and differentiation of normal cells, when performing PDT it is necessary to apply a suitable concentration of QD–conjugates in order that the receptors in normal cells are targeted only to a minimal extent, whilst the overexpressed receptors in cancer cells are targeted predominantly.

8.8.2
In Vivo Targeting and Imaging Cancer with QDs

The bright fluorescence and remarkable photostability of QDs represent two major benefits for *in vivo* fluorescence imaging. Although the basic principle underlying *in vitro* targeting may be applied *in vivo*, the biodistribution of QDs remains a clear challenge for *in vivo* cancer imaging and PDT. Today, the most widely used method for the *in vivo* delivery of QD conjugates is systemic intravenous injection, although local administration via subcutaneous and intramuscular injection may also be used when targeting peripheral cancers. In contrast to the subcutaneous and intramuscular injection of QDs, multiple factors – notably interactions with the blood components and the immune response – must be paid strict attention when QDs are injected directly into the blood. With the introduction of QD-conjugates capable of specifically recognizing cancer cells, the applications of bioconjugated QDs gradually moved in the areas of *in vivo* imaging and PDT of cancer.

The *in vivo* application of QDs was first assessed by injecting CdSe/ZnS QDs coated with peptides into the tail vein of a mouse [56], after which the injected QDs were seen to distribute preferentially into endothelial cells lining the blood vessels of the lungs. QDs conjugated to peptides with an affinity for cancer cells and the tumor vasculature also bound preferentially bound to the tumors, as indicated by the *ex vivo* fluorescence microscopic imaging of tissue sections. A variety of QDs bioconjugated with cancer markers, such as an antibody to PSMA [37], RGD peptide [53], alpha-fetoprotein [57] and anti-Her2 antibody [47], have been tested *in vivo* in mouse models. Intravenously injected QD-PSMA antibody conjugates were shown to localize efficiently in human prostate cancer implanted subcutaneously in mice [37], as characterized by *in vivo* imaging of the whole animal (Figure 8.10a). The distribution of QD–PSAM antibody conjugates in the tumor milieu may be controlled by specific binding between the antigen and antibody. As noted in Section 8.8.1, the RGD peptide binds selectively with $\alpha_v\beta_3$ integrin overexpressed in human breast cancer cells (MDA-MB-435) and glioblastoma cells (U87MG). Thus, the NIR QD–RGD conjugate, when administered intravenously to mice bearing subcutaneous U87MG human glioblastoma, was efficiently localized in the tumor [53]. Figure 8.10b shows the signal-to-background ratio of NIR QD–RGD conjugates targeted in the tumor. The targeting of glioblastoma with QD–RGD conjugate is characterized by both *in vivo* fluorescence imaging of the whole animal, and *ex vivo* fluorescence imaging of the tumor. The *in vivo* imaging of human hepatocarcinoma cells, implanted subcutaneously in mice, using a QD conjugated with an antibody to alpha-fetoprotein (anti-AFP), is another example [57]. AFP, a major component of mammalian serum, is an important marker protein for liver cancer. QD–anti-AFP conjugates introduced systemically by tail vein injection in mice accumulate efficiently in subcutaneously grown HCCLM6 human hepatocarcinoma cells. In contrast to these antibodies and peptides, immunoliposomes developed for both *in vivo* targeting cancer and chemotherapy are multifunctional with respect to imaging and drug delivery. For example, liposomes conjugated with NIR QDs and anti-Her2 antibody, and

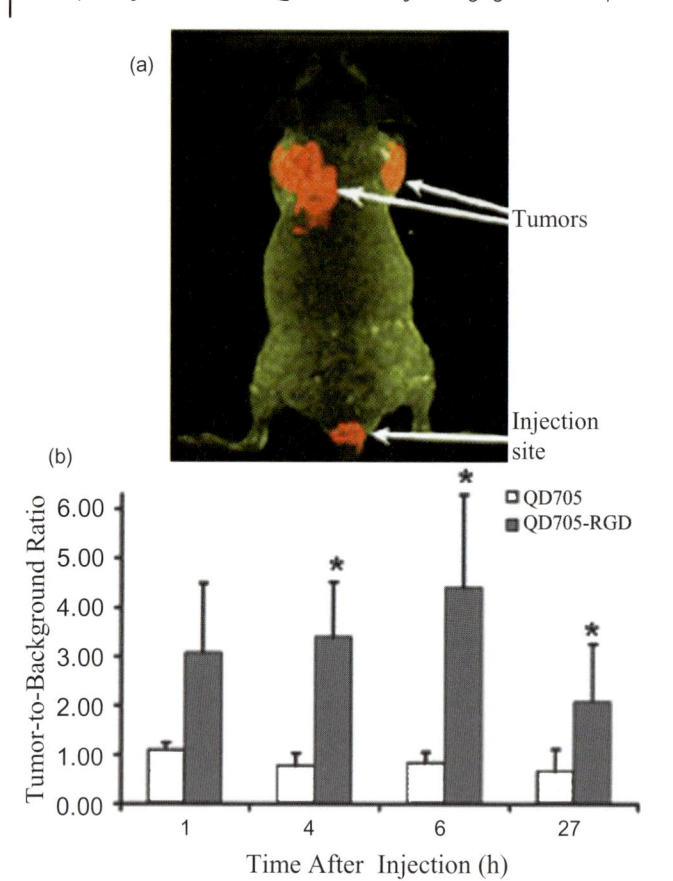

Figure 8.10 (a) Fluorescence image of human prostate cancer implanted in a mouse. Here, the tumor is targeted with anti-PSMA antigen-conjugated CdSe/ZnS QDs; (b) Histogram of fluorescence signal from U87MG tumor-bearing mice injected with NIR QD–RGD peptide. Reprinted by permission from Macmillan Publishers Ltd: Nature Biotechnology Ref. [37], copyright (2004) (a), and reprinted with permission from Ref. [53] (b).

encompassed with doxorubicin (see Section 8.8.1.1) were shown to target MCF-7/Her2 xenografts implanted in nude mice. This targeting method has a great potential to be extended for the targeted PDT of cancer.

8.8.3
In Vivo Targeting of Tumor Vasculature and Lymph Nodes with QDs

Targeting of the tumor vasculature and lymph nodes with bioconjugated QDs represents a promising approach for the *in vivo* imaging and PDT of cancer. Recently, applications of bioconjugated for vasculature imaging have emerged demonstrating the two-photon excited fluorescence imaging of capillaries in adipose tissue and skin in live mice [19]. For this, water-soluble CdSe/ZnS QDs are administered by tail vein injection, followed by NIR excitation and imaging. The large two-photon absorption cross-section of QDs allowed effective excitation of the QDs with NIR light. Two-photon excited fluorescence imaging was later applied for imaging the tumor vasculature using PEG–phosphatidylethanolamine-labeled CdS/ZnS and CdSe/ZnCdS QDs in C3H mice bearing isogenic mouse adenocarcinoma (MCaIV) tumor implants [58]. The above-mentioned QD–RGD peptide conjugate is another example for imaging the tumor vasculature.

Lymph node metastasis is an important symptom of cancers; thus, the imaging lymph nodes and lymphatic drainage may be of great help when staging cancers. The fluorescence visualization of sentinel lymph nodes also helps physicians to locate and dissect samples for biopsy. On this basis, bioconjugated QDs have become widely used for the fluorescence imaging of lymph nodes in animal models. For example, sentinel lymph nodes in the mouse and pig may be imaged *in vivo* using phosphine-coated NIR CdTe/CdSe QDs [59], with lymph nodes 1 cm below the skin being successfully visualized under NIR excitation. Lymph nodes may also be imaged by using QDs without any specific surface functional group. For example, lymph nodes in mice are visualized fluorescently by injecting CdSe/ZnS and CdSe/CdTe/ZnS QDs coated with negative-, positive-, and neutral-terminated PEG molecules [60]. These PEG-coated QDs are injected either intravenously or directly into subcutaneously grown tumors. All three types of QD drain towards the inguinal node, and this can be visualized though the skin under both visible and NIR excitation. Importantly, carboxylic acid-conjugated QDs are drained selectively into the right inguinal and axillary lymph nodes in the tumor-bearing side (Figure 8.11). Despite the efficient *in vivo* visualization of lymph nodes associated with tumors, QDs are retained in the first draining lymph nodes due to their large size. Unfortunately, this may cause problems when imaging distant lymph node metastases for deciding the boundaries for PDT and surgery.

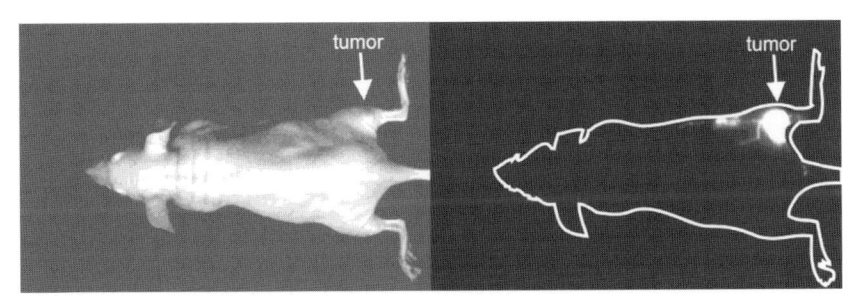

Figure 8.11 Nude mouse bearing M21 melanoma, dorsal view 3 min after injection into the tumor using 655 nm PEG 5k–COOH QDs. Left, visible light; right, fluorescence at 655 nm. Reprinted with permission from Ref. [60]. Copyright (2007) American Chemical Society.

8.9
Quantum Dots for Photodynamic Therapy of Cancer

Bioconjugated QDs bearing cancer-selective antibodies and peptides are well-demonstrated agents for targeting and visualizing cancer *in vivo*. However, the photosensitized production of ROIs at high efficiency is the primary requirement for PDT, for which the efficient absorption of light (especially in the NIR region), a long-lived excited state, photostability and biocompatibility are necessary. Thus, the bright fluorescence and incredible photostability of QDs, along with *in vivo* cancer targeting, promise the development of QDs for the image-guided PDT of cancer. The broad absorption band and large two-photon absorption cross-section of QDs, as well as the availability of NIR QDs, are added advantages of QDs to become the standard photosensitizers for PDT.

8.9.1
Quantum Dot Alone for PDT

The potential of QDs alone for PDT depends on the efficiency of energy transfer from photoexcited QDs to molecular oxygen, and the production of ROIs. The direct photoactivation of QDs in both organic and aqueous phases produces singlet oxygen, the efficiency of which is extremely low (<5%). For example, hydrophobic-capped CdSe QDs with a 65% fluorescence quantum efficiency produce only ~5% singlet oxygen in toluene [61]. Phosphorescence emission ~1270 nm is the standard for detecting singlet oxygen. The low efficiency of QDs for singlet oxygen production is due to short-lived excited states (<100 ns). In addition, other factors such as how the size, surface chemistry, and organic and polymer capping of QDs interfere with collisional quenching of the excited state by oxygen, remain unknown. On the other hand, conventional photosensitizers such as porphyrins and phthalocyanines, which have small size and long-lived triplet states (microseconds to milliseconds), produce ROIs at much higher efficiencies (up to 75%). Despite QDs producing singlet oxygen at low efficiency, the incredible photostability of QDs offers prolonged photoactivation and a persistent production of singlet oxygen and other ROIs. Thus, QDs can offer cumulative effects in PDT. For example, the prolonged photoactivation of CdSe/ZnS QDs conjugated to DNA results in strand breakage and nucleobase damage [62]. Such breakage and damage of DNA are due to the production of ROIs such as singlet oxygen and hydroxyl radicals, followed by the abstraction of hydrogen atoms from the bases or the pyranose ring in DNA. The abstraction of hydrogen atom creates radical centers in DNA and results in the breakage and damage of DNA. ROI production by photoativated QDs, and the subsequent breakage and damage of DNA, are shown schematically in Figure 8.12. The photosensitized DNA breakage and damage by QDs is promising for nucleus-targeted PDT if combined with the targeting nucleus of cancer cells. The photoactivation of QDs delivered in cancer cells also results in the production of superoxide and peroxynitrite, and the breakage of lysosomes

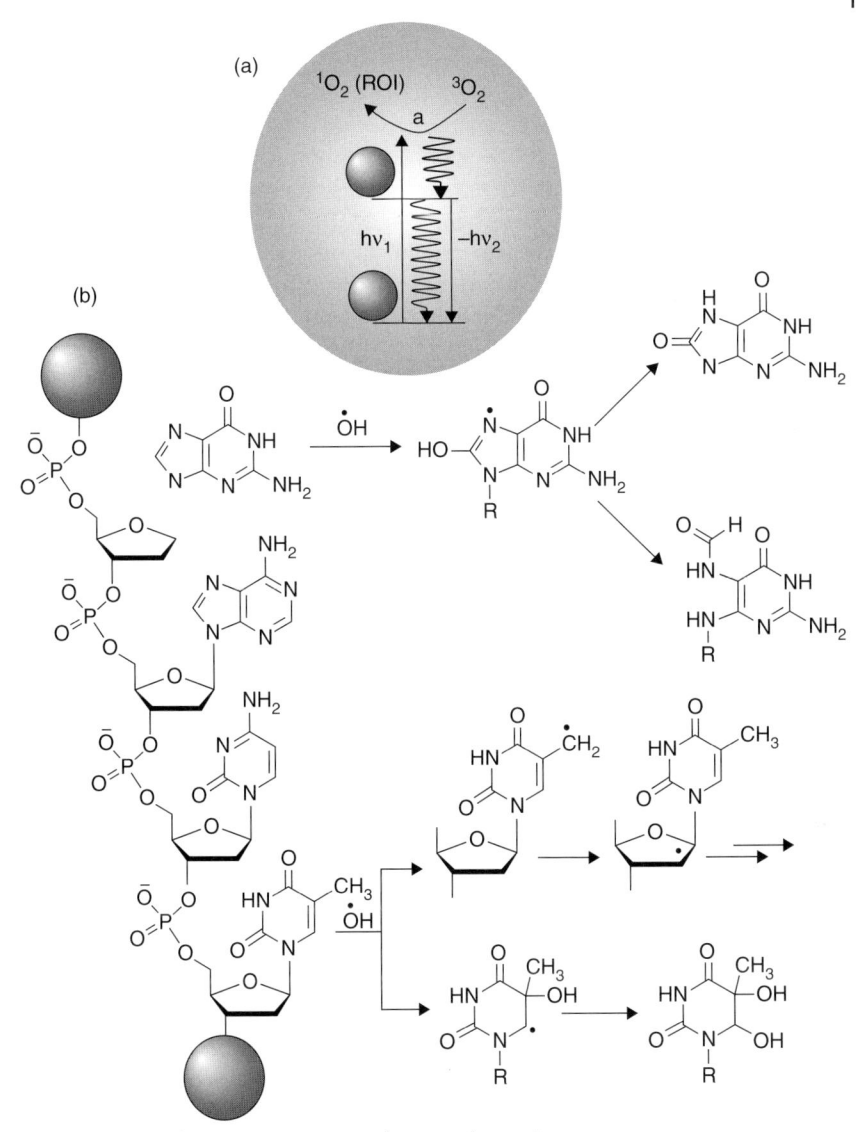

Figure 8.12 (a) Schematic presentation of ROI production by a QD; (b) Reactions of a DNA molecule with hydroxyl radical and subsequent nucleobase damage and strand breakage in DNA. Reprinted with permission from Ref. [62]. Copyright (2008) American Chemical Society.

[63]. Thus, the properties of QDs such as photostability, photosensitized production of reactive oxygen and nitrogen intermediates, and the damage and breakage of DNA and lysosomes, demonstrate the potential of QDs alone as photosensitizers for PDT.

8.9.2
Potentials of QD–Photosensitizer Conjugates for PDT

There are many advantages and limitations for both conventional photosensitizers and QDs when separately applied for PDT. For example, the unique optical properties – especially the NIR absorption, large two-photon absorption cross-section, broad absorption band and photostability – of QDs show promise for PDT, but not the efficiency of ROI production. On the other hand, the optical properties of conventional photosensitizers, such as the narrow absorption band and poor photostability, are not attractive for PDT, but rather for the high efficiency of ROI production. In other words, the advantages of QDs complement the limitations of photosensitizers, and *vice versa*. Therefore, a combination of QD and conventional photosensitizers would serve as an ideal drug for modern PDT. In such coupled QD–photosensitizer systems, the photosensitizers are indirectly excited by nonradiative energy transfer (also called Förster resonance energy transfer; FRET) from photoactivated QDs. Thus, the FRET efficiency becomes critical for the photosensitized production of ROIs. The FRET efficiency varies inversely with the sixth power of the distance between the energy donor (QD) and the energy acceptor (photosensitizer), such that close conjugation (typically within 10 nm) of the photosensitizer on the QD is necessary for efficient energy transfer and ROI production. QD-based FRET systems are widely applied as biosensors [64], while several QD–photosensitizer systems composed of CdSe, CdSe/ZnS, and CdTe as energy donors and porphyrins, phthalocyanines, Rose Bengal, and metal complexes as energy acceptors, have also been developed for PDT applications. The photoactivation of a QD–photosensitizer conjugate, the energy transfer from QD to photosensitizers, and the production of ROI are shown in Figure 8.13.

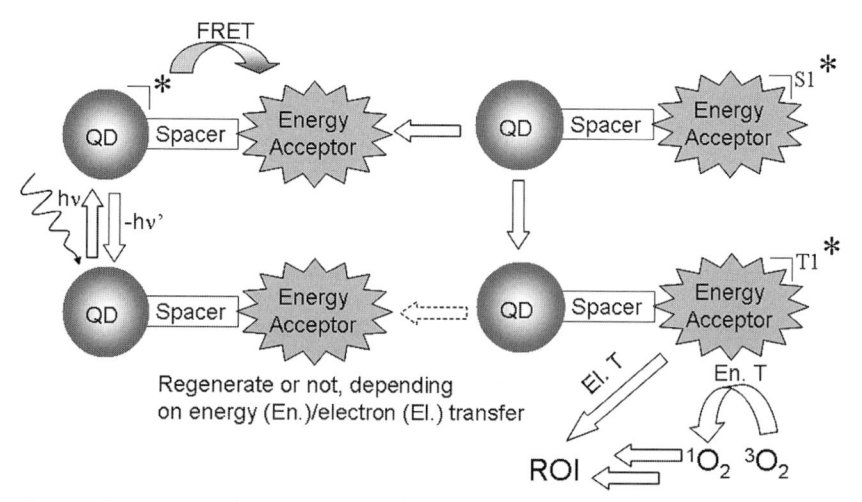

Figure 8.13 Energy-transfer processes in a photoactivated QD–photosensitizer system, and the production of ROIs. Reprinted with permission from Ref. [3].

The advantages of such QD–photosensitizer conjugates are many. For example, indirect photoactivation avoids photobleaching of the photosensitizers, while the large surface area and biocompatibility of QDs offer a platform for conjugating multiple photosensitizer molecules that are otherwise hydrophobic and insoluble in the aqueous phase. The photosensitizer, when excited by energy transfer from the QDs, relaxes to its long-lived triplet state and initiates ROI production (see Section 8.2). Thus, a higher efficiency for ROI production and photostability of the sensitizer can be achieved for QD–photosensitizer systems.

One example of a QD–photosensitizer system is a noncovalent mixture of CdSe QDs and an aluminum phthalocyanine [65]. The *in vitro* PDT of leukemic cells was assessed by labeling the cells with a QD-anti-CD90 conjugate, followed by illumination with UV light in the presence of aluminum phthalocyanine. Here, although sensitization of leukemic cells was observed, energy transfer from QDs on the cell wall to photosensitizers either in the culture medium or adsorbed onto the cell wall, was not warranted. The photoactivation of cells with UV light, which directly damages DNA and other vital biomolecules, is undesirable for PDT.

The first example of a QD–photosensitizer conjugate is CdSe QDs conjugated with silicon phthalocyanine (Pc4) [61]. Figure 8.14 shows QD–Pc4 conjugates with varying donor-to-acceptor distances, and their steady-state fluorescence spectra. These CdSe–Pc4 conjugates have been investigated extensively, and their physical and chemical parameters for FRET optimized. The energy transfer efficiency in these conjugates depends on the number of Pc4 molecules on the surface of a single CdSe QD, and the length of the spacer between CdSe and Pc4. Maximum energy transfer efficiency is obtained when three Pc4 are conjugated to one CdSe QD. Photoactivation of the QD–Pc4 system resulted in an efficient energy transfer from QDs to Pc4, identified by transient absorption measurements. QD–Pc4 was also found to be capable of producing singlet oxygen. The major concern regarding the QD–Pc4 system for PDT is its poor water-solubility. Examples of water-soluble QD–photosensitizer conjugates are CdTe QDs conjugated with *meso*-tetra(4-sulfonatophenyl)porphine dihydrochloride [66] and sulfonated aluminum phthalocyanine [67], and CdSe/CdS/ZnS QDs conjugated with Chlorin e6 and Rose Bengal [68]. Among these systems, QD–Chlorin e6 and QD–Rose Bengal are attractive for PDT due to their higher efficiency (>30%) for singlet oxygen production. The QD–Chlorin e6 and QD–Rose Bengal systems are prepared by covalently conjugating the photosensitizer to a phytochelatin peptide, followed by coating of the surface of QDs with the peptide–photosensitizer conjugates. A typical example of water-soluble QD–phthalocyanine conjugates is CdTe–aluminum phthalocyanine tetrasulfonate, which is prepared by conjugating aluminum phthalocyanine tetrasulfonate to CdTe QDs capped with thioglycolic acid [67]. This QD–photosensitizer system provides ~15% singlet oxygen quantum efficiency. Although neither *in vitro* nor *in vivo* PDT experiments using QD–photosensitizer have yet been investigated in detail, singlet oxygen production by QDs and QD–photosensitizer systems and the photostability of QDs demonstrates the potential of QDs for applications in the imaging and PDT of cancer.

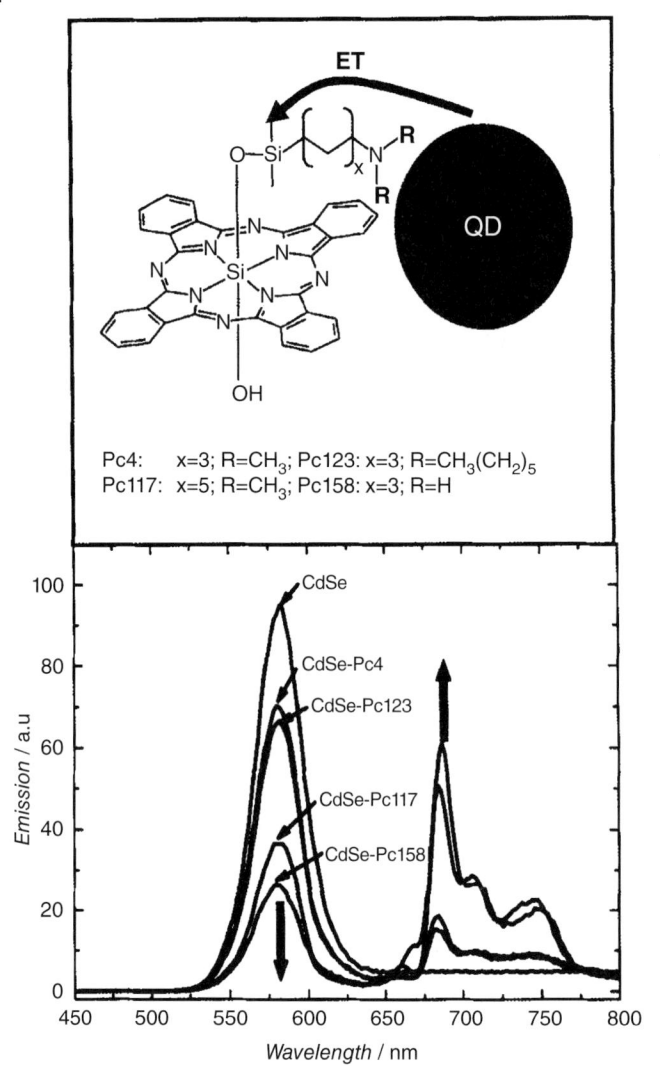

Pc4:   x=3; R=CH$_3$; Pc123: x=3; R=CH$_3$(CH$_2$)$_5$
Pc117: x=5; R=CH$_3$; Pc158: x=3; R=H

Figure 8.14 (Top) Energy transfer from QD to Pc4 with varying QD-to-Pc4 spacer length. Here, the FRET efficiency varies from 23 to 77%. (Botttom) Fluorescence spectra of QD–Pc4 conjugates with varying spacer length. Reprinted with permission from Ref. [69] with permission from Springer Science and Business Media.

8.10
Toxicity of QDs

Several issues remain concerning the extension of QDs to *in vivo* applications in humans, notably those of toxicity and pharmacokinetic. The most attractive QDs for *in vivo* imaging and PDT contain cadmium, which is known to not only accumulate in vital organs such as the liver and kidneys, but also to damage DNA and protein molecules. In addition, as the half-life of cadmium in humans is approximately 20 years, cadmium-based core-only QDs, and core–shell QDs based on cadmium shells, are less attractive for *in vivo* imaging and PDT. A number of controversial reports have been made relating to the toxicity of cadmium-based QDs. For example, mercaptoacetic acid-coated CdSe QDs were shown to release Cd^{2+} as a result of both air oxidation (126 ppm) and UV illumination (82 ppm) [70]; such levels of Cd^{2+} are highly toxic and cause cell death. As the release of Cd^{2+} by QDs is due to oxidation-induced surface etching, the cytotoxicity of QDs can be avoided by coating the surface with suitable protecting shells. For example, CdSe QDs coated with ZnS shells essentially do not release Cd^{2+} ions under UV illumination [70]. In general, QDs shelled with ZnS, silica and polymers are rather safe for *in vitro* applications. However, an additional mechanism of toxicity – that is, the precipitation of QDs onto the cell membrane – exists for polymer-shelled QDs at high concentrations [71].

The scenario of the *in vivo* toxicity of QDs is different from that of *in vitro* cytotoxicity. Whether QDs accumulate in vital organs or are cleared from the body depends on their hydrodynamic size; notably, the plasma half-lives of QDs with large hydrodynamic sizes are extended, which not only results in an increased accumulation of QDs in the vital organs but also affects their pharmacokinetics. In general, the longer the half-life of QDs in the body, the greater are the chances for biochemical degradation of the protection shells and consequent release of Cd^{2+}. As an example, when QDs were administered systemically to rats they accumulated in the sinusoid edges of liver, the red pulp of the spleen, the subcapsular sinus in the lymph nodes, and vascular sinus periphery in the bone marrow [72]. The glomerular barrier and small pore size of the mammalian vasculature are limiting factors for the clearance of larger QDs from the body. A typical example is the renal clearance of smaller QDs of 4.36 nm diameter into the urinary bladder, and the distribution of larger QDs of 8.65 nm diameter into vital organs such as the liver, spleen, and lungs [73]. Thus, the *in vivo* administration of cadmium- or other toxic metal-based QDs with a hydrodynamic diameter >5 nm is not recommended. Although, larger particles are removed from the body by phagocytosis, this mechanism is not considered "safe" for QDs, due to possible degradation and subsequent release of metal ions under the highly oxidative conditions of phagocytosis. Thus, for advanced *in vivo* imaging and PDT it is necessary to develop QDs without toxic metals. In this case, one promising candidate is InP QDs, the biological applications of which are currently under active investigation.

8.11
Conclusions

Semiconductor QDs are attractive materials for the *in vivo* imaging and PDT of cancer, due to their unique optical properties such as size-tunable absorption and fluorescence bands, exceptional photostability, large two-photon absorption cross-section, and broad absorption and narrow fluorescence bands. These unique properties, combined with the good availability of NIR QDs, continue to show promise for the replacement of conventional organic dyes and second-generation photosensitizers. One notable advantage of QDs is their large surface area for bioconjugation, and consequently their surface, polymer and bioconjugate chemistries have been efficiently exploited, with many biocompatible QDs having visible and NIR absorption and fluorescence having been prepared. On the basis of such biocompatibility, QD use has begun to infiltrate cancer research, and many bioconjugates for imaging cancer cells, tumors, tumor vasculature and lymph nodes have been developed. These bioconjugates include QD–antibody and QD–peptide conjugates which can efficiently target prostate, breast, pancreatic, liver and tongue cancers, both *in vitro* and *in vivo*. In parallel, the photosensitized generation of singlet oxygen and other ROIs (both of which are essential elements of PDT) by QDs alone, by mixtures of QDs and photosensitizers, and by conjugates of QDs and photosensitizers, have been identified. Such photostability and ROI production make QDs potential candidates for PDT. Thus, an interface between the targeted delivery of QDs in cancers and the efficient generation of ROIs is under active investigation for advancing the *in vivo* imaging and PDT of cancer. Recent developments, such as self-illuminating QDs, afterglow nanoparticles and X-ray-activated fluorescence from QDs, have shown great promise for the imaging and PDT of deep cancers that cannot be accessed by visible and NIR light. Despite these advantages of QDs and bioconjugated QDs, it remains necessary to optimize their *in vivo* applications, notably to remove concerns regarding their cytotoxicity and pharmacokinetics, before their application to humans.

Abbreviations

CPP	Cell-penetrating peptide
CTB	Cholera toxin subunit B
DHLA	Dihydrolipoic acid
EGFR	Epidermal growth factor receptor
FRET	Förster resonance energy transfer
FTP	Alpha-fetoprotein
IGF1R	Insulin-like growth factor receptor
IgG	Immunoglobulin G
NIR	Near-infrared
PDT	Photodynamic therapy
PEG	Polyethylene glycol
PI-3K	Phosphoinositide-3OH kinase
PS	Photosensitizer
PSCA	Prostate stem cell antigen
PSMA	Prostate-specific membrane antigen
QDs	Quantum dots
RGD	Arginine-glycine-aspartic acid

ROI   Reactive oxygen intermediates
TGA   Thioglycolic acid
UV   Ultraviolet

References

1 Dougherty, T.J., Grindey, G.B., Fiel, R., Weishaupt, K.R. and Boyle, D.G. (1975) Photoradiation therapy. II. Cure of animal tumors with hematoporphyrin and light. *Journal of the National Cancer Institute*, **55**, 115–21.

2 Nseyo, U.O., Dehaven, J., Dougherty, T.J., Potter, W.R., Merrill, D.L., Lundahl, S.L. and Lamm, D.L. (1998) Photodynamic therapy (PDT) in the treatment of patients with resistant superficial bladder cancer: a long-term experience. *Journal of Clinical Laser Medicine and Surgery*, **16**, 61–8.

3 Biju, V., Itoh, T., Anas, A., Sujith, A. and Ishikawa, M. (2008) Semiconductor quantum dots and metal nanoparticles: syntheses, optical properties, and biological applications. *Analytical and Bioanalytical Chemistry*, **391**, 2469–95.

4 Dabbousi, B.O., Rodriguez-Viejo, J., Mikulec, F.V., Heine, J.R., Mattoussi, H., Ober, R., Jensen, K.F. and Bawendi, M.G. (1997) (CdSe)ZnS core-shell quantum dots: synthesis and characterization of a size series of highly luminescent nanocrystallites. *Journal of Physical Chemistry, B*, **101**, 9463–75.

5 Murray, C.B., Norris, D.J. and Bawendi, M.G. (1993) Synthesis and characterization of nearly monodisperse CdE (E = sulfur, selenium, tellurium) semiconductor nanocrystallites. *Journal of the American Chemical Society*, **115**, 8706–15.

6 Bowen Katari, J.E., Colvin, V.L. and Alivisatos, A.P. (1994) X-ray photoelectron spectroscopy of cdse nanocrystals with applications to studies of the nanocrystal surface. *Journal of Physical Chemistry*, **98**, 4109–17.

7 Qu, L., Peng, Z.A. and Peng, X. (2001) Alternative routes toward high quality CdSe nanocrystals. *Nano Letters*, **1**, 333–7.

8 Biju, V., Makita, Y., Nagase, T., Yamaoka, Y., Yokoyama, H., Baba, Y. and Ishikawa, M. (2005) Subsecond luminescence intensity fluctuations of single CdSe quantum dots. *Journal of Physical Chemistry, B*, **109**, 14350–5.

9 Bharali, D.J., Lucey, D.W., Jayakumar, H., Pudavar, H.E. and Prasad, P.N. (2005) Folate-receptor-mediated delivery of InP quantum dots for bioimaging using confocal and two-photon microscopy. *Journal of the American Chemical Society*, **127**, 11364–71.

10 Rogach, A.L., Franzl, T., Klar, T.A., Feldmann, J., Gaponik, N., Lesnyak, V., Shavel, A., Eychmüller, A., Rakovich, Y.P. and Donegan, J.F. (2007) Aqueous synthesis of thiol-capped CdTe nanocrystals: state-of-the-art. *Journal of Physical Chemistry, C*, **111**, 14628–37.

11 Peng, X., Schlamp, M.C., Kadavanich, A.V. and Alivisatos, A.P. (1997) Epitaxial growth of highly luminescent CdSe/CdS core/shell nanocrystals with photostability and electronic accessibility. *Journal of the American Chemical Society*, **119**, 7019–29.

12 Kim, S., Fisher, B., Eisler, H.-J. and Bawendi, M. (2003) Type-II quantum dots: CdTe/CdSe(core/shell) and CdSe/ZnTe(core/shell) heterostructures. *Journal of the American Chemical Society*, **125**, 11466–7.

13 Talapin, D.V., Rogach, A.L., Kornowski, A., Haase, M. and Weller, H. (2001) Highly luminescent monodisperse CdSe and CdSe/ZnS nanocrystals synthesized in a hexadecylamine-trioctylphosphine oxide-trioctylphospine mixture. *Nano Letters*, **1**, 207–11.

14 Underwood, D.F., Kippeny, T. and Rosenthal, S.J. (2001) Ultrafast carrier dynamics in CdSe nanocrystals determined by femtosecond fluorescence upconversion spectroscopy. *Journal of Physical Chemistry. B*, **105**, 436–43.

15 Manoj, N. and Brus, L. (1999) Luminescence photophysics in

semiconductor nanocrystals. *Accounts of Chemical Research*, **32**, 407–14.

16 Wu, X.Y., Liu, H.J., Liu, J.Q., Haley, K.N., Treadway, J.A., Larson, J.P., Ge, N.F., Peale, F. and Bruchez, M.P. (2003) Immunofluorescent labeling of cancer marker Her2 and other cellular targets with semiconductor quantum dots. *Nature Biotechnology*, **21**, 41–6.

17 Biju, V., Kanemoto, R., Matsumoto, Y., Ishii, S., Nakanishi, S., Itoh, T., Baba, Y. and Ishikawa, M. (2007) Photoinduced photoluminescence variations of CdSe quantum dots in polymer solutions. *Journal of Physical Chemistry, C*, **111**, 7924–32.

18 Blanton, S.A., Hines, M.A., Schmidt, M.E. and Guyot-Sionnest, P. (1996) Two-photon spectroscopy and microscopy of II-VI semiconductor nanocrystals. *Journal of Luminescence*, **70**, 253–68.

19 Larson, D.R., Zipfel, W.R., Williams, R.M., Clark, S.W., Bruchez, M.P., Wise, F.W. and Webb, W.W. (2003) Water-soluble quantum dots for multiphoton fluorescence imaging *in vivo*. *Science*, **300**, 1434–6.

20 Medintz, I.L., Uyeda, H.T., Goldman, E.R. and Mattoussi, H. (2005) Quantum dot bioconjugates for imaging, labelling and sensing. *Nature Materials*, **4**, 435–46.

21 Chan, W.C.W. and Nie, S. (1998) Quantum dot bioconjugates for ultrasensitive nonisotopic detection. *Science*, **281**, 2016–18.

22 Uyeda, H.T., Medintz, I.L., Jaiswal, J.K., Simon, S.M. and Mattoussi, H. (2005) Synthesis of compact multidentate ligands to prepare stable hydrophilic quantum dot fluorophores. *Journal of the American Chemical Society*, **127**, 3870–8.

23 Vu, T.Q., Maddipati, R., Blute, T.A., Nehilla, B.J., Nusblat, L. and Desai, T.A. (2005) Peptide-conjugated quantum dots activate neuronal receptors and initiate downstream signaling of neurite growth. *Nano Letters*, **5**, 603–7.

24 Dubertret, B., Skourides, P., Norris, D.J., Noireaux, V., Brivanlou, A.H. and Libchaber, A. (2002) *In vivo* imaging of quantum dots encapsulated in phospholipid micelles. *Science*, **298**, 1759–62.

25 Dahan, M., Levi, S., Luccardini, C., Rostaing, P., Riveau, B. and Triller, A. (2003) Diffusion dynamics of glycine receptors revealed by single-quantum dot tracking. *Science*, **302**, 442–5.

26 Shah, L.S., Clark, P.A., Moioli, E.K., Stroscio, M.A. and Mao, J.J. (2007) Labeling of mesenchymal stem cells by bioconjugated quantum dots. *Nano Letters*, **7**, 3071–9.

27 Howarth, M., Takao, K., Hayashi, Y. and Ting, A.Y. (2005) Targeting quantum dots to surface proteins in living cells with biotin ligase. *Proceedings of the National Academy of Sciences of the United States of America*, **102**, 7583–8.

28 Tokumasu, F. and Dvorak, J. (2003) Development and application of quantum dots for immunocytochemistry of human erythrocytes. *Journal of Microscopy*, **211**, 256–61.

29 Rosenthal, S.J., Tomlinson, A., Adkins, E.M., Schroeter, S., Adams, S., Swafford, L., Mcbride, J., Wang, Y.Q., Defelice, L.J. and Blakely, R.D. (2002) Targeting cell surface receptors with ligand-conjugated nanocrystals. *Journal of the American Chemical Society*, **124**, 4586–94.

30 Lidke, D.S., Nagy, P., Heintzmann, R., Arndt-Jovin, D.J., Post, J.N., Grecco, H.E., Jares-Erijman, E.A. and Jovin, T.M. (2004) Quantum dot ligands provide new insights into erbB/HER receptor-mediated signal transduction. *Nature Biotechnology*, **22**, 198–203.

31 Jaiswal, J.K., Mattoussi, H., Mauro, J.M. and Simon, S.M. (2003) Long-term multiple color imaging of live cells using quantum dot bioconjugates. *Nature Biotechnology*, **21**, 47–51.

32 Chu, T.C., Shieh, F., Lavery, L.A., Levy, M., Richards-Kortum, R., Korgel, B.A. and Ellington, A.D. (2006) Labeling tumor cells with fluorescent nanocrystal-aptamer bioconjugates. *Biosensors & Bioelectronics*, **21**, 1859–66.

33 Minet, O., Dressler, C. and Beuthan, J. (2004) Heat stress induced redistribution of fluorescent quantum dots in breast tumor cells. *Journal of Fluorescence*, **14**, 241–7.

34 van Tilborg, G.A.F., Mulder, W.J.M., Chin, P.T.K., Storm, G., Reutelingsperger, C.P., Nicolay, K. and Strijkers, G.J. (2006)

Annexin A5-conjugated quantum dots with a paramagnetic lipidic coating for the multimodal detection of apoptotic cells. *Bioconjugate Chemistry*, **17**, 865–8.

35 Zhou, M., Nakatani, E., Gronenberg, L.S., Tokimoto, T., Wirth, M.J., Hruby, V.J., Roberts, A., Lynch, R.M. and Ghosh, I. (2007) Peptide-labeled quantum dots for imaging GPCRs in whole cells and as single molecules. *Bioconjugate Chemistry*, **18**, 323–32.

36 Sukhanova, A., Devy, M., Venteo, L., Kaplan, H., Artemyev, M., Oleinikov, V., Klinov, D., Pluot, M., Cohen, J.H.M. and Nabiev, I. (2004) Biocompatible fluorescent nanocrystals for immunolabeling of membrane proteins and cells. *Analytical Biochemistry*, **324**, 60–7.

37 Gao, X.H., Cui, Y.Y., Levenson, R.M., Chung, L.W.K. and Nie, S.M. (2004) In vivo cancer targeting and imaging with semiconductor quantum dots. *Nature Biotechnology*, **22**, 969–76.

38 Derfus, A.M., Chan, W.C.W. and Bhatia, S.N. (2004) Intracellular delivery of quantum dots for live cell labeling and organelle tracking. *Advanced Materials*, **16**, 961–4.

39 Duan, H. and Nie, S. (2007) Cell-penetrating quantum dots based on multivalent and endosome-disrupting surface coatings. *Journal of the American Chemical Society*, **129**, 3333–8.

40 Ruan, G., Agarwal, A., Marcus, A.I. and Nie, S. (2007) Imaging and tracking of tat peptide-conjugated quantum dots in living cells: new insights into nanoparticle uptake, intracellular transport, and vesicle shedding. *Journal of the American Chemical Society*, **129**, 14759–66.

41 Delehanty, J.B., Medintz, I.L., Pons, T., Brunel, F.M., Dawson, P.E. and Mattoussi, H. (2006) Self-assembled quantum dot–peptide bioconjugates for selective intracellular delivery. *Bioconjugate Chemistry*, **17**, 920–7.

42 Rozenzhak, S.M., Kadakia, M.P., Caserta, T.M., Westbrook, T.R., Stone, M.O. and Naik, R.R. (2005) Cellular internalization and targeting of semiconductor quantum dots. *Chemical Communications*, 2217–19.

43 Biju, V., Muraleedharan, D., Nakayama, K., Shinohara, Y., Itoh, T., Baba, Y. and Ishikawa, M. (2007) Quantum dot-insect neuropeptide conjugates for fluorescence imaging, transfection, and nucleus targeting of living cells. *Langmuir*, **23**, 10254–61.

44 Anas, A., Okuda, T., Kawashima, N., Nakayama, K., Itoh, T., Ishikawa, M. and Biju, V. (2009) Clathrin-mediated endocytosis of quantum dot-peptide conjugates in living cells. *ACS Nano*, **3**, 2419–2429.

45 Tan, W.B., Huang, N. and Zhang, Y. (2007) Ultrafine biocompatible chitosan nanoparticles encapsulating multi-coloured quantum dots for bioapplications. *Journal of Colloid and Interface Science*, **310**, 464–70.

46 Schroeder, J.E., Shweky, I., Shmeeda, H., Banin, U. and Gabizon, A. (2007) Folate-mediated tumor cell uptake of quantum dots entrapped in lipid nanoparticles. *Journal of Controlled Release*, **124**, 28–34.

47 Weng, K.C., Noble, C.O., Papahadjopoulos-Sternberg, B., Chen, F.F., Drummond, D.C., Kirpotin, D.B., Wang, D., Hom, Y.K., Hann, B. and Park, J.W. (2008) Targeted tumor cell internalization and imaging of multifunctional quantum dot-conjugated immunoliposomes in vitro and in vivo. *Nano Letters*, **8**, 2851–7.

48 Chakraborty, S.K., Fitzpatrick, J.A.J., Phillippi, J.A., Andreko, S., Waggoner, A.S., Bruchez, M.P. and Ballou, B. (2007) Cholera toxin B conjugated quantum dots for live cell labeling. *Nano Letters*, **7**, 2618–26.

49 Yoo, J., Kambara, T., Gonda, K. and Higuchi, H. (2008) Intracellular imaging of targeted proteins labeled with quantum dots. *Experimental Cell Research*, **314**, 3563–9.

50 Zhang, H., Sachdev, D., Wang, C., Hubel, A., Gaillard-Kelly, M. and Yee, D. (2009) Detection and downregulation of type I IGF receptor expression by antibody-conjugated quantum dots in breast cancer cells. *Breast Cancer Research and Treatment*, **114**, 277–85.

51 Yong, K.-T., Ding, H., Roy, I., Law, W.-C., Bergey, E.J., Maitra, A. and Prasad, P.N.

(2009) Imaging pancreatic cancer using bioconjugated InP quantum dots. *ACS Nano*, **3**, 502–10.

52 Kaul, Z., Yaguchi, T., Kaul, S.C., Hirano, T., Wadhwa, R. and Taira, K. (2003) Mortalin imaging in normal and cancer cells with quantum dot immunoconjugates. *Cell Research*, **13**, 503–7.

53 Cai, W.B., Shin, D.W., Chen, K., Gheysens, O., Cao, Q.Z., Wang, S.X., Gambhir, S.S. and Chen, X.Y. (2006) Peptide-labeled near-infrared quantum dots for imaging tumor vasculature in living subjects. *Nano Letters*, **6**, 669–76.

54 Bharali, D.J., Lucey, D.W., Jayakumar, H., Pudavar, H.E. and Prasad, P.N. (2005) Folate-receptor-mediated delivery of InP quantum dots for bioimaging using confocal and two-photon microscopy. *Journal of the American Chemical Society*, **127** (32), 11364–71.

55 Bagalkot, V., Zhang, L., Levy-Nissenbaum, E., Jon, S., Kantoff, P.W., Langer, R. and Farokhzad, O.C. (2007) Quantum dot – aptamer conjugates for synchronous cancer imaging, therapy, and sensing of drug delivery based on bi-fluorescence resonance energy transfer. *Nano Letters*, **7**, 3065–70.

56 Akerman, M.E., Chan, W.C.W., Laakkonen, P., Bhatia, S.N. and Ruoslahti, E. (2002) Nanocrystal targeting *in vivo*. *Proceedings of the National Academy of Sciences of the United States of America*, **99**, 12617–21.

57 Yu, X., Chen, L., Li, K., Li, Y., Xiao, S., Luo, X., Liu, J., Zhou, L., Deng, Y., Pang, D. and Wang, Q. (2007) Immunofluorescence detection with quantum dot bioconjugates for hepatoma in vivo. *Journal of Biomedical Optics*, **12**, 014008.

58 Stroh, M., Zimmer, J.P., Duda, D.G., Levchenko, T.S., Cohen, K.S., Brown, E.B., Scadden, D.T., Torchilin, V.P., Bawendi, M.G., Fukumura, D. and Jain, R.K. (2005) Quantum dots spectrally distinguish multiple species within the tumor milieu *in vivo*. *Nature Medicine*, **11**, 678–82.

59 Kim, S., Lim, Y.T., Soltesz, E.G., Grand, A.M.D.E., Lee, J., Nakayama, A., Parker, J.A., Mihaljevic, T., Laurence, R.G., Dor, D.M., Cohn, L.H., Bawendi, M.G. and Frangioni, J.V. (2004) Near-infrared fluorescent type II quantum dots for sentinel lymph node mapping. *Nature Biotechnology*, **22**, 93–7.

60 Ballou, B., Ernst, L.A., Andreko, S., Harper, T., Fitzpatrick, J.A.J., Waggoner, A.S. and Bruchez, M.P. (2007) Sentinel lymph node imaging using quantum dots in mouse tumor models. *Bioconjugate Chemistry*, **18**, 389–96.

61 Samia, A.C.S., Chen, X. and Burda, C. (2003) Semiconductor quantum dots for photodynamic therapy. *Journal of the American Chemical Society*, **125**, 15736–7.

62 Anas, A., Akita, H., Harashima, H., Itoh, T., Ishikawa, M. and Biju, V. (2008) Photosensitized breakage and damage of DNA by CdSe-ZnS quantum dots. *Journal of Physical Chemistry, B*, **112**, 10005–11.

63 Juzenas, P., Generalov, R., Asta, J., Juzeniene, A. and Moan, J. (2008) Generation of nitrogen oxide and oxygen radicals by quantum dots. *Journal of Biomedical Nanotechnology*, **4**, 450–6.

64 Medintz, I.G. and Mattoussi, H. (2008) Quantum dot-based resonance energy transfer and its growing application in biology. *Physical Chemistry Chemical Physics*, **11**, 17–45.

65 Bakalova, R., Ohba, H., Zhelev, Z., Nagase, T., Jose, R., Ishikawa, M. and Baba, Y. (2004) Quantum dot anti-CD conjugates: are they potential photosensitizers or potentiators of classical photosensitizing agents in photodynamic therapy of cancer? *Nano Letters*, **4**, 1567–73.

66 Shin, L., Hernandez, B. and Selke, M. (2006) Singlet oxygen generation from water-soluble quantum dot–organic dye nanocomposites. *Journal of the American Chemical Society*, **128**, 6278–9.

67 Idowu, M., Chen, J.-Y. and Nyoko, T. (2008) Photoinduced energy transfer between water-soluble CdTe quantum dots and aluminium tetrasulfonated phthalocyanine. *New Journal of Chemistry*, **32**, 290–6.

68 Tsay, J.M., Trzoss, M., Shi, L., Kong, X., Selke, M., Jung, M.E. and Weiss, S. (2007) Singlet oxygen production by peptide-coated quantum dot-photosensitizer conjugates. *Journal of the American Chemical Society*, **129**, 6865–71.

69 Samia, A.C.S., Dayal, S. and Burda, C. (2006) Quantum dot-based energy transfer: perspectives and potentials for applications in photodynamic therapy. *Photochemistry and Photobiology*, **82**, 617–25.

70 Derfus, A.M., Chan, W.C.W. and Bhatia, S.N. (2004) Probing the cytotoxicity of semiconductor quantum dots. *Nano Letters*, **4**, 11–18.

71 Kirchner, C., Liedl, T., Kudera, S., Pellegrino, T., Javier, A.M., Gaub, H.E., Stölzle, S., Fertig, N. and Parak, W.J. (2005) Cytotoxicity of colloidal CdSe and CdSe/ZnS nanoparticles. *Nano Letters*, **5**, 331–8.

72 Fischer, H.C., Liu, L., Pang, K.S. and Chan, W.C.W. (2006) Pharmacokinetics of nanoscale quantum dots: *in vivo* distribution, sequestration, and clearance in the rat. *Advanced Functional Materials*, **16**, 1299–305.

73 Choi, S.H., Liu, W., Misra, P., Tanaka, E., Zimmer, J.P., Ipe, B.I., Bawendi, M.G. and Frangioni, J.V. (2007) Renal clearance of quantum dots. *Nature Biotechnology*, **25**, 1165–70.

74 Biju, V., Mundayoor, S., Omkumar, R.V., Anas, A. and Ishikawa, M. (2009) Bioconjugated quantum dots for cancer research present status, prospects and remaining issues. *Biotechnology Advances*, **28**, 0000.

Part III
Synthesis, Characterization, and Toxicology

9
Type-I and Type-II Core–Shell Quantum Dots: Synthesis and Characterization

Dirk Dorfs, Stephen Hickey and Alexander Eychmüller

9.1
Introduction

The physical and optical properties of semiconductor quantum dots (QDs) are interesting for a variety of applications. However, due to their very large surface/volume ratio, semiconductor QDs can also lose some of their desirable properties due to, for example, corrupted surfaces. Thus, the protection and passivation of the QD surface is of major importance for almost all of their possible applications. In most cases, this passivation is performed through the coating of a given semiconductor material with another material having a larger bulk bandgap than the core substance and a relative position of the bands, such that both charge carriers, the electrons, and the holes are confined to the core. This energetic situation is called type-I. In contrast, in type-II structures, the electrons and holes are separated so that one is located in the core material the other in the shell material. Both systems have been synthesized via various routes in recent years, and will be reviewed in the first part of the chapter. Subsequently, data will be provided relating to spherical semiconductor nanoheterostructures exhibiting more than one shell. The section describing ultraviolet-visible (UV-Vis) active nanomaterials concludes with a review on nonspherical core–shell systems, namely elongated rod-like structures, and a brief outline on the characterization of the aforementioned different nanoheterostructures. The timely consideration of steady developments presently being made in the area of infrared (IR) active materials is also addressed. Hence, a brief synopsis of the progress that has been achieved in the field to date is followed by a discussion of the synthesis and characterization of a number of near-infrared (NIR) nanomaterials is presented. Finally, a more in-depth discussion of the resultant type-I and type-II core–shell NIR materials is undertaken.

9.2
Core–Shell UV-Vis Nanoparticulate Materials

9.2.1
Type-I Core–Shell Structures

One of the main differences of semiconductor QDs in comparison with the corresponding bulk materials is the strongly increased band gap fluorescence observed in QDs. It could be shown that the quantum yield of a QD depends mainly on the surface properties of the nanocrystal. The quantum yield is high if the surface is well passivated, which means that all surface valences are saturated, for example, by organic ligands which are present in most QD systems as a consequence of the synthetic process.

However, these ligands may possibly be weakly bound, resulting in a dynamic equilibrium of adsorbing and desorbing ligands, and they can also be sensitive to all types of post-synthetic treatments of the QDs. Thus, a more stable method to saturate – and thus passivate – the nanoparticle surface was required in order to obtain QDs with high quantum yields and which also retained their fluorescence properties under various post-synthetic conditions.

The so-called core–shell structures were synthesized in order to passivate these surface traps, and thus increase the quantum efficiencies of the QD. Pioneering studies were conducted by Spanhel et al., who reported a significant enhancement of the band gap fluorescence of CdS QDs synthesized in water when treated with OH^- in the presence of excessive Cd^{2+} [1]. This increase in quantum yield was attributed to a passivating shell of $Cd(OH)_2$ which formed around the QD and effectively hindered the free charge carriers from reaching the particle surface (and hence the trap states located at the particle surface).

Nowadays, most core–shell QDs are synthesized in high-boiling organic solvents based on the CdE (E = S, Se, Te) synthesis of Murray et al. [2]. This synthesis yields samples with a much lower degree of polydispersity than the older water-based synthesis. The first example of a successful shell growth of another semiconductor material (ZnS in this case) onto CdSe QDs synthesized in organic solvents was reports by Guyot-Sionnest et al. in 1996 [3], and another very detailed study on the same system followed soon after by Bawendi et al. [4], while a systematic study of the growth of CdS onto CdSe QDs was reported by Peng et al. [5].

Both systems – CdSe/ZnS and CdSe/CdS core–shell nanocrystals – were subsequently studied in great detail, and are today standard systems when QDs with a high fluorescence quantum yield are required. The choice of CdSe as one of the most popular core materials for core–shell structures originates from its bulk band gap of 1.7 eV (corresponding to a photon wavelength of 728 nm), which allows the covering of almost the entire visible range of the spectrum with this material using QDs in a size regime of 1.7 nm to 6 nm in diameter. The choice of ZnS and CdS as popular shell materials for CdSe QDs originates from two different physical properties. While ZnS exhibits a wide band gap of 3.7 eV, which allows the free charge carriers to be effectively confined to the core of the core–shell particles even

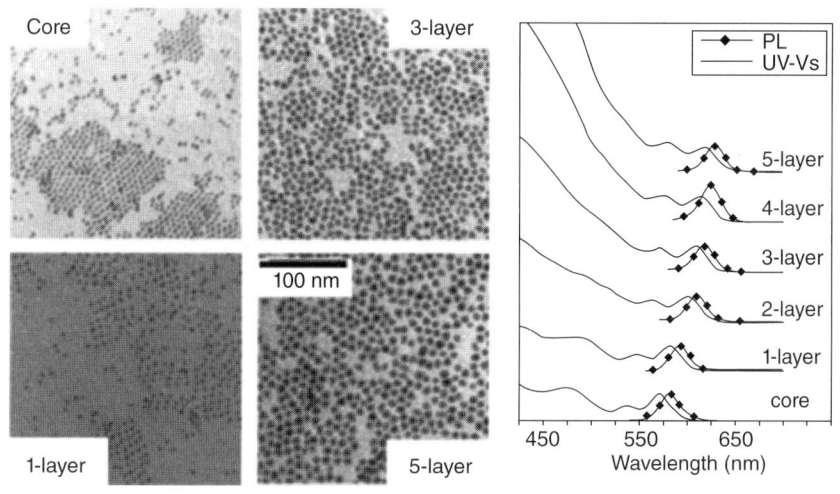

Figure 9.1 Left: Transmission electron microscopy images of CdSe nanocrystals and CdSe/CdS core–shell particles with one, three, and five ionic monolayers of a CdS shell; Right: The corresponding absorption and emission spectra [6].

with quite thin shells, it has a significant lattice mismatch with CdSe, which makes an epitaxial growth more challenging than for materials with a smaller lattice mismatch. CdS as a shell material for CdSe nanocrystals has a comparatively low lattice mismatch, allowing the epitaxial growth of thicker shells; however, the band gap of this material (2.5 eV) is smaller than that of ZnS, which causes the electronic passivation to be less effective at a given shell thickness.

Absorption and emission spectra of CdSe/CdS core–shell structures are shown in Figure 9.1. The main findings from these spectra are: (i) the first absorption maximum is shifted towards a lower energy with an increase in CdS shell thickness; and (ii) the fluorescence quantum yield is greatly enhanced in the core–shell QDs compared to the pure CdSe QDs. These findings can be explained as follows: the shift towards lower energies originates from a relaxation of the wavefunctions of both charge carriers into the CdS shell (however, the contribution of the electron should be much larger than that of the hole, as its effective mass is smaller and the band offsets between the conduction bands of both materials is smaller than the offset in the valance bands, both of which allow the electron to tunnel "more easily" into the shell than the hole), while the increase in quantum efficiencies originates from the passivation of surface trap states of the original QD and the much lower amplitude of the probability density function at the surface of the core–shell particle. For ZnS coating, which is the other very common coating material for CdSe, the observations are similar but also show two main differences: (i) the shift of the first absorption maximum towards lower energies is less pronounced; and (ii) the fluorescence quantum yield passes through a maximum at a thickness of the ZnS coating of 1.8 monolayers [4], while thicker coatings will cause the quantum yield to drop again. The less-pronounced shift of the first electronic transition can be explained with the very high band gap of ZnS compared to the band gap of CdS – hence, the relaxation of the charge carrier wavefunctions into the ZnS shell is much weaker than in the CdS case. The maximum in the fluorescence quantum yield can be explained by the large lattice mismatch between CdSe and ZnS, which causes considerable lattice strain and thus can cause crystallographic defects upon thicker ZnS coating.

The experimental procedures available for the colloidal synthesis of core–shell structures have improved considerably during the past decade. The first attempts dealt with aqueous media, and yielded materials with comparably high polydispersities. Subsequently, the pioneering studies of Guyot-Sionnest demonstrated the possibility of synthesizing core–shell structures in high-boiling organic solvents, and these showed a considerably lower polydispersity than their aqueous counterparts. The synthesis in high-boiling solvents was then further developed, for example, by Peng et al. [6], and this gave rise to an easy control of the shell thickness. Today, a variety of type-I core–shell structures can be synthesized in organic solvents, including CdSe/CdS [5], CdSe/ZnS [3, 4], CdS/ZnS [7], and InP/ZnS [8].

Due to their high fluorescence quantum yields and high photo- and chemical stabilities, type-I core–shell structures represent extremely interesting alternatives to common organic dye molecules with regards to labeling processes (e.g., "biolabeling," which is the *in vitro* or *in vivo* labeling of a bio molecule of interest which is then traced by detecting the photoluminescence (PL) of the QD [9]). Compared to the organic dyes in particular, the strongly enhanced photostability is important as well as the much narrower line width of the emission lines, which enables multicolor labeling.

9.2.2
Type-II Core–Shell Structures

Whilst in type-I semiconductor heterostructures, the band gap alignment leads to a confinement of the charge carriers in one compartment of the heterostructure, the opposite is true for type-II nanoheterostructures.

The electron and hole are localized in two different compartments of the heterostructure in a type-II structure, and this usually leads to a significantly slowed down recombination rate of the charge carriers and thus to an extended luminescence lifetime.

While type-I nanoheterostructures are usually superior compared to homogeneous QDs in terms of their fluorescence quantum yield and photostability, type-II heterostructures have other major differences compared to type-I structures and homogeneous QDs. First, the staggered band alignment gives rise to a spatially indirect transition which occurs at lower energies than both of the band gaps of the two materials used (when not taking into account the additional quantization energy). Hence, type-II core–shell QDs can be emitters at wavelengths that cannot be achieved with any of the two materials alone. Furthermore, the luminescence lifetime of type-II heterostructures is strongly increased.

Another, recently much-discussed, application of type-II core–shell QDs is the lasing of nanoparticles. Due to the type-II band alignment and lower overlap of the charge carrier wave functions, the exciton–exciton interaction becomes repulsive in type-II structures whilst being attractive in type-I systems. As demonstrated by Klimov et al., this behavior may prove advantageous when using QDs for amplified stimulated emission (ASE), which might open the way towards QD-based lasers.

Among the first colloidal type-II systems to be studied were CdTe/CdSe core–shell particles [10]. In these structures, the band offsets of the two materials cause the electron to be located in the CdSe shell, while the hole is confined to the core of the nanocrystals. The possibility of a spatially indirect transition from the CdTe valence band to the CdSe conduction band causes the absorption and emission wavelength of these structures to be strongly shifted towards lower energies compared to the pure CdTe nanoparticles, and even makes accessible wavelengths which would not be accessible with one of the two materials alone. Thus, the emission wavelength in these core–shell structures can be tuned up to 1000 nm.

Another system that was intensively studied, especially in the context of ASE from nanoparticles, was the CdS/ZnSe core–shell nanoparticle [11], with the group of Klimov et al. successfully demonstrating their use for ASE. Figure 9.2 shows the emission spectra of CdS/ZnSe core–shell nanocrystals with two different CdS core sizes (1.6 and 2.4 nm) and different ZnSe shell thicknesses [12]. The first finding here is that the emission wavelengths are smaller than the bulk band gaps of both materials (CdS = 2.5 eV; ZnSe = 2.7 eV), and thus can be assigned to a

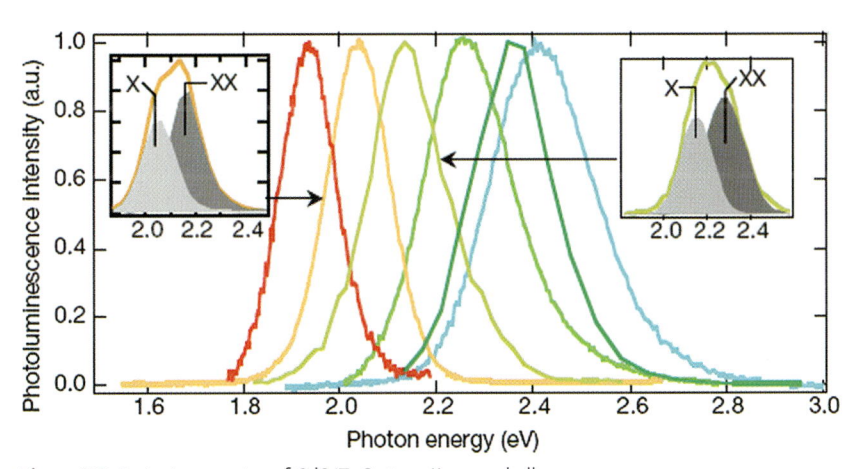

Figure 9.2 Emission spectra of CdS/ZnSe type-II core–shell nanocrystals (different core diameters and different shell thicknesses). The insets show emission spectra measured immediately after excitation at high pump intensities, which can be deconvoluted into a long-living part of the single excited particles and a very fast decaying part from double-excited particles [12].

spatially indirect recombination of a hole in the ZnSe valence band with an electron in the CdS conduction band (in the bulk, this transition would correspond to an energy of 1.8 eV).

The insets in Figure 9.2 show early time emission spectra measured immediately after the excitation at high pump energies. The main finding here is that the early time emission spectra can be deconvoluted into two signals, which can be assigned to a very fast decaying component originating from doubly excited particles (labeled as XX) and to a long-living component (labeled as X). The shift of the XX transition compared to the X transition has a magnitude of approximately 100 meV, and the XX transition is shifted towards higher energies. Both findings are related to the type-II structure of the nanocrystals. In type-I nanocrystals where the electron(s) and hole(s) are located in the same area of the nanocrystal, the exciton–exciton interactions are usually weak and attractive; thus, the XX transition would be shifted by small value towards lower energies. However, a different behavior is observed for type-II systems as, due to the spatial separation of electron(s) and hole(s), the exciton–exciton interaction becomes strong and repulsive, causing a much larger shift of the XX transition towards higher energies.

This behavior has interestingly an advantageous consequence when examining the ASE behavior of the nanocrystals. As can be seen in Figure 9.3 (left), ASE in type-I systems can only be observed at the wavelength of the XX transition, and hence from doubly excited particles. In the case of the type-II systems, however, ASE can be observed also from singly excited particles, and thus at the center of the ensemble emission wavelength. This behavior causes the pump energy threshold for ASE to be much lower in case of type-II systems than with type-I systems (see Figure 9.3, right). This behavior is responsible for type-II nanoheterostructures being very promising candidates for QD (rod)-based lasers.

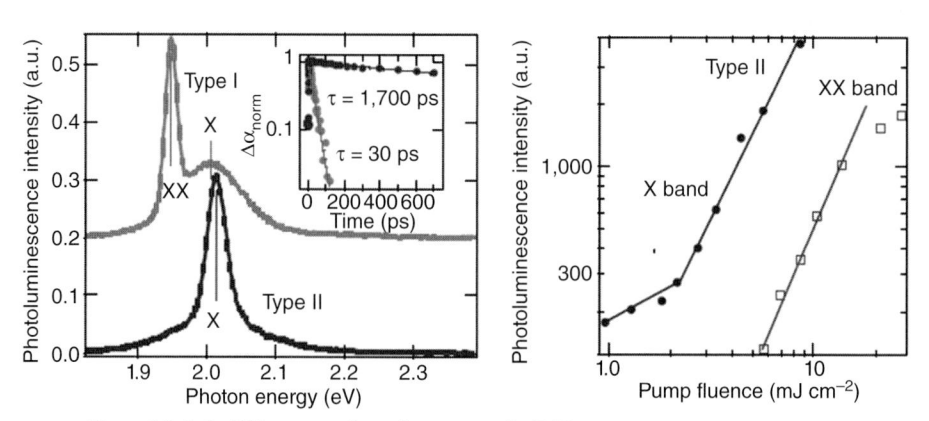

Figure 9.3 Left: ASE spectra of type-I nanocrystals (CdSe nanocrystals) and of type-II nanocrystals (CdS/ZnSe core–shell nanocrystals); Right: ASE intensity as a function of pump energy for type-I and type-II systems [12].

9.2.3
Multiple Shell Structures

A more complicated structure, namely the QD quantum well system, which consists of CdS particles with an embedded layer of HgS, was first synthesized in 1993 [13]. These particles were examined with different characterization techniques, including static and time-resolved photoluminescence, transient photobleaching, and high-resolution electron microscopy. The results obtained were compared with theoretical data, and the findings on this model system are summarized.

Likewise, research progress into different multishell nanocrystals will also be surveyed, describing three types of nanostructure: (i) Ternary core–shell shell systems with an intermediate layer as a "lattice adapter"; (ii) double QD quantum well (double-QDQW) systems; and (iii) an "inversed QDQW" system. Sorting multilayered nanocrystals into these three categories is justified by the different potential steppings of the semiconductor materials involved. Figure 9.4 shows the principal potential steppings of the valence and conduction bands in these three types of multilayered system.

The potential stepping on the left in Figure 9.4 causes both charge carriers (electron and hole) to be confined in the core of the nanocrystal (as above: type-I situation). In the case of QDQW systems (middle), the electron and hole are confined in the potential well, which is embedded in the QD. In the case of the "inversed" QDQW systems (right), the charge carriers are located in the core of the nanocrystal and in the outer shell. Depending on the different potential steppings, each of these systems will show unique properties (see below).

Based on polyphosphate-stabilized water-soluble CdS nanocrystals, the QDQW system consisting of CdS nanocrystals with an embedded layer of HgS (a QW within a QD) became a model system for a number of fundamental studies.

The synthetic concept was developed by Mews et al. in 1993 [13, 14]. To a solution of CdS QDs, Hg^{2+}-ions were added; this resulted in a substitution reaction on the particle surface, where the outermost layer of Cd-ions was replaced by Hg-ions and the Cd-ions were released into the solution (cf. Figure 9.5). By analyzing the concentration of the free Cd- and Hg-ions in solution it was shown that, for an excessive addition of Hg-ions, no further substitution reaction took place, as no

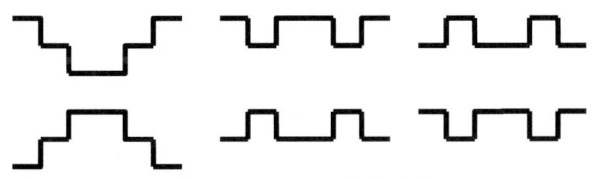

Figure 9.4 Potential stepping in core–shell–shell (CSS) nanocrystals (left), quantum dot quantum well (QDQW) nanocrystals (middle) and "inverse" QDQW nanocrystals (right).

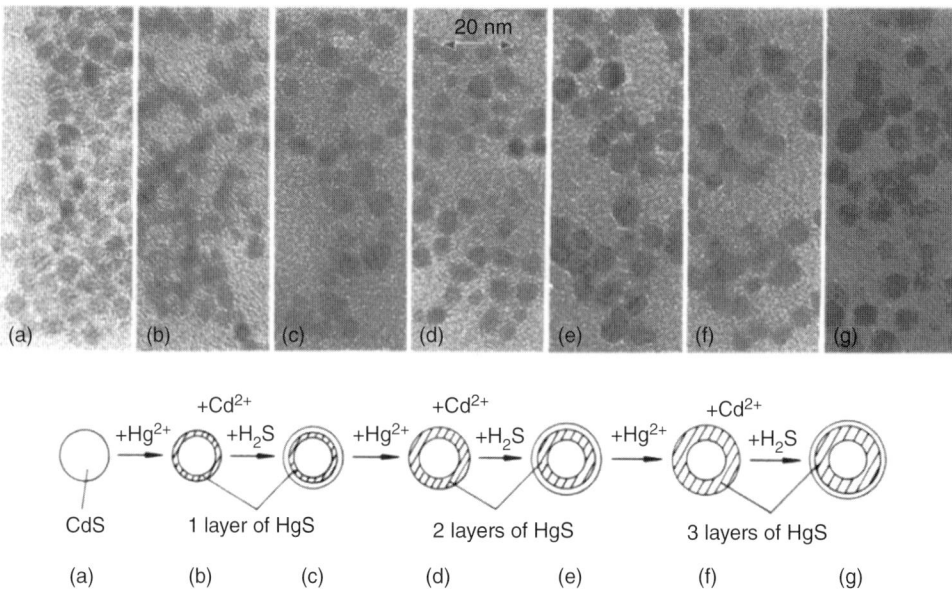

Figure 9.5 Scheme of the synthesis of CdS/HgS/CdS QDQWs and TEM images at various stages of the synthesis.

further increase in Cd-ion concentration occurred while the Hg-ion concentration suddenly began to rise. This indicated that only one ionic monolayer had been substituted. Subsequently, the Cd-ions released into solution could be reprecipitated onto the particles by the addition of H_2S. The emerging colloidal particles consisted of a CdS core surrounded by a monolayer of HgS, and almost one monolayer of CdS as the outermost shell. From this point the preparation was seen to divide into two branches: (i) either increasing the HgS layer thickness; or (ii) increasing the thickness of the outermost CdS layer. The thickening of the HgS layer was achieved simply by repeating the substitution and reprecipitation steps described above. An increase in the CdS layer thickness was performed independently of the thickness of the formerly prepared HgS layer by the further addition of Cd-ions and precipitation of these onto the nanocrystals with H_2S.

The spectral evolution in the course of the further preparation is depicted in Figure 9.6. The major finding was a strong shift of the absorption onset towards lower energies, with an increase in the thickness of the HgS well. This behavior could be explained with the small band gap of HgS (0.5 eV) with respect to CdS (2.5 eV), and a localization of the charge carriers in the HgS wells. One remarkable finding was that the CdS capping of the particles also led to a significant shift of the absorption onset towards lower energies, even though CdS itself cannot absorb in this spectral region.

Theoretical calculations for these structures were performed by the same group in 1994 [15], applying the "particle in the box" model with the effective approximation resulting in calculated values for the first electronic transition and the cor-

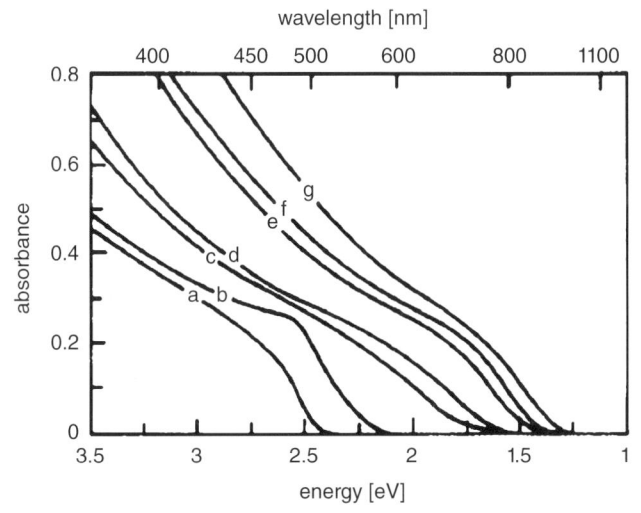

Figure 9.6 Absorption spectra of the colloidal solutions of (a) CdS, (b) a + 8 × 10⁻⁵ M Hg²⁺, (c) b + H₂S, (d) c + 8 × 10⁻⁵ M Hg²⁺, (e) d + H₂S, (f) e + 8 × 10⁻⁵ M Hg²⁺, and (g) f + H₂S (see Figure 9.5).

responding wave functions for the charge carriers. The presented results were in good agreement with the measured optical data. Further theoretical treatments were carried out by Bryant *et al.* after 1995 [16, 17].

Transient photobleaching experiments have been performed on these structures [18], with the most interesting result being that the photobleaching followed (spectrally) the newly evolving 1s-1s electronic transition of the composite particles. Some considerations concerning the charge carrier dynamics in the novel QDQWs were also outlined by these authors.

The subpicosecond photoexcitation of CdS/HgS/CdS QDQW nanoparticles at wavelengths shorter than their interband absorption (390 nm) leads to a photobleach spectrum at longer wavelengths (440–740 nm) [19]. The photobleach spectrum changes, and its maximum red-shifts with increased delay time. These results may be explained by the rapid quenching of the initially formed laser-excited excitons by two types of energy acceptor (traps); the first trap is proposed to be due to CdS molecules at the CdS/HgS interface, while the second trap is that present in the CdS/HgS/CdS well. The results of excitation at longer wavelengths, as well as the formation and decay of the bleach spectrum at different wavelengths, strongly support this description.

The homogeneous absorption and fluorescence spectra of the CdS/HgS QDQW system were also investigated using transient hole burning and fluorescence line-narrowing spectroscopy. Again, these photophysical measurements provided evidence for a charge-carrier localization within the HgS well [20].

High-resolution transmission microscopy (HRTEM) studies conducted with the CdS/HgS/CdS QDQW system have been performed to identify the details of the system's crystallography [21]. The HRTEM images of CdS nanocrystals showed triangular features, with the spacings and angles between the lattice planes showing alignment along the (110) axis of the zinc-blende crystal structure of CdS. The decrease in contrast when moving from the apex to the base implies a decrease in thickness which, in turn, suggests that the nanocrystal is a tetrahedron terminated in (111) surfaces. Corresponding HRTEM simulations agreed with this interpretation of the experimental image, although only a small fraction of the crystallites were aligned along the proper crystallographic axis to allow the shape to be discerned. The basic morphology was preserved in the next step of the synthesis, in which the surface cadmium ions of the CdS crystallites were exchanged with mercury. The final coating of the particles was carried out by adding excess cadmium ions to the solution and growing CdS on top of the HgS layer, via a slow H_2S injection. The close match of the CdS and HgS lattice parameters (a_{HgS} = 5.852 Å, a_{CdS} = 5.818 Å) and the presence of faceted crystallites with only one exposed plane favored this growth mode.

The influence of the crystallography of the interface between the CdS and the HgS well on the optical properties of the CdS/HgS/CdS system was the subject of further characterization which employed optical detected magnetic resonance spectroscopy (ODMR) [22].

In the field of QDQW structures, a variety of new material combinations have also been used to prepare these structures. El-Sayed first reported on a ZnS/CdS QDQW system [23] in which the structures were characterized optically and the results compared with theoretical calculations. A QDQW system which consisted of CdS nanocrystals with an embedded monolayer of CdSe was presented by Battaglia et al. [24]. For this, the QDQW structures were prepared in high-boiling organic solvents using the SILAR technique (as introduced by Peng et al., cf. Ref. [6]), yielding almost monodisperse QDQW systems with high emission quantum efficiencies.

In 2001, the first results were reported by the present authors of an extended CdS/HgS QDQW with two embedded HgS wells [25], and also by the group of El-Sayed [26]. When further characterization of these structures were conducted using X-ray photoelectron spectroscopy [27] as a depth profiling technique, all results obtained were in good agreement with the predicted structure.

These structures are of interest as they allow studies to be made of the distance-dependent interaction between two QWs within one QD. In a later report, the spectroscopic properties of these double-well QDs were also compared with theoretical calculations in the frame of the effective mass approximation [28]. Basically, the synthetic procedure for the double-well QD nanocrystals is the same as for the normal QDQW systems, with the outermost ion monolayer of the CdS nanocrystals being substituted with HgS by the addition of $Hg(ClO_4)_2$. The Cd^{2+}-ions released into solution are then reprecipitated onto the particles by the addition of H_2S. Different double-well QD systems can be obtained by a sequence of growing of CdS shells and substituting them by HgS. For ease of discussion, a nomenclature for the CdS/HgS/CdS/HgS/CdS samples is introduced according to Figure 9.7 as CdS/HgS-ABCD, where each letter reflects the thickness of the correspond-

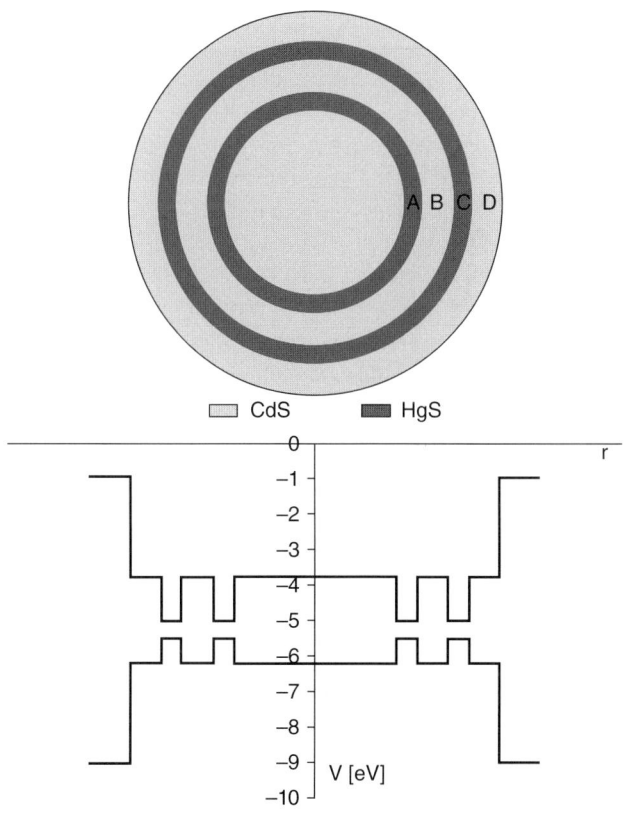

Figure 9.7 Idealized picture of the double-well QD structure with introduction of the CdS/HgS-ABCD nomenclature, and the corresponding radial potential for electron and hole [28].

ing layer in the monolayers. For example, CdS/HgS-1213 relates to a nanocrystal consisting of a CdS core, followed by one monolayer of HgS, two monolayers of CdS, one monolayer of HgS, and again three monolayers of CdS as the outermost capping. CdS/HgS/CdS QDQWs are analogously named as CdS/HgS-AB. Of course, whilst this nomenclature refers to an idealized situation, in reality the particles will have variations in the thickness of all layers and are expected to exhibit inhomogeneities within the different layers.

Figure 9.8 displays the calculated radial probabilities of presence for the electron and the hole for the CdS/HgS-1x13 series of nanocrystals, named according to the nomenclature described above, where x is varied from $x = 0$ to $x = 7$. These nanocrystals all contain the same CdS core – two wells each consisting of one monolayer of HgS, and three outer cladding layers of CdS. The only difference here is that the distance between the two HgS wells can be varied from zero to seven monolayers of CdS. As might be expected, the probability of presence has a maximum within the HgS wells for both the electron and the hole, by this giving rise to a

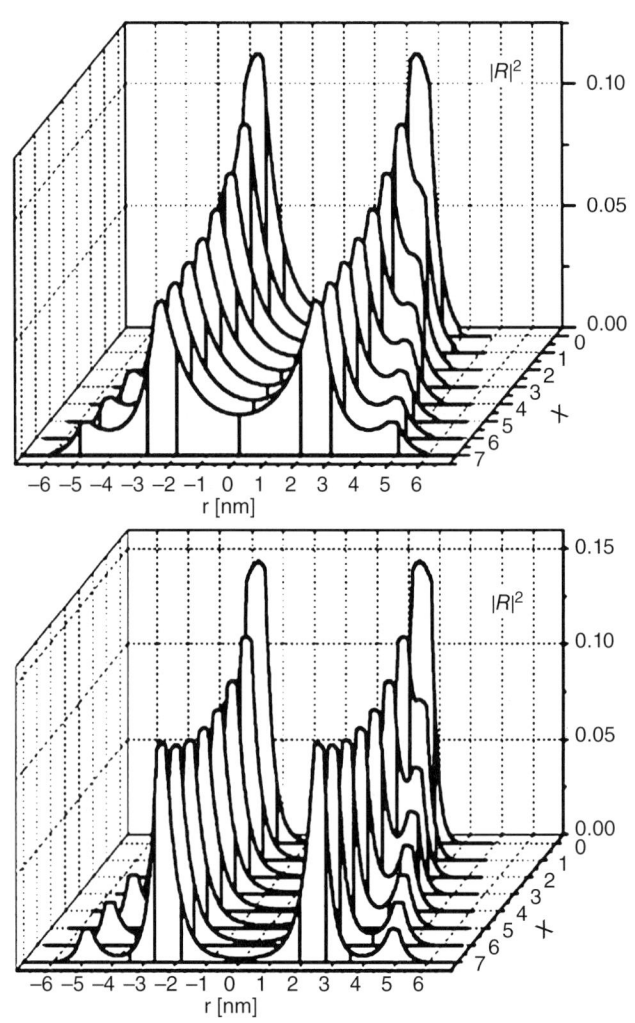

Figure 9.8 Radial probability of presence in the CdS/HgS-1x13 systems (x = 0–7) for the electron (top) and the hole (bottom), where r is the radial distance from the particle center [28].

spatial overlap of the two wavefunctions within the HgS wells. In both materials the effective mass of the hole is larger than that of the electron, and therefore the localization is much stronger for the hole than for the electron in the same systems. Increasing the distance between the two HgS monolayers results in a more pronounced separation of the two maxima of the probability of presence; however, this separation becomes increasingly smeared out with a decrease in the distance of the two HgS layers. A similar behavior can be observed for the electron, although due to the smaller effective mass the maxima are not as well separated,

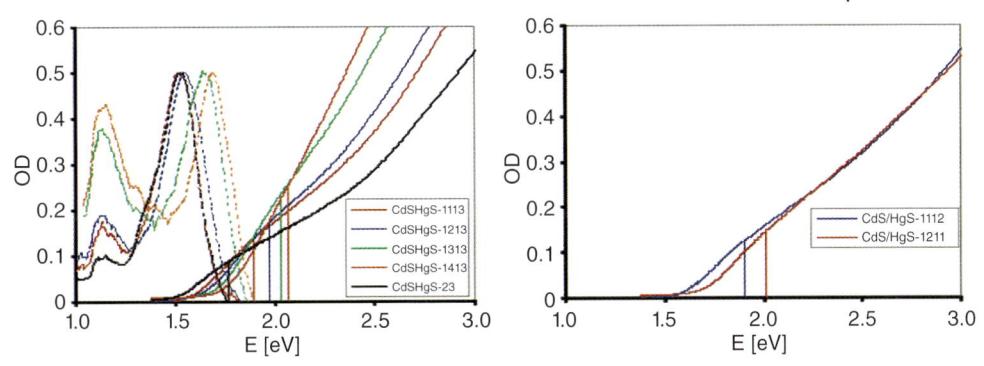

Figure 9.9 UV/Visible absorption and emission spectra (lines and dotted lines, respectively) of the CdS/HgS-1x13 systems, together with the calculated first electronic transition (left) and the comparison of two systems with the same amount of CdS and HgS but different layer structures (right). Lines of the same color refer to the same sample in the left figure [28].

even for the system with the largest distance between the two HgS layers (i.e., CdS/HgS-1713).

Figure 9.9 (left) shows the UV-Vis absorption spectra of the CdS/HgS-1x13 series of nanocrystals, with x ranging from 0 to 4. Here, the vertical bars represent the calculated first electronic transition (E_{gap}) for the corresponding idealized systems, while the point of maximum curvature is considered to represent the first electronic transition of the sample. For the samples CdS/HgS-23, CdS/HgS-1113, and CdS/HgS-1213, this point matches quite well with the calculated transition energies. In the case of the samples CdS/ HgS-1313 and CdS/HgS-1413 such a point is difficult to discern, but the absorption onset of those samples is still shifted towards higher energies, and is in good agreement with the calculated values (cf. the vertical bars). This may be interpreted by a decreasing interaction between the two HgS layers, but with an increasing distance between them. The high energy absorption (above 2.5 eV) depends mainly on the total amount of absorbing material. The absorption behavior between 1.8 and 2.2 eV is not easily explained because this simple theoretical treatment is not suitable for explaining the oscillator strengths in different regions of the spectra. The dotted lines in Figure 9.9 (left) are the corresponding normalized emission spectra of the particles in these samples. Here, each sample shows an emission close to the band gap absorption onset; moreover, each sample shows a second emission at 1.15 eV which is likely to be "trap emission." These traps are most probably related to stacking faults at the interfaces of the layers, as shown previously using ODMR measurements on the CdS/HgS/CdS systems [22]. In order to demonstrate that the absorption onsets of those systems are not only affected by the molar ratio of CdS to HgS, but do indeed depend on the layer structure, a comparison of two structures with the same molar ratio but a different layer structure (namely, CdS/HgS-1112 and CdS/HgS-1211) is shown in Figure 9.9. The structure in which the HgS wells are separated by two monolayers of CdS, and with a capping layer of one monolayer, displays an absorption onset at a higher energy than that with a separation of one layer of CdS and two capping layers of CdS. In both cases, the agreement with the calculated transition energy is satisfactory.

The first report of a ternary core–shell/shell (CSS) structure was made by Reiss et al. in 2003, and included the details of a synthetic procedure that yielded CdSe/ZnSe/ZnS CSS particles [29].

In 2004, Talapin synthesized similar particles [30], whereby CdSe nanocrystals capped with TOP (n-trioctylphospine)/TOPO (n-trioctylphosphinoxide) were coated with a shell of either CdS or ZnSe. In both cases, the outermost shell was composed of ZnS, the main purpose of which was to avoid charge carrier migration towards the surface of the particles (ZnS is a good candidate due to its large bandgap of 3.7 eV). The large lattice mismatch between CdSe and ZnS was problematic, but this was overcome by the use of an intermediate ZnSe or CdS shell as a "lattice adapter."

Early reports on simple CdSe/ZnS core–shell particles showed that the quantum yield of the nanocrystals as a function of the ZnS layer thickness passed through a maximum at a ZnS layer thickness of approximately two monolayers [4]. The reason given for this behavior was the strain induced in the system by the lattice mismatch of ZnS and CdSe.

Figure 9.10 displays the basic structure of the CSS structures (panel a), together with the potential stepping of the valence and conduction band edges (panel b), and the band gaps of the materials as a function of the lattice spacing (panels c and d). It can be seen, that not only the band gaps but also the lattice spacing of CdS and ZnSe (the materials used as buffer layers) were between those of CdSe and ZnS and, therefore, both materials were well suited to act as "lattice adapters."

Figure 9.11 shows the development of the absorption and emission spectra during the coating procedure, together with the development of the quantum efficiency as a function of the shell thickness for different shell compositions. Remarkably, the shift to lower energies of the first absorption signal was stronger for the ZnSe coating than for the ZnS coating, which was interpreted as a stronger "leakage" of the wavefunction of the exciton into the ZnSe shell compared to the ZnS shell. According to these authors, another significant finding was that the quantum efficiencies of the CSS structures did not fall as strongly for an increased ZnS layer thickness as they did for "normal" CdSe/ZnS core–shell structures. This was attributed to a higher crystallinity as a consequence of stress release of the CSS structures compared to the CdSe/ZnS structure.

Another identified benefit of the CSS structures was their greatly enhanced photostability. For example, when single-particle luminescence images of CdSe/ZnS core–shell structures and CdSe/ZnSe/ZnS CSS structures were recorded, after illumination with a laser beam under ambient conditions, almost all of the CSS structures were still luminescent after 10 min, whereas most of the core–shell particles had already been extinguished. This situation was assigned to a higher stability against photo-oxidation of the CSS structures in comparison with the

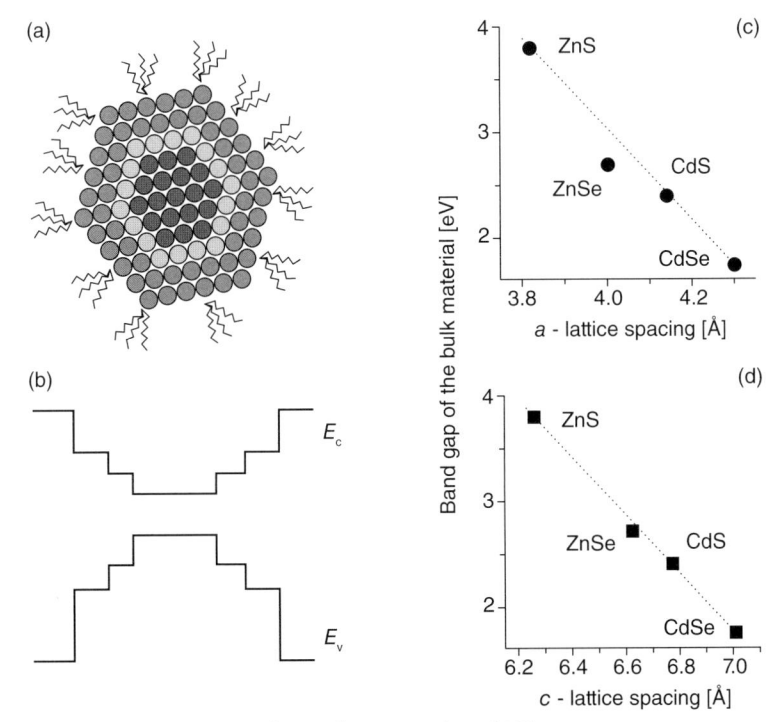

Figure 9.10 CSS nanocrystal. (a) Schematic outline; (b) The schematic energy level diagram; (c, d) Relationship between band gap energy and lattice parameter of bulk wurzite phase CdSe, ZnSe, CdS, and ZnS [30].

core–shell particles. Recently, Jun et al. presented a simplified one-step procedure for the synthesis for CdSe/CdS/ZnS CSS nanocrystals [31].

In 2005, another example of CSS structures with a lattice-adapting layer was reported by Xie et al. [32]; in this case, the new structure was a CdSe/CdS/ZnS CSS structure (as presented above). However, the authors reported the possibility of including an alloyed layer of $Zn_{0.5}Cd_{0.5}S$ into the structure, resulting in CdSe/CdS/$Zn_{0.5}Cd_{0.5}$S/ZnS multishell particles; they also reported on very high quantum efficiencies for the multishell particles. In addition, their results with regards to stability versus photo-oxidation showed the same trend as those noted by others [30].

Transmission electron microscopy (TEM) images presented elsewhere [32] (Figure 9.12) demonstrated the growth of the particles, and showed clearly that the particles had retained a very narrow size distribution throughout the coating procedure.

Recently, the group of Banin reported an example of a CSS structure with InAs, and thus a Group III–V core material [33]; these InAs cores were covered with an intermediate layer of CdSe and an outermost layer of ZnSe. As noted elsewhere for pure Group II–VI CSS particles, very high emission quantum efficiencies (up

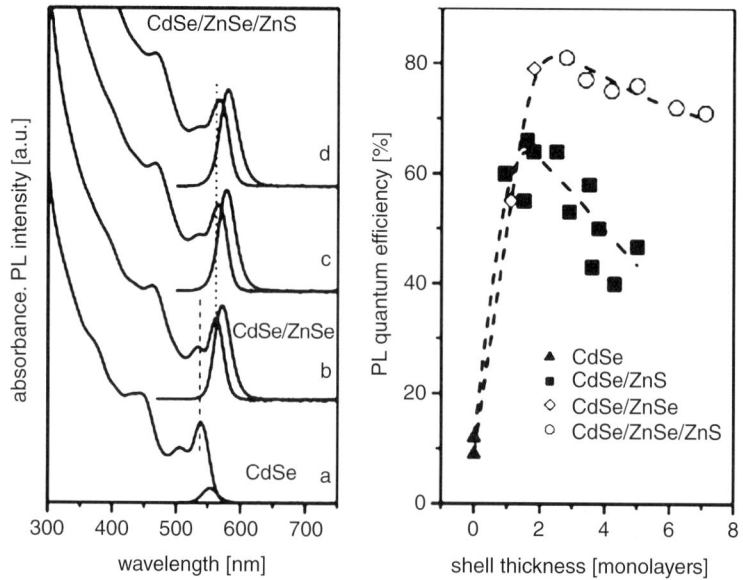

Figure 9.11 Left: Absorption and PL spectra of (a) CdSe cores, (b) CdSe/ZnSe core–shell nanocrystals (thickness of the ZnSe shell ~2 monolayers); (c, d) CdSe/ZnSe/ZnS nanocrystals with the thicknesses of the ZnS shell ~2 monolayers (c) and ~4 monolayers (d). Right: Room-temperature PL quantum yields of CdSe, CdSe/ZnSe, and CdSe/ZnSe/ZnS nanocrystals dissolved in chloroform. For comparison, the dependence of the PL quantum yield on the shell thickness for various samples of CdSe/ZnS nanocrystals is displayed [30].

to 70%) were reported. However, due to the use of InAs as the core material, in this case the emission wavelength was tunable throughout the NIR region of the spectrum (800–1600 nm). This report showed clearly that the properties of the CSS nanocrystals can neither be obtained with simple InAs/CdSe, nor with InAs/ZnSe core–shell nanocrystals. Figure 9.13 shows the evolution of the quantum yield as a function of the coating for InAs/CdSe core–shell nanocrystals, InAs/ZnSe nanocrystals, and the CSS system (where the first layer added is from CdSe and the outer shells are composed of ZnSe). In the case of the InAs/CdSe system, a clear maximum of the quantum yield was observed for one monolayer of CdSe, whilst for the InAs/ZnSe particles the quantum efficiencies increased continuously but reached only about 15%. In the CSS system, the quantum efficiencies increased continuously to about 50%.

In summary, the CSS structures with a lattice-adapting intermediate layer have been proven to be superior systems with regards to their luminescence quantum yields and stability versus photo-oxidation.

Another interesting structure developed by Peng and coworkers was the CdSe/ZnS/CdSe CSS system [34]. As in this case the intermediate layer is composed of the high-band-gap material ZnS, these are not QDQW systems in the truest sense but, nevertheless, were of superior fundamental interest (cf. also Figure 9.12). This

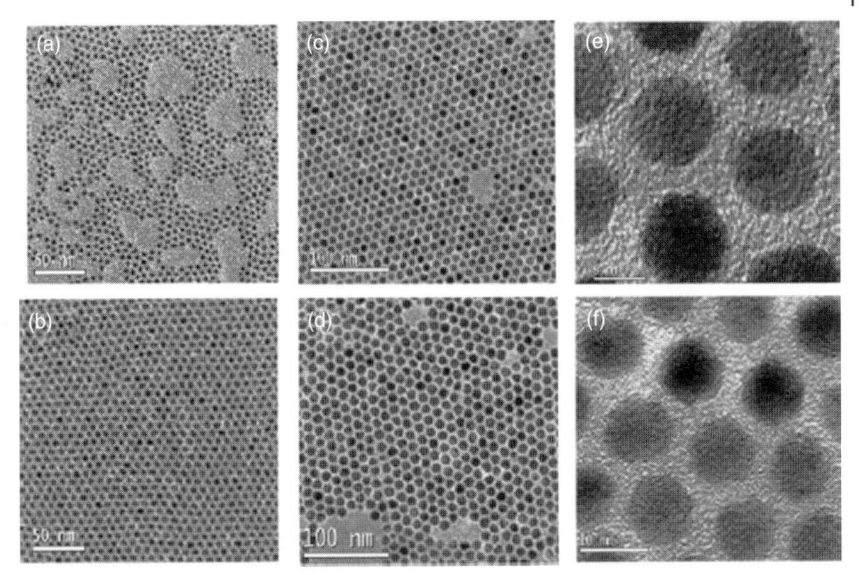

9.2 Core–Shell UV-Vis Nanoparticulate Materials

Figure 9.12 Transmission electron microscopy (TEM) images of the plain CdSe cores and core–shell nanocrystals obtained under typical reaction conditions. (a) TEM images of CdSe cores (before injection of Cd^{2+} solution); (b) (a) plus 2 monolayers of CdS; (d and e) (b) plus 3.5 monolayers of $Zn_{0.5}Cd_{0.5}S$; (d and f) (c) plus 2 monolayers of ZnS [32].

report was the first to describe two distinct emission signals from one nanocrystal, where none of the signals was trap-related. The embedded ZnS layer seemed to be thick enough to completely decouple the outer shell of CdSe from the inner CdSe core of the particle. Figure 9.14 shows the evolution of the emission signal in this type of inversed QDQW. For particles with a ZnS layer of one monolayer thickness, an increase in the outermost CdSe QW led to a shift of the emission signal to lower energies, but not yet to the occurrence of a second emission signal. In contrast, at a ZnS layer thickness of three monolayers, no shift of the emission signal could be observed; rather, a second emission signal occurred at higher energies which was shifted to lower energies with increasing thickness of the outer CdSe well. This observation was explained as the phenomenon of "coupled and decoupled quantum systems in one semiconductor nanocrystal."

The shift towards lower energies for thin ZnS buffer layers can be interpreted as an interaction between the core and the outer shell; this is comparable to the interaction between the two QWs described above and referred to as a "double-well QD structure." The two distinct emission signals observed with thick intermediate ZnS layers are interpreted as a complete electronic decoupling of the outer CdSe QW and the CdSe core. These structures were subjected to further spectroscopic investigations [35] and also investigated as possible sources of white light emission [36]. However, as the presented measurements were ensemble in nature, it could not be excluded without doubt that in some, or all, cases the single nanocrystals

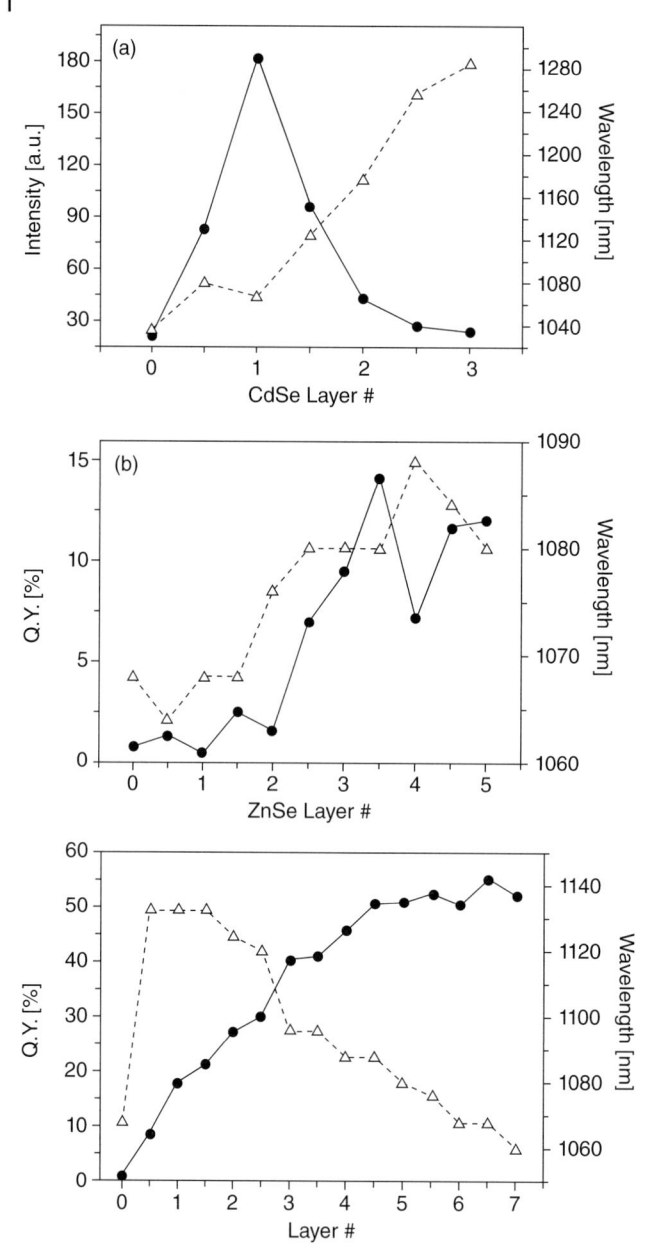

Figure 9.13 Quantum yields (black dots with full lines) and emission wavelengths (white triangles with dotted lines) as a function of the shell thickness in monolayers for InAs/CdSe core–shell nanocrystals (top), InAs/ZnSe core–shell nanocrystals (middle) and InAs/CdSe/ZnSe CSS nanocrystals (bottom, the first layer is CdSe, all other layers are ZnSe) [33].

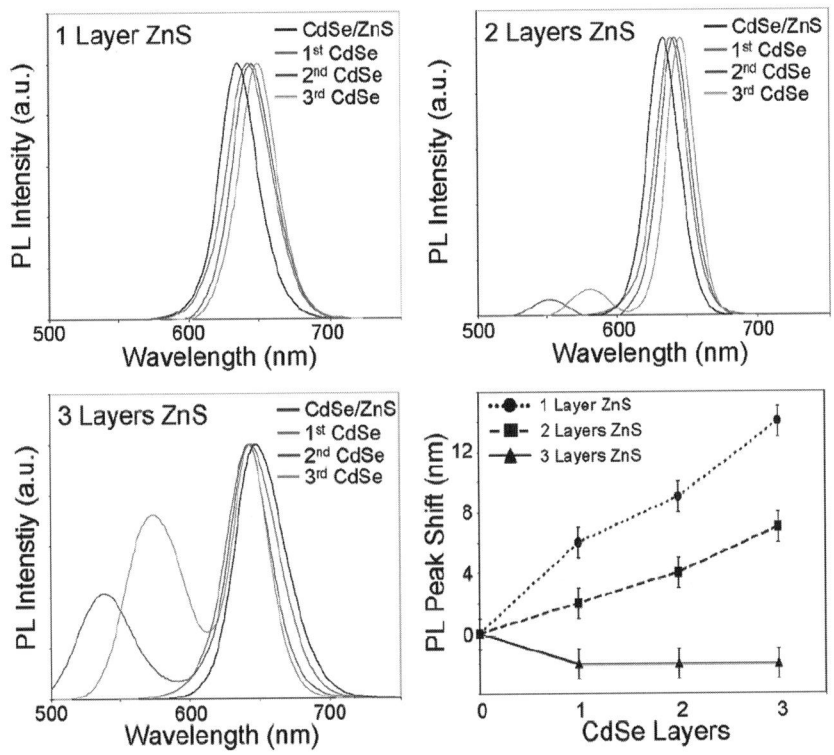

Figure 9.14 Emission spectra of CdSe/ZnS/CdSe nanocrystals with a ZnS barrier layer having the thickness of one (top left), two (top right) and three (bottom left) monolayers. All spectra are shown for the simple CdSe/ZnS core–shell structure and for an outer shell of one, two, and three monolayers of CdSe. The variations of the positions of the emission maxima is summarized in the lower right figure [34].

had only an emitting core or an emitting shell, and not both. Only single-particle spectroscopy studies performed on these structures could ultimately resolve this question.

9.2.4
Nonspherical Nanoheterostructures

Following the development of synthetic routes towards an efficient shape control of homogeneous quantum rods [37], the development of nonspherical nanocrystals composed of more than one material was a logical step. The simplest approach here would be to overcoat a nanorod with an inorganic shell; this would result in a type-I band alignment and thus increased luminescence quantum yields (whilst retaining the polarized emission of the anisotropic quantum rod). An example would be CdSe quantum rods covered with ZnS [33], where the increase in fluorescence quantum yield could be explained in similar fashion to the case of spherical core–shell structures (reducing the probability of the presence of any charge carriers at the particle surface).

Other nonspherical heterostructures include compartment-like rods (e.g., CdTe/CdSe and others [39]), and another class of nonspherical heterostructures which were started from QDs and the seeds used for the growth of a quantum rod. Following the pioneering studies of Talapin *et al.* [40], who synthesized elongated CdSe/CdS core–shell structures, this approach has recently been further redeveloped by Talapin [41] and Manna [38], and has resulted in highly monodisperse, size- and aspect ratio-tunable CdSe seeded CdS nanorods (see Figure 9.15 for emission and absorption spectra and TEM images). One interesting aspect about this type of structure is that, especially for high-aspect-ratio rods, the absorption spectra of these heterostructures are dominated by the absorption of the CdS (since it is in 10- to 40-fold excess compared to CdSe), while the emission arises only from the CdSe core (as the interband relaxation is much faster than the radiative recombination). Taking into account the fact that the photoluminescence excitation spectra also almost match the absorption spectra, it can be concluded that energy transfer from the rod towards the core is very efficient, and that the rod can be regarded as an antenna which absorbs light, while the emission of the light occurs always from the seed. This also means that light emitted by these seeded rods is much less reabsorbed than, for example, the case of pure CdSe rods. Consequently, such rods may attract interest for those applications where comparatively thick layers of nanocrystals are used, when the reabsorption should be limited as much as possible. The large-scale assembly of these seeded rods was also shown to be possible; this might be of special interest due to the capability of these structures to emit polarized light [38]. The seeded growth approach was recently applied to the development of other nonspherical heterostructures, such as ZnSe seeded CdS rods [42] (hence an anisotropic type-II structure with a charge carrier separation along the rod axes). These type-II nanorods show not only extended luminescence lifetimes but also a clearly spatially indirect transition with transition energies of around 2.1 eV (thus much smaller than both bulk band gaps of CdS and ZnSe, which were 2.5 eV and 2.7 eV, respectively). A variety of seeded tetrapods was also synthesized using the seeded growth approach [43].

9.3
Characterization of Nanoheterostructures

Although, today, nanoheterostructures are widely accessible (e.g., CdSe/ZnS core–shell particles are available commercially from several sources), the characterization of these structures remains a major challenge for chemists and physicists. Whilst homogenous QDs can usually be sufficiently characterized with standard methods such as TEM, X-ray diffraction (XRD) and UV-Vis/NIR absorption and emission spectroscopy, further investigations are normally required for the (complete) characterization of heterogeneous quantum structures. Especially in the

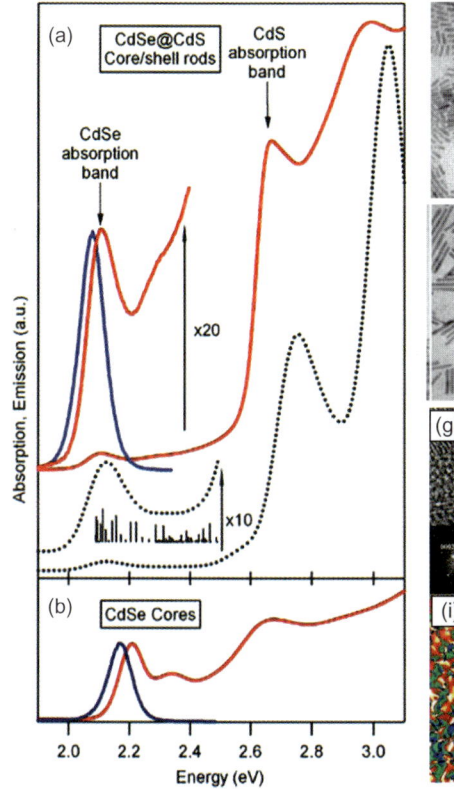

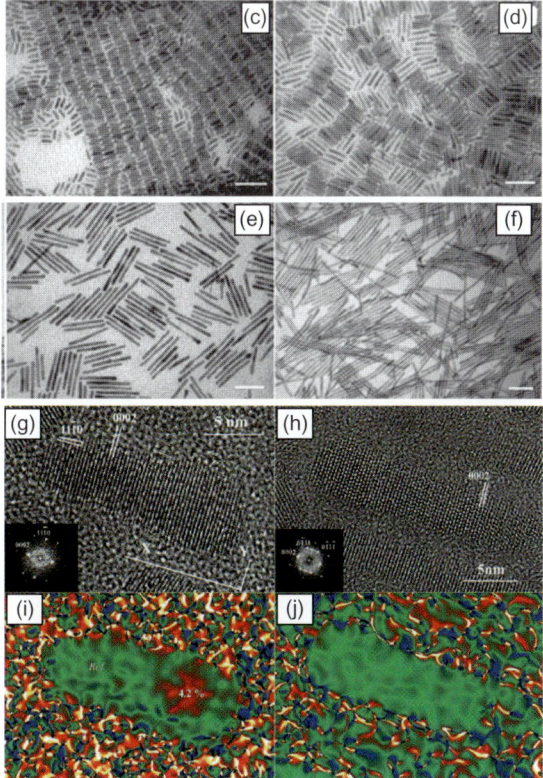

Figure 9.15 (a) Absorption (red) and emission (blue) spectra of CdSe-seeded CdS rods. The dotted line is a calculated spectrum; (b) Absorption and emission spectra of the CdSe seeds; (c–f) Transmission electron microscopy (TEM) images of CdSe-seeded CdS rods of different length and aspect ratios; (g) High-resolution TEM images of a CdSe-seeded CdS nanorod; (h) High-resolution TEM (HRTEM) images of pure CdS nanorod; (i, j) Corresponding "mean dilatation" images. This technique allows a mean dilatation mapping from high-resolution TEM images or, indeed, any type of lattice image. It uses a color scale for displaying variations in the periodicity of the HRTEM contrast. Areas of the same color are regions with the same periodicity. The mean dilatation image of the CdSe/CdS rod (i) shows an area with lattice parameters altered by 4.2% with respect to the reference area, situated at the opposite tip of the rod. For comparison, the same analysis is performed on "CdS-only" rods, and no variation of the lattice parameters over the whole length of the nanorod can be observed (j) [38].

case of epitaxially grown shells, information about shell thicknesses and shell composition that can be obtained with crystallographic methods such as XRD, and to some extent also with HRTEM, are limited. A preliminary concept of the composition of a core-shell particle can be obtained by simply measuring the diameter of the original core particles and their diameter after shell growth. However, it must not be taken for granted that the original core particles do not grow by themselves under the shell growth conditions (e.g., via Ostwald ripening), and consequently carefully performed reference experiments and additional elemental analyses will be required to interpret the results. Even if diameter measurements, when combined with elemental analysis, might allow determination of the shell thickness, neither approach can solve the question of the degree of alloying in a core-shell structure. Bearing in mind that typical core-shell structures have shell thicknesses of around 0.5–2 nm, the question of how much such a shell is alloyed (or not) seems almost impossible to answer. Further information might be obtained, however, by applying X-ray photoelectron spectroscopy. As the escape depth of the photoelectrons generated is typically in the range of just a few nanometers, thus is an extremely surface-sensitive technique. When using an energy-tunable X-ray source (e.g., a synchrotron), this escape depth can even be tuned to some degree and hence a depth profiling of a given core-shell structure is possible. Borchert et al. successfully applied the method not only to the characterization of simple CdSe/ZnS core-shell structures [44], but also to more sophisticated CdS/HgS QDQW structures, as described above (cf. Ref. [27]).

9.4
Core-Shell Infrared Nanoparticulate Materials

The International Commission on Illumination (CIE) has recommended the division of longer-wavelength radiation into the following three bands: IR-A (700–1400 nm); IR-B (1400–3000 nm); and IR-C (3000 nm to 1 mm).

The following is a commonly used working scheme:

- **Near-infrared (NIR, IR-A DIN):** 0.75–1.4 µm in wavelength, defined by the water absorption, and commonly used in fiber optic telecommunications because of low attenuation losses in the SiO_2 glass (silica) medium. Applications utilizing this spectral region include image intensifiers, such as night-vision goggles.

- **Short-wavelength infrared (SWIR, IR-B DIN):** 1.4–3 µm, water absorption increases significantly at 1450 nm. The 1530 to 1560 nm range is the dominant spectral region employed for long-distance telecommunications.

- **Mid-wavelength infrared (MWIR, IR-C DIN) also called intermediate infrared (IIR):** 3–8 µm. In guided missile technology the 3–5 µm portion of this band is the atmospheric window in which the homing heads of passive IR "heat-seeking" missiles are designed to work, homing on to the IR signature of the target aircraft, typically the jet engine exhaust plume.

- **Long-wavelength infrared (LWIR, IR-C DIN):** 8–15 µm. This is the "thermal imaging" region, in which sensors can obtain a completely passive picture of the outside world based on thermal emissions only and requiring no external light or thermal source such as the sun, moon or infrared illuminator. Forward-looking infrared (FLIR) systems use this area of the spectrum. It is sometimes also referred to as the "far-infrared".

- **Far-infrared (FIR):** 15–1000 µm.

Interest in NIR-active materials is primarily due to the fact that water is transparent across a number of wavelength ranges in this spectral region and, as a result, atmospheric transmission technologies and a number of biological applications can greatly benefit if these wavelengths are employed. The fact that there are no efficient and stable dyestuffs that can be used to address this spectral region had given extra impetus to the development of inorganic possibilities, and semiconductor nanocrystals have been demonstrated to adequately fulfill this role. Semiconductor nanocrystalline core materials exist that may be used directly to access this region, such as the Group IV–VI materials lead sulfide, lead selenide and lead telluride, as well as the Group III–V material InAs [45]. However, as the surface of nanoparticles is very sensitive to their immediate environment – a fact that usually results in a deterioration of the optical properties and, in particular, of the emission efficiency – it is as a rule beneficial to add a surface coating.

In order to achieve optical responses in the NIR, or to extend the spectral range of the nanocrystals further into the red, those materials of which the bulk bandgap is already reasonably narrow have been the main focus of investigation. A number of such systems have been reported, and have in general concentrated on the following as core materials: CdSe [46, 47], CdTe [48–54] or HgTe [55, 56] and the mixed ternary compounds $Cd_xHg_{1-x}Te$ [54] as Group II–VI materials, InP [57–59], InAs [33, 60–63] and GaAs [64–67] as Group III–V materials, and PbS [68], PbSe [69–76] and PbTe [77, 78] as Group IV–VI materials. Some mention of CSS systems [33, 46, 49, 60, 63] and different geometries such as tetrapods [47] and wires [64–67] have also been reported. It is worth noting at this point that there are a number of other core–shell material sets involving phosphors [79], transition metal oxides [80], noble metals [81] and their chalcogenides [82, 83], as well as various approaches to system design [84] and mixed nanoparticle-dyestuff compositions [85]. For example, although a CdSe/CdS core–shell system is not inherently an IR emitter, NIR emission can be achieved by coupling these nanocrystals to IR fluorescent dyestuffs. However, at this point attention will be focused on core–shell systems where the core is a QD composed of one of the Group II–VI, III–V, and IV–VI materials.

9.5
Type I Core–Shell Infrared Structures

Amongst the materials with which CdTe cores have been coated to produce type-I structures are ZnS [54] and InP/ZnS [49]. In addition, CdHgTe/ZnS core–shell nanocrystals which are highly luminescent, stable for months in butanol, have a higher resistance to photobleaching compared to cores, and which were synthesized using a hybrid approach (i.e., an aqueous-based synthesis, followed by transfer to organic solvents) have been reported by Tsay et al. These authors reported a high quantum yield (20–50%) in the NIR wavelength region (>700 nm), a region which is attractive for various biological applications because of the reduced autofluorescence background, improved penetration into scattering tissues, and enhanced photostability. In order to demonstrate their biological utility, the

CdHgTe/ZnS core–shells were coated with phospholipid micelles. This coating is known to suppress the toxicity of the nanocrystals (NCs) and also to render them water-soluble, so as to allow bioactivity and keep the quantum yields high [86]. Studies on IR-active core–shell HgTe/Hg$_x$Cd$_{1-x}$Te(S) particles, the objectives of which were to elucidate the nature of the recombination emission being associated with a Cd–Hg mixed site and structural information on the NC core–shell interface, have also been reported [55].

Indium phosphide nanoparticles have also been a popular and productive choice of material for study, as this Group III–V material has a bulk bandgap of 1.35 eV (~920 nm) [57] and a reduced toxicity in comparison to the Groups II–VI and IV–VI materials. However, to date InP (in fact, all of the Group III–V materials) have remained synthetically very challenging since, even when proving possible to synthesize, they usually possessed a low quantum yield that could be increased by coating the material with a wider band gap material. The coating materials of choice have been the Group II–VI semiconductors ZnS and ZnSe, and also with CdSe to result in a type-II structure [57–59], due to their favorable band offsets and lattice mismatches (see Figure 9.16). In studies conducted by Shu and colleagues, a simple method was reported for the synthesis of highly luminescent InP NCs with quantum yields of up to 30%, and InP/ZnS NCs with quantum yields of up to 60%. This method allowed the preparation of both InP and InP/ZnS NCs within 20 min, and no size-selection process was required. The NCs were

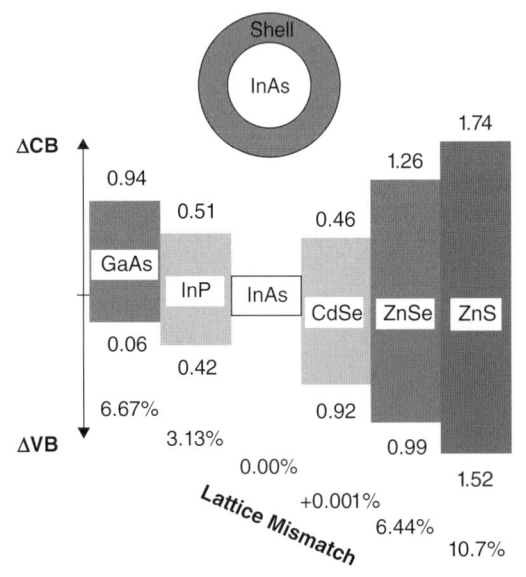

Figure 9.16 Summary of the band offsets (in eV) and lattice mismatch (in %) between the core InAs and a number of III–V semiconductor shells (left side), and II–VI semiconductor shells (right side). CB = conduction band; VB = valence band. Reproduced from Ref. [59].

obtained by a reacting InCl$_3$ complex and tris(trimethylsilyl)phosphine in octadecene in the presence of zinc undecylenate. The photoluminesence spectra of differently sized InP/ZnS core–shell nanocrystals were presented that displayed peak maxima from 480 to 735 nm. In the studies of Langhof and coworkers, a similar synthetic strategy was employed, using both InCl$_3$ and tris(trimethylsilyl) phosphine, but the solvent mixture was a more coordinating solvent mixture that consisted of TOP/TOPO. The ZnS shell was achieved by transferring the NCs to pyridine and overcoating with the ZnS in this solution at 100 °C. The ZnS shell was synthesized by the reaction of diethylzinc and hexylmethyldisilathiane in tributylphosphine.

In this study, linearly polarized photoluminesence measurements showed that the InP/ZnS NCs studied had an elongated shape and an ellipsoidal eccentricity of about 0.6. Optical pumping photoluminesence decay curves revealed the spin relaxation time to be substantially shorter than the radiative lifetime of an exciton in InP/ZnS NCs, while the magnetic field-induced circularly polarized photoluminesence measurements supplied information about the g-factor of the exciton, electron, and hole. In studies performed by Banin *et al.*, core–shell semiconductor nanocrystals with InAs cores were synthesized, and onto these Group III–V semiconductor shells (InP and GaAs), and Group II–VI semiconductor shells (CdSe, ZnSe, and ZnS) had been overgrown on cores of various radii, employing a two-step procedure. During initial attempts to carry out growth of the Group III–V semiconductor shell materials (InP, GaAs) it was found that these shells could be grown only at higher temperatures (>240 °C). In contrast to the Group III–V semiconductor shells, however, growth of the Group II–VI semiconductor shells (CdSe, ZnSe, ZnS) was observed at the comparatively lower temperature of approximately 150 °C.

Yet, it was found that when growing Group II–VI semiconductor shells, a higher temperature was needed to increase the fluorescence quantum yield. For shell growth at 260 °C, the maximum fluorescence quantum yield of the products was about fourfold larger than that obtained at 160 °C. In addition, the comparison was noted between core–shell nanocrystals with InAs and CdSe cores. InAs/CdSe and CdSe/CdS, the core–shells were analogous in their electronic structure and hence, in both of these cases, because of the relatively low conduction band offsets, the electron wave function extended into the shell and to the nanocrystal surface; this in turn caused the spectrum to red-shift upon shell growth. Additionally, InAs/ZnSe and InAs/ZnS core–shell materials were analogous to CdSe/ZnS. In this study, the band gap was barely changed upon shell growth, on account of the large band offsets between the core and the shell semiconductors. An overview of the absorbance and emission trends for the coating materials is shown in Figure 9.17.

9.6
Type II Core–Shell Infrared Structures

Included amongst the shells employed in the case of CdSe are CdTe [47] and CdTe/ZnTe [46]. In most cases, coating with CdTe is reported to yield a type-II

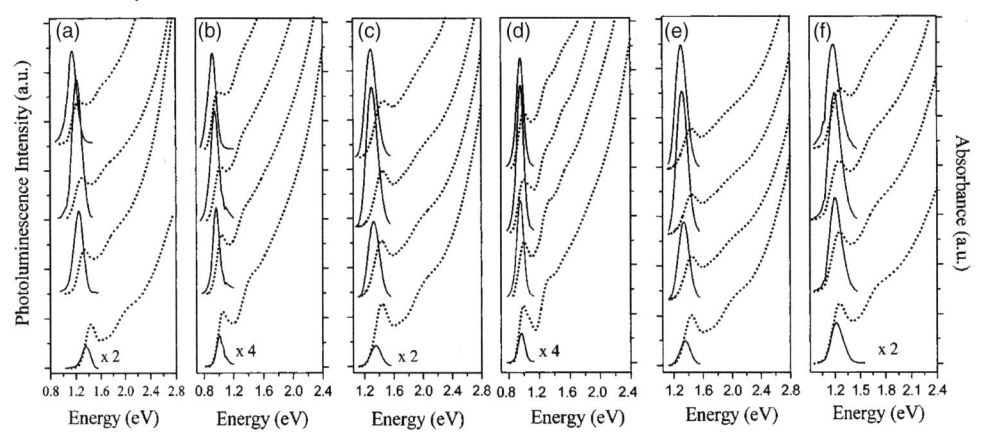

Figure 9.17 Evolution of absorption (dotted lines), and photoluminescence (solid lines) for growth of core–shells. The PL spectra are given on a relative scale for comparison of the enhancement of QY with shell growth. (a) InAs/CdSe with an initial core radius of 1.2 nm. The shell thickness (in number of monolayers, ML) and QY for the traces from bottom to top were respectively: 0, 1.2%; 0.6, 13%; 1.2, 21%; 1.8, 18%. (b) InAs/CdSe with initial core radius of 2.5 nm. The shell thickness (in number of ML) and QY for the traces from bottom to top were respectively: 0, 0.9%; 0.7, 11%; 1.2, 17%; 1.6, 14%; (c) InAs/ZnSe with an initial core radius of 1.2 nm. The shell thickness (in number of ML) and QY for the traces from bottom to top were respectively: 0, 1.2%; 0.6, 9%; 1.5, 18%; 2.5, 14%; (d) InAs/ZnSe with an initial core radius of 2.8 nm. The shell thickness (in number of ML) and QY for the traces from bottom to top were respectively: 0, 0.9%; 0.7, 13%; 1.3, 20%; 2.2, 15%; (e) InAs/ZnS with an initial core radius of 1.2 nm. The shell thickness (in number of ML) and QY for the traces from bottom to top were respectively: 0, 1.2%; 0.7, 4%; 1.3, 8%; 1.8, 7%; (f) InAs/ZnS with an initial core radius of 1.7 nm. The shell thickness (in number of ML) and QY for the traces from bottom to top were respectively: 0, 1.1%; 0.6, 5%; 1.3, 7.1%; 2.2, 6.3%. Reproduced from Ref. [59].

system, but in a number of studies [87, 88] this was reported as type-I. This fact may be explained by a relative shift in the bands as the thickness of the coating layer was increased. Interestingly, in the CdSe/CdTe/ZnTe system, when the CdTe was first used as a coating material, an emission signal at 1027 nm was observed; yet, when it was employed as a sandwich layer the emission was observed to shift to between 1415 and 1470 nm, depending on the thickness of the ZnTe layer (1.3 nm to 1.8 nm) (see Figure 9.18). In a second approach, the CdSe core size was altered from 3.4 to 5.7 nm, while the shell thicknesses of both CdTe and ZnTe remained unchanged (within experimental uncertainty); as a result, the emission red-shifted from 1415 to 1518 nm. These results correlated well with the band offsets of the ZnTe valence and the CdSe conduction band edges, and the resultant decrease of the CdSe → ZnTe transition. It was also noted that, in the case of the CdSe/CdTe/ZnTe structure (3.4/1.8/1.3 nm), a lifetime of up to 150 ns was observed for the CdSe → ZnTe 1415 nm emission. This result further indicated a very long radiative lifetime of ≈10 ms, due to the spatial separation of electron and hole by the CdTe intermediate layer.

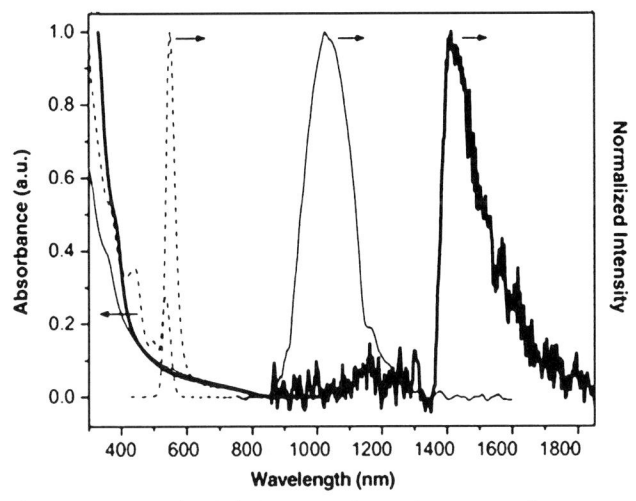

Figure 9.18 Normalized absorption and emission spectra of CdSe core (3.4 nm, dashed gray line), CdSe/CdTe (3.4/1.8 nm) core–shell (solid gray line), and CdSe/CdTe/ZnTe (3.4/1.8/1.3 nm) core–shell–shell (solid black line) QDs in toluene. Reproduced from Ref. [46].

Included amongst the materials that CdTe cores have been coated with to produce type-II structures are CdS [48] and CdSe [50–53]. For CdS-coated nanoparticles, the size distribution was observed to remain quite narrow upon coating with shell thicknesses of one to five monolayers. However, with increasing shell thickness the exciton lifetimes were seen to become longer (the average lifetime of the CdTe cores was 11.9 ns). For CdTe/CdS structures with two- and four-monolayer-thick shells, the time was increased to 23.2 and 26.8 ns respectively, while the slightly nonexponential decay curve observed for the bare CdTe core shifted to a monoexponential exciton decay, and was also accompanied by an increase in the photoluminescence quantum yields. In studies conducted by Kairdolf *et al.* on CdTe/CdSe core–shell nanocrystals, a one-pot procedure for the synthesis of CdTe core QDs, prepared using multidentate polymer ligands *in situ*, resulted in a series of monodisperse CdTe QDs that showed a bright fluorescence from 515 to 655 nm (green to red). Fluorescence emission spectra showing the transition from CdTe cores to CdTe/CdSe core–shell QDs, the emission of which was red-shifted from the original QD core emission of 650 nm to the NIR at 810 nm were presented. The absorbance spectra showed the red-shifting and eventual loss of the first exciton peak as the CdSe shell was grown on the CdTe core, typical of type-II QDs. Elsewhere [51], an aqueous route to CdTe/CdSe materials with moderate fluorescence quantum yields (10–20%) has been presented in which the uncoated CdTe displays an emission maximum at just below 650 nm, while the CdSe-coated cores have an emission peak maximum at approximately 760 nm. The lifetimes were also found to show more monoexponential decays and longer lifetimes as the CdSe shell was increased in thickness. These authors demonstrated the ability of the core–shell materials to selectively sense Cu(II) ions in the presence of other physiologically relevant cations in solution, by the quenching of the NIR fluorescence. Seo et al. observed that the presence of the alkyl metal precursors ethyl zinc or ethyl aluminum during the coating procedure led to rectangular shapes of CdTe/CdSe type-II dots, but spherically shaped materials could not be obtained. It was suggested that this synthesis proceeded by a mechanism in which the selenide anion, which is activated by ethyl zinc or ethyl aluminum, subsequently proceeded to react with excess cadmium cations. The absorption and emission spectra of two differently sized CdTe cores and their nonspherical core–shells, red-shifted from 692 nm and 751 nm to 760 nm and 802 nm, respectively, which was to be expected for type-II dots. In subsequent studies based on femtosecond upconversion techniques, Chou and coworkers reported the first observation of early relaxation dynamics on CdTe/CdSe type-II QDs interband emission. CdTe/CdSe QDs, in which the core (core–shell) sizes, as measured with TEM size histograms were calculated as 5.3 (6.3), 6.1 (7.1), and 6.9 (7.8) nm, were used in the study. Upon an increase in diameter of the CdTe cores from 5.3 to 6.9 nm, the CdTe emission peak wavelength-shifted from 690 nm to 737 nm, while the emission of CdTe/CdSe, coated with a similar thickness (1.0 nm) of CdSe, showed a similar systematic bathochromic shift from 1025 nm (core 5.3 nm/shell 1.0 nm) to 1061 nm (core 6.9 nm/shell 0.9 nm). The relaxation results presented indicated that the electron-separation rate decreased as the size of the cores increased. In the absence of any observation of coherent optical phonon modes for both CdTe core (in CdTe/CdSe) emission and CdTe/CdSe interband emission, it is possible that the finite rate of charge separation might be due to the small electron–phonon coupling, causing a weak coupling between the initial and charge-separated states. The study results indicated that, in particular, the degree of control of the rate of electron transfer might prove useful in applications where a rapid carrier separation followed by charge transfer into a matrix or electrode was important, such as in photovoltaic devices.

An example of $InAs_xP_{1-x}/InP/ZnSe$ III–V alloyed core–shell QDs, the bandgap of which has been engineered for applications in the NIR, has been reported by the group of Bawendi. The synthetic procedure for the alloyed core used the indium acetate, tris(trimethylsilyl)phosphine, and tris(trimethylsilyl)arsine as indium, phosphorus, and arsenic precursors, respectively. The resultant alloyed dot cores were overcoated with a shell of InP so as to increase their size and quantum yield. This shell was grown by injecting a mixture of indium acetate and tris(trimethylsilyl)phosphine to a solution of alloyed dots at 140 °C, which was low enough to avoid nucleation. The temperature was then raised to 180 °C so as to initiate shell growth. Two successive injections of In and P precursors, both at 140 °C, to a solution of $InAs_{0.82}P_{0.18}$ (738 nm fluorescence peak) QDs resulted in core–shell particles with an emission at 765 nm after the first injection (full-width, half-maximum; FWHM 103 nm) and 801 nm after the second injection (FWHM 119 nm) (Figure 9.19), and a tripling of the QY.

These NIR-emitting CSS $InAs_xP_{1-x}/InP/ZnSe$ were then subsequently successfully used in a sentinel lymph node (SLN) mapping experiment. When injected

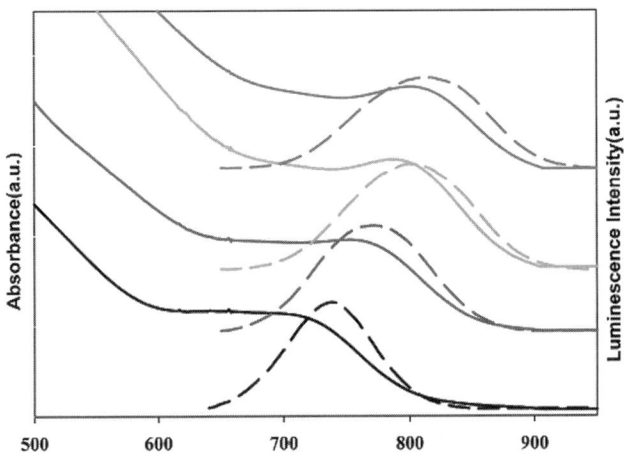

Figure 9.19 Absorbance (solid line) and corresponding photoluminescence (dashed line) spectra of core $InAs_{0.82}P_{0.18}$ QDs (emission 738 nm, FWHM 86 nm), after a first shell of InP (emission 765 nm, FWHM 103 nm), after a second shell of InP (emission 801 nm, FWHM 119 nm), and after the final ZnSe shell (emission 815 nm, FWHM 120 nm), from bottom to top, respectively. Reproduced from Ref. [63].

intradermally into the paw of a mouse, the small amount and low concentration of NIR QDs injected could not be seen on the color video images. However, the fluorescence image revealed the fine detail of lymphatic flow from the injection site to the SLN. As the background autofluorescence from the tissue was low in this spectral region, this permitted a high signal-to-background ratio.

The Group IV–VI lead chalcogenide semiconductor nanocrystals PbS, PbSe, and PbTe all form type-II core–shell structures with one another, and also have the advantage of possessing narrow bulk bandgaps (0.41, 0.278, and 0.31 eV at room temperature and 0.29, 0.17, and 0.19 at 4 K, respectively) [45]. The band gap of PbS may be tuned to 800 nm in the absorbance [89], resulting in a set of materials that can be engineered to address the spectral range between 800 nm (quantum-confined PbS) to beyond 7000 nm (bulk PbTe). The applications (as opposed to academic) interest in coating in the case of lead chalcogenides was primarily one of protecting the core, as the complete NIR region is accessible to these materials without any requirement for coating. Elsewhere, PbS has been coated with PbSe [68], and PbSe with PbS [70–76].

In the latter cases, PbS would be expected to be more robust to atmospheric-induced oxidation than would the PbSe core. The SILAR (Successive Ion Layer Addition and Reaction) technique has been applied to add up to five monolayers, with a consequent shift in absorbance of 85 nm for the coating of a 6 nm core. In the group of Lifshitz, the synthesis of both core–shell and core-alloyed shell has been undertaken, and the materials evaluated as optical components in Q-switching lasers. In this way, the synthesis of the core materials in the size range

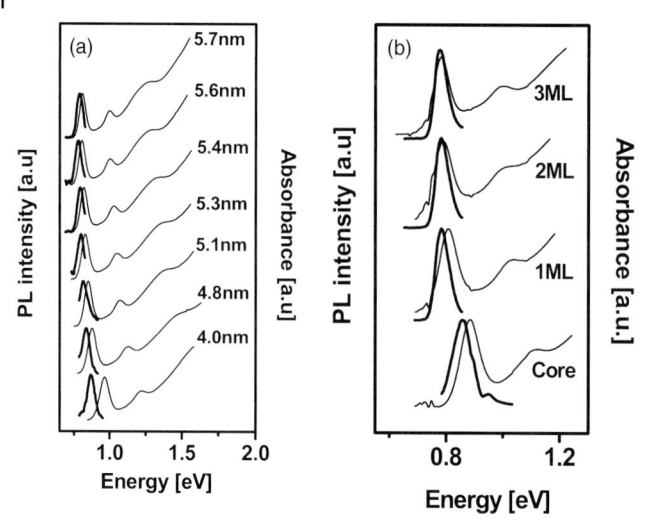

Figure 9.20 (a) Absorbance (thin lines) and corresponding PL (bold lines) spectra of core PbSe nanocrystals with NC diameters as indicated; (b) Absorbance (thin lines) and corresponding PL spectra (bold lines) of 4.9 nm PbSe core (lowest curves) and the corresponding PbSe/PbS core–shell (top three curves) nanocrystals with n monolayers ($n = 1, 2$ and 3) of a PbS shell. Reproduced from Ref. [76].

of 2.3 to 7.0 nm was demonstrated; spectral evidence was also provided for a decrease in the Stokes shift with increasing particle size (Figure 9.20).

The preparation of PbSe/PbSe$_x$S$_{1-x}$ core-shells structures could be achieved by using a single injection of the precursors into a single round flask. For this, a stock solution of Se, S was prepared by mixing 0.15 g Se dissolved in 1.4 ml TOP, with 0.03–0.10 g S dissolved in 0.3 ml trioctylphosphine. The amount of S in the new stock solution corresponded to a stoichiometric amount of one to two monolayers of the PbS compound. Thus, the molar ratio of the precursors Pb:Se:S ranged between 10:9:5 and 7:6:14. The periodic removal of aliquots during the reaction was followed by purification and examination of the species by HR-TEM and absorption spectroscopy. The results suggested a rapid nucleation of the PbSe core, followed by a slower precipitation of the PbSe$_x$S$_{1-x}$ shell, with a gradual change in the chemical composition. This single process is reported to produce materials of a narrow size distribution (5%) and high quantum efficiencies (55%) when compared to conventional core and core–shell materials. The claim was also made that these systems, PbSe/PbS and PbSe/PbSe$_x$S$_{1-x}$ core–shell nanocrystal QDs, were air-stable and therefore possessed the advantage that no further processing would be required to protect the surfaces from atmospheric oxidation for applications [73]. However, in studies conducted by Stouwdam et al., who investigated the photostability of colloidal PbSe and PbSe/PbS core–shell nanocrystals in solution and in the solid state, no differences were observed in stability between the coated NCs as compared to uncoated NCs in solution upon irradiation with a xenon lamp [72]. The possible reason given for this was that either the PbS shell was unable to confine the charge carriers, or there was incomplete shell passivation within the PbSe core and, as a result, the core–shell NCs had comparable stability. These findings highlighted and typified a number of problems encountered when attempting to characterize the structure and stability of these systems. PbSe has also been coated with Group II–VI semiconductors to controllably synthesize PbSe/CdSe type-I core–shell QDs that are stable against fading and spectral shifting [69]. These core–shells can then further undergo additional shell growth to produce PbSe/CdSe/ZnS CSS QDs that represent the initial steps toward bright, biocompatible, NIR optical labels.

The coating of PbTe has been carried out using controlled oxidation in the presence of amines [77], and by cation exchange to form PbTe/CdTe. In the former case the oleylamine induces the formation of $Pb(OH)_2$ or lead oxide at the surface of the PbTe nanocrystals. If no treatment is involved to remove the excessive oleylamine from the surfaces of the freshly prepared PbTe after PbTe NC synthesis, the Te^{2-} in PbTe nanocrystals can be partially replaced by OH^- from the trace amount of water present. This reaction is facilitated by the presence of oleylamine, and initially produces a shell of amorphous $Pb(OH)_2$. The shell may subsequently be transformed into crystalline PbO after an annealing treatment, while retaining the particle shape. The material retains its narrow size distribution and forms supercrystals – this is an interesting observation, given that the nanoparticles are cubes and not spheres. Hence, through an anion-exchange mechanism in the presence of oleylamine, the PbTe nanocubes can be converted into core–shell building blocks by "shrinking" themselves into a truncated octahedral or a near-spherical core with an amorphous but quasi-cubic shell (PbTe@Pb(OH)$_2$ and PbTe@PbO). The cubic core–shell particles can still be packed into a two-dimensional pattern or a super crystal, although the long-range order was lost due to a slight truncation of the corners of the NCs (Figure 9.21).

In a cation-exchange reaction, Lambert et al. took the PbS/CdS and PbSe/CdSe core–shell formation by cation exchange (as reported by Pietryga and coworkers) as a starting point and extended it to PbTe/CdTe. They also showed that a combination of the PbTe rocksalt structure and the CdTe zinc blende structure allowed for direct observation of the core and shell with HRTEM. This enabled a direct visualization of the crystallographic properties of the PbTe/CdTe QDs, and an evaluation of the cation-exchange reaction. A seamless match was also observed between the PbTe and CdTe crystal lattices. It was concluded that, in the case of lead chalcogenides, cationic exchange represented a straightforward technique for the formation of lead chalcogenide/cadmium chalcogenide core–shell materials.

9.7
Summary and Conclusions

In this chapter, progress in the synthesis of colloidal nanoheterostructures was summarized. Depending on their electronic structure (type-I or type-II) and their

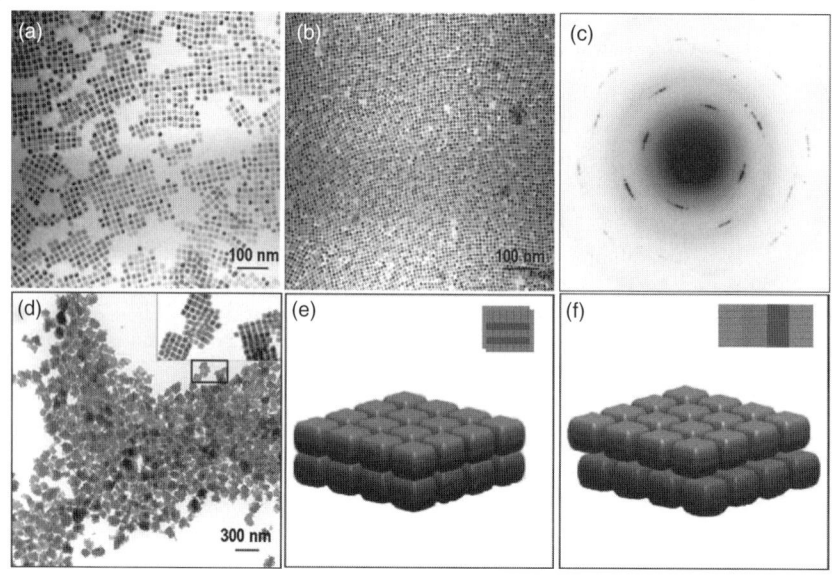

Figure 9.21 Electron microscopy images of PbTe nanocubes and simple cubic (SC). (a) TEM image of PbTe monolayer assemblies; (b) TEM image of PbTe monolayer assembly; (c) Selected area electron diffraction pattern of monolayer assembly (negative pattern); (d) TEM image of PbTe multilayer assemblies; (e) Model of stacking structure of two-layer assembly ("simple cubic SC"); (f) Model of stacking structure of two-layer assembly ("1D-shifted SC"). Reproduced from Ref. [77].

geometry (spherical or nonspherical), the properties and application potential of these structures was discussed. Type-I structures were introduced as potent dyes with very high fluorescence quantum yields and high photostability, while type-II heterostructures may be used for extended band gap engineering. The latter continue to show promising properties with regards to amplified stimulated emission.

References

1 Spanhel, L., Haase, M., Weller, H. and Henglein, A. (1987) Photochemistry of colloidal semiconductors. 20. Surface modification and stability of strong luminescing CdS particles. *Journal of the American Chemical Society*, **109**, 5649–55.
2 Murray, C.B., Norris, D.J. and Bawendi, M.G. (1993) Synthesis and characterization of nearly monodisperse CdE (E = sulfur, selenium, tellurium) semiconductor nanocrystallites. *Journal of the American Chemical Society*, **115**, 8706–15.
3 Hines, M.A. and Guyot-Sionnest, P. (1996) Synthesis and characterization of strongly luminescing ZnS-Capped CdSe nanocrystals. *Journal of Physical Chemistry*, **100**, 468–71.
4 Dabbousi, B.O. et al. (1997) (CdSe)ZnS core-shell quantum dots: synthesis and characterization of a size series of highly luminescent nanocrystallites. *Journal of Physical Chemistry B*, **101**, 9463–75.
5 Peng, X.G., Schlamp, M.C., Kadavanich, A.V. and Alivisatos, A.P. (1997) Epitaxial growth of highly luminescent CdSe/CdS

core/shell nanocrystals with photostability and electronic accessibility. *Journal of the American Chemical Society*, **119**, 7019–29.

6 Li, J.J. *et al.* (2003) Large-scale synthesis of nearly monodisperse CdSe/CdS core/shell nanocrystals using air-stable reagents via successive ion layer adsorption and reaction. *Journal of the American Chemical Society*, **125**, 12567–75.

7 Steckel, J.S. *et al.* (2004) Blue luminescence from (CdS)ZnS core-shell nanocrystals. *Angewandte Chemie, International Edition*, **43**, 2154–8.

8 Haubold, S., Haase, M., Kornowski, A. and Weller, H. (2001) Strongly luminescent InP/ZnS core-shell nanoparticles. *ChemPhysChem*, **2**, 331–4.

9 Bruns, O.T. *et al.* (2009) Real-time magnetic resonance imaging and quantification of lipoprotein metabolism in vivo using nanocrystals. *Nature Nanotechnology*, **4**, 193–201.

10 Kim, S., Fisher, B., Eisler, H.J. and Bawendi, M. (2003) Type-II quantum dots: CdTe/CdSe(core/shell) and CdSe/ZnTe(core/shell) heterostructures. *Journal of the American Chemical Society*, **125**, 11466–7.

11 Ivanov, S.A. *et al.* (2007) Type-II core/shell CdS/ZnSe nanocrystals: synthesis, electronic structures, and spectroscopic properties. *Journal of the American Chemical Society*, **129**, 11708–19.

12 Klimov, V.I. *et al.* (2007) Single-exciton optical gain in semiconductor nanocrystals. *Nature*, **447**, 441–6.

13 Eychmüller, A., Mews, A. and Weller, H. (1993) A quantum-dot quantum-well – CdS/HgS/CdS. *Chemical Physics Letters*, **208**, 59–62.

14 Mews, A., Eychmüller, A., Giersig, M., Schooss, D. and Weller, H. (1994) Preparation, characterization, and photophysics of the quantum-dot quantum-well system CdS/HgS/CdS. *Journal of Physical Chemistry*, **98**, 934–41.

15 Schooss, D., Mews, A., Eychmüller, A. and Weller, H. (1994) Quantum-dot quantum-well CdS/HgS/CdS – theory and Experiment. *Physical Review B*, **49**, 17072–8.

16 Bryant, G.W. (1995) Theory for quantum-dot quantum wells: pair correlation and internal quantum confinement in nanoheterostructures. *Physical Review B*, **52**, 16997–7000.

17 Jaskolski, W. and Bryant, G.W. (1998) Multiband theory of quantum-dot quantum wells: dim excitons, bright excitons, and charge separation in heteronanostructures. *Physical Review B*, **57**, R4237–40.

18 Eychmüller, A., Vossmeyer, T., Mews, A. and Weller, H. (1994) Transient photobleaching in the quantum-dot quantum-well CdS/HgS/CdS. *Journal of Luminescence*, **58**, 223–6.

19 Kamalov, V.F., Little, R., Logunov, S.L. and El-Sayed, M.A. (1996) Picosecond electronic relaxation in CdS/HgS/CdS quantum dot quantum well semiconductor nanoparticles. *Journal of Physical Chemistry*, **100**, 6381–4.

20 Banin, U., Mews, A., Kadavanich, A.V., Guzelian, A.A. and Alivisatos, A.P. (1996) Homogeneous optical properties of semiconductor nanocrystals. *Molecular Crystals and Liquid Crystals Science and Technology. Section A: Molecular Crystals and Liquid Crystals*, **283**, 1–10.

21 Mews, A., Kadavanich, A.V., Banin, U. and Alivisatos, A.P. (1996) Structural and spectroscopic investigations of CdS/HgS/CdS quantum-dot quantum wells. *Physical Review B*, **53**, 13242–5.

22 Lifshitz, E. *et al.* (1999) Optically detected magnetic resonance study of CdS/HgS/CdS quantum dot quantum wells. *Journal of Physical Chemistry, B*, **103**, 6870–5.

23 Little, R.B., El-Sayed, M.A., Bryant, G.W. and Burke, S. (2001) Formation of quantum-dot quantum-well heteronanostructures with large lattice mismatch: ZnS/CdS/ZnS. *Journal of Chemical Physics*, **114**, 1813–22.

24 Battaglia, D., Li, J.J., Wang, Y.J. and Peng, X.G. (2003) Colloidal two-dimensional systems: CdSe quantum shells and wells. *Angewandte Chemie, International Edition*, **42**, 5035–9.

25 Dorfs, D. and Eychmüller, A. (2001) A series of double well semiconductor quantum dots. *Nano Letters*, **1**, 663–5.

26 Braun, M., Burda, C. and El-Sayed, M.A. (2001) Variation of the thickness and number of wells in the CdS/HgS/CdS quantum dot quantum well system.

Journal of Physical Chemistry, A, **105**, 5548–51.

27 Borchert, H. et al. (2003) Photoemission study of onion like quantum dot quantum well and double quantum well nanocrystals of CdS and HgS. *Journal of Physical Chemistry, B*, **107**, 7486–91.

28 Dorfs, D., Henschel, H., Kolny, J. and Eychmüller, A. (2004) Multilayered nanoheterostructures: theory and experiment. *Journal of Physical Chemistry, B*, **108**, 1578–83.

29 Reiss, P., Carayon, S., Bleuse, J. and Pron, A. (2003) Low polydispersity core/shell nanocrystals of CdSe/ZnSe and CdSe/ZnSe/ZnS type: preparation and optical studies. *Synthetic Metals*, **139**, 649–52.

30 Talapin, D.V. et al. (2004) CdSe/CdS/ZnS and CdSe/ZnSe/ZnS core-shell-shell nanocrystals. *Journal of Physical Chemistry, B*, **108**, 18826–31.

31 Jun, S., Jang, E. and Lim, J.E. (2006) Synthesis of multi-shell nanocrystals by a single step coating process. *Nanotechnology*, **17**, 3892–6.

32 Xie, R.G., Kolb, U., Li, J.X., Basche, T. and Mews, A. (2005) Synthesis and characterization of highly luminescent CdSe-Core CdS/Zn0.5Cd0.5S/ZnS multishell nanocrystals. *Journal of the American Chemical Society*, **127**, 7480–8.

33 Aharoni, A., Mokari, T., Popov, I. and Banin, U. (2006) Synthesis of InAs/CdSe/ZnSe core/shell1/shell2 structures with bright and stable near-infrared fluorescence. *Journal of the American Chemical Society*, **128**, 257–64.

34 Battaglia, D., Blackman, B. and Peng, X.G. (2005) Coupled and decoupled dual quantum systems in one semiconductor nanocrystal. *Journal of the American Chemical Society*, **127**, 10889–97.

35 Dias, E.A., Sewall, S.L. and Kambhampati, P. (2007) Light harvesting and carrier transport in core/barrier/shell semiconductor nanocrystals. *Journal of Physical Chemistry, C*, **111**, 708–13.

36 Sapra, S., Mayilo, S., Klar, T.A., Rogach, A.L. and Feldmann, J. (2007) Bright white-light emission from semiconductor nanocrystals: by chance and by design. *Advanced Materials*, **19**, 569–72.

37 Peng, X.G. et al. (2000) Shape control of CdSe nanocrystals. *Nature*, **404**, 59–61.

38 Carbone, L. et al. (2007) Synthesis and micrometer-scale assembly of colloidal CdSe/CdS nanorods prepared by a seeded growth approach. *Nano Letters*, **7**, 2942–50.

39 Milliron, D.J. et al. (2004) Colloidal nanocrystal heterostructures with linear and branched topology. *Nature*, **430**, 190–5.

40 Talapin, D.V. et al. (2003) Highly emissive colloidal CdSe/CdS heterostructures of mixed dimensionality. *Nano Letters*, **3**, 1677–81.

41 Kraus, R.M. et al. (2007) Room-temperature exciton storage in elongated semiconductor nanocrystals. *Physical Review Letters*, **98**, 017401-1–017401-4.

42 Dorfs, D., Salant, A., Popov, I. and Banin, U. (2008) ZnSe quantum dots within CdS nanorods: a seeded-growth type-II system. *Small*, **4**, 1319–23.

43 Fiore, A. et al. (2009) Tetrapod-shaped colloidal nanocrystals of II–VI semiconductors prepared by seeded growth. *Journal of the American Chemical Society*, **131**, 2272–82.

44 Borchert, H. et al. (2003) High resolution photoemission study of CdSe and CdSe/ZnS core-shell nanocrystals. *Journal of Chemical Physics*, **119**, 1800–7.

45 Rogach, A.L., Eychmüller, A., Hickey, S.G. and Kershaw, S.V. (2007) Infrared-emitting colloidal nanocrystals: synthesis, assembly, spectroscopy, and applications. *Small*, **3**, 536–57.

46 Chen, C.Y. et al. (2005) Type-II CdSe/CdTe/ZnTe (core-shell-shell) quantum dots with cascade band edges: the separation of electron (at CdSe) and hole (at ZnTe) by the CdTe layer. *Small*, **1**, 1215–20.

47 Peng, P. et al. (2005) Femtosecond spectroscopy of carrier relaxation dynamics in type-II CdSe/CdTe tetrapod heteronanostructures. *Nano Letters*, **5**, 1809–13.

48 Wang, J., Long, Y.T., Zhang, Y.L., Zhong, X.H. and Zhu, L.Y. (2009) Preparation of highly luminescent CdTe/CdS core/shell quantum dots. *ChemPhysChem*, **10**, 680–5.

49 Kim, S. et al. (2009) Bandgap engineered reverse type-I CdTe/InP/ZnS core-shell nanocrystals for the near-infrared. *Chemical Communications*, 1267–9.

50 Kairdolf, B.A., Smith, A.M. and Nie, S. (2008) One-pot synthesis, encapsulation, and solubilization of size-tuned quantum dots with amphiphilic multidentate ligands. *Journal of the American Chemical Society*, **130**, 12866–7.

51 Xia, Y.S. and Zhu, C.Q. (2008) Aqueous synthesis of type-II core/shell CdTe/CdSe quantum dots for near-infrared fluorescent sensing of copper(II). *Analyst*, **133**, 928–32.

52 Seo, H. and Kim, S.W. (2007) In situ synthesis of CdTe/CdSe core-shell quantum dots. *Chemistry of Materials*, **19**, 2715–17.

53 Chou, P.T. et al. (2006) Spectroscopy and femtosecond dynamics of type-II CdTe/CdSe core-shell quantum dots. *ChemPhysChem*, **7**, 222–8.

54 Tsay, J.M., Pflughoefft, M., Bentolila, L.A. and Weiss, S. (2004) Hybrid approach to the synthesis of highly luminescent CdTe/ZnS and CdHgTe/ZnS nanocrystals. *Journal of the American Chemical Society*, **126**, 1926–7.

55 Lifshitz, E. and Eychmuller, A. (2007) Spectroscopic investigations on II–VI-semiconductor nanocrystals and their assemblies. *Journal of Cluster Science*, **18**, 5–18.

56 Harrison, M.T. et al. (2000) Wet chemical synthesis of highly luminescent HgTe/CdS core/shell nanocrystals. *Advanced Materials*, **12**, 123–5.

57 Xu, S., Ziegler, J. and Nann, T. (2008) Rapid synthesis of highly luminescent InP and InP/ZnS nanocrystals. *Journal of Materials Chemistry*, **18**, 2653–6.

58 Langof, L. et al. (2004) Colloidal InP/ZnS core-shell nanocrystals studied by linearly and circularly polarized photoluminescence. *Chemical Physics*, **297**, 93–8.

59 Cao, Y. and Banin, U. (2000) Growth and properties of semiconductor core/shell nanocrystals with InAs cores. *Journal of the American Chemical Society*, **122**, 9692–702.

60 Ben-Lulu, M., Mocatta, D., Bonn, M., Banin, U. and Ruhman, S. (2008) On the absence of detectable carrier multiplication in a transient absorption study of InAs/CdSe/ZnSe core/shell1/shell2 quantum dots. *Nano Letters*, **8**, 1207–11.

61 Zimmer, J.P. et al. (2006) Size series of small indium arsenide-zinc selenide core-shell nanocrystals and their application to in vivo imaging. *Journal of the American Chemical Society*, **128**, 2526–7.

62 Millo, O., Katz, D., Cao, Y. and Banin, U. (2001) Tunneling and optical spectroscopy of InAs and InAs/ZnSe core/shell nanocrystalline quantum dots. *Physica Status Solidi, (B) Basic Research*, **224**, 271–6.

63 Kim, S.W. et al. (2005) Engineering $InAs_xP_{1-x}$/InP/ZnSe III–V alloyed core/shell quantum dots for the near-infrared. *Journal of the American Chemical Society*, **127**, 10526–32.

64 Hua, B., Motohisa, J., Kobayashi, Y., Hara, S. and Fukui, T. (2009) Single GaAs/GaAsP coaxial core-shell nanowire lasers. *Nano Letters*, **9**, 112–16.

65 Prete, P. et al. (2008) Luminescence of GaAs/AlGaAs core-shell nanowires grown by MOVPE using tertiary butylarsine. *Journal of Crystal Growth*, **310**, 5114–18.

66 Titova, L.V. et al. (2006) Temperature dependence of photoluminescence from single core-shell GaAs–AlGaAs nanowires. *Applied Physics Letters*, **89**, 2865–7.

67 Noborisaka, J., Motohisa, J., Hara, S. and Fukui, T. (2005) Fabrication and characterization of freestanding GaAs/AlGaAs core-shell nanowires and AlGaAs nanotubes by using selective-area metalorganic vapor phase epitaxy. *Applied Physics Letters*, **87**, 093109-1–093109-3.

68 Koktysh, D.S., McBride, J.R., Dixit, S.K., Feldman, L.C. and Rosenthal, S.J. (2007) PbS/PbSe structures with core-shell type morphology synthesized from PbS nanocrystals. *Nanotechnology*, **18**, 495607-1–495607-4.

69 Pietryga, J.M. et al. (2008) Utilizing the lability of lead selenide to produce heterostructured nanocrystals with bright, stable infrared emission. *Journal of the American Chemical Society*, **130**, 4879–85.

70 Cui, D., Xu, J., Paradee, G., Xu, S.Y. and Wang, A.Y. (2007) Developing PbSe/PbS core-shell nanocrystals quantum dots toward their potential heterojunction applications. *Journal of Experimental Nanoscience*, **2**, 13–21.

71 Xu, J. et al. (2006) Synthesis and surface modification of PbSe/PbS core-shell nanocrystals for potential device applications. *Nanotechnology*, **17**, 5428–34.

72 Stouwdam, J.W. et al. (2007) Photostability of colloidal PbSe and PbSe/PbS core/shell nanocrystals in solution and in the solid state. *Journal of Physical Chemistry, C*, **111**, 1086–92.

73 Lifshitz, E. et al. (2006) Air-stable PbSe/PbS and PbSe/PbSe$_x$S$_{1-x}$ core-shell nanocrystal quantum dots and their applications. *Journal of Physical Chemistry, B*, **110**, 25356–65.

74 Brumer, M. et al. (2006) Nanocrystals of PbSe core, PbSe/PbS, and PbSe/PbSe$_x$S$_{1-x}$ core/shell as saturable absorbers in passively Q-switched near-infrared lasers. *Applied Optics*, **45**, 7488–97.

75 Kigel, A., Brumer, M., Sashchiuk, A., Amirav, L. and Lifshitz, E. (2005) PbSe/PbSe$_x$S$_{1-x}$ core-alloyed shell nanocrystals. *Materials Science & Engineering C–Biomimetic and Supramolecular Systems*, **25**, 604–8.

76 Brumer, M. et al. (2005) PbSe/PbS and PbSe/PbSe$_x$S$_{1-x}$ core/shell nanocrystals. *Advanced Functional Materials*, **15**, 1111–16.

77 Zhang, J. et al. (2008) Simple cubic super crystals containing PbTe nanocubes and their core-shell building blocks. *Journal of the American Chemical Society*, **130**, 15203–9.

78 Lambert, K., De Geyter, B., Moreels, I. and Hens, Z. (2009) PbTe/CdTe core/shell particles by cation exchange: A HRTEM study. *Chemistry of Materials*, **21**, 778–80.

79 Lim, M.A., Il Seok, S., Chung, W.J. and Hong, S.I. (2008) Near infrared luminescence properties of nanohybrid film prepared from LaPO$_4$:Er^{3+}/LaPO$_4$ core/shell nanoparticles and silica-based resin. *Optical Materials*, **31**, 201–5.

80 Zhang, M.F. et al. (2008) Preparation and characterization of near-infrared luminescent bifunctional core/shell nanocomposites. *Journal of Physical Chemistry, C*, **112**, 2825–30.

81 Lee, J.S., Shevchenko, E.V. and Talapin, D.V. (2008) Au–PbS core-shell nanocrystals: plasmonic absorption enhancement and electrical doping via intra-particle charge transfer. *Journal of the American Chemical Society*, **130**, 9673–5.

82 Schwartzberg, A.M., Grant, C.D., van Buuren, T. and Zhang, J.Z. (2007) Reduction of HAuCl$_4$ by Na$_2$S revisited: the case for Au nanoparticle aggregates and against Au$_2$S/Au Core/Shell particles. *Journal of Physical Chemistry, C*, **111**, 8892–901.

83 Norman, T.J. et al. (2002) Near infrared optical absorption of gold nanoparticle aggregates. *Journal of Physical Chemistry, B*, **106**, 7005–12.

84 Paltiel, Y., Aharoni, A., Banin, U., Neuman, O. and Naaman, R. (2006) Self-assembling of InAs nanocrystals on GaAs: the effect of electronic coupling and embedded gold nanoparticles on the photoluminescence. *Applied Physics Letters*, **89**, 033108-1–033108-3.

85 Xuan, Y. et al. (2007) Near infrared light-emitting diodes based on composites of near infrared dye, CdSe/CdS quantum dots and polymer. *Semiconductor Science and Technology*, **22**, 1021–4.

86 Dubertret, B. et al. (2002) In vivo imaging of quantum dots encapsulated in phospholipid micelles. *Science*, **298**, 1759–62.

87 Ablyazov, N.N., Areshkin, A.G., Melekhin, V.G., Suslina, L.G. and Fedorov, D.L. (1986) Fluctuation-induced broadening of exciton reflection spectra in A^{II}B^{VI} solid-solutions. *Physica Status Solidi B–Basic Research*, **135**, 217–25.

88 Niles, D.W. and Hochst, H. (1990) Band offsets and interfacial properties of cubic CdS grown by molecular-beam epitaxy on CdTe(110). *Physical Review B*, **41**, 12710–19.

89 Hines, M.A. and Scholes, G.D. (2003) Colloidal PbS nanocrystals with size-tunable near-infrared emission: observation of post-synthesis self-narrowing of the particle size distribution. *Advanced Materials*, **15**, 1844–9.

10
Nanowire Quantum Dots

Thomas Aichele, Adrien Tribu, Gregory Sallen, Catherine Bougerol, Régis André, Jean-Philippe Poizat, Kuntheak Kheng and Serge Tatarenko

10.1
Introduction

Over the past few decades, the use of semiconductor quantum dots (QDs) has become widespread in the development of new light sources, such as lasers, light-emitting diodes (LEDs), and single-photon sources, for nanoelectronic devices, and also in chemistry and the life sciences, where they act as nanoscopic probes and labels. Many of the initial experiments were conducted on either epitaxially grown self-assembled QDs or chemically synthesized, colloidal nanocrystal QDs. Self-assembled systems have certain advantages, due to their compatibility with semiconductor technologies and their longer operational times, whereas nanocrystals can be integrated flexibly into biological systems or linked to specific molecules. Nonetheless, over the years additional requirements for QD systems have emerged such that, today, QD samples of low density and at well-defined locations have become increasingly important for quantum optical technologies. Although many techniques have been developed to achieve this, these are restricted to specific materials or geometries. Self-assembly processes allow QD growth only within certain parameters (this applies especially to Group II–VI materials), while colloidal QDs suffer from bleaching and blinking effects that drastically reduce their efficiency. Thus, QDs in nanowires may offer an interesting alternative since, due to the narrow lateral size, any strain that is built up in heterostructure compositions can be relaxed directly onto the sidewall of the nanowire. Consequently, restrictions on self-assembly processes can be lifted, and heterostructures such as QDs can be included with a high degree of freedom. Additional parameters, such as the nanowire diameter can also often be controlled within large ranges. Although nanowire formation is specific to the materials and growth methods employed, nanowire growth has been reported for every important semiconductor material developed to date. In addition, by using nanomanipulation techniques nanowires can be detached from the as-grown substrate and positioned on secondary samples with nanometer precision. The absence of a wetting layer, which typically appears in self-assembled systems, means that nanowire QDs are in many respects more similar to nanocrystals. Nevertheless, blinking or bleaching are typically not reported.

The aim of this chapter is to highlight the selected current trends of research into nanowire QDs, notably with epitaxially prepared systems. After introducing the general concept of QDs, techniques for the epitaxial growth of QDs, nanowires and included heterostructures are reviewed, with the growth of CdSe/ZnSe nanowires using molecular beam epitaxy being described in detail. The applications of nanowire QDs as probes and sensors for biological molecules, as optically and electronically active devices in nanoelectronics, and as sources for single photons, applied as transmitters for quantum information, are then outlined.

10.2
Quantum Dots

It is well known that electrons, when confined in a potential trap with extensions that are similar to or smaller than those of the electron wave-function, exhibit a discrete energy spectrum. QDs, as intermediate systems in the evolution from a single atom to a solid, are semiconductor structures with very small spatial dimensions, surrounded by higher band-gap material. In such a system, electrons in the conduction band and holes in the valence band are strongly confined in all three spatial dimensions (see Figure 10.1). This leads to a discretization of the energy level scheme, which means that QDs are in many ways more similar to atoms than to bulk semiconductors.

In order to calculate the electronic states in QDs, several schemes have been developed with different levels of sophistication [2]. As an example, one of the simplest models, an electron (effective mass m^*) that is confined in a cubic potential of size L with infinite barriers possesses energy eigen-values of the form:

$$E_{lmn} = \frac{\pi^2 \hbar^2}{2m^* L^2}(l^2 + m^2 + n^2), \quad \text{for } l,m,n = 1,2,3,\ldots. \tag{10.1}$$

It can be seen that only discrete energies are allowed, comparable to the situation in atoms. Higher-dimensional structures as quantum wires [two-dimensional (2-D) confinement] and quantum wells [one-dimensional (1-D) confinement] also have quantized **k**-vectors along the confinement direction. For finite barriers and small sizes, only the first few states are considered, whilst above the potential barrier a continuum of energy levels is present. Although, here, many simplifications have been made compared to more realistic dot geometries, this model is suitable to provide a qualitative understanding and to demonstrate the discrete energy scheme. For more realistic potentials (see Ref. [2]), the degeneracy of the levels might be changed or lifted.

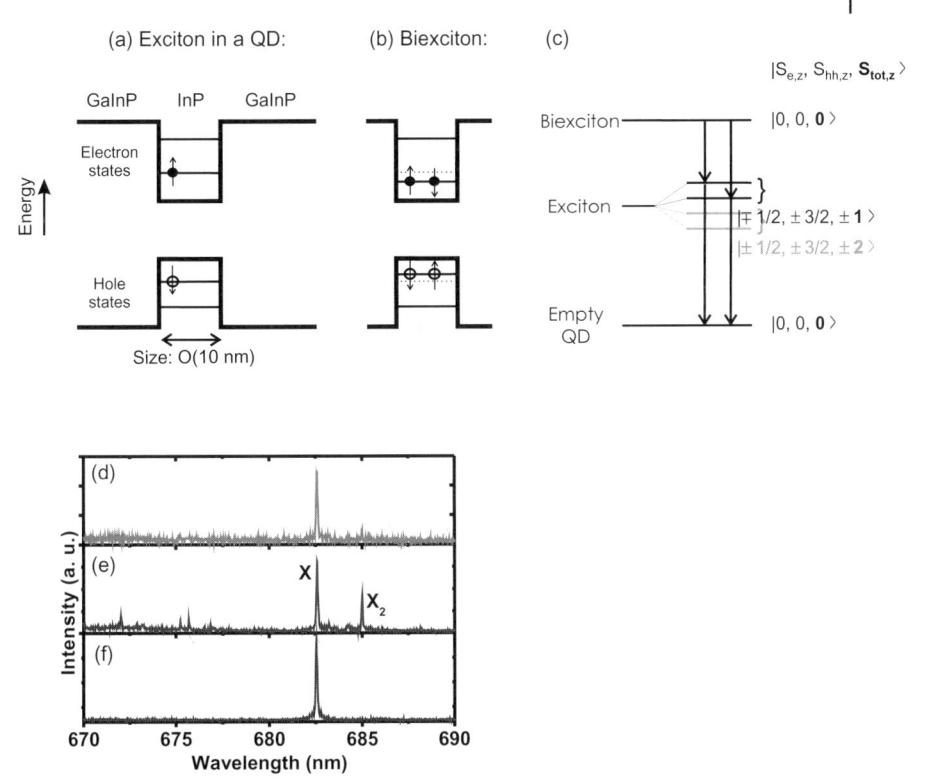

Figure 10.1 Excitations in a quantum dot. (a) Exciton formed by an electron-hole pair; (b) The biexciton containing two electron-hole pairs with generally a different energy than the exciton; (c) Schematic of the exciton and biexciton decay cascade. The two dark excitons are indicated by gray lines; (d–e) Photoluminescence spectra taken on a single InP quantum dot. Image (d) was acquired at low excitation power, and (e) at a higher excitation power. X and X_2 indicate the exciton and biexciton spectral line, respectively. Spectrum (f) was taken through a narrow bandpass filter. From Ref. [1].

When the QD is occupied with several quasi-free charge carriers (electrons or holes), the coulombic interaction must be taken into account. While equally charged carriers suffer a repulsion, for an electron-hole pair the energy of the system is lowered and an *exciton* is formed (Figure 10.1a). The recombination of the exciton leads to the emission of a single photon, with two typical regimes being distinguishable:

- If the extension of the QD clearly exceeds the exciton Bohr radius, the center-of-mass motion is quantized by the confinement potential, while the relative carrier motion is dominated by the coulombic interaction; this case is termed the *weak-confinement regime*.

- By contrast, in the *strong-confinement regime* the dot radius is smaller than the exciton Bohr radius, and the kinetic energy, due to size quantization, is the dominant energy contribution.

In the same way, two electron-hole pairs form a *biexciton*, but within a general different energy due to coulombic interaction (Figure 10.1b). When decaying, one electron-hole pair first recombines; this leads to the emission of one photon and to a remaining exciton in the QD, which in turn can lead to a second emission of a photon with a different wavelength (Figure 10.1c–f).

The excitation of a QD can be performed either electrically (see Refs [3, 4]) or optically. If the ground state of the exciton is pumped resonantly, then the absorbed and emitted light will have the same wavelength, which makes it impossible to separate luminescence and stray light in continuous-wave (cw) experiments. For excitation of the exciton close to resonance, an electron-hole pair is created in the higher settled states directly inside the QD. Such an excited exciton state relaxes very quickly and nonradiatively to its ground state. When exciting at energies higher than the band gap of the barrier material, free charges are created directly in the conduction and valence band of the bulk, respectively, and are finally captured by the QD, where they relax quickly to the exciton ground state. Figure 10.2

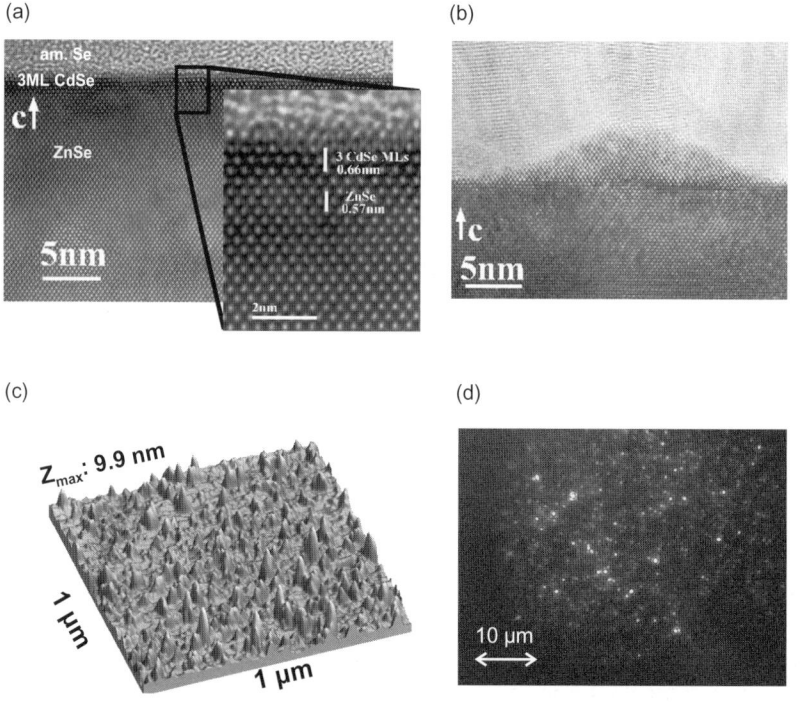

Figure 10.2 (a) Transmission electron microscopy image of a CdSe layer on ZnSe below the critical thickness (3 monolayers); (b) Above the critical thickness the layer relaxes by forming a QD; (c) Atomic force microscopy image of CdSe QDs distributed on a ZnSe surface; (d) Microphotoluminescence image of InP quantum dots in GaInP. The image was taken through a bandpass filter to suppress excitation stray light. Panels (a–c) are from Ref. [5]; panel (d) is from Ref. [6].

shows a transmission electron microscopy (TEM), atomic force microscopy (AFM) and photoluminescence (PL) images of self-assembled QD systems. A second type of epitaxial QD, where the heterostructures are included in nanowires, will be reviewed in greater detail in the next sections.

Due to the Pauli exclusion principle, each electron state can be occupied by at most two electrons corresponding to the two spin states $S_{e,z} \pm 1/2$. In low-dimensional systems, the dispersion relation is changed with respect to a bulk crystal, so that heavy holes (with $S_{hh,z} = \pm 3/2$) possess a lower energy than light holes ($S_{lh,z} = \pm 1/2$). Thus, in the lowest state, the QD can also be occupied by only two holes, leading to four combinations of the electron and hole spin in the exciton ground state with total spin $S_{tot,z} = (\pm 1, \pm 2)$. In the biexciton ground state, both spin states of the electrons and holes are occupied, resulting in a single state with $S_{tot,z} = 0$. As for radiative transitions, the change of spin must be carried away by the photon, two of the exciton states (with total spin ± 2) are dark states and participate neither in the biexciton nor the exciton decay, as symbolized in Figure 10.1c. A more extended overview of the level scheme of QD excitons can be found in Ref. [7].

10.3
Growth of Quantum Dots, Nanowires and Nanowire Heterostructures

A variety of techniques is available to achieve zero-dimensional QD structures. One of the first realizations of QDs were nanocrystal inclusions in glass [8] or in colloidal solutions (e.g., Ref. [9]), which emit at room temperature and are available for the whole visible spectrum. The main drawback here is a susceptibility for photobleaching due to chemical destruction, and for blinking. The latter condition describes the effect of interrupted emission even on large time scales, due to the presence of long-lived dark states.

A variety of production methods also exists by which QDs and quantum wires have been prepared, starting from higher-dimensional semiconductor heterostructures, such as etching pillars in quantum well (QW) systems or forming intersections of QWs or quantum wires [10]. The growth of nanostructures on patterned substrates, such as grooves and pyramids, has also led to successful QD and quantum wire formation [11, 12].

These fabrication methods allow a high degree of control over the positioning, and this is especially advantageous if there is a need to couple the QD to microcavities and photonic waveguides.

The vast majority of studies with QDs were conducted with self-assembled systems, with the QDs being fabricated by epitaxial growth of one type of crystal on top of another. If the lattice constants of the two materials are very close (such as AlAs and GaAs), then large thicknesses of the two layers can be achieved. However, if the lattice constants differ noticeably (e.g., Jn(Ga)As/GaAs), then small islands of the top material will be formed to minimize the strain [13]. The self-assembly process will compete with other strain-relaxation processes, and this

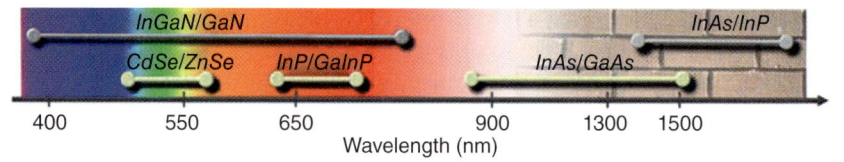

Figure 10.3 Schematic representation of the wavelength ranges accessible with different Stranski–Krastanow QD material systems. The region inaccessible to silicon detectors is indicated by a brick wall.

limits the achievable ranges for size, shape, and density of QD islands. A thin layer (the "wetting layer") will remain and completely cover the substrate. In this growth mode (known as Stranski–Krastanov growth), the wetting layer forms a QW which normally shows photoluminescence below the QD emission wavelength. Finally, the islands become overgrown by the substrate material to form QDs. Depending on the underlying semiconductor materials, self-assembled QDs can cover a broad spectral range, from ultraviolet to the infrared regime (see Figure 10.3). Moreover, they typically do not show any blinking and bleaching effects, and they allow Fourier-limited light emission [14], which is an important criterion in quantum information processing.

There are several important advantages for QDs in nanowires, compared to self-assembled systems. First, the strain of the heterostructure can relax directly onto the sidewalls, due to the narrow dimensions of the nanowire. Thus, there is no longer any the self-assembly process, and an increased freedom can be gained in composing the longitudinal size and composition of materials along the nanowire. It should be noted that nanowire QDs are not surrounded by a wetting layer that may introduce undesired nonradiative decay channels for excitations in the QD. In addition, the position of a QD along the nanowire axis can be defined by controlling the growth times of the according material. Finally, it is very easy to detach nanowires from the as-grown sample into a solvent, from which emitters with an arbitrary low density can be transferred to another substrate. Similar to other nanoparticles, positioning along the surface with nanometer precision can be achieved by using scanning probe microscopy techniques [15, 16], while electrical contacting results in electronically active nanowires [17–20]. The different growth principles of semiconductor nanowires are reviewed in the following sections.

One of the most frequently employed techniques is that of vapor–liquid–solid (VLS) growth, which was originally developed by Wagner and Ellis during the 1960s to produce micrometer-sized whiskers [21]. Since the 1990s, however, this method has been used to create nanowires and nanorods from a rich variety of materials, including elemental semiconductors [22, 23], Group III–V semiconductors [24–27], Group II–VI systems [28–30], and oxides [31–33]. The growth techniques used comprise diverse epitaxial methods, such as chemical vapor deposition (CVD) [23, 25, 26], molecular beam epitaxy (MBE) [28, 34], laser ablation [27, 31], or physical vapor transport [22]. In the VLS method, nanometer-sized metal particles are first deposited on the surface. During subsequent growth the substrate is heated above the melting point of the metal nanoparticles, to a temperature at which it forms a eutectic phase with one of the epitaxial semiconductor reactants. Continued feeding of the semiconductor atoms into the liquid droplet causes the eutectic phase to become supersaturated. This alloy then acts as a reservoir of reactants and favors growth at the solid–liquid interface such that a 1-D nanowire is formed with the alloy droplet remaining on the top. The size of the metal particle also affects the diameter of the nanowire and its rate of growth [31]. In addition to VLS growth, a catalyst-free growth of nanowire systems has also been reported, for both oxide systems [35, 36] and nitrides [37, 38]. An exhaustive overview of epitaxial and other developmental processes for nanowires is provided by Xia et al. [39] (and citations therein).

Figure 10.4 shows the growth process of MBE-grown ZnSe nanowires [40]. Following the preparation of a 0.5 nm gold film on a GaAs substrate, the sample was transferred into the ultrahigh vacuum MBE chamber, where the growth process was supervised using reflecting high-energy electron refraction (RHEED). Initially, sputtering of the gold film resulted in an amorphous layer. After annealing the sample to 580 °C for 5 min, the film is dewetted by forming small gold particles with a size distribution depending on the original film thickness (see inset of Figure 10.4b). This roughening of the surface is reflected by a point-like RHEED image. When starting the growth process, a Zn–Au alloy is formed that introduces the nanowire growth. Due to the mixed orientation of the nanowires in this example, circles appear and soon dominate the RHEED pattern.

After preparation of the gold particles, the samples were introduced into the MBE chamber and ZnSe growth was started. The scanning electron microscopy (SEM) and TEM images in Figure 10.5 show the sensitivity of the nanowire formation with respect to the growth conditions employed. When growing under an excess of Se (for the depicted sample, a Zn (Se) flux of $2.5(7.5) \times 10^{-7}$ torr was used) and a sample temperature of 350–450 °C, a dense carpet of narrow nanowires with high aspect ratios was seen to cover the substrate. The nanowires had a uniform diameter of 20–50 nm and a length up to 2 μm after growth for 1 h (Figure 10.5a). The according growth rate was approximately one monolayer per second, which clearly exceeded that of epitaxial 2-D growth at the same fluxes (typically 0.2 monolayer per second). When, on the other hand, growing at a low temperature (300 °C) or with an inverted Zn–Se flux ratio, needle-shaped nanowires were formed (Figure 10.5b). (Hereafter, these nanostructures will be referred to as nanoneedles, to distinguish them from the narrow nanowires described before.) The nanoneedles had a wide base (ca. 80 nm diameter) and a sharp tip (5–10 nm). Details of these studies have been reported elsewhere [40]. The formation of nanoneedles rather than nanowires can be explained by the slower adatom mobility expected at low temperatures, or at a low Se flux. The slower mobility promotes nucleation on the sidewalls before reaching the gold particle at the nanoneedle tip. Similar observations have been reported by Colli et al. [41]. The detailed shape in which nanowires are formed under specific conditions is, however, specific to both the material and the underlying growth technique.

(a) deposition of a gold film

(b) annealing (580 °C), dewetting of the gold film

(c) beginning of MBE growth

(d) inclusion of a CdSe QD

(e) end of growth

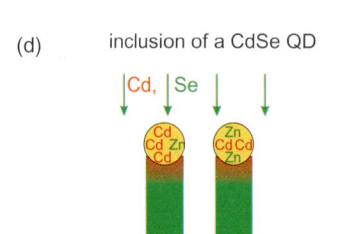

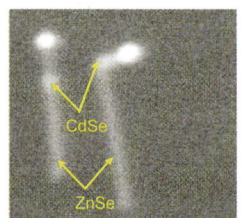

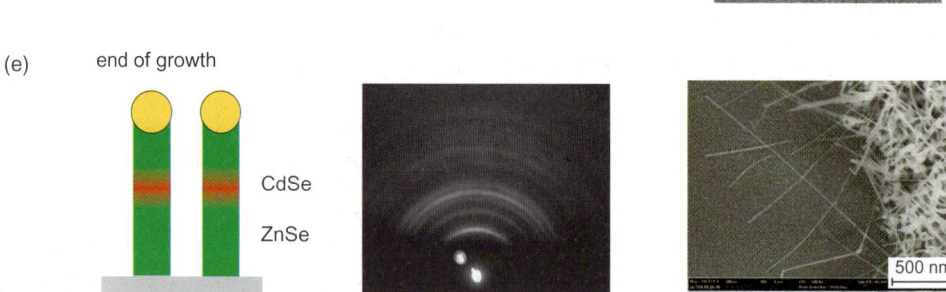

Figure 10.4 Principle of the VLS growth on the example of ZnSe MBE growth at 450 °C. (a) Deposition of a gold film; (b) Annealing and dewetting of the gold film; (c) Beginning of MBE growth; (d) Inclusion of a QD; (e) End of growth. Left column: Schematics of the growth; Central column: RHEED images obtained during growth; Right column: Scanning electron microscopy images of the sample after the respective growth step. The inset in (b) shows the size distribution of the gold particles after annealing of a 0.5 nm film.

10.3 Growth of Quantum Dots, Nanowires and Nanowire Heterostructures | 375

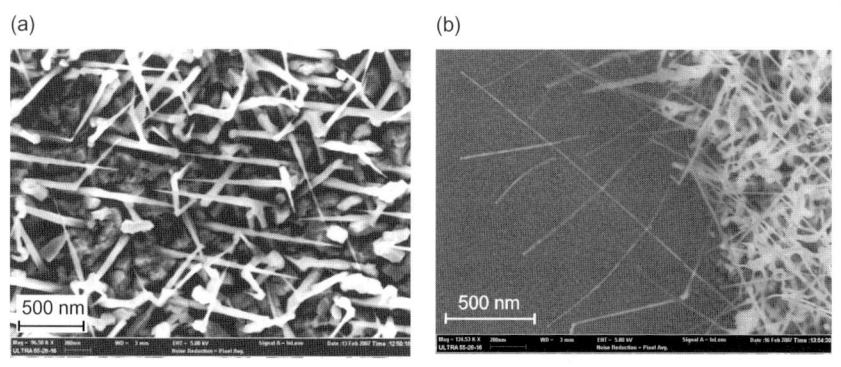

(a)   (b)

Figure 10.5 Scanning electron microscopy images of (a) ZnSe nanoneedles grown at 300 °C in an excess of Se, and (b) nanowires grown at 350 °C in an excess of Se.

When including heterostructures, such as a small QD zone, the VLS growth mechanism may introduce certain obstacles that must be overcome. For example, when replacing the element that is bound in the gold particle (e.g., Zn) by another material (e.g., Cd to create a CdSe QD), the initial material will not be immediately removed from the gold particle, but rather will cause a gradient of material composition (see Figure 10.4d,e). In this way, a smooth transition from ZnSe to CdSe is created instead of an abrupt transition from ZnSe to CdSe, which also smears out the QD potential and reduces the confinement strength of the charge carriers within the QD. Possible ways to avoid this effect may be to include heterostructures that replace the nonbound adatoms instead (e.g., Zn*Te* in Zn*Se*), or by previously emptying the gold particles from the Zn atoms by temporarily interrupting the Zn flux.

A further problem can be introduced due to stacking fault defects, such as localized changes of the crystalline structures, that are often observed in nanowires [25, 26, 40, 42]. Such defects can introduce nonradiative escape channels for the QD excitons, or they may act as local traps for charge carriers, that eventually radiatively decay and cause an additional spectral background that overlaps with the quantum emission [40, 42].

For the case of the above-described ZnSe nanowires, this circumstance is demonstrated in Figure 10.6. The spectra in the right-hand column were obtained using a photoluminescence set-up (as described later in Figure 10.9). The samples were excited by a continuous-wave diode laser at 405 nm, and the light was collected with an microscope objective and investigated with a grating spectrograph. Ensemble spectra were recorded on the as-grown sample, while results on single objects were obtained for nanostructures prepared on a separate substrate. Figure 10.6a,b shows the results for nanowires and nanoneedles taken from the samples depicted in Figure 10.5. The TEM images demonstrate the predominance of a wurtzite crystal structure, but with many intersections of regions in the zinc-blende phase. The presence of both wurtzite and zinc-blende shows that, under

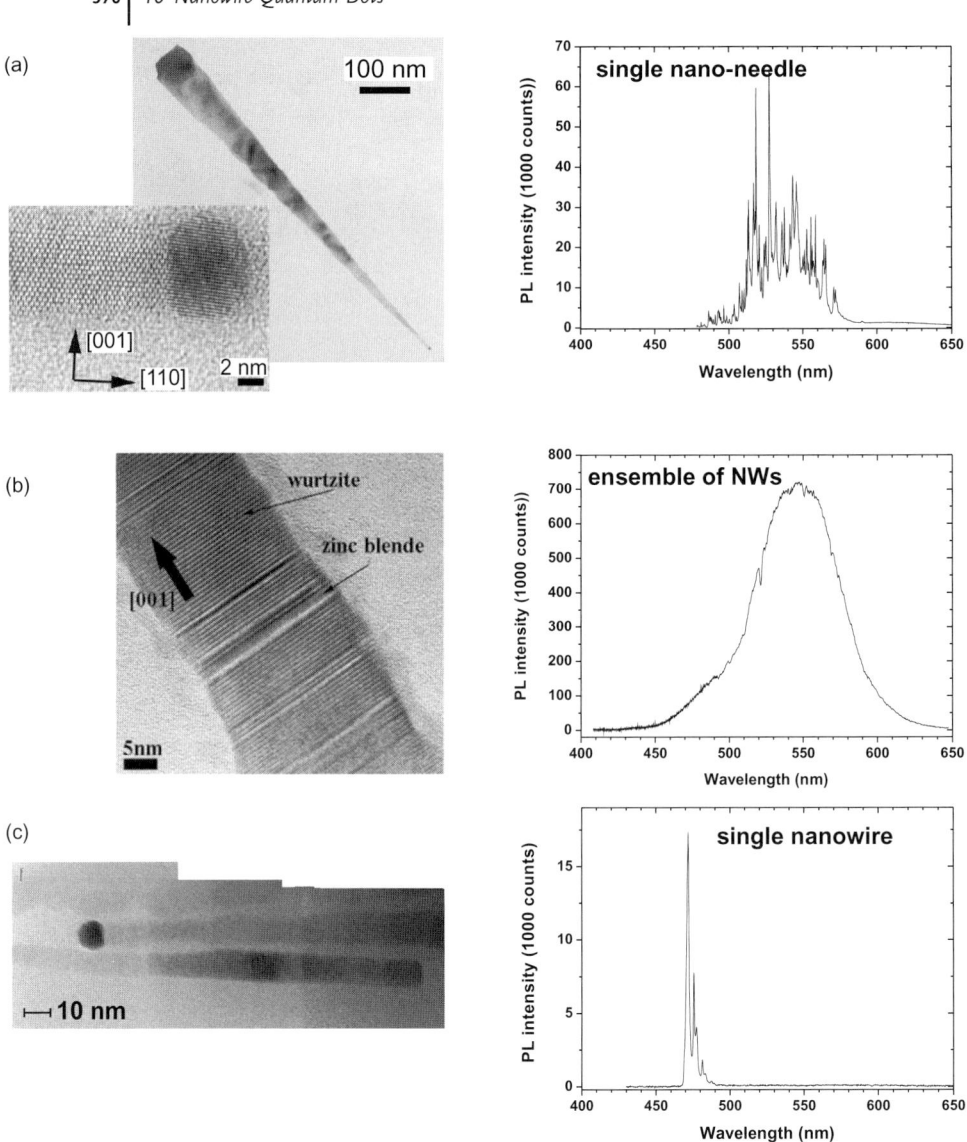

Figure 10.6 Transmission electron microscopy (TEM) images (left) and photoluminescence (PL) spectra (right) of (a) nanoneedles, (b) a nanowire, and (c) a nanowire grown on top of a nanoneedle (for details, see the text). The small TEM image in (a) shows a close-up into the nanoneedle tip region. The spectrum in (a) was obtained from an individual nanoneedle, and that in (b) from a nanowire ensemble.

the growth conditions used, both phases would be allowed. Although, the observation of wurtzite structure is in contrast to the zinc-blende that naturally occurs in bulk ZnSe, it is not an uncommon behavior for NWs (as discussed in Ref. [43]). Accordingly, the photoluminescence shows an intense emission at 500–600 nm, even in the case of individual nanoneedles. This suggests that point defects effectively capture the excited charge carriers and quench the band edge emission [42]. In contrast to the long nanowires, the defect planes in the nanoneedles (see Figure 10.6b) are here disoriented with respect to the nanoneedle axis. It seems that this disorientation hinders the propagation of defects in the growth direction, especially for lower diameters. Defects zones are rapidly blocked on the side walls, providing a high structural quality towards the nanoneedle tip. The observation of a decreasing defect density from the base towards the top in these nanoneedles inspired the following growth recipe. Initially, growth was started with an excess of Zn for 30 min, which led to the formation of nanoneedle structures. The Zn- and Se-flux was then inverted and nanowires were grown for another 30 min on top of the nanoneedles. In this way, growth at the sidewalls was aborted and regrowth started on the defect-free and strain-relaxed nanoneedle tips, where the high structural quality of the crystal lattice could be preserved along the narrow nanowire that is formed in this second growth step. As expected, the TEM images showed the nanowires to have a greatly reduced density of defects. An example of a nanowire with a monocrystalline structure over a large area is shown in Figure 10.6c. This was also reflected by the fluorescence spectra, where nanowires with almost no emission at wavelengths above 500 nm were identified.

The suppression of these undesired defects are a crucial precondition for the observation and investigation of QD fluorescence. In a next step, samples with a small region of CdSe inside the nanowires were prepared by interrupting the ZnSe growth, changing to CdSe for a short time, and then continuing the ZnSe growth (see SEM image in Figure 10.4d). From the CdSe growth time, the height of the CdSe slice was estimated as 1.5 to 4 nm, while the diameter of the nanowire was approximately 10 nm. These sizes were both on the order of the Bohr diameter of excitons in bulk CdSe ($2a_B = 11$ nm), which means that the carriers in these QDs were in the strong confinement regime. When observing individual QDs under microphotoluminescence, spectra as shown in Figure 10.7 may be recorded; these showed very few discrete spectral lines, similar to the spectra observed for self-assembled QD systems. The power-dependent scaling (Figure 10.7c), together with a comparison of the relative energy positions with known emission lines in the spectra of self-assembled CdSe/ZnSe QDs, suggests that these lines correspond to exciton (X), biexciton (XX) and charged exciton (CX) transitions. The X–CX (X–XX) energy splitting was found to be approximately 10 meV (20 meV), as compared to 15–22 meV (19–26 meV) for self-assembled CdSe/ZnSe QDs. The mean exciton energy was also similar (2.25 ± 0.08 meV) compared to 2.45 ± 0.2 meV for self-assembled CdSe/ZnSe QDs. Unambiguous proof for the assignment of these lines has been provided by using photon correlation spectroscopy [44]. These data demonstrate the successful implementation of CdSe QDs into ZnSe.

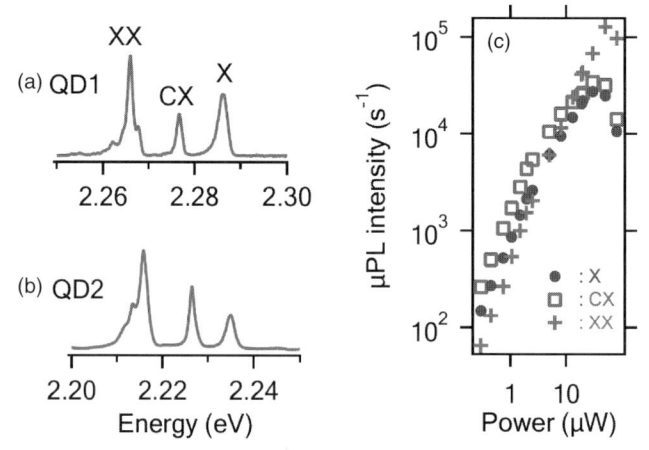

Figure 10.7 (a, b) Micro-photoluminescence spectra of two CdSe/ZnSe nanowire QDs. X: exciton, XX: biexciton, CX: charged exciton transition, respectively; (c) Line intensities of quantum dot QD2 as a function of excitation power.

10.4
Applications for Nanowires and Nanowire Quantum Dots

10.4.1
Nanowires and Quantum Dots in the Life Sciences

Whilst the original interest in semiconductor nanostructures arose from the physical and chemical sciences, attention was very soon also received from the areas of biology and medicine. The most important point here is the very small sizes of these structures, which typically are on the same order as proteins. By taking such structures as the starting point, much larger assemblies can be organized that approach the order of biological systems, such as viruses.

In biology, chemically synthesized nanocrystal QDs are used as imaging labels. Compared to organic dye molecules, nanocrystals have several advantages, including a higher photostability and a higher quantum yield. The latter parameter allows photoexcitation at lower intensities, thus reducing unwanted interactions with the specimen. Moreover, the emission wavelength of nanocrystals can be tuned over a large spectral range, simply by adjusting their size. The chemical synthesis process can be performed with high size uniformity since, due to the specific chemical synthesis process, the nanocrystal surface can be coated with organic ligands (e.g., trioctylphosphine oxide, TOPO) [45]. When used in biological labeling, bioconjugates for specific proteins or nucleic acids can be attached to the ligands [46]. In recent studies, nanocrystal labels have been used *in situ* and *in vivo* as labels in a variety of biological systems, including specific genomic sequences, antigens, plasma membrane proteins, and cytoplasmic proteins (see Ref. [46] and citations therein). A further promising area is the development of nanocrystal-based biosensors, in which the presence of specific molecules quenches the photoemission and thus allows their detection [47].

In another fascinating application, semiconducting nanowires can be integrated into miniaturized electronic systems that allow the sensing and electrical readout of single biological nano-objects, the basic building block of which is the nanowire field-effect transistor (FET). In a standard FET, a doped semiconductor is connected to metal electrodes (called the source and drain), through which a current is injected. The conductance of the semiconductor between the source and drain can then be controlled by a gate electrode which is coupled via a thin dielectric layer. Nanowire FETs, where the semiconductor is formed by a nanowire, have the capacity to outperform conventional FETs (see also Section 10.4.2). In FET sensors, charged or polar biological or chemical species take over part of the gate voltage. Here, nanowires are of special interest, as they offer a large surface-to-volume ratio and, by linking surface receptors to the nanowire surface, the device can be made sensitive to specific molecules. When such a sensor is placed in a solution with accordingly charged macromolecules, binding to the receptors will lead to a change in conductivity through the transistor that can be detected in real time [48]. At their very limit, nanowire sensors can detect *single* objects, as demonstrated recently for a single virus [49]. Consequently, by fabricating arrays of different nanowires sensors, small-scale, multiplexed sensors for a wide range of molecules can be created [50] that may eventually serve as a powerful diagnostic tool for medical applications. A comprehensive review on this topic has been produced by Patolski *et al.* [51].

10.4.2
Nanowire Electronic Devices

The most prominent applications of QDs and nanowires are their use as light-emitting devices and laser-active materials. A QD laser from an ensemble of (ideally identical) QDs has very interesting properties, such as ultra-low threshold lasing, a reduced-temperature sensitivity of the threshold, and high material and differential gain [52, 53]. Free-standing semiconductor nanowires represent attractive building blocks for creating electrically driven lasers, because their defect-free structures exhibit the superior electrical transport of high-quality planar inorganic devices. If the ends of the nanowire are cleaved, they can act as two reflecting mirrors that define a Fabry-Pérot optical resonator, and can thus function as a stand-alone optical cavity and gain medium for laser activity. Such nanowire lasing was first realized with optically pumped ZnO nanowires by Huang *et al.* [54], and for electrically driven CdS nanowires by Duan *et al.* [55].

The possibility of achieving a broad wavelength regime via the correct choice of material and size of both QDs and nanowires (see Figure 10.3) also makes these materials attractive for the fabrication of LEDs of diverse colors. The very high light extraction efficiency, in addition to the fact that high-quality nanowires can be grown on cheap substrates such as silicon, makes nanowire LEDs very attractive for the lighting/display industry [17, 19, 56, 57]. The use of a single QD nanowire LED was demonstrated by Minot et al. [20].

Semiconducting nanowires are also potential alternatives to planar metal-oxide semiconductor field-effect transistors (MOSFETs). For example, Xiang et al. [58] have demonstrated a nanowire FET made from a Ge/Si core-shell nanowire, the transconductance and on-current of which substantially exceeded that of planar silicon MOSFETs by a factor of three to four.

10.4.3
Single-Photon Sources

10.4.3.1 Single-Photon Generation

Photons are ideal carriers to transmit quantum information over large distances, due to their low interaction with the environment. In 1984, Bennett and Brassard proposed a secret key-distribution protocol [59] that used the single-particle character of a photon to avoid any possibility of eavesdropping on an encoded message (for a review, see Ref. [60]). The implementation of efficient quantum gates based on photons and linear optics was also proposed [61] and demonstrated [62, 63]. In addition, photons have been proposed for use as information carriers in larger networks between the processing nodes of stationary qubits [64].

The application of linear optics in quantum information processing requires the reliable generation of single- or few-photon states, on demand. However, due to their bosonic character, photons tend to appear in bunches, a characteristic which hinders the implementation of classical sources particularly in quantum cryptographic systems, as an eavesdropper may gain partial information via a beam splitter attack. Similar obstacles occur for linear optics quantum computation, where photonic quantum gates [61], quantum repeaters [65], and quantum teleportation [66] require the preparation of single- or few-photon states *on demand* in order to obtain reliability and high efficiency.

The single-photon character of a light source can be tested by measuring the normalized second-order correlation function $g^{(2)}(t_1, t_2)$ via detecting the light intensity $\langle \hat{I}(t) \rangle$ at two points in time. For stationary fields, it reads

$$g^{(2)}(\tau = t_1 - t_2) = \frac{\langle :\hat{I}(0)\hat{I}(\tau): \rangle}{\langle \hat{I}(0) \rangle^2} \tag{10.2}$$

where: : denotes normal operator ordering. For classical fields, $g^{(2)}(0) \geq 1$ and $g^{(2)}(0) \geq g^{(2)}(\tau)$ hold, which prohibits values smaller than unity (for a comprehensive discussion of light field characteristics, see Ref. [67]). For thermal light sources, there is an increased probability of detecting a photon shortly after another, and this bunching phenomenon leads to $g^{(2)}(0) \geq 1$ (Figure 10.8a). For coherent states, photons appear independent of each other in time, such that $g^{(2)}(\tau) = 1$ for all τ, according to a Poisson photon number distribution (Figure 10.8b). A single-photon state shows the anti-bunching effect of a sub-Poisson

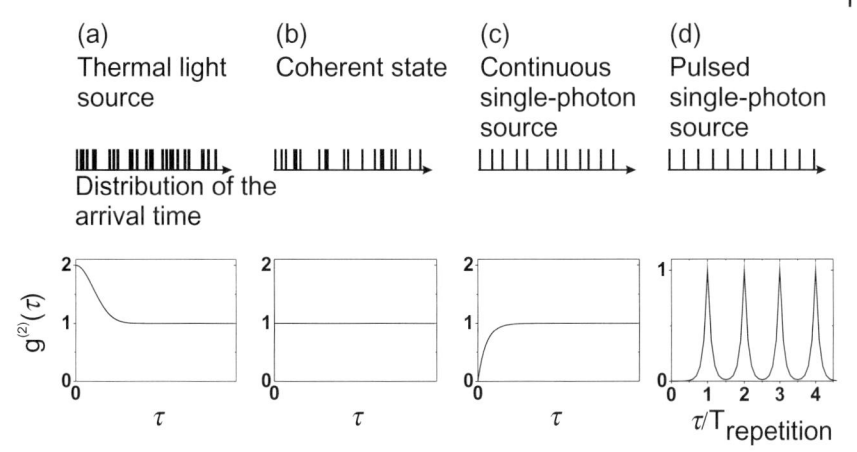

Figure 10.8 $g^{(2)}$ function of (a) a thermal light source, (b) coherent light, (c) a continuously driven single-photon source, and (d) a pulsed single-photon source. The top row symbolizes the different arrival times of photon detection events.

distribution with $g^{(2)}(0) \leq 1$ (Figure 10.8c). This represents the fact, that the appearance of a photon reduces the probability for a second photon within a time window. Ideally, for a single-photon state $g^{(2)}(0) = 0$, while higher values account for the statistically increased contribution of additional photons. The pulsed excitation of a single-photon source leads to a peaked structure in $g^{(2)}(\tau)$, with a missing peak at zero delay indicating triggered single-photon emission (Figure 10.8d).

In order to circumvent the detector dead time (e.g., ≈50 ns for avalanche single-photon detectors), the second-order correlation function is measured in a Hanbury–Brown–Twiss arrangement [68] as depicted in Figure 10.9, which consists of two photo detectors and a 50:50 beam splitter. A large number of time intervals between detection events is measured and binned together in a histogram.

10.4.3.2 High-Temperature Single-Photon Emission from Nanowire QDs

A promising process for single-photon generation is the spontaneous emission from a single quantum emitter. Numerous emitters have been used to demonstrate single-photon emission [69], though single atoms or ions represent the most fundamental systems [70–72]. Other systems capable of single-photon generation are single molecules and single nanocrystals [73, 74], although their drawback is a susceptibility for photobleaching and blinking [75]. Stable alternatives are nitrogen-vacancy defect centers in diamond [76, 77], but these show broad optical spectra together with comparably long lifetimes.

Single semiconductor QDs allow the stable generation of single photons over a wide spectral range, as demonstrated elsewhere [1, 78–81]. Moreover, they are compatible with current semiconductor manufacturing technologies. For commercial applications – and especially quantum cryptography – compact and easy-to-handle devices that function at room temperature are desired. However, in contrast

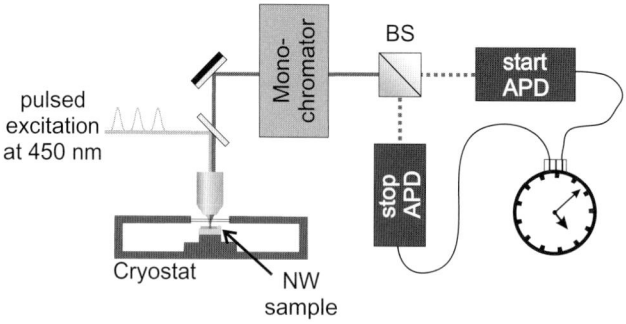

Figure 10.9 Sketch of the micro-photoluminescence set-up. A pulsed laser excites the nanowire (NW) QDs inside a helium cryostat through a microscope objective. The fluorescence is filtered by a monochromator and projected over a beamsplitter (BS) on two single-photon detectors (APDs), from which time-dependent coincidences were measured.

to diamond color centers and nanocrystals, this has not yet been achieved, and represents the major drawback of QD systems.

Alternatively, a QD formed inside a semiconductor nanowire may allow the design of well-positioned and size-controlled QDs with no wetting layer, and without the need for self-organization mechanisms. Single-photon emission from a GaAsP/GaP nanowire QD was first reported at cryogenic temperature by Borgström et al. [82], who prepared the nanowires by using MOCVD and transferred them to a patterned SiO_2 substrate before optical characterization. Both, exciton and biexciton transitions were identified, and a profound pulsed anti-bunching at 5.2 K was achieved which indicated the emission of only one photon per laser pulse. Moreover, a high quantum yield with photon count rates exceeding one million per second was observed. The brightness of these QDs was explained due to the fact that, unlike self-assembled QDs, they were not suffering from being embedded in a high-refractive index material, although this effect was restricted to low temperature. In the same study, the stacking of several QDs in one nanowire was also explored, which established a first step towards the design of QD molecules or chains, where coupling between charges (spins) in essentially identical QDs could be mediated via either tunneling or dipole–dipole interactions [82]. Furthermore, the possibility of electrical excitation represents a very realistic and promising perspective for semiconducting nanowires [20].

For high-temperature operation, Group II–VI materials are especially interesting, as they may exhibit large biexciton binding energies and strong carrier confinements [80]. High-temperature single-photon emission up to 220 K from a single CdSe quantum inside a ZnSe nanowire was recently reported [83], and is reviewed in the following section.

The growth of these nanowires was described above (see Figure 10.6c and Section 10.3) when, in order to prepare QDs, a small region of CdSe was inserted into the ZnSe nanowire. When studying single nanowires, the sample was placed in a methanol ultrasonic bath for 30 s; this caused some nanowires to break from the substrate into the solution. Droplets of the solution were then placed on a clean substrate, leaving behind a low density of individual nanowires. Metal markers on the substrate prepared by optical lithography allowed orientation on the sample surface.

The sample was mounted on a variable-temperature cryostat which allowed experiments to be conducted over a temperature range from 4 K to room temperature. The nanowire emission was efficiently collected using a microscope objective and then dispersed with a monochromator. The sample was excited with a 200 fs-pulsed, frequency-doubled Ti:Sa laser. The second-order correlation function $g^{(2)}(\tau)$ was obtained using a Hanbury-Brown–Twiss set-up, in which the photoluminescence signal (filtered through a monochromator) is split on a 50:50 beamsplitter and projected on two avalanche single-photon detectors. Coincidences as a function of the delay time were detected in a time correlation electronics, based on a time-to-amplitude converter.

The typical spectra of the QDs (as shown in Figure 10.7) showed discrete spectral lines from the exciton, biexciton, and charged exciton transitions. One characteristic feature of such nanowire structures is their polarization behavior. As seen in Figure 10.10a, the excitation efficiency and luminescence are both heavily polarization-dependent. The photoluminescence is highly polarized with a contrast of 80–90% (the nanowire was excited here with a circularly polarized laser light). Conversely, the photoluminescence intensity has a sine-like variation as a function of the linear laser polarization. In this case, the emitted light was highly polarized with the same direction as the preferred excitation polarization. Previously, the

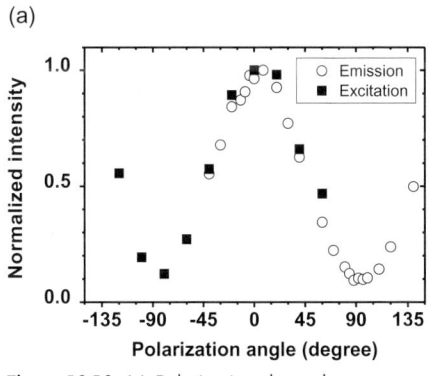

Figure 10.10 (a) Polarization dependence. Solid squares indicate the variation of the total photoluminescence intensity as a function of the polarization of the excitation laser. Open circles indicate the photoluminescence intensity observed through an analyzing polarizer while the excitation laser was circularly polarized; (b) Radiative lifetime of an ensemble of nanowire structures as a function of the sample temperature. The solid (dashed) curve is a linear (exponential) fit to the data points below (above) 100 K. Inset: Ensemble spectrum observed during the time-resolved measurement at 4 K. The vertical shaded area is the spectral range within which the lifetimes were evaluated. From Ref. [83].

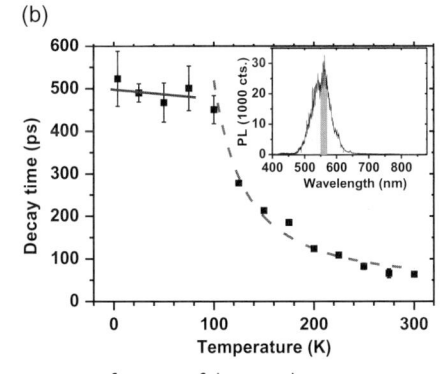

polarization has been found to be highly oriented along the nanowire emission [84, 85]. This striking polarization anisotropy of absorption can be explained by the dielectric contrast between the nanowire material and the surrounding environment; the polarization of the emitted light results from a competition between these electromagnetic effects and the orientation of the dipole within the QD [85, 86].

Time-resolved measurements of the exciton lifetimes were performed using a streak camera (with a resolution of ca. 1 ps) on an ensemble of nanowires. Figure 10.10b shows a flat temperature-dependency of the decay time at low temperature, followed by an exponential decay above 100 K. The flat temperature-dependency is characteristic of 3-D confinement (confined excitons), while the exponential decay is characteristic of a nonradiative recombination regime. In the radiative regime, the decay time was approximately 500 ps and dominated up to 100 K; this was slightly larger than is observed in self-assembled CdSe/ZnSe QDs (ca. 300 ps [87]), but may be due to a piezoelectric field resulting from the wurtzite structure in the nanowires, which separates the electron and hole wave functions and thus reduces the oscillator strength.

Figure 10.11b shows the spectra from a nanowire quantum at different temperatures. The spectral line can be attributed to a charged-exciton (trion) transition.

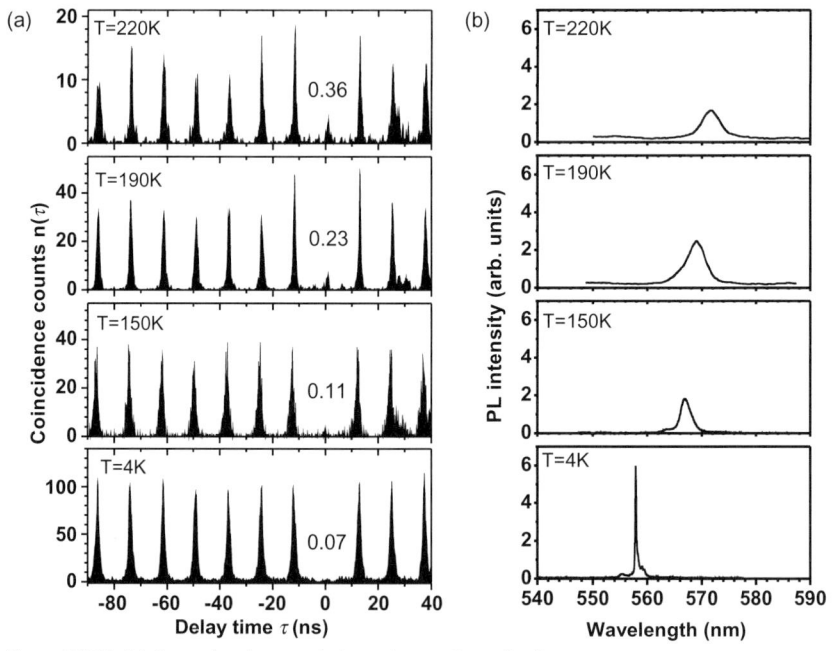

Figure 10.11 (a) Second-order correlation taken under pulsed excitation at temperatures between 4 and 220 K. The numbers in the graphs are measured values of $g^{(2)}(0)$ (i.e., the area under the peak at $\tau = 0$ relative to the peaks at $|\tau| > 0$); (b) Spectra from the same nanowire QD taken during the same experimental run as Figure 10.3a. From Ref. [83].

With higher excitation intensity, an exciton (X) and biexciton transition (XX) also become visible, whilst above 150 K the spectral lines are significantly broadened. Finally, the autocorrelation measurements in Figure 10.11 characterize the single-photon character of the studied nanowire QD. These graphs are the raw histograms of measured coincidences, without any correction for background count events. The area under each peak at time $\tau = 0$. was normalized with respect to the average area under the peaks at $\tau > 0$. Each peak area was calculated by integrating the coincidences within 12 ns windows (the repetition time of the excitation laser). The correlation functions were taken at different temperatures between 4 K and 220 K. At 4 K, the peak at $\tau = 0$ is suppressed to a normalized value of 7%, showing the high quality of the single-photon generation. With increasing temperature, this value increases only slightly to finally reach 36% at 220 K. This value is far below 50%, and the emitted light field is thus clearly distinguished from states with two or more photons. Therefore, even without correcting for background events, these emitters can be used directly as a high-quality, single-photon device even when operating at high temperature, with a strongly suppressed probability for two-photon events. At this temperature, Peltier cooling becomes an alternative to liquid helium or nitrogen cooling, which makes these emitters an interesting candidate for the development of compact, stable and cost-efficient quantum devices operating near room temperature. More details of these autocorrelation measurements as a function of the temperature are reported in Ref. [83]. For a comparison, high-temperature experiments from individual self-assembled QDs were reported before from CdSe/ZnSe [80] and from GaN/AlN QDs [81]. Both of these experiments showed photon anti-bunching up to 200 K, but with normalized dip values above 0.5 (81% and 53%, respectively).

Another promising material system for producing single photons from nanowire QDs at high temperature are GaN QDs. Single GaN QDs may also be used to demonstrate single-photon emission at room temperature, due to large band offsets between GaN and AlN. Such GaN/AlGaN heterostructures present a fundamental interest due to their peculiarities: when grown along the c-axis, they exhibit a huge internal electric field as a consequence of the wurtzite structure. Furthermore, the large band offsets and the large hole effective mass lead respectively to strong confinement and localization effects. In QDs grown in the Stranski–Krastanow mode, the QD density is difficult to master, and defects in the structure lead to residual emission in the QD emission spectral region, and also to strong spectral diffusion effects. Concurrently, nanowire geometry appears to be a solution for achieving a high crystal quality in III–N material. Therefore, defect-free GaN QDs based on nanowire heterostructures are expected to be achieved, and to overcome the limitations occurring in the case of Stranski–Krastanow QDs. The narrow emission (down to 1 meV) of single GaN/AlN QDs in nanowires, and the identification of excitonic and biexcitonic recombinations, have been recently reported by Renard et al. [34]. In order to create GaN/AlN insertions, GaN nanowires are first grown on a Si(111) substrate under a nitrogen-rich atmosphere by using plasma-assisted MBE. The nanowires grew along the c-axis, which is the polar direction of the wurtzite structure. After the growth of GaN nanowires, an AlN barrier was deposited, followed successively by a GaN

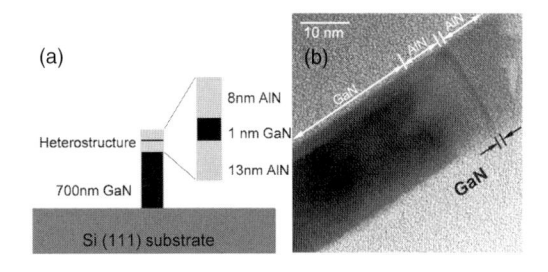

Figure 10.12 (a) Schematic illustration of GaN/AlN inclusions on the top of GaN nanowires grown on Si (111) substrate; (b) High-resolution TEM image of a single GaN/AlN inclusion; (c and d) Power-dependent spectra of QD emissions at 5 K. Each spectrum has been normalized to its maximum intensity. Excitation powers (increasing from the bottom): 10, 50, 100, 200, 300, and 500 μW for (a), and 25, 100, 120, 170, 200, 250, and 300 μW for (b). From Ref. [34].

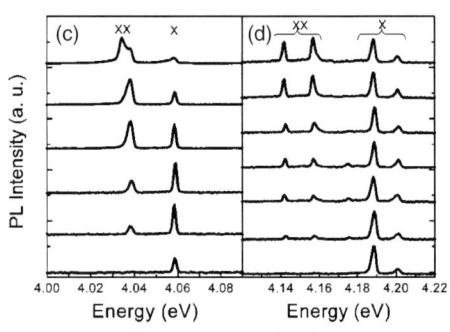

insertion and a top AlN barrier (see Figure 10.12a,b). Again, both excitonic and biexcitonic transitions were observed (Figure 10.12c,d). Although, single-photon emission has not yet been confirmed in this system due to a large spectral background emission, these results described, for the first time, a large positive biexciton binding energy in GaN/AlN QDs, consistent with the negligible effect of the electric field in these thin GaN QDs.

10.5
Conclusions

In this chapter, the growth of ZnSe nanowires and of CdSe/ZnSe nanowire heterostructures using MBE has been described. The formation of nanowires and nanoneedles has been observed, depending on the detailed growth parameters, and a route was described towards the suppression of stacking faults that appear in these structures. In this way, ZnSe nanowires with a mostly monocrystalline tip can be created, in which perturbing emission from trapped charge carriers are strongly reduced. The growth techniques of other QDs and nanowire systems were also reviewed, and the application of nanowires and nanowire QDs in biology, medicine, nanoelectronics and photonics was discussed. Particular attention was paid to the development of single-photon generation, with single-photon emission from CdSe/ZnSe nanowire QDS systems up to a temperature of 220 K being observed in particular. In this regime, compact and cost-efficient single-photon devices based on epitaxial QDs should become feasible. The ease with which low-density samples can be prepared, together with the possibility of electric contacting, underlines the suggestion that nanowires possess great promise for future use in devices in the life sciences, electronics, and quantum photonics.

References

1. Aichele, T., Zwiller, V. and Benson, O. (2004) Visible single-photon generation from semiconductor quantum dots. *New Journal of Physics*, **6**, 90.
2. Bimberg, D., Grundmann, M. and Ledentsov, N. (1998) *Quantum Dot Heterostructures*, John Wiley & Sons, Inc., New York.
3. Yuan, Z., Kardynal, B.E., Stevenson, R.M., Shields, A.J., Lobo, C.J., Cooper, K., Beattie, N.S., Ritchie, D.A. and Pepper, M. (2002) Electrically driven single-photon source. *Science*, **295** (5552), 102–5.
4. Xu, X., Williams, D.A. and Cleaver, J.R.A. (2004) Electrically pumped single-photon sources in lateral p-i-n junctions. *Applied Physics Letters*, **85** (15), 3238–40.
5. Robin, I., André, R., Bougerol, C., Aichele, T. and Tatarenko, S. (2006) Elastic and surface energies: two key parameters for CdSe quantum dot formation. *Applied Physics Letters*, **88** (23), 233103.
6. Aichele, T., Scholz, M., Ramelow, S. and Benson, O. (2006) Non-classical light states from single quantum dots, in *Advances in Atomic, Molecular, and Optical Physics*, chapter 1, vol. 1 (eds G. Rempe and M. Scully), Academic Press, USA.
7. Dekel, E., Gershoni, D., Ehrenfreund, E., Garcia, J.M. and Petroff, P.M. (2000) Carrier-carrier correlations in an optically excited single semiconductor quantum dot. *Physical Review B (Condensed Matter and Materials Physics)*, **61** (16), 11009–20.
8. Eldmov, A.I., Onushchenko, A.A. and Tsekhornskii, V.A. (1980) Exciton light absorption by CuCl microcrystals in glass matrix. *Soviet Journal of Glass Physics and Chemistry*, **6**, 511–12.
9. Murray, C.B., Norris, D.J. and Bawendi, M.G. (1993) Synthesis and characterization of nearly monodisperse CdE (E = sulfur, selenium, tellurium) semiconductor nanocrystallites. *Journal of the American Chemical Society*, **115** (19), 8706–15.
10. Schedelbeck, G., Wegscheider, W., Bichler, M. and Abstreiter, G. (1997) Coupled quantum dots fabricated by cleaved edge overgrowth: from artificial atoms to molecules. *Science*, **278**, 1792.
11. Jetter, M., Pérez-Solórzano, V., Gröning, A., Ubl, M., Gräbeldinger, H. and Schweizer, H. (2004) Selective growth of GaIN quantum dot structures. *Journal of Crystal Growth*, **272**, 204.
12. Merano, M., Sonderegger, S., Crottini, A., Collin, S., Renucci, P., Pelucchi, E., Malko, A., Baier, M.H., Kapon, E., Deveaud, B. and Ganiere, J.-D. (2005) Probing carrier dynamics in nanostructures by picosecond cathodoluminescence. *Nature*, **438**, 479.
13. Notzel, R. (1996) Self-organized growth of quantum-dot structures. *Semiconductor Science and Technology*, **11** (10), 1365–79.
14. Santori, C., Fattal, D., Vuckovic, J., Solomon, G.S. and Yamamoto, Y. (2002) Indistinguishable photons from a single-photon device. *Nature*, **419** (6907), 594–7.
15. Junno, T., Deppert, K., Moritelius, L. and Samuelson, L. (1995) Controlled manipulation of nanoparticles with an atomic force microscope. *Applied Physics Letters*, **66** (26), 3627–9.
16. Wang, Y., Zhang, Y., Li, B., Lü, J. and Hu, J. (2007) Capturing and depositing one nano-object at a time: single particle dip-pen nanolithography. *Applied Physics Letters*, **90** (13), 133102.
17. Haraguchi, K., Katsuywna, T., Hiruma, K. and Ogawa, K. (1992) GaAs p-n junction formed in quantum wire crystals. *Applied Physics Letters*, **60** (6), 715–47.
18. Kim, H., Cho, Y., Lee, H., Kim, S.I., Ryu, S.R., Kim, D.Y., Kang, T.W. and Chung, K.S. (2004) High-brightness light-emitting diodes using dislocation-free indium gallium nitride/gallium nitride multiquantum-well nanorod arrays. *Nano Letters*, **4** (6), 1059–62.
19. Bao, J., Zimmler, M.A., Capasso, F., Wang, X. and Ren, Z.F. (2006) Broadband ZnO single-nanowire light-emitting diode. *Nano Letters*, **6** (8), 1719–22.
20. Minot, E.D., Kelkensberg, F., van Kouwen, I.V.I., van Dam, J.A., Kouwenhoven, L.P., Zwiller, V., Borgström, M.T., Wunnicke, O.,

Verheijen, M.A. and Bakkers, E.P.A.M. (2007) Single quantum dot nanowire LEDs. *Nano Letters*, **7** (2), 367–71.

21 Wagner, R.S. and Ellis, W.C. (1964) Vapor-liquid-solid mechanism of single crystal growth. *Applied Physics Letters*, **4** (5), 89–90.

22 Wu, Y. and Yang, P. (2000) Germanium nanowire growth via simple vapor transport. *Chemistry of Materials*, **12** (3), 605–7.

23 Westwater, J., Gosain, D.P., Tomiya, S., Usui, S. and Ruda, H. (1997) Growth of silicon nanowires via gold/silane vapor-liquid-solid reaction. *Journal of Vacuum Science & Technology B: Microelectronics and Nanometer Structures*, **15** (3), 554–7.

24 Yeh, C.-C. and Chen, C.-C. (2000) Large-scale catalytic synthesis of crystalline gallium nitride nanowires. *Advanced Materials*, **12** (10), 738–41.

25 Hiruma, K., Yazawa, M., Katsuyama, T., Ogawa, K., Haraguchi, K., Koguchi, M. and Kakibayashi, H. (1995) Growth and optical properties of nanometer-scale GaAs and InAs whiskers. *Journal of Applied Physics*, **77** (2), 447–62.

26 Yazawa, M., Koguchi, M., Muto, A. and Hiruma, K. (1993) Semiconductor nanowhiskers. *Advanced Materials*, **5**, 577–80.

27 Shi, W.S., Zheng, Y.F., Wang, N., Lee, C.S. and Lee, S.T. (2001) Synthesis and microstructure of gallium phosphide nanowires. *Journal of Vacuum Science & Technology. B: Microelectronics and Nanometer Structures*, **19** (4), 1115–18.

28 López-López, M., Guillén-Cervantes, A., Rivera-Alvarez, Z. and Henrnández-Calderón, I. (1998) Hillocks formation during the molecular beam epitaxial growth of ZnSe on gags substrates. *Journal of Crystal Growth*, **193** (1), 528–34.

29 Wang, Y., Zhang, L., Liang, C., Wang, G. and Peng, X. (2002) Catalytic growth and photoluminescence properties of semiconductor single-crystal ZnSe nanowires. *Chemical Physics Letters*, **357** (3-4), 314–18.

30 Wang, Y., Zhang, L. and Zhang, J. (2002) Catalytic growth of large-scale single-crystal CdS nanowires by physical evaporation and their photoluminescence. *Chemistry of Materials*, **14** (4), 1773–7.

31 Morales, A.M. and Lieber, C.M. (1998) A laser ablation method for the synthesis of crystalline semiconductor nanowires. *Science*, **279** (5348), 208–11.

32 Huang, M.H., Wu, Y., Feick, H., Tran, N., Weber, E. and Yang, P. (2001) Catalytic growth of zinc oxide nanowires by vapor transport. *Advanced Materials*, **13**, 113–16.

33 Chen, Y., Li, J., Han, Y., Yang, X. and Dai, J. (2002) The effect of nig vapor source on the formation of mgo whiskers and sheets. *Journal of Crystal Growth*, **245** (1–2), 163–70.

34 Renard, J., Songmuang, R., Bougerol, C., Daudin, B. and Gayral, B. (2008) Exciton and biexciton luminescence from single GaN/AlN quantum dots in nanowires. *Nano Letters*, **8** (7), 2092–6.

35 Ristic, J., Sánchez-García, M.A., Calleja, E., Sanchez-Páramo, J., Calleja, J.M., Jahn, U. and Ploog, K.H. (2002) AlGaN nanocolumns grown by molecular beam epitaxy: optical and structural characterization. *Physica Status Solidi A*, **192** (1), 60–6.

36 Zhang, Y., Jia, H. and Yu, D. (2004) Metal-catalyst-free epitaxial growth of aligned ZnO nanowires on silicon wafers at low temperature. *Journal of Physics D: Applied Physics*, **37** (3), 413–15.

37 Bartness, K.A., Sanford, N.A., Barker, J.M., Schlager, J.B., Rosllho, A., Davydov, A.V. and Levin, I. (2006) Catalyst-free growth of GaN nanowires. *Journal of Electronic Materials*, **35** (4), 576.

38 Kills, Y.H., Lee, J.Y., Lee, S.-H., Oh, J.-E. and Lee, H.S. (2005) Synthesis of aligned GaN nanorods on Si(111) by molecular beam epitaxy. *Applied Physics A: Materials Science and Processing*, **80** (8), 1635–9.

39 Xia, Y., Yang, P., Sun, Y., Wu, Y., Mayers, B., Gates, B., Yin, Y., Kim, F. and Yan, H. (2003) One-dimensional nanostructures: synthesis, characterization, and applications. *Advanced Materials*, **15** (5), 353–89.

40 Aichele, T., Tribu, A., Bougerol, C., Kheng, K., André, R. and Tatarenko, S. (2008) Defect-free ZnSe nanowire and nanoneedle nanostructures. *Applied Physics Letters*, **93** (14), 143106.

41 Colli, A., Hofmann, S., Ferrari, A.C., Ducati, C., Martelli, F., Rubini, S., Cabrini, S., Franciosi, A. and Robertson, J. (2005) Low-temperature synthesis of ZnSe nanowires and nanosaws by catalyst-assisted molecular-beam epitaxy. *Applied Physics Letters*, **86** (15), 153103.

42 Philipose, U., Xu, T., Yang, S., Sun, P., Buda, H.E., Wang, Y.Q. and Kavanagh, K.L. (2006) Enhancement of band edge luminescence in ZnSe nanowires. *Journal of Applied Physics*, **100** (8), 084316.

43 Glas, F., Harmand, J. and Patriarche, G. (2007) Why does wurtzite form in nanowires of III–V zinc blende semiconductors? *Physical Review Letters*, **99** (14), 146101.

44 Sallen, G., Tribu, A., Aichele, T., André, R., Bougerol, C., Tatarenko, S., Kheng, K. and Poizat, J. Ph. (2009) Exciton dynamics of a single quantum dot embedded in a nanowire. *Physical Review B*, **80**, 085310.

45 Talapin, D.V., Rogach, A.L., Kornowski, A., Haase, M. and Weller, H. (2001) Highly luminescent monodisperse CdSe and CdSe/ZnS nanocrystals synthesized in a hexadecylamine-trioctylphosphine oxide-trioctylphospine mixture. *Nano Letters*, **1** (4), 207–11.

46 Smith, A.W., Gao, X. and Nie, S. (2004) Quantum dot nanocrystals for *in vivo* molecular and cellular imaging. *Photochemistry and Photobiology*, **80** (3), 377–85.

47 Medintz, I.L., Clapp, A.R., Mattoussi, H., Goldman, E.R., Fisher, B. and Mauro, J.M. (2003) Self-assembled nanoscale biosensors based on quantum dot FRET donors. *Nature Materials*, **2** (9), 630–8.

48 Cui, Y., Wei, Q., Park, H. and Lieber, C.M. (2001) Nanowire nanosensors for highly sensitive and selective detection of biological and chemical species. *Science*, **293** (5533), 1289–92.

49 Patolsky, F., Zheng, G., Hayden, O., Lakadamyali, M., Zhuang, X. and Lieber, C.M. (2004) Electrical detection of single viruses. *Proceedings of the National Academy of Sciences of the United States of America*, **101** (39), 14017–22.

50 Zheng, G., Patolsky, F., Cui, Y., Wang, W.U. and Lieber, C.M. (2005) Multiplexed electrical detection of cancer markers with nanowire sensor arrays. *Nature Biotechnology*, **23** (10), 1294–301.

51 Patolsky, F., Zheng, G. and Lieber, C.M. (2006) Nanowire sensors for medicine and the life sciences. *Nanomedicine*, **1** (1), 51–65.

52 Arakawa, Y. and Sakaki, H. (1982) Multidimensional quantum well laser and temperature dependence of its threshold current. *Applied Physics Letters*, **40** (11), 939–41.

53 Asada, M., Miyamoto, Y. and Suematsu, Y. (1986) Gain and the threshold of three-dimensional quantum-box lasers. *IEEE Journal of Quantum Electronics*, **22** (9), 1915.

54 Huang, M.H., Mao, S., Feick, H., Yan, H., Wu, Y., Kind, H., Weber, E., Russo, R. and Yang, P. (2001) Room-temperature ultraviolet nanowire nanolasers. *Science*, **292** (5523), 1897–9.

55 Duan, X., Huang, Y., Agarwal, R. and Lieber, C.M. (2003) Single-nanowire electrically driven lasers. *Nature*, **421** (6920), 241–5.

56 Huang, Y., Duan, X. and Lieber, C.M. (2005) Nanowires for integrated multicolor nanophotonics, *Small*, **1**, 142–7.

57 Qian, F., Gradecak, S., Li, Y., Wen, C. and Lieber, C.M. (2005) Core/multishell nanowire heterostructures as multicolor, high-efficiency light-emitting diodes. *Nano Letters*, **5** (11), 2287–91.

58 Xiang, J., Lu, W., Hu, Y., Wu, Y., Yan, H. and Lieber, C.M. (2006) Ge/Si nanowire heterostructures as high-performance field-effect transistors. *Nature*, **441** (7092), 489–93.

59 Bennett, C.H. and Brassard, G. (1984) Quantum cryptography: quantum key distribution and coin tossing. Proceedings of the IEEE International Conference on Computers, Systems and Signal Processing, Bangalore, India, pp. 175–9.

60 Gisin, N., Ribordy, G., Tittel, W. and Zbinden, H. (2002) Quantum cryptography. *Review of Modern Physics*, **74** (1), 145–95.

61 Knill, E. Laflamme, R. and Milburn, G.J. (2001) A scheme for efficient quantum computation with linear optics. *Nature*, **409** (6816), 46–52.

62 Gasparoni, S., Pan, J., Walther, P., Rudolph, T. and Zeilinger, A. (2004) Realization of a photonic controlled-not gate sufficient for quantum computation. *Physical Review Letters*, **93** (2), 020504.

63 Kiesel, N., Schmid, C., Weber, U., Tóth, G., Gühne, O., Ursin, R. and Weinfurter, H. (2005) Experimental analysis of a four-qubit photon cluster state. *Physical Review Letters*, **95** (21), 210502.

64 Tian, L., Rabl, P., Blatt, R. and Zoller, P. (2004) Interfacing quantum-optical and solid-state qubits. *Physical Review Letters*, **92** (24), 247902.

65 Briegel, H.J., Dür, W., Cirac, J.I. and Zoller, P. (1998) Quantum repeaters: the role of imperfect local operations in quantum communication. *Physical Review Letters*, **81** (26), 5932–5.

66 Bouwmeester, D., Pan, J., Mattle, K., Eibl, M., Weinfurter, H. and Zeilinger, A. (1997) Experimental quantum teleportation. *Nature*, **390** (6660), 575–9.

67 Mandel, L. and Wolf, E. (1995) *Optical Coherence and Quantum Optics*, Cambridge University Press, Cambridge.

68 Hanbury Brown, R. and Twiss, R.Q. (1956) A test of a new type of stellar interferometer on Sirius. *Nature*, **178** (4541), 1046–8.

69 Grangier, P., Sanders, B. and Vuckovic, J. (eds) (2004) Special issue on single-photon sources. *New Journal of Physics*, **6**, 85–163.

70 Kuhn, A., Hennrich, M. and Rempe, G. (2002) Deterministic single-photon source for distributed quantum networking. *Physical Review Letters*, **89** (6), 067901.

71 Keller, M., Lange, B., Hayasaka, K., Lange, W. and Walther, H. (2004) Continuous generation of single photons with controlled waveform in an ion-trap cavity system. *Nature*, **431** (7012), 1075–8.

72 McKeever, J., Boca, A., Boozer, A.D., Miller, R., Buck, J.R., Kuzmich, A. and Kimble, H.J. (2004) Deterministic generation of single photons from one atom trapped in a cavity. *Science*, **303** (5666), 1992–4.

73 Brunel, C., Lounis, B., Tamarat, P. and Orrit, M. (1999) Triggered source of single photons based on controlled single molecule fluorescence. *Physical Review Letters*, **83** (14), 2722–5.

74 Michler, P., Imamoglu, A., Mason, M.D., Carson, P.J., Strouse, G.F. and Buratto, S.K. (2000) Quantum correlation among photons from a single quantum dot at room temperature. *Nature*, **406** (6799), 968–70.

75 Nirmal, M., Dabbousi, B.O., Bawendi, M.G., Macklin, J.J., Trautman, J.K., Harris, T.D. and Brus, L.E. (1996) Fluorescence intermittency in single cadmium selenide nanocrystals. *Nature*, **383** (6603), 802–4.

76 Kurtsiefer, C., Mayer, S., Zarda, P. and Weinfurter, H. (2000) Stable solid-state source of single photons. *Physical Review Letters*, **85** (2), 290–3.

77 Beveratos, A., Kühn, S., Brouri, R., Gacoin, T., Poizat, J. and Grangier, P. (2002) Room temperature stable single-photon source. *European Physical Journal D – Atomic, Molecular, Optical and Plasma Physics*, **18** (2), 191–6.

78 Michler, P., Kiraz, A., Becher, C., Schönfeld, W.V., Petroff, P.M., Zhang, L., Hu, E. and Imamoglu, A. (2000) A quantum dot single-photon turnstile device. *Science*, **290** (5500), 2282–5.

79 Santori, C., Pelton, M., Solomon, G.S., Dale, Y. and Yamamoto, Y. (2001) Triggered single photons from a quantum dot. *Physical Review Letters*, **86** (8), 1502–5.

80 Sebald, K., Michler, P., Passow, T., Hommel, D., Bacher, G. and Forchel, A. (2002) Single-photon emission of CdSe quantum dots at temperatures up to 200 K. *Applied Physics Letters*, **81** (16), 2920–2.

81 Kako, S., Santori, C., Hoshino, K., Götzinger, S., Yamamoto, Y. and Arakawa, Y. (2006) A gallium nitride single-photon source operating at 200 K. *Nature Materials*, **5** (11), 887–92.

82 Borgström, M.T., Zwiller, V., Müller, E. and Imamoglu, A. (2005) Optically bright quantum dots in single nanowires. *Nano Letters*, **5** (7), 1439–43.

83 Tribu, A., Sallen, G., Aichele, T., André, R., Poizat, J., Bougerol, C., Tatarenko, S. and Kheng, K. (2008) A high-temperature single-photon source from nanowire quantum dots. *Nano Letters*, **8**, 12.

84 Wang, J., Gudiksen, M.S., Duan, X., Cui, Y., Lieber, C.M. (2001) Highly polarized photoluminescence and

photodetection from single indium phosphide nanowires. *Science*, **293** (5534), 1455–7.

85 Lan, A., Giblin, J., Protasenko, V. and Kuno, M. (2008) Excitation and photoluminescence polarization anisotropy of single CdSe nanowires. *Applied Physics Letters*, **92** (18), 183110.

86 Niquet, Y. and Mojica, D.C. (2008) Quantum dots and tunnel barriers in InAs/InP nanowire heterostructures: electronic and optical properties. *Physical Review B (Condensed Matter and Materials Physics)*, **77** (11), 115316.

87 Bacher, G., Weigand, R., Seufert, J., Kulakovskii, V.D., Gippius, N.A., Forchel, A., Leonardi, K. and Hommel, D. (1999) Biexciton versus exciton lifetime in a single semiconductor quantum dot. *Physical Review Letters*, **83** (21), 4417–20.

11
Quantum Dot–Core Silica Glass–Shell Nanomaterials: Synthesis, Characterization, and Potential Biomedical Applications

Norio Murase

11.1
Introduction

The size of semiconductor quantum dots (QDs), between cluster and bulk materials, ranges from 2 to 10 nm, while the number of their constituent atoms is typically in the range of 10^2 to 10^4. Because of the rapid decay of the photoluminescence (PL) of QDs, a bright PL would be expected with a tunable wavelength, that would be determined also by the QD's size. Although one of the most promising applications for semiconductor QDs is for biological tagging, their use as electronic materials represents another important application.

In recent years, many reports have been made on semiconductor QDs. In this highly competitive field, the materials investigated have mostly included CdSe QDs prepared in organic solution, and CdTe QDs prepared in aqueous solution. Despite the properties of QDs having been reviewed comprehensively [1, 2], one further important aspect that should also be considered with regards to their applications is that of coating. As QDs possess a high surface-to-volume ratio, any surface defects will cause their original narrow spectral features to be quickly degraded. However, as glass has a greater chemical robustness, a greater light resistance and also a low ionic permeability when compared to polymers, the applicability of QDs could be extended if they were to be coated with glass. Moreover, the addition of a glass matrix would create new optical phenomena by providing a distinctive reaction field.

At this point, mention should be made of the nomenclature used in this field. Although sol–gel-derived glass contains a significant amount of organic molecules together with water, it is usually referred to as "glass," "silica," "silica glass," or just SiO_2. This is because there are no network-modifying metal ions in the matrix. For simplicity in this chapter, these sol–gel-derived matrices are referred to as glass, while the topics discussed refer to "glass"-coated QDs and the preparation of water-dispersible QDs.

The chapter first provides an historic overview of QDs in glass matrices with regards to biological tagging using either single molecules or particles. The incorporation of QDs into small glass beads, with their diverse applications, can be achieved by either of two schemes, namely the Stöber and the reverse micelle methods. The glass not only protects and disperses the QDs, but also plays an active role in creating new optical properties by forming composite structures. Subsequently, the glass beads can be conjugated to biological molecules in readiness for biomedical applications. Whilst glass is able to protect against the QDs interacting with their surroundings, its role is limited, and cadmium-free QDs may be required for use, especially in biological areas. When sol–gel-derived glass is used for such incorporation, the water dispersibility of QDs is advantageous; hence, the recent development of hydrophilic QDs (mainly green- to red-emitting InP QDs with blue-emitting $ZnSe_xTe_{1-x}$ QDs) is also described.

11.2
Historical Overview

11.2.1
Quantum Dots and their Incorporation into Glass Matrices

When the optical properties of QDs are compared with their corresponding bulk counterparts, the most prominent effect is the blue shift of the absorption spectrum. This property was reported for CdS QDs in a glass matrix as early as 1926 [3], and since then glass has been used as a matrix to grow QDs by annealing normally at over 500 °C for several tens of hours after oversaturation of the matrix [4]. Both, melting glass [5] and sol–gel-derived glass [6] were used for this method. As the prepared transparent glass can stably disperse a high concentration of QDs, it is used for optical materials with high third-order nonlinear susceptibilities [7]. In an advanced study, the glass matrix was even used as a reaction field to treat the surface of QDs in glass [8], although it proved difficult to control the reaction to reduce only the surface. Furthermore, the prepared glass inevitably disperses the ions that are required to create the QDs, and which disturb any further surface treatment following the preparation of QDs. Although glass is an attractive proposition as a matrix for QDs, there is seemingly no way in which the reaction of the QDs within it can be precisely controlled.

In contrast, when QDs are prepared in solution they quickly reach equilibrium under relatively mild conditions; moreover, the removal of the required materials from a reaction mixture is straightforward. In 1997, researchers noticed clearly that colloidally prepared QDs show size-dependent, narrow, and efficient PL when the surfaces are well passivated [9, 10]. It is, therefore, a good way to incorporate QDs after colloidal preparation into glass matrices. Various sol–gel techniques can be used for this purpose; however, the surface of QDs must be protected during processing.

A theoretical approach clarified the importance and delicate aspect of the surface condition to obtain a high PL efficiency [11]. If one QD is considered, with a radius r and surface energy μ, the growth speed of the QD can be written under the appropriate assumptions as follows:

$$\frac{dr}{dt} = K\mu\left(\frac{1}{r_{cr}} - \frac{1}{r}\right) \tag{11.1}$$

where K and r_{cr} are constants determined by solution temperature, diffusion coefficient, solubility, and so on. According to this equation, QDs smaller than r_{cr} will dissolve and disappear eventually with time, whereas QDs larger than r_{cr} will grow larger by utilizing the dissolved matter.

Equation 11.1 further indicates that QDs with a radius close to r_{cr} do not grow or dissolve on average (zero-growth rate), and when the colloidal solutions are inspected carefully the QDs close to r_{cr} have the highest PL efficiency. This was attributed to the fact that QDs with a zero-growth rate have the smoothest surface, with the lowest degree of disorder [12].

The incorporation of colloidally prepared QDs into a sol–gel-derived glass matrix was first reported in 2001 [13], although the material seemed not to be a solid. Furthermore, the PL efficiency was decreased to 5–10% when QDs were incorporated into the "glass" matrix. As the importance of preserving of the surface was deemed to maintain the initial high PL efficiency in colloidal solution, the following five protocols were considered for creating emitting glass matrices.

- For incorporation into sol–gel-derived glass, aqueously prepared QDs have an advantage in maintaining the initial efficiency. Thioglycolic acid (TGA)-stabilized CdTe QDs should, therefore, be selected.

- As carboxylic acid groups are situated on the surface of the QDs, a silicon alkoxide with the amino group (3-aminopropyltrimethoxysilane; APS, a type of silane coupling agent) is appropriate for the stable dispersion of QDs.

- Reaction in alkaline region is required to maintain the dispersibility of the QDs. This pH condition also accelerates the sol–gel reaction.

- To prevent surface deterioration during preparation, the constituent ions (e.g., Cd^{2+}) are dispersed in the solution; this effectively diminishes the dissolution rate of the QDs.

- To prevent surface deterioration and agglomeration during preparation, the QDs are added into the sol only after it shows some viscosity. This drastically reduces the degree to which QDs can be dissolved into the matrix.

The aqueously prepared CdTe QDs with high PL efficiency [14] were thus incorporated into a glass matrix without changing their spectral properties, as shown in Figure 11.1, and the PL efficiency was maintained at about 40%. The PL color images of the glass are shown in Figure 11.2 [15]. To improve this method, the ultimately high concentration of QDs in glass was attained by the adsorption of QDs onto a glass layer [16]. An estimation method of the precise PL efficiency quickly helped in evaluating the preparation [17]. Knowledge obtained in these studies was subsequently used to prepare glass beads that were several tens of nanometers in size and impregnated with emitting QDs.

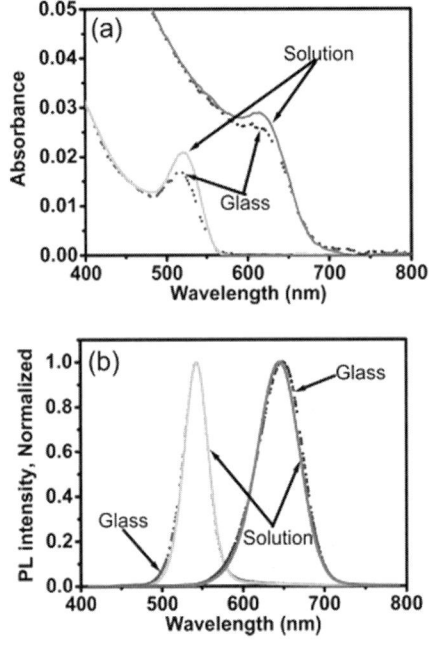

Figure 11.1 (a) Absorption and (b) phospholuminescence spectra of CdTe QDs in dilute solution and glass matrices [15].

11.2.2
Single-Particle Spectroscopy and its Application to Biological Systems

During investigations into a type of wavelength-multiplexed recording to obtain an ultrahigh recording density in an optical memory, it was noted that the signal from an individual dye molecule could be detected and analyzed spectroscopically [18, 19]. Although, initially, such spectroscopy was considered to be applicable only in basic research fields, it rapidly became used to tag biomolecules or viruses [20] for their detection as well as monitoring their amount, distribution, and even movement. Consequently, the types of molecules that could be used for such detection purposes increased rapidly in number.

During the same period, it was also noted that luminescent QDs could also be used for such detection, by estimating the PL decay time, efficiency, and Stokes' shift [21, 22]. The advantage of the detection method using QDs is its robustness. Since the dye molecules degrade quickly [23], the observation time is very limited. QDs have a great advantage because they usually emit photons roughly two orders or more before degradation than with dyes. By using CdSe QDs, a stepwise movement of a type of motor protein (dynein and kinesin) activated by ATP (adenosine triphosphate) was observed with a spatial resolution of 1 nm [24].

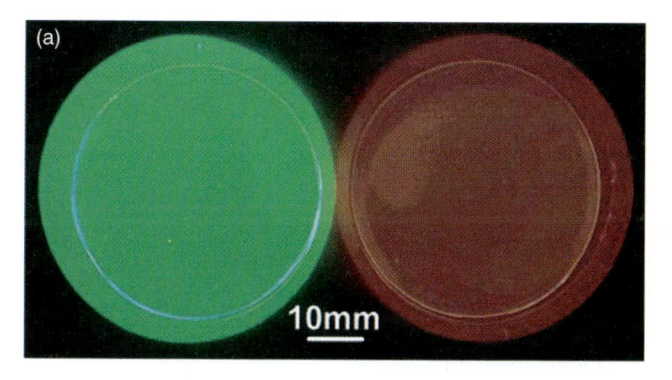

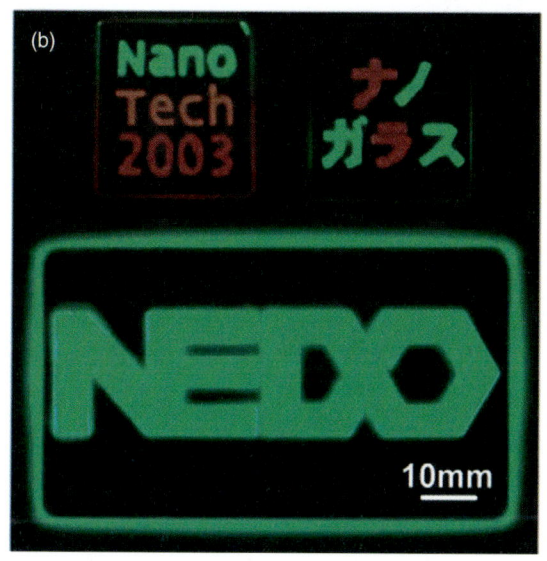

Figure 11.2 Color fluorescent images of the prepared glass excited by an UV lamp at 365 nm. (a) Phosphor prepared in a Teflon chalet; (b) Phosphor adhered to patterned grooves on the surface of the glass substrates [15].

Moreover, when the number of photons emitted in unit time is increased, the position of the emitter can be more precisely determined [25]. Thus, the appearance of an innovative emitter with high brightness, such as a glass bead impregnated with many QDs, is anticipated.

11.3
Incorporation of Quantum Dots in Glass Beads for Bioapplications

As noted above, two methods are mainly be used to incorporate QDs into small glass beads, namely the Stöber and reverse micelle methods, and these are described in the following sections. Yet, several studies have been conducted to achieve the incorporation of a single QD into one glass bead [26, 27, 28].

11.3.1
The Stöber Method

In 1968, Stöber et al. first showed that silica spheres, ranging in size from 50 nm to 2 μmm, and with a narrow size distribution, could be created by the hydrolysis and condensation of a silicon alkoxide (tetraethoxysilane, TEOS) under highly alkaline conditions [29]. The method was shown to be easily reproducible, and was subsequently improved upon by many groups who clarified the mechanism. Ultimately the method became known as the Stöber method, and was recognized as the definitive approach to creating small silica spheres in solution.

More recently, several research groups have used this method to create a glass layer on seeds prepared in advance. When the seeds were Au nanocrystals (NCs) dispersed in water, their surfaces were first activated using a silicon alkoxide with an amino group (APS; see Section 11.2) which made the surfaces vitreophilic when dispersed in ethanol. The Stöber method was then applied to the ethanol solution by adding sodium silicate solution, such that a glass shell, with thickness ranging from 10 to 83 nm, was deposited on the surface of a Au particle of diameter 15 nm [30].

As with the synthesis of glass-coated QDs, 40- to 80-nm diameter seeds containing QDs, such as CdTe, CdSe or CdSe (core)/CdS (shell), were first prepared by adding a thiol-containing alkoxide (2-mercaptopropyltrimethoxysilane, MPS), with the subsequent addition of sodium silicate. The Stöber process was then adapted to these seeds for the preparation of glass beads of 100–700 nm in size. However, in this study the initial PL efficiency of the QDs (4–5%) was decreased significantly (ca. 20-fold) following addition of the silicate [31]. Another report described the deposition of a layer (ca. 100 nm thick) of glass-coated CdSe/ZnS or CdS/ZnS QDs onto the surface of monodisperse silica microspheres (ca. 550 nm). Here, the glass coating was carried out by exchanging the surface from tri-n-octylphosphine oxide (TOPO) to two types of amino compound, so as to impart dispersibility in ethanol and an affinity towards the silica surface. The resultant PL efficiency was decreased from 38% to 13% [32].

Any of these methods – at least for the present time – will significantly reduce the PL efficiency because of the required changes in the QD surfaces that will allow the deposition of silica molecules. It should be noted that alcohol, which normally is indispensable for this method, is generally used to precipitate both aqueously and organically prepared QDs. In other words, QDs tend to lose their stability when this method is applied.

11.3.2
Reverse Micelle Method

A reverse micelle, namely a water-in-oil microemulsion, is formed after mixing oil and a small amount of water together with amphiphilic molecules as surfactant.

The size of the water droplet will depend on the ratio of water to surfactant. When alkoxide molecules such as TEOS are added to this microemulsion, they first disperse in the oil phase, but are then gradually transferred to the water phase of the alkaline region due to hydrolysis that most likely occurs on the surface of the water phase. Glass spheres which reflect the initial size of the water droplets are thus formed after condensation in the water droplet [33].

When QDs are hydrophilic, they become distributed in the water droplets in a microemulsion. Then, accompanied by the hydrolysis and condensation of alkoxide in the water droplet, glass beads impregnated with many emitting QDs will be formed. In this preparation procedure, which was first devised in 2001 [34] but initially reported in 2004 [35], TGA-capped CdTe QDs (hydrophilic) were located only on the surface of glass beads with a rather low PL efficiency (ca. 5%). As this method is seemingly straightforward, several reports have been made. For example, when the same water-dispersible CdTe QDs with a surfactant of TGA and 1-thioglycerol were used, beads impregnated with QDs were formed only in the central region. This effect was ascribed qualitatively to the electrostatic interaction of both negatively charged QDs and the hydrolyzed silica molecules. The PL efficiency of these beads was estimated at 7% [36].

For hydrophobic CdSe or CdSe/ZnS QDs, the above method must be slightly modified. Thus, in one study the reverse micelle method was applied after changing their surfaces to APS, and the PL efficiencies of the resultant glass beads were in the range of 10–20%. The detailed mechanism of the incorporation of QDs into glass beads was not discussed, however [37]. The most comprehensive study on the incorporation mechanism of hydrophobic QDs was presented by Meijerink's group [38]. For this, the alkylamine-capped CdSe/ZnS QDs (to be more precise, a CdSe core with two CdS shells + three $Cd_{0.5}Zn_{0.5}S$ shells + two ZnS shells) were dispersed in cyclohexane, after which the PL intensity of the solution was monitored during the addition of TEOS, surfactant (amphiphilic molecule), and aqueous ammonia. It was concluded that the surface amine molecules were quickly exchanged by these molecules, as shown in Scheme 11.1. This surface exchange accompanies the rapid reduction in the PL efficiency of 90% in the worst case (Figure 11.3). At the expense of this reduction, the QDs acquire water dispersibility and transfer into water droplets, where silica growth takes place. The PL efficiencies of the glass beads were carefully evaluated to obtain 35% for the best case.

Although, at present, it is better to select originally hydrophilic QDs to maintain the PL efficiency in glass beads, it is still necessary to protect the surface during preparation, mainly because of the dissolution of QDs (as noted in Section 11.2). Therefore, a two-step synthesis using a modified Stöber method and subsequently also a reverse micelle method (see Scheme 11.2) succeeded in maintaining the initial colloidal PL efficiency [39].

In Step 1 of the modified Stöber method, TGA-capped CdTe QDs were coated with a thin SiO_2 layer by using diluted ammonia and TEOS at pH ~10. The thickness of the layer was estimated to be 0.6 nm. "Modified" here means that alcohol is not added. This plays a role in reducing the hydrolysis speed of TEOS for the purpose of preventing the hydrolyzed silica molecules from nucleation. We also know alcohol always precipitates any QDs in solution as well. By measuring the

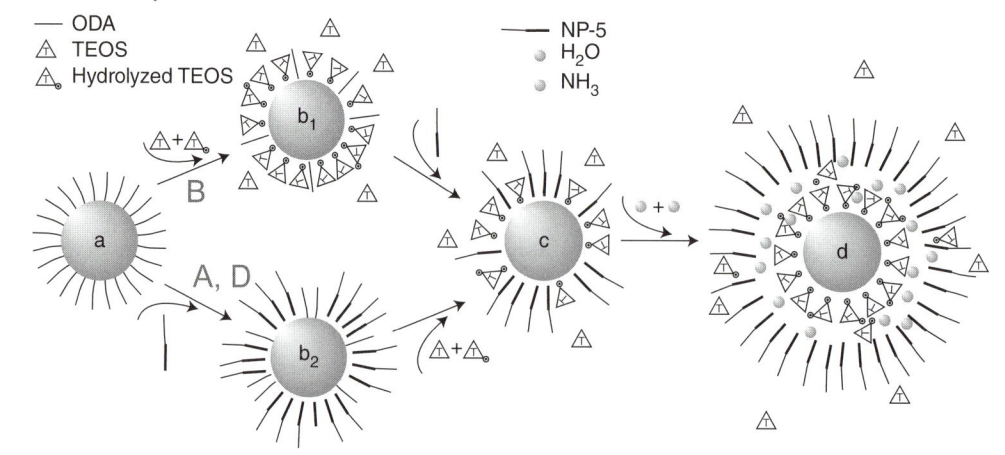

400 | *11 Quantum Dot–Core Silica Glass–Shell Nanomaterials*

Scheme 11.1 Schematic representation of the incorporation mechanism of hydrophobic QDs (in orange) in glass beads by the reverse micelle method. ODA (octadecylamine) is a type of alkylamine. NP-5 is the amphiphilic molecule (poly(5) oxyethylene-4-nonylphenyl-ether) [38].

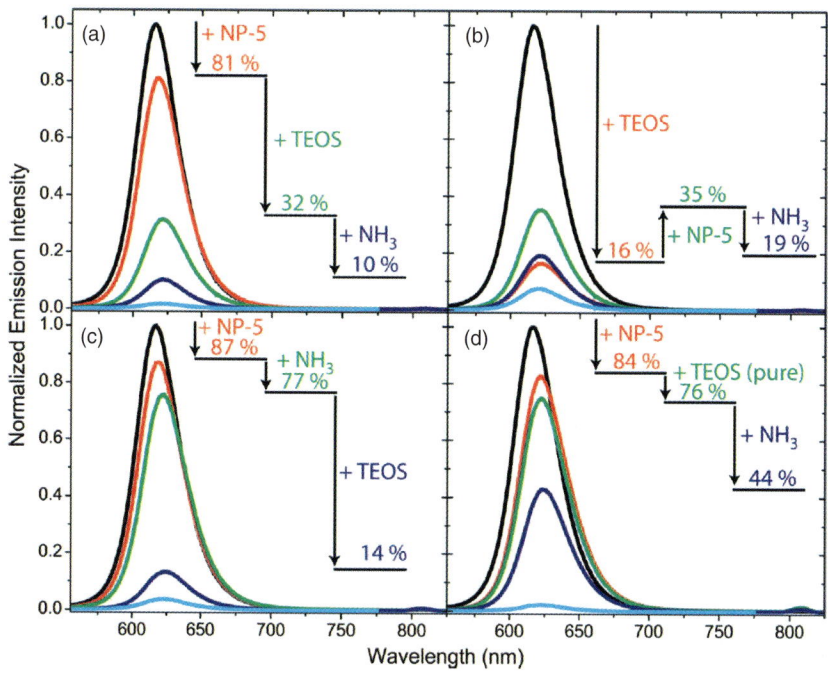

Figure 11.3 Phospholuminescence spectra of CdSe QDs (ODA-coated) at different stages of the reverse microemulsion synthesis (λ_{exc} = 406 nm). (a–c) NP-5, TEOS, and NH$_3$ were added in different orders, as indicated by the arrows in each panel; (d) This synthesis was identical to that synthesis in panel (a), except for the use of ultra pure and water-free TEOS. The PL spectra were corrected for the significant dilution factor upon addition of NP-5, and percentages represent the integrated emission intensity. The emission spectra in light blue were measured one week after the synthesis [38].

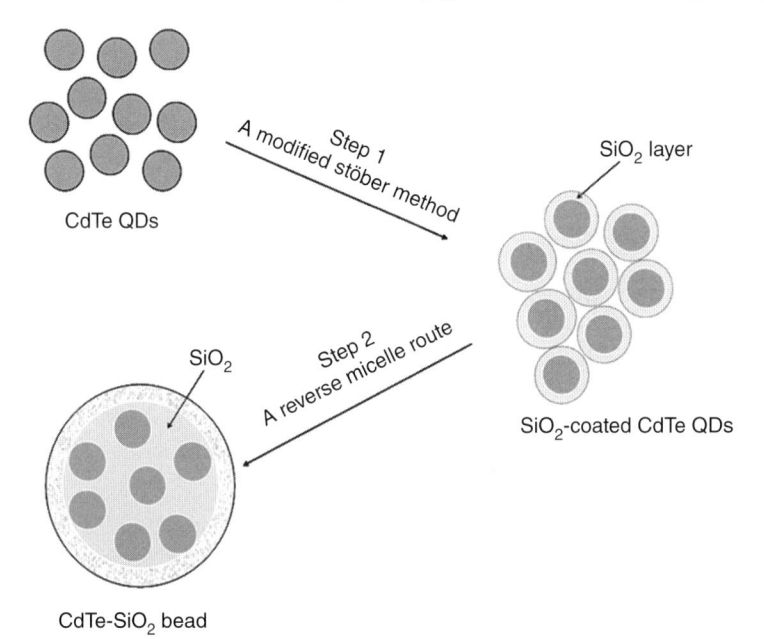

11.3 Incorporation of Quantum Dots in Glass Beads for Bioapplications

Scheme 11.2 Encapsulation of emitting CdTe QDs within a glass bead via a modified Stöber synthesis (Step 1) and a reverse micelle route (Step 2) [39].

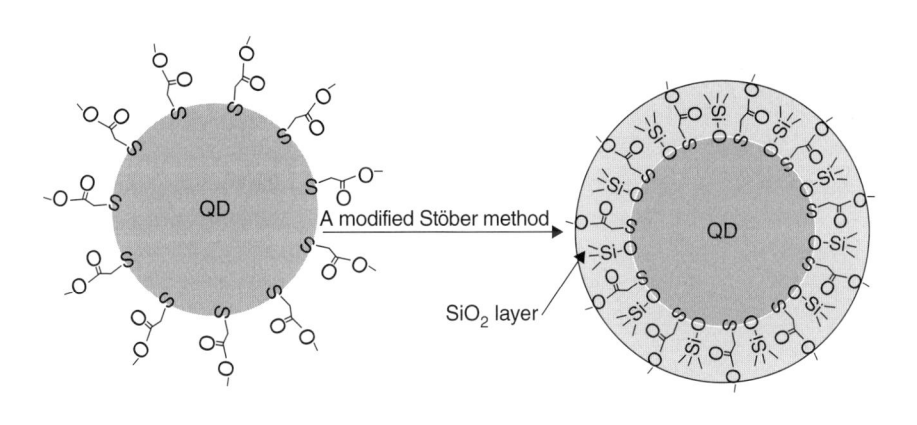

Scheme 11.3 Formation mechanism of SiO_2-coated CdTe QDs during Step 1 [39].

molar ratio of S/Cd and zeta-potential just after this step, the TGA molecules on the surface were found still to be retained after the Stöber process (see Scheme 11.3). As ammonia accelerates the reaction, this step is completed within 3 h, and contributes to maintaining the surface of the QDs (see below).

Table 11.1 Effect of water-to-oil (W/O) and oil-to-surfactant (O/S) ratios and injection speed of H_2O phase on glass bead size [39].

Glass bead size (mean)	W/O (molar ratio)[a]	O/S (molar ratio)[b]	Injection speed (ml min^{-1})
500 nm	0.51	37	0.09
1 µm	0.51	37	0.30
2 µm	0.51	37	1.40
30 nm	0.50	30	0.05
40 nm	0.40	35	0.10
1 µm	0.56	30	0.40
2 µm	0.75	30	0.50

a) Molar ratio of water to oil in microemulsion.
b) Molar ratio of oil to surfactant in microemulsion.

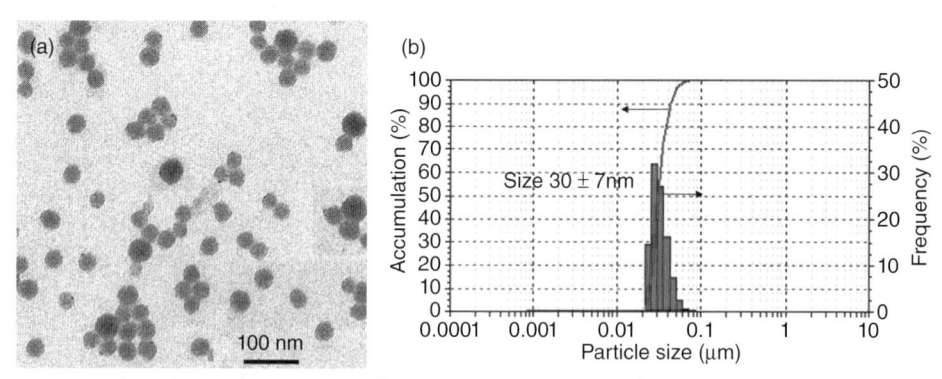

Figure 11.4 (a) Transmission electron microscopy image and (b) size distribution of red-emitting glass beads impregnated with CdTe QDs [39].

In Step 2, the Stöber beads prepared in Step 1 were used as a starting material. For this, a reverse micelle with Igepal CO-520 (poly(5)oxyethylene-4-nonylphenyl-ether) as a surfactant, and cyclohexane as an oil phase, was prepared. Then, in order to encapsulate many CdTe QDs into a glass bead, TEOS was injected into the microemulsion, followed by the addition of diluted ammonia; the microemulsion solution was then stirred for 4 h.

The size of the glass beads could be controlled over a wide range, from 30 nm to 2 µm in diameter, by altering the molar ratio of water, oil, surfactant, and the injection rate of the water phase (solution after Step 1). An example of these relationships is listed in Table 11.1, while the typical size distribution is shown in Figure 11.4 for those beads of 30 nm diameter. An elemental analysis using inductively coupled plasma (ICP) spectrometry after dissolution of the beads showed that 14 of the CdTe QDs had been incorporated into one bead of this size. Figure 11.5 shows the absorption and PL spectra of the green- and red-emitting glass

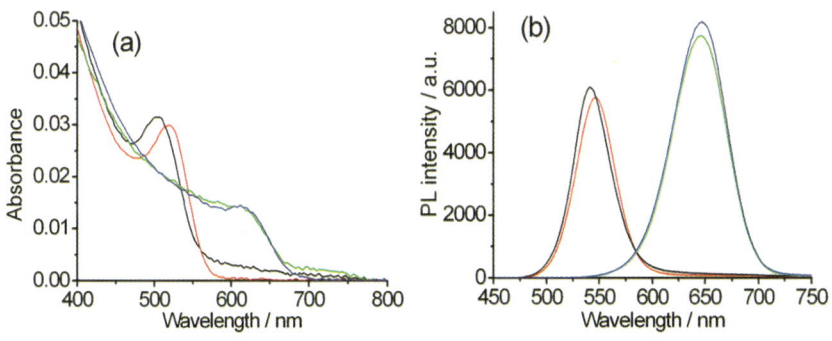

Figure 11.5 (a) Absorption and (b) phospholuminescence spectra of glass beads in microemulsions and CdTe QDs in aqueous solutions. Black, red, green, and blue curves are from green-emitting beads, green-emitting QDs, red-emitting beads, and red-emitting QDs, respectively [39].

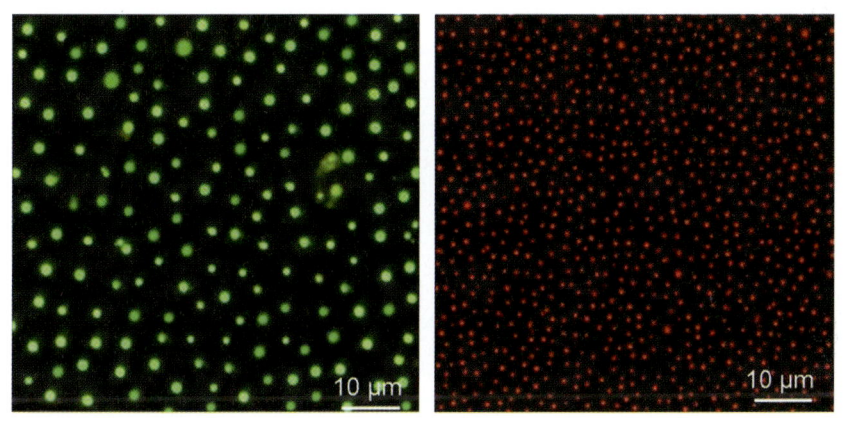

Figure 11.6 Color images of green- and red-emitting glass beads by irradiation of UV light (365 nm). The mean sizes of the green- and red-emitting beads were 2 ± 0.8 and $1 \pm 0.5\,\mu m$, respectively [39].

beads of this size. The first absorption and PL peaks of the red-emitting beads were very similar to those of the colloidal solution, whereas a small blue shift was observed for the green-emitting beads. The PL efficiencies of the prepared green- and red-emitting beads were estimated as 27% and 65%, respectively. This indicates that the initial PL efficiencies of the colloidal solutions were maintained for both green- and red-emitting QDs. When the beads were of micrometer size, however, emission from individual beads could be observed by using fluorescence microscopy (Figure 11.6).

Until recently, the encapsulation of QDs into glass beads has always been accompanied by a reduction in PL efficiencies. Therefore, when considering the reasons for maintaining the efficiencies of the present study, three important aspects can be highlighted:

1. The thin silica layer that is created in Step 1 (as explained clearly in this section), and which prevents the removal of surface cap molecules (TGA) during Step 2. Although the thickness of the silica shell prepared in the short reaction time in Step 1 is thin, it is sufficient to prevent the surface from deteriorating during Step 2. As noted in Section 11.2, the movement of molecules and ions in the glass matrix is effectively prohibited by the glass networks. Therefore, the modified Stöber synthesis with a short reaction period is effective in reducing PL degradation of the QDs. Additionally, this layer increases the concentration of the QDs in the glass bead. It is known that surface deterioration by exchanging or removing capping molecules will always reduce the PL.

2. Proper pH during incorporation of the QDs in Step 2. To obtain a high PL efficiency of QDs in glass beads, it is important to retain the stability of the QDs during incorporation. It has been pointed out that TGA-capped CdTe QDs in aqueous solution exhibit a high PL efficiency and good stability only in a pH range of 6–10, due to the dissociation constant (pK_a = 3.67) of the carboxyl group in TGA [40]. In the preparation of the glass beads, an alkaline solution of pH 8–10 was selected during both the Stöber synthesis (Step 1) and the reverse micelle method (Step 2).

3. A reduced reaction time during the two-step synthesis. A conventional Stöber synthesis is used to prepare glass beads in an ethanol solution with a long reaction period, which leads to a reduction in the PL of QDs such as Rogach's results [31]. In that report, the formation of a silica coating on CdTe QDs resulted in a drastic reduction in PL efficiency because of a long reaction time (five days). However, such surface deteriorations were not observed in the present study because of the two-step synthesis with a short reaction time (ca. 7 h).

Although the PL efficiency is maintained with this method after incorporation, the remaining issues – such as control of the size distribution of the beads and the number of QDs in one bead while maintaining the initial PL efficiency – must be overcome for bioapplications.

Bioconjugation to the glass surface, where the biomolecules are either covalently or electrostatically bonded to the glass surface through linkers, was first introduced using silane coupling agents such APS and MPS by Alivisatos *et al.* [41]. A comprehensive review was provided for such a conjugation technique of several NCs [42], while glass-coated CdSe QDs and Fe_2O_3 NCs were conjugated with the cell membranes [43].

11.3.3
Bifunctional Glass Beads Derived from the Reverse Micelle Method

In this section, three topics are introduced relating to the creation of new structures and morphologies with novel PL properties, using the peculiarities of the glass matrix and sol–gel synthesis. According to progress in studies of the reverse

(a) MNCs and QDs → SiO₂ coating

(b) SiO₂ bead + MNCs, QDs, TEOS, H₂O, NH₄OH →

○ QD
● MNC (Magnetic Nanocrystals)

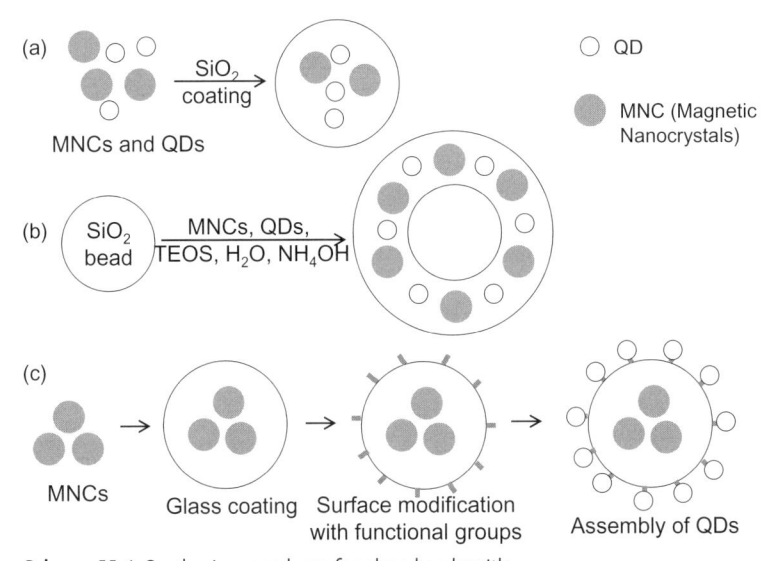

(c) MNCs → Glass coating → Surface modification with functional groups → Assembly of QDs

Scheme 11.4 Synthesis procedures for glass beads with magnetic nanocrystals (MNCs) and luminescent QDs [53].

micelle method, a procedure was created that would provide new structural glass beads with better functionalities [44], notably with magnetic and PL properties.

Until recently, several sol–gel approaches had been used to synthesize magnetic-luminescent glass beads, as shown in Scheme 11.4 [45–50], but the direct attachment of magnetic and luminescent QDs tends to reduce the PL efficiency of QDs. The situation is similar when a polymer matrix is used. For example, a threefold decrease in PL efficiency was reported for magnetic-luminescent beads with an average diameter of 30 nm [a composite of a polymer-coated Fe_2O_3 core with a shell of QDs (CdSe/ZnS)] [51]. In particular, a PL efficiency of about 15% was reported for γ-Fe_2O_3/QD (CdSe/ZnS) beads in a liquid crystal phase with a diameter of approximately 25 nm, whereas the initial efficiency of QDs in $CHCl_3$ was 41% [52].

Recently, a new method has been presented to incorporate multiple Fe_3O_4 and CdTe QDs in one glass bead, using both the Stöber and reverse micelle methods. The details of this method are shown in Scheme 11.5 [53]. In this case, instead of adding the aqueous phase in one batch (as is usually carried out for the reverse micelle method), an aqueous solution of a thin glass-coated CdTe QDs prepared using the Stöber method (first aqueous phase) is first added at the start of the reverse micelle method. Then, when an aqueous Fe_3O_4 NC solution (second aqueous phase) with a pH higher than that of the initial aqueous phase is subsequently added in Step 2 (after proper condensation of the hydrolyzed silicon alkoxide supplied by the first aqueous phase), the Fe_3O_4 NCs are incorporated into the hollow space created by the second aqueous phase in the beads, which is separated from the glass matrix where the CdTe QDs are dispersed. This two-step aqueous phase addition results in a high PL efficiency (68%) for CdTe QDs, mainly

Step 1: Preparation of first aqueous phase through modified Stöber method

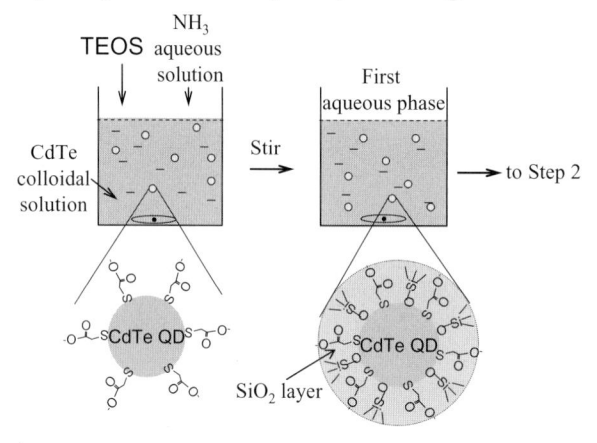

Step 2: Preparation of glass beads through modified reverse micelle route

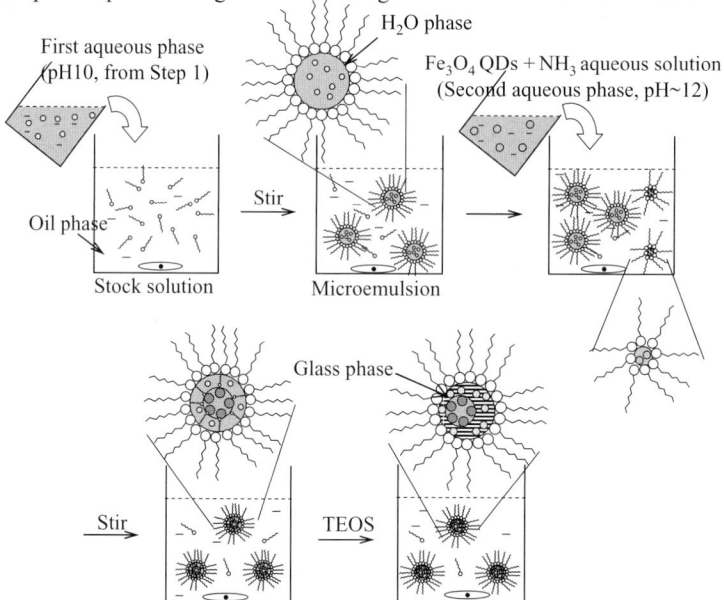

Scheme 11.5 Preparation and formation process for glass beads encapsulating CdTe QDs and Fe_3O_4 NCs by two-step synthesis [53].

because neither the QDs nor the NCs are mixed in the glass beads. The fact that the pH of the second aqueous phase containing Fe_3O_4 NCs is higher than that of the first phase represents a crucial factor in this procedure.

Figure 11.7a–c show the transmission electron microscopy (TEM) images of prepared glass beads (these are the same sample on one grid, but at different

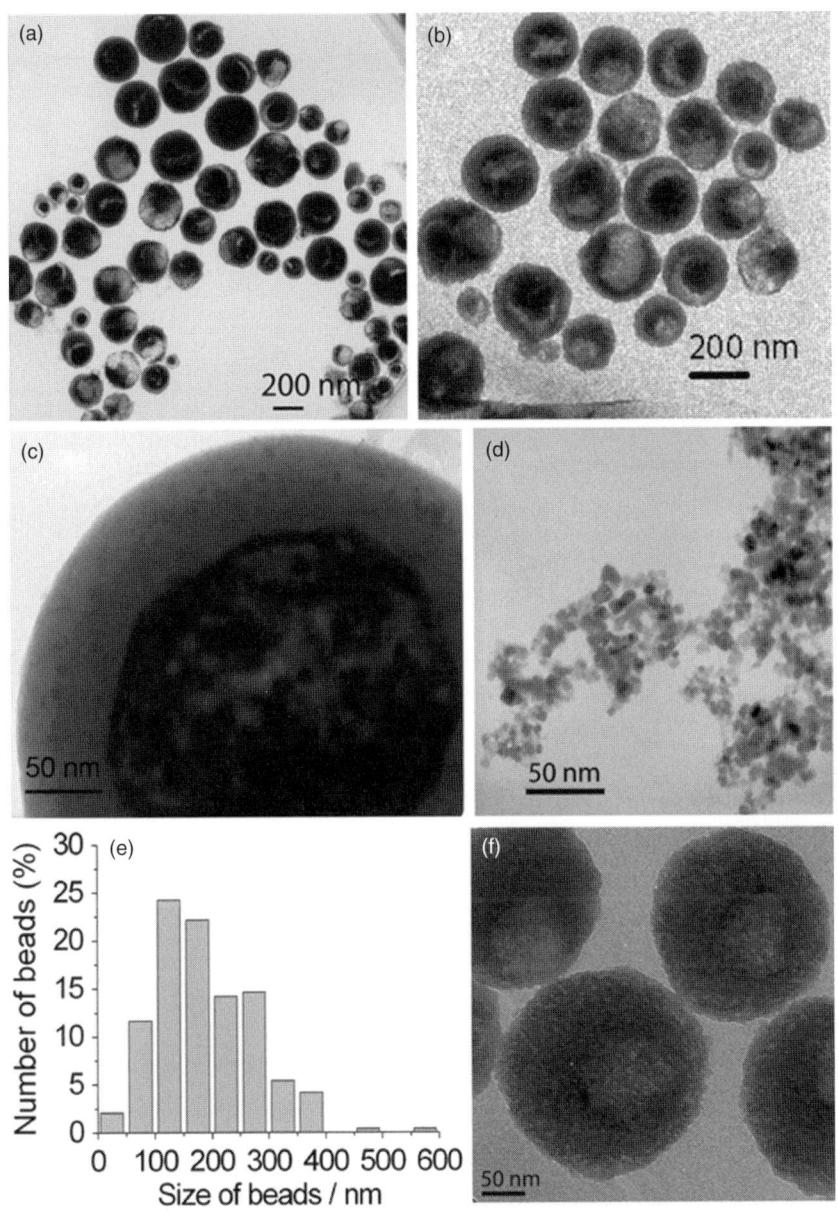

Figure 11.7 (a–c) Transmission electron microscopy (TEM) images of glass beads encapsulating Fe_3O_4 nanocrystals (NCs) and CdTe QDs; (d) TEM image of Fe_3O_4 NCs before encapsulation. In panel (c) the Fe_3O_4 NCs are seen to be located in the middle part of the beads, while the CdTe QDs are dispersed within the solid part of the bead (glass shell). Because of the high magnification, the originally hollow part in panel (c) was modified by the focused electron beam (300 KeV, ca. 8 µA) during the observation; (e) Size distribution of the beads. The histogram was plotted by measuring the size of hundreds of beads in the TEM images; (f) TEM image of hollow glass beads, shown for comparison [53].

positions), while Figure 11.7d shows the morphology of the Fe_3O_4 NCs (ca. 8 nm) used in the preparation. The beads shown in Figure 11.7a,b exhibit a hollow structure inside. Figure 11.7c shows that the CdTe QDs (ca. 4 nm) are dispersed in the solid part of the bead, whereas the Fe_3O_4 NCs are dispersed as larger black dots in the middle of the bead. The histogram shown in Figure 11.7e was plotted by measuring the size of hundreds of glass beads in the TEM images. For comparison, the TEM image of the hollow glass beads is shown in Figure 11.7f, where the second water phase does not include any nanocrystals. These empty beads contained only CdTe QDs in a glass matrix.

This formation mechanism may require further explanation. When the second aqueous phase is added in Step 2, newly formed small water droplets are encapsulated in the initial droplets created by a first addition of the aqueous phase. As the pH of the second aqueous phase (solution dispersing Fe_3O_4 NCs, pH 12) is higher than that of the first aqueous phase (just after Step 1, pH 10), a rapid condensation of hydrolyzed TEOS at the interface between the two aqueous phases occurs, when the second aqueous phase enters the first aqueous phase. This creates a glass wall on the surface of the second aqueous phase in the droplet of the first aqueous phase. This wall becomes a barrier that separates the two types of NCs (or QDs) in one glass bead. As the second aqueous phase does not contain alkoxide, this component creates the hollow part when the water is finally removed from the hollow part, during the separation process. Because the aqueously prepared Fe_3O_4 NCs are positively charged, but the hydrolyzed TEOS is negatively charged, the NCs tend to attach to the glass wall created at the interface. The remainder of the NCs in the second aqueous phase are subsequently deposited when dried. As a consequence, dark areas consisting of many Fe_3O_4 NCs in the middle of the beads were observed in the TEM images.

Figure 11.8 shows the X-ray diffraction (XRD) pattern of the prepared beads. The angles of six diffraction peaks match well with the reported values for a cubic spinel structure of magnetite. Three other diffraction peaks were assigned to a cubic CdTe. A vibrating sample magnetometer indicated that both glass beads and pure Fe_3O_4 NCs revealed typical superparamagnetic behavior at room temperature.

Both, annular dark-field scanning transmission electron microscopy (ADF-STEM) and energy-dispersive X-ray (EDX) analysis were used to further analyze the beads. Figure 11.9a shows an ADF-STEM image of a glass bead, while Figure 11.9b,c show the EDX analysis results obtained from the indicated detection area in the glass bead. The EDX analysis showed that the bead contained Cd, Te, and Fe in the middle. It was noted that the Cd and Te signals derived from the surrounding glass shell on both sides of the detection area, while Si, Cd, and Te were detected in the solid part (glass shell) of the bead. This means that the Fe_3O_4 NCs were dispersed only in the middle. A wet chemical analysis of these beads indicated a molar ratio for Si/Fe/Cd/Te/S of 1/0.017/0.027/0.011/0.017, such that the average concentration of these QDs in the beads was estimated at approximately 0.0017 M. This corresponds to approximately 3700 CdTe QDs in one bead with a diameter of 200 nm, and a hollow part with a size of 100 nm.

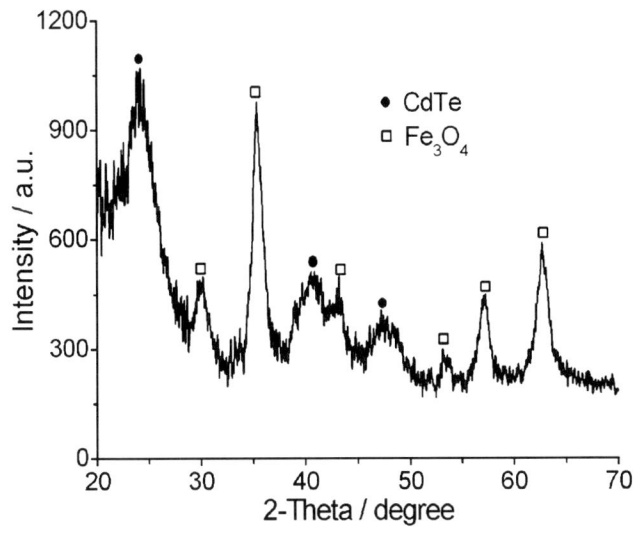

Figure 11.8 The X-ray diffraction pattern of the as-prepared beads. All of the diffraction peaks were indexed to cubic Fe_3O_4 or CdTe [53].

The absorption and PL properties of these beads with those of the initial CdTe colloidal solution are shown in Figure 11.10. A PL efficiency of 68% with a slight blue shift of the PL peaks was observed for the beads, whereas the PL efficiency of the initial CdTe colloidal solution was 71%. To the best of the present authors' knowledge, this PL efficiency of 68% is the highest reported value for two types of NC in one glass bead. The process introduced here should open a new approach for preparing glass beads with dual functions by dispersing other materials in the hollow part, such as metal NCs, oxide NCs, and dye molecules.

11.3.4
Complex Structures Created in Small Glass Beads with Novel Photoluminescent Properties

A glass matrix can supply a distinctive reaction field to create a new structure, whereas other materials such as solutions and polymers do not [54].

When the glass-coated CdTe QDs prepared by the Stöber method were refluxed at approx. 100 °C in an aqueous solution dispersing Cd^{2+} and TGA, their PL color changed drastically from green to red with time (see Figure 11.11a), while the inset of the figure shows the PL of the samples. Figure 11.11b indicates the time evolution of PL efficiency and width (full width, half-maximum; FWHM) of the spectrum upon refluxing. The efficiency was increased from 23% to 78%, with a narrowing of the spectral width from 50 nm to 42 nm in the best case. When the glass layer on the surface of the QDs was removed chemically, it was evident that the size of the QDs had remained unchanged during this reflux. Furthermore, it

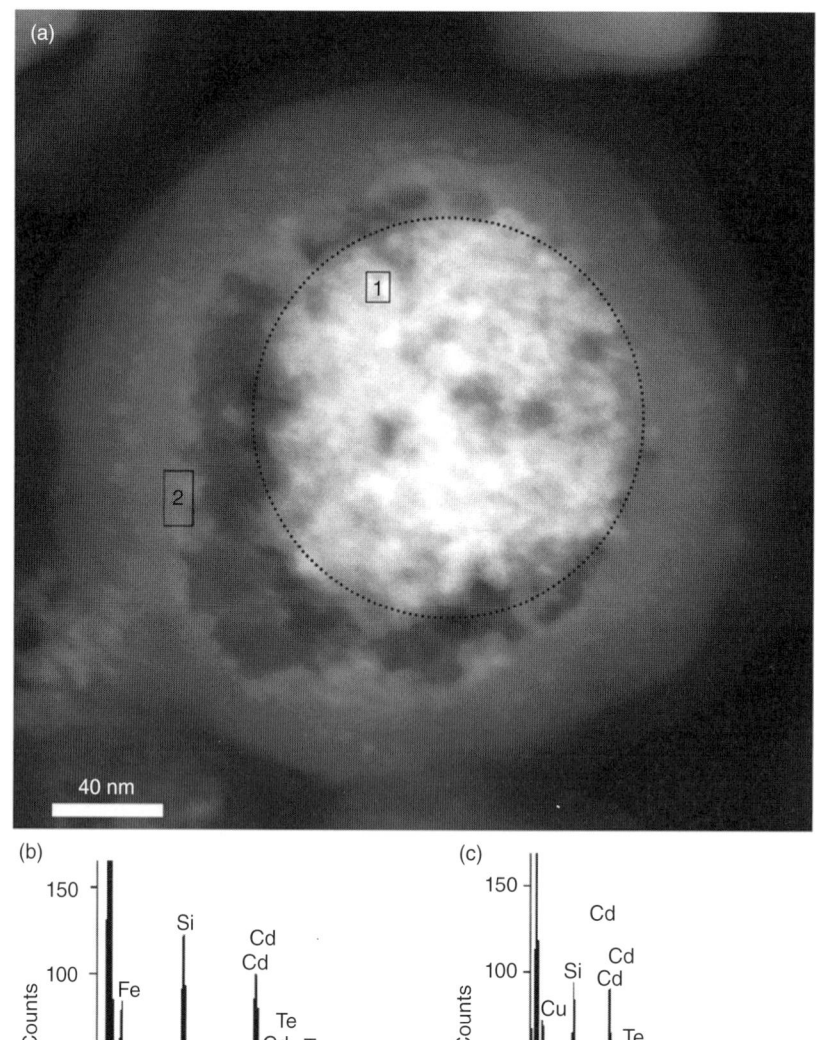

Figure 11.9 (a) ADF-STEM image of a luminescent hollow glass bead encapsulating Fe_3O_4 NCs; (b) EDX data from area 1 in panel (a), where an Fe signal was detected; (c) EDX data from area 2 in panel (a), where no Fe signal was detected. An Fe signal was detected only within the area shown by the black dotted circle [53].

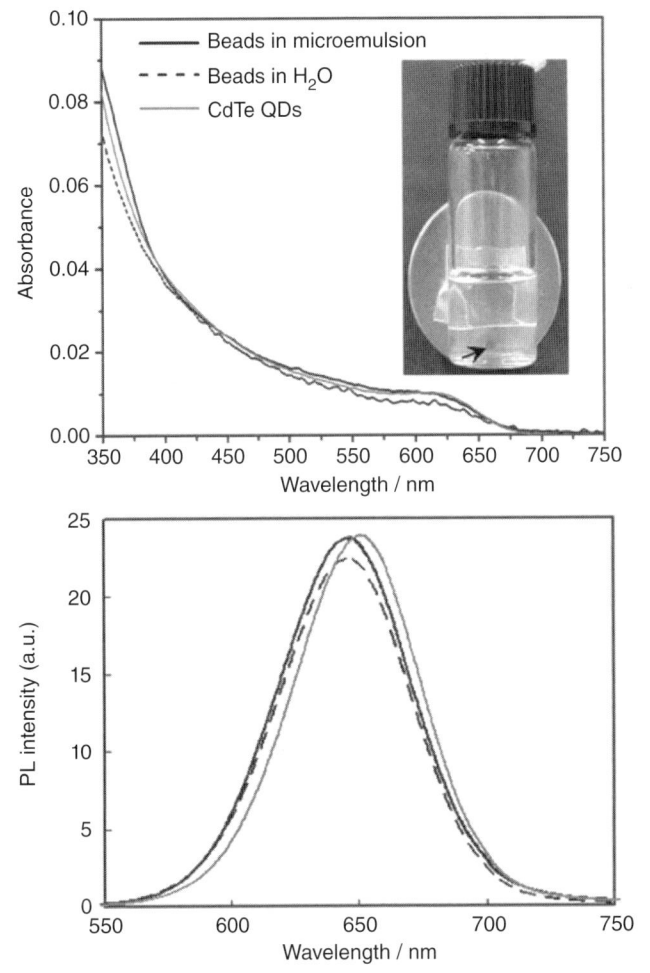

Figure 11.10 Absorption (upper) and phospholuminescence (PL) (lower) spectra of luminescent glass beads encapsulating Fe_3O_4 NCs in a microemulsion (black line; size distribution shown in Figure 11.7e) and in H_2O (dashed black line, <200 nm diameter). The beads were easily collected with a permanent magnet (insert, black arrow). The absorption and PL spectra of a CdTe colloidal solution are shown for comparison (gray line). The PL efficiency of the beads in the microemulsion was 68%, while that of the CdTe colloidal solution was 71%. After redispersion in pure H_2O, the beads (<200 nm diameter) revealed a PL efficiency of 40% [53].

was noted that only 2 h of reflux was sufficient to bring the PL to red, despite it normally taking up to 100 h to obtain red-emitting CdTe QDs from their green-emitting counterparts by normal colloidal preparation [14]. The molar extinction coefficient (ε) of the QDs at the first absorption peak wavelength was different from that of the conventionally prepared QDs [55]. Furthermore, the absorption coefficients of QDs smaller than approximately 400 nm were increased after

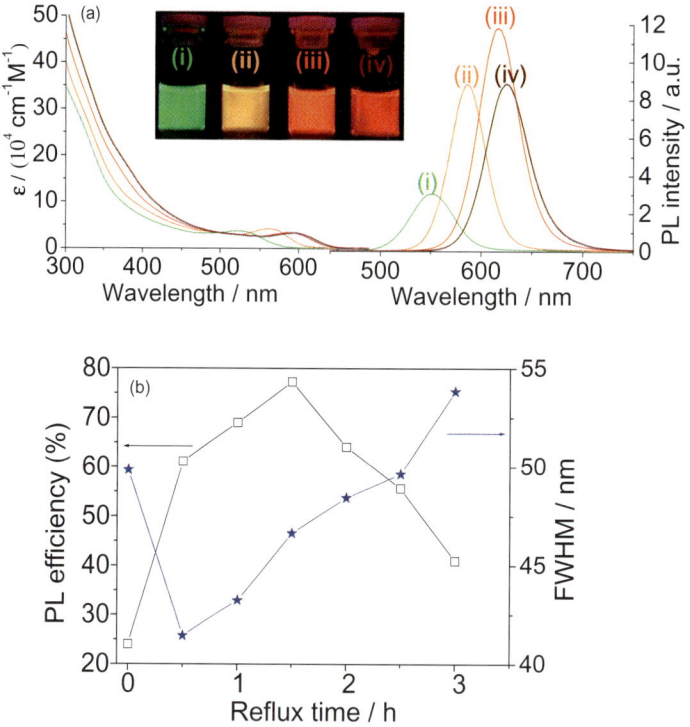

Figure 11.11 Spectral properties of glass-coated CdTe QDs prepared using a green-emitting CdTe colloidal solution. (a) Tunable absorption and PL spectra before and after refluxing: before refluxing (i) and after 0.5 (ii), 1.5 (iii), and 2.5 h (iv) of refluxing. The inset shows images of samples (i) to (iv) under UV light; (b) PL efficiency and spectral width (full width at half-maximum, FWHM) versus reflux time. The PL efficiency increased dramatically, followed by a steady decline after refluxing for 1.5 h [54].

reflux, this being ascribed to the formation of clusters in the glass layer during reflux.

When TGA is used as the surfactant, its hydrolysis during reflux produces S^{2-} ions [56, 57]. In the present study, the released S^{2-} ions reacted with the free Cd^{2+} ions distributed in the glass matrix, forming CdS-like clusters. If this were to occur in aqueous solution, the product would be easily incorporated into QDs via diffusion. However, in the present study, this diffusion was greatly reduced by the glass network around the QDs, such that only the ions were sufficiently small to disperse actively enough for clusters to form and grow.

In order to support the formation of clusters quantitatively, quantum mechanical calculations were performed using the theory presented by Eychmüller *et al.* [58]. Figure 11.12a shows a schematic diagram of a green-emitting QD (2.6 nm diameter), and of a QD with a glass layer containing Cd^{2+} and TGA before and after reflux. Reflux produces CdS-like clusters of size *d* in the glass layer. In Figure 11.12a, ③ illustrates the complex structure used, where the clusters are formed inside the glass layer. The lower part of Figure 11.12a shows the radial probability

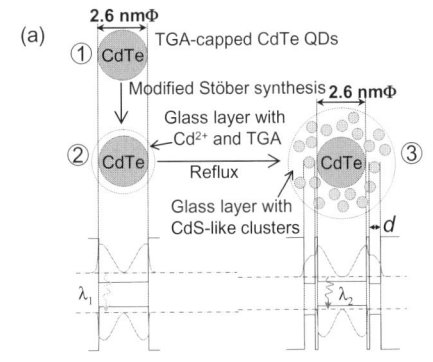

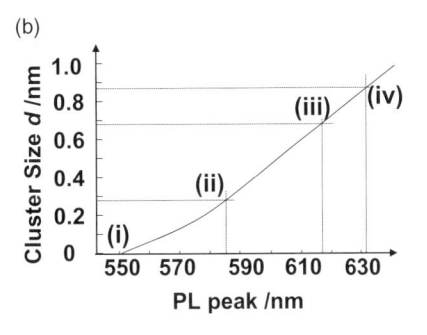

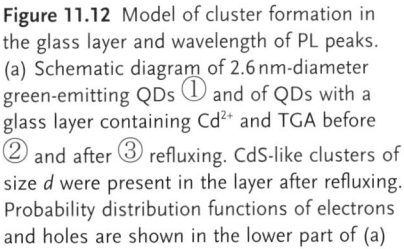

11.3 Incorporation of Quantum Dots in Glass Beads for Bioapplications

Figure 11.12 Model of cluster formation in the glass layer and wavelength of PL peaks. (a) Schematic diagram of 2.6 nm-diameter green-emitting QDs ① and of QDs with a glass layer containing Cd^{2+} and TGA before ② and after ③ refluxing. CdS-like clusters of size d were present in the layer after refluxing. Probability distribution functions of electrons and holes are shown in the lower part of (a) for ② and ③, together with PL wavelengths (λ_1 and λ_2); (b) PL peak wavelengths (abscissa) shown in Figure 11.11a (i–iv) and calculated wavelengths (solid curve) as a function of cluster size (ordinate) based on the cluster model shown in (a). The distance between clusters and QDs was set to 0.1 nm [54].

distribution functions of the electrons and holes in the excitons for ② and ③. The probability distribution of electrons was confined to the core for ②, while the distribution extended into the surrounding clusters through the tunneling effect for ③. As shown in the middle of the figure, this extension resulted in a red shift of the PL peak from λ_1 to λ_2 through a reduction of the quantum confinement effect. Figure 11.12b shows the results of this calculation. When the distance between the clusters and QDs was 0.1 nm, the calculated sizes of the clusters leading to the observed red-shift (samples i–iv in Figure 11.11b) are shown in the graph. The sizes increased upon reflux; for example, the PL peak shifted to 617 nm (sample iii) when the cluster size became 0.68 nm.

The main reason for the increased PL efficiency in the present study was attributed to the absence of an interface between the core (QD) and shell (clusters). As the clusters were not chemically bonded directly to the QD surfaces, a lattice constant mismatch between the core and shell and lattice defects in the shell, due to a distortion at the interface between the core and shell, had no effect. This can also be attributed to a narrowing of the PL spectrum, because the absence of these interfaces and distortions reduced the inhomogeneity of the total QD ensemble.

The observed PL enhancement can be regarded as a new type of surface passivation that is different from others [9, 57, 59, 60]. The same phenomenon was observed with initially red-emitting rather than green-emitting CdTe QDs (as explained here), together with ZnSe QDs having a similar glass shell containing ZnS-like clusters.

With regards to stability, the PL efficiency of the QDs in glass beads for this study after reflux was maintained in a phosphate-buffered saline (PBS) solution for orders of magnitude longer than that of normal QDs. This stability is crucial for the bioapplication of this type of QD.

11.3.5
Emitting Glass Fibers Created by Self-Assembly of the Complex Structured Beads

It was found that the glass-coated CdTe QDs, when dispersed in solution, would self-assemble to create fibers during reflux. The morphologies of these fibers were of three types – tubular, belt-like, and solid – depending on the ratio of chemicals in the solution. When the green-emitting QDs were used as the starting material, the PL color of the fibers could be tuned from yellow to red by controlling the duration of the reflux [61].

Figure 11.13 shows the evolution of the PL spectra of these fibers (solid curves) and colloidally dispersed glass-coated CdTe QDs (dashed curves) during reflux. These fibers exhibited almost the same PL peak wavelength and spectral width as did the glass-coated QDs obtained in solution (complex structured glass beads, as explained above). This means that the glass-coated QDs incorporated in these fibers had grown to the same complex structure as those in the solution during reflux. The PL efficiency of these fibers was 16%. To the best of the present authors' knowledge, there have been no reports of the PL efficiency of luminescent fibers containing QDs, most probably because of the low values obtained. From the absorbance at the first absorption peak, and so on, the concentration of QDs in the fibers was estimated at 0.002 M; this corresponded to approximately 20% of the ultimate value (ca. 0.01 M), when concentration quenching is observed [16].

Scanning electron microscopy (SEM) images of the fibers are shown in Figure 11.14, where a tubular morphology is shown in Figure 11.14a1,b, and the diameter of the hollow portion was approximately 150 nm. Powder XRD patterns of these luminescent fibers revealed that they had a crystal structure, although not all of the diffraction peaks could be indexed to a compound in the XRD database. Similar diffraction peaks were observed for a composite of Cd^{2+}, TGA, and polymer [62].

Figure 11.15 shows the SEM images of luminescent fibers of the other two morphologies (belt-like and solid) after reflux. As shown in the lower part of the figure, these fibers exhibited a bright PL, similar to those in Figure 11.13.

11.4
Preparation of Cadmium-Free Quantum Dots With Water Dispersibility

It is crucial that biologically applied QDs do not contain highly toxic elements, despite evidence indicating that cadmium ions are present in strongly emitting QDs (such as CdSe and CdTe). Likewise, when dealing with glass coating through sol–gel techniques, water dispersibility represents another very important issue. Consequently, in this section attention is focused on state-of-the-art technologies for preparing cadmium-free, hydrophilic QDs.

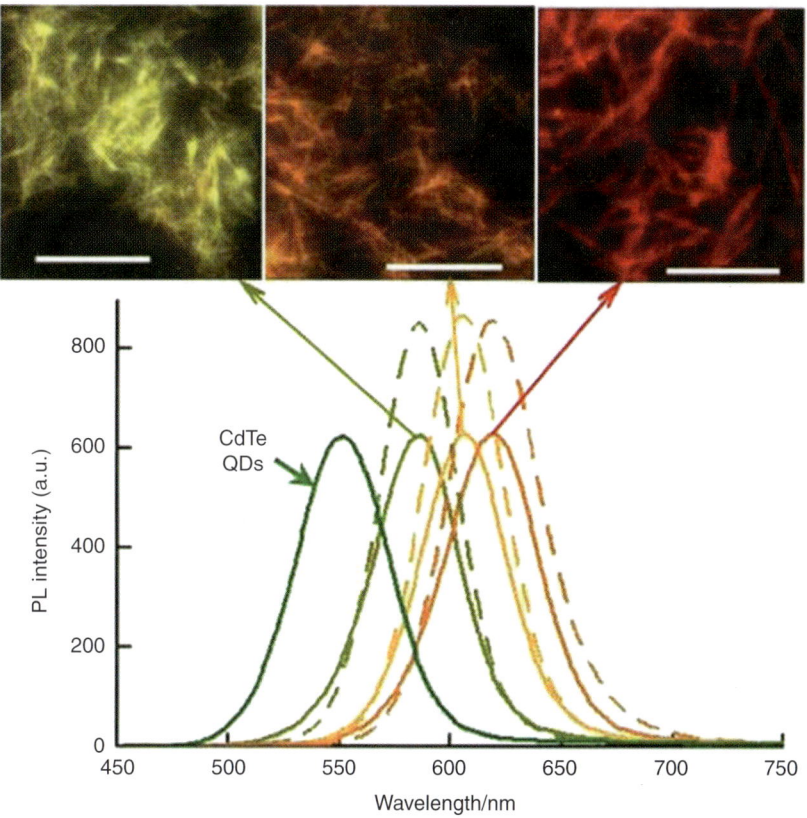

Figure 11.13 Evolution of PL spectra of luminescent fibers (solid curves) and complex structured glass beads (dashed curves) in solution during refluxing. The reflux times were 0.5, 1.0, and 2.0 h, from left to right. The peak wavelength of PL and width of spectra were almost the same as those of the glass beads. PL peak wavelengths after refluxing for 0.5, 1.0, and 2.0 h were 585, 606, and 620 nm, respectively. The initial PL spectrum of green-emitting CdTe QDs is shown on the left for comparison. The color images above are of corresponding fibers, and were captured under 365 nm UV light. Scale bar = ca. 50 μm [61].

The band gaps of Group II–VI and III–V compound bulk semiconductors are shown in Figure 11.16. Other than these, chalcopyrite crystals are known. However, their PL mostly originates from defects with longer decay times. The data in Figure 11.16 show that QDs containing heavier elements generally have a narrower bandgap, although when the QDs becomes smaller the quantum–size effect makes the corresponding gap larger. Therefore, the number of materials of QD that demonstrate the PL wavelength over a wide visible range are limited. Among these materials, InP becomes the only major candidate when those containing Cd and As are eliminated; consequently, the preparation of InP QDs has undergone intensive investigation during recent years.

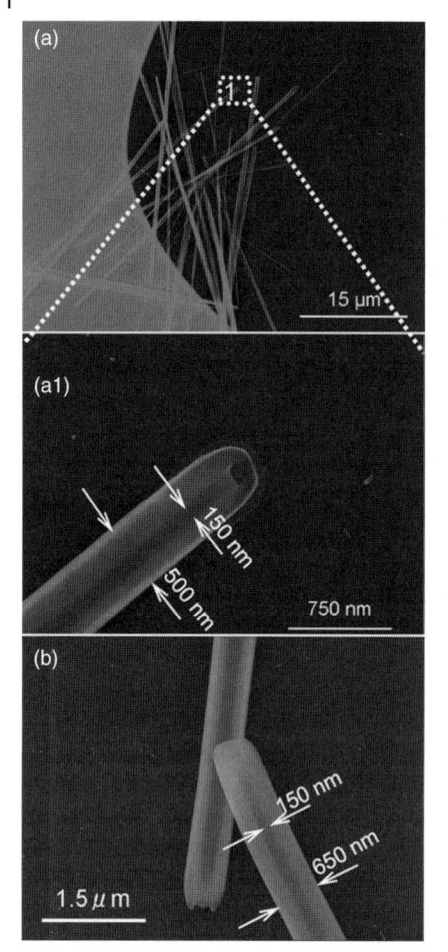

Figure 11.14 (a) Scanning electron microscopy image of luminescent fibers after refluxing for 2 h; a(1) Magnified image of area 1 in (a). The hollow part of tube is clearly evident; (b) Magnified image of fibers with different outside diameters. As shown in a(1) and (b), even though the outside diameters differed, the diameters of the hollow parts were almost the same [61].

The most frequently used method for the synthesis of InP QDs employs the dehalosilylation reaction between indium chloride ($InCl_3$) and tris(trimethylsilyl) phosphine ($P(Si(CH_3)_3)_3$), in the presence of a mixture of trioctylphosphine oxide (TOPO) and trioctylphosphine (TOP) or dodecylamine (DDA) as coordinating solvents. This reaction requires a high temperature (>200 °C), an oxygen-free atmosphere, and several days of reaction time [63–68]. As the phosphor source used is both expensive and explosive, many investigations have been conducted to improve the synthesis by eliminating complex reactions and using less-toxic

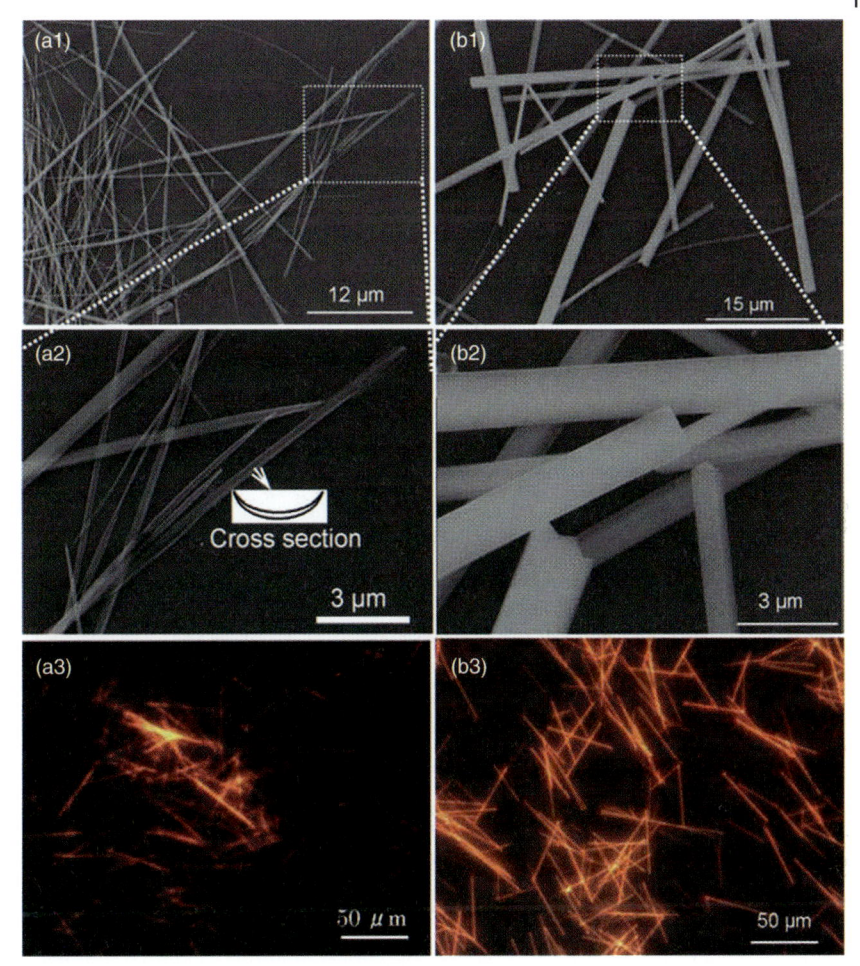

11.4 Preparation of Cadmium-Free Quantum Dots With Water Dispersibility | 417

Figure 11.15 Scanning electron microscopy images of luminescent fibers after refluxing for 2 h showing (a[1], a[2]) belt-like and (b[1], b[2]) solid morphologies. a(3) and b(3) correspond to color images under 365 nm UV light [61].

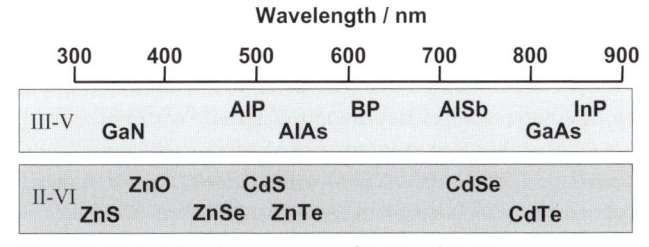

Figure 11.16 Bulk bandgap energies of II–VI and III–V compound semiconductors. Note: the energy is converted to wavelength.

and less-flammable precursors [69–71]. Recently, Maenosono et al. reported the use of a readily available phosphorus source (tris(dimethylamino) phosphine, $P(N(CH_3)_2)_3$) [69]. A solvothermal ("hydrothermal" when the solvent is water) synthesis also proved to be effective when reacting at a relatively low temperature [70, 71], although the quality of the prepared QDs was inadequate.

All of the as-prepared InP QDs described above have been synthesized in organic solution, and showed almost no PL immediately after preparation. The most widely used method for acquiring PL is to coat the core with a shell such as ZnS; the PL efficiency of InP/ZnS QDs prepared in organic solutions was 23–40% [72, 73]. Another way of increasing the PL efficiency would be to treat the QDs with hydrogen fluoride (HF), which removes the surface traps caused by phosphorus atoms. The PL efficiency of these QDs was reported to be 25–40% in organic solution [65, 74].

The surfaces of QDs prepared in organic solution may be exchanged for hydrophilic ones in order to attain water dispersibility, although such a phase transfer would normally cause a significant reduction in PL efficiency. Parasad et al. transferred the InP/ZnS QDs into aqueous solution by exchanging the surfactants for TGA, but the PL efficiency in water was not reported [75]. Bawendi et al. prepared water-dispersible InAs/ZnSe core–shell QDs of various sizes with PL efficiencies of 6–9% [76]. More recently, Peng et al. reported the preparation of water-dispersible InP/ZnS QDs with PL efficiencies of 40% [73]. These methods for obtaining water-dispersible III–V QDs involve exchanging the hydrophilic surfactants prepared using organometallic methods, rather than a direct formation of a hydrophilic layer on the surface of bare QDs in water.

Post-preparative irradiation has been reported to effectively form a ZnS shell on II–VI cores in aqueous solution [77–80], and water-dispersible blue-emitting, cadmium-free QDs ($ZnSe_xTe_{1-x}$) were successfully obtained in this way [79]. This method was also found to be applicable for coating those InP QDs that originally were in a hydrophilic condition. The synthesis procedure is divided into three steps, as shown in Scheme 11.6a; schematic representations of the QDs at each of the steps are shown in Scheme 11.6b [81].

In Step 1, the InP core is prepared via a solvothermal synthesis in organic solution [82]. In a typical synthesis, $InCl_3$, $P(N(CH_3)_2)_3$, and dodecylamine (DDA) are mixed well with toluene and placed in an autoclave. The temperature is then raised to 70 °C for 1 h to dissolve the $InCl_3$ and DDA in the toluene solvent, and subsequently maintained at 180 °C for 24 h for solvothermal synthesis; it is then reduced to room temperature. The DDA acts as a good stabilizing agent for the preparation of QDs [83], as it also reduces the temperature required for QD formation [84]. The $P(N(CH_3)_2)_3$ decomposes at a relatively low temperature to produce phosphor; then, when the byproducts have been precipitated and discarded, the crude solution is subjected to size-selective precipitation. The quality of the QDs is confirmed by measuring the PL properties after photoetching with HF.

In Step 2, the as-prepared QDs dispersed in organic solution are transferred into aqueous solution. After redispersion of the prepared QDs in a mixture of organic solvents (hexane/butanol, 67/33, v/v), the mixture is further mixed with

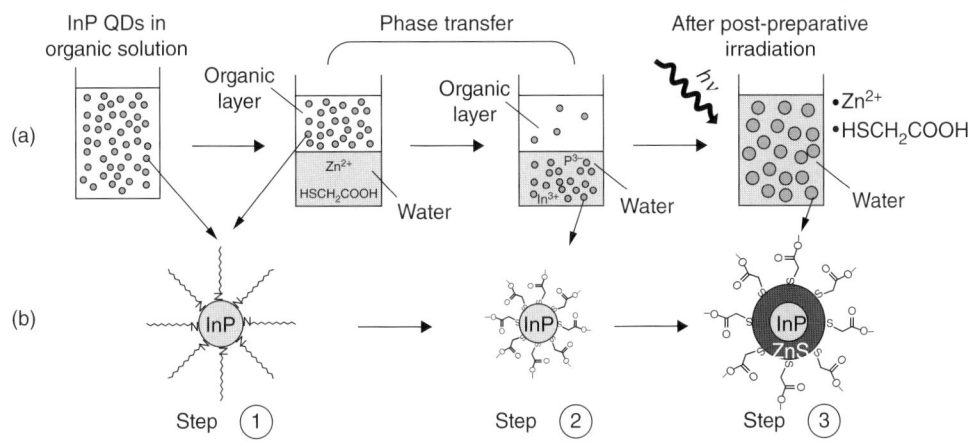

Scheme 11.6 (a) Illustration of synthesis of highly luminescent core–shell InP QDs in aqueous solution[a]; (b) Schematic drawing of InP QDs at different synthesis Stages [81]. [a]Step 1: solvothermal preparation of InP QDs in organic solution. Step 2: phase transfer of InP QDs from organic to aqueous solution. Step 3: creation of ZnS shell on surface of InP QDs by postpreparative UV irradiation.

a water phase containing TGA and Zu^{2+} ions, and stirred. This procedure efficiently transfers the QDs in organic solution to the water phase. As butanol is miscible with water, the effective surface area between the butanol and water is significantly larger, and this is probably the main reason for the efficient transfer. When the QDs are transferred, the color of the organic phase quickly fades, accompanied by a coloration of the water phase. Raising the temperature of the solution will increase the speed of this transfer.

In Step 3, a ZnS shell is created via a postpreparative UV-irradiation in water. As noted above, UV irradiation creates a thickness-controlled ZnS shell on the surface of II–VI QDs. However, surface dissolution of the InP QDs in water, under conditions where bulk InP does not dissolve due to its covalent bond nature, competed with formation of the ZnS layer at the start of UV irradiation. Thus, in the present study it was necessary to increase the power of the UV light to $4\,W\,cm^{-2}$ so as to quickly create a ZnS shell.

The absorption and PL spectral evolution of InP QDs from Steps 1 to 3 are shown in Figure 11.17. QDs of approximately 3.1 nm were used for this experiment. At Step 1, the QDs showed a very weak PL on the lower energy side, with the first excitonic absorption peak at 570 nm. However, when the QDs were transferred into the aqueous solution (Step 2), the first excitonic peak shifted to a shorter wavelength by approximately 30 nm, this being ascribed to a decrease in the particle size. From an elemental analysis, approximately one monolayer of InP was dissolved from the surface during this transfer, whilst the PL efficiency of InP QDs was increased from approximately 0% to 2% at this stage. This increment

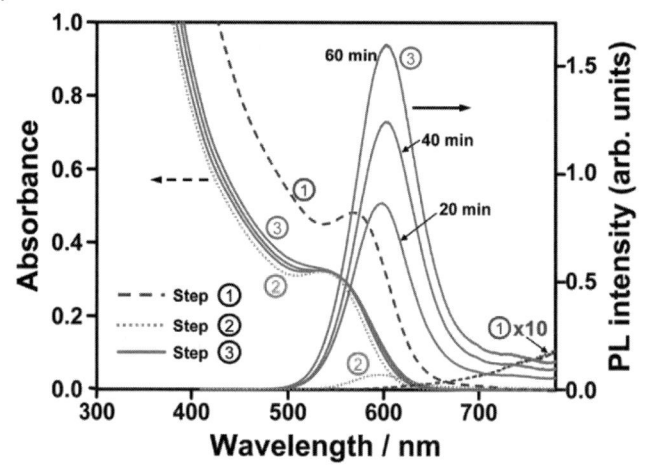

Figure 11.17 Absorption and PL spectra of QDs in organic solution (Step 1), after phase transformation (Step 2), and after UV irradiation (Step 3). The times in the figure indicate the duration of UV irradiation [81].

was caused by the creation of a thin ZnS shell on the surface of the QDs, as clarified by XRD and elemental analysis. The phase transfer introduced here differed a great deal from the conventional method for obtaining water-soluble III–V QDs by the simple exchange of surfactants for hydrophilic ones. In the present study, the surface of the InP QDs was dissolved, and a hydrophilic layer formed on the surface simultaneously. Then, following irradiation in Step 3, the PL intensities were increased dramatically. Although small red-shifts of both the excitonic absorption and PL peak were observed, these were much smaller than had been previously reported for $ZnSe_{1-x}Te_x/ZnS$ QDs [79]. This difference was attributed to the larger band offset between the InP core and ZnS shell, which causes the carrier wave functions to be more confined within the core, as indicated quantitatively by quantum mechanical calculations.

Figure 11.18a shows a series of absorption and PL spectra of the prepared QDs; photographic images of the prepared QDs under room and UV light are shown in Figure 11.18b,c. When the preparation conditions were optimized, a high PL efficiency (30–68%) was obtained for the emission from green to red in water, with sufficient stability. The main component of the PL decay time was in the range 60–70 ns, regardless of the size. It was confirmed that there was no lifetime component longer than 1 μs in any of the samples.

To clarify the origin of this PL increment during UV irradiation, the wavefunctions of the electrons and holes of the QDs at each step of the preparation were calculated, together with the 1s-1s transition energy. This calculation was simplified by assuming that the QDs consisted of an InP core covered with a concentric shell of ZnS, and were without an alloying layer at the interface. Figure 11.19 shows the radial distribution function of electrons and holes in the QDs at each

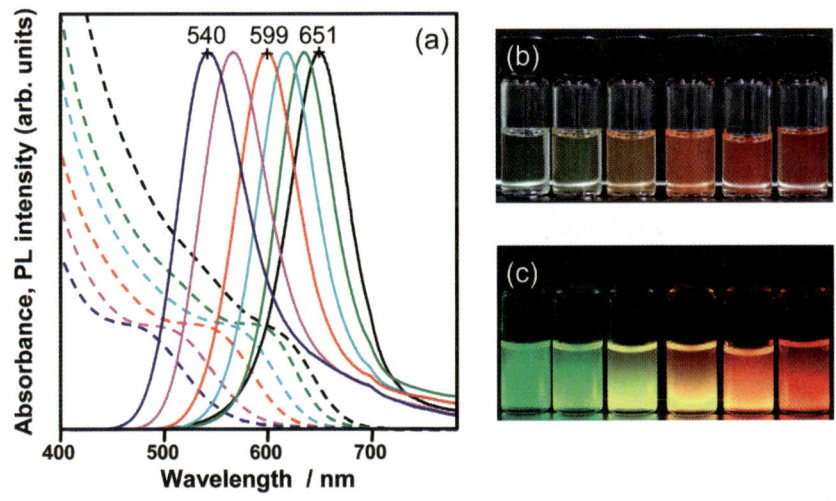

Figure 11.18 (a) Absorption and PL spectra of InP/ZnS QDs in the aqueous solution. The QDs were size-selectively precipitated at Step 2 and irradiated by UV light at Step 3. Fluorescence images of the prepared QDs placed under the room light (b) and excited by UV lamp at 365 nm (c) [81].

step. The size parameters used for the calculation (InP cores: 3.1, 2.5, and 2.5 nm, ZnS shell: 0.25 and 1.5 nm) were taken from TEM observations and an elemental analysis. In Step 1, the electrons and holes were spread over the particles, and tunneled into the surrounding organic solvent (Figure 11.19a). Then, when the InP QDs were transferred to the aqueous solution in Step 2, a thin shell of ZnS (0.25 nm) formed on the surface of the QDs. The electrons tunneled into the shell, and slightly further into the surrounding water. In contrast, the holes tunneled only slightly into the shell but did not reach the surrounding water to any substantial degree (Figure 11.19b). In Step 3, a thicker ZnS shell (1.5 nm) completely confined the electron inside the particle (Figure 11.19c). The calculations showed that a thicker ZnS shell (>1 nm) would be required to confine the electrons completely inside the NCs. This would explain why a high PL efficiency was obtained in the study, compared to when only a thin ZnS layer (0.5 nm) was created on the surface of the InP core by epitaxial growth [72, 75].

The calculated energies of the 1s-1s transition of the InP QDs are shown in Figure 11.19d, together with the experimental values, which were in good agreement with the calculations. Figure 11.19e shows the evolution of PL efficiency during UV irradiation. When the thickness of the ZnS shell was sufficient to confine the electrons after an irradiation of approximately 40 min under these experimental conditions, the PL efficiency was almost saturated. The calculation shown here explains quite well the drastic increase in PL efficiency due to the formation of a thick shell. The incorporation of this hydrophilic QD into a glass matrix is most likely to be the next topic for future investigations.

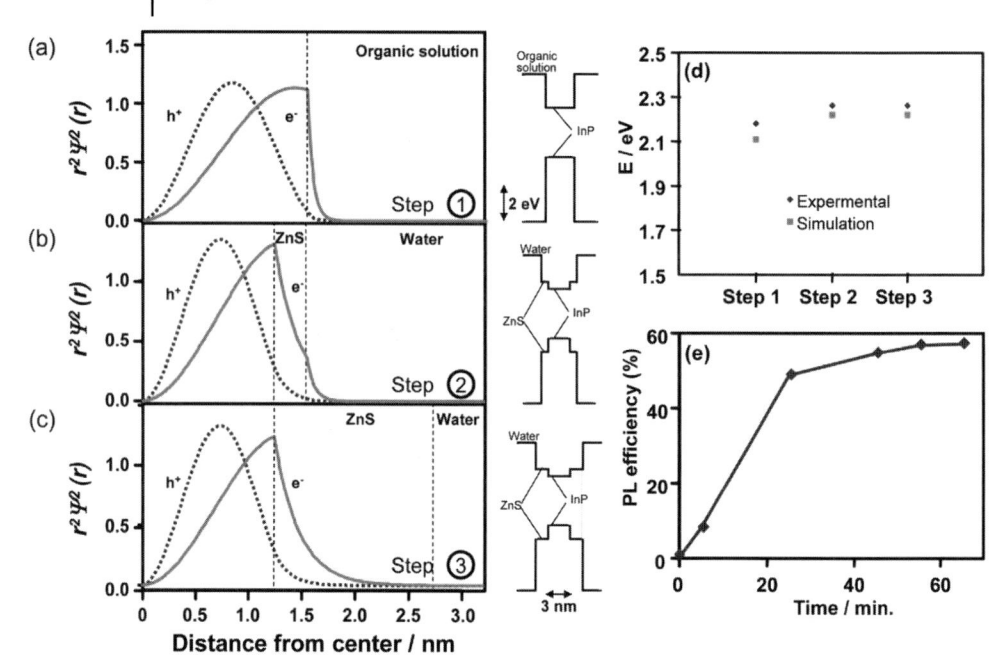

Figure 11.19 Radial probabilities for presence of electrons (e⁻) and holes of exciton (h⁺) in QDs in organic solution (a), after phase transformation (b), and after postpreparative irradiation (c). Thicknesses of ZnS shell created on core after phase transformation and postpreparative irradiation were 0.25 and 1.5 nm, respectively. Schematic illustrations at the center show band structures used for calculations; (d) Calculated 1s-1s transition energy for QDs at different stages, together with experimental data obtained from absorption spectra; (e) PL efficiency of QDs as a function of irradiation time. The particle size was increased with the irradiation time [81].

11.5
Summary and Future Perspective

The history of technology, when closely examined, often appears to develop as a result of a tight competition between two incompatible routes. In the present study, the case always considered was whether glass is better than polymer. As noted in the Introduction, coatings on QDs are indispensable for a variety of applications and, indeed, with polymer-coated QDs having been investigated from an early stage, many are already in practical use today [85]. At the outset of these studies, the advantages of glass over polymers, notably their robustness and light resistance, were well recognized. Yet, other than protective and passive roles, glass was known to play an active part in creating original structures with novel optical properties. Additionally, it was necessary to prepare QDs that did not contain highly toxic elements, whilst as long as sol–gel-derived glass matrices were used, the water dispersibility of the QDs was assured. It is hoped that, in the near future, these highly creative approaches using novel QDs will provide momentous information, especially in the areas of biomedicine and diagnostics. One clear target here would be the highly sensitive and specific detection of the viruses that currently threaten the human race.

Acknowledgments

The author acknowledges his colleagues, Drs M. Ando, C. L. Li and P. Yang, for their dedicated effort and continuous discussions. Previously, these studies were supported by the Nano-glass project from NEDO; they are currently supported by the Creation and Support Program for Start-ups from Universities from JST.

References

1 Efros, A.L., Lockwood, D.J. and Tsybeskov, L. (eds) (2003) *Semiconductor Nanocrystals from Basic Principles to Applications*, Kluwer Academic/Plenum Publishers, ISBN: 0-306-47751-3.

2 Schmid, G. (ed.) (2004) *Nanoparticles from Theory to Application*, Wiley-VCH Verlag GmbH, ISBN: 987-3-527-30507-0.

3 Jaeckel, G. (1926) Über einige neuzeitliche Absorptionsgläser. *Zeitschrift für Technische Physik*, **6**, 301–4.

4 Ekimov, A. (1996) Growth and optical properties of semiconductor nanocrystals in a glass matrix. *Journal of Luminescence*, **70**, 1–20.

5 Ekimov, A.I., Efros, A.L. and Onushchenko, A.A. (1985) Quantum size effect in semiconductor microcrystals. *Solid State Communications*, **56**, 921–4.

6 Nogami, M., Nagaska, K. and Kato, E., (1990) Preparation of small-particle-size, semiconductor CdS-doped silica glasses by the sol-gel process. *Journal of the American Ceramic Society*, **73**, 2097–9.

7 Jain, R.K. and Lind, R.C. (1983) Degenerate four-wave mixing in semiconductor-doped glasses. *Journal of the Optical Society of America*, **73**, 647–53.

8 Murase, N. and Yazawa, T. (2001) Partially reduced cuprous oxide nanoparticles formed in porous glass reaction fields. *Journal of the American Ceramic Society*, **84**, 2269–73.

9 Dabbousi, B.O., Rodriguez-Viejo, J., Mikulec, F.V., Heine, J.R., Mattoussi, H., Ober, R., Jensen, K.F. and Bawendi, M.G. (1997) (CdSe)ZnS core-shell quantum dots: synthesis and characterization of a size series of highly luminescent nanocrystallites. *Journal of Physical Chemistry, B*, **101**, 9463–75.

10 Peng, X., Schlamp, M.C., Kadavanich, A.V. and Alivisatos, A.P. (1997) Epitaxial growth of highly luminescent CdSe/CdS core/shell nanocrystals with photostability and electronic accessibility. *Journal of the American Chemical Society*, **119**, 7019–29.

11 Talapin, D.V., Haubold, S., Rogach, A.L., Kornowski, A., Haase, M. and Weller, H. (2001) A novel organometallic synthesis of highly luminescent CdTe nanocrystals. *Journal of Physical Chemistry, B*, **105**, 2260–3.

12 Talapin, D.V., Rogach, A.L., Shevchenko, E.V., Kornowski, A., Haase, M. and Weller, H. (2002) Dynamic distribution of growth rates within the ensembles of colloidal II–VI and III–V semiconductor nanocrystals as a factor governing their photoluminescence efficiency. *Journal of the American Chemical Society*, **124**, 5782–90.

13 Selvan, S.T., Bullen, C., Ashokkumar, M. and Mulvaney, P. (2001) Synthesis of tunable, highly luminescent QD-glasses through Sol-Gel processing. *Advanced Materials*, **13**, 985–8.

14 Li, C.L. and Murase, N. (2005) Surfactant-dependent photoluminescence of CdTe nanocrystals in aqueous solution. *Chemistry Letters*, **34**, 92–3.

15 Li, C.L. and Murase, N. (2004) Synthesis of highly luminescent glasses

incorporating CdTe nanocrystals through sol-gel processing. *Langmuir*, **20**, 1–4.

16 Yang, P., Li, C.L. and Murase, N. (2005) Highly photoluminescent multilayer QD–glass films prepared by LbL self-assembly. *Langmuir*, **21**, 8913–17.

17 Murase, N. and Li, C.L. (2008) Consistent determination of photoluminescence quantum efficiency for phosphors in the form of solution, plate, thin film, and powder. *Journal of Luminescence*, **128**, 1896–903.

18 Moerner, W.E. and Kador, L. (1989) Optical detection and spectroscopy of single molecules in a solid. *Physical Review Letters*, **62**, 2535–8.

19 Orrit, M. and Bernard, J. (1990) Single pentacene molecules detected by fluorescence excitation in a p-terphenyl crystal. *Physical Review Letters*, **65**, 2716–19.

20 Seisenberger, G., Ried, M.U., Endreß, T., Büning, H., Hallek, M. and Bräuchle, C. (2001) Real-time single-molecule imaging of the infection pathway of an adeno-associated virus. *Science*, **294**, 1929–2932.

21 Nirmal, M., Dabbousi, B.O., Bawendi, M.G., Macklin, J.J., Trautman, J.K., Harris, T.D. and Brus, L.E. (1996) Fluorescence intermittency in single cadmium selenide nanocrystals. *Nature*, **383**, 802–4.

22 Empedocles, S.A., Norris, D.J., and Bawendi, M.G. (1996) Photoluminescence spectroscopy of single CdSe nanocrystallite quantum dots. *Physical Review Letters*, **77**, 3873–6.

23 Toomre, D. and Manstein, D.J. (2001) Lighting up the cell surface with evanescent wave microscopy. *Trends in Cell Biology*, **11**, 298–303.

24 Toba, S., Watanabe, T.M., Yamaguchi-Okimoto, L., Yano Toyoshima, Y. and Higuchi, H. (2006) Overlapping hand-over-hand mechanism of single molecular motility of cytoplasmic dynein. *Proceedings of the National Academy of Sciences of the United States of America*, **103**, 5741–5.

25 Yildiz, A., Forkey, J.N., McKinney, S.A., Ha, T., Goldman, Y.E. and Selvin, P.R. (2003) Myosin V walks hand-over-hand: single fluorophore imaging with 1.5-nm localization. *Science*, **300**, 2061–5.

26 Nann, T. and Mulvaney, P. (2004) Single quantum dots in spherical silica particles. *Angewandte Chemie, International Edition*, **43**, 5393–6.

27 Darbandi, M., Thomann, R. and Nann, T. (2005) Single quantum dots in silica spheres by microemulsion synthesis. *Chemistry of Materials*, **17**, 5720–5.

28 Gerion, D., Pinaud, F., Williams, S.C., Parak, W.J., Zanchet, D., Weiss, S. and Alivisatos, A.P. (2001) Synthesis and properties of biocompatible water-soluble silica-coated CdSe/ZnS semiconductor quantum dots. *Journal of Physical Chemistry, B*, **105**, 8861–71.

29 Stöber, W., Fink, A. and Bohn, E. (1968) Controlled growth of monodisperse silica spheres in the micron size range. *Journal of Colloid and Interface Science*, **26**, 62–9.

30 Liz-Marzán, L.M., Giersig, M. and Mulvaney, P. (1996) Synthesis of nanosized gold-silica core-shell particles. *Langmuir*, **12**, 4329–35.

31 Rogach, A.L., Nagesha, D., Ostrander, J.W., Giersig, M. and Kotov, N.A. (2000) "Raisin bun"-type composite spheres of silica and semiconductor nanocrystals. *Chemistry of Materials*, **12**, 2676–85.

32 Chan, Y., Zimmer, J.P., Stroh, M., Steckel, J.S., Jain, R.K. and Bawendi, M.G. (2004) Incorporation of luminescent nanocrystals into monodisperse core-shell silica microspheres. *Advanced Materials*, **16**, 2092–7.

33 Yamauchi, H., Ishikawa, T. and Kondo, S. (1989) Surface characterization of ultramicro spherical particles of silica prepared by w/o microemulsion method. *Colloids and Surfaces*, **37**, 71–80.

34 Murase, N. and Yazawa, T. (2001) Japanese Patent 3677538.

35 Selvan, S.T., Li, C.L., Ando, M. and Murase, N. (2004) Formation of luminescent CdTe-silica nanoparticles through an inverse microemulsion technique. *Chemistry Letters*, **33**, 434–5.

36 Yang, Y. and Gao, M.Y. (2005) Preparation of fluorescent SiO2 particles with Single CdTe nanocrystal cores by the reverse microemulsion method. *Advanced Materials*, **17**, 2354–7.

37 Selvan, S.T., Tan, T.T. and Ying, J.Y. (2005) Robust, non-cytotoxic, silica-coated CdSe quantum dots with efficient

photoluminescence. *Advanced Materials*, **17**, 1620–5.
38. Koole, R., van Schooneveld, M.M., Hilhorst, J., Donegá, C.M., Hart, T.D.C., van Blaaderen, A., Vanmaekelbergh, D. and Meijerink, A. (2008) On the incorporation mechanism of hydrophobic quantum dots in silica spheres by a reverse microemulsion method. *Chemistry of Materials*, **20**, 2503–12.
39. Yang, P., Ando, M. and Murase, N. (2007) Encapsulation of emitting CdTe QDs within silica beads to retain initial photoluminescence efficiency. *Journal of Colloid and Interface Science*, **316**, 420–7.
40. Ando, M., Li, C.L. and Murase, N. (2004) Photoluminescence properties and zeta potential of water-dispersible CdTe nanocrystals. *Materials Research Society Symposium Proceedings*, **789**, 123–8.
41. Bruchez, M. Jr., Moronne, M., Gin, P., Weiss, S. and Alivisatos, A.P. (1998) Semiconductor nanocrystals as fluorescent biological labels. *Science*, **281**, 2013–16.
42. Niemeyer, C.M., Nanoparticles, P. and Acids, N. (2001) Biotechnology meets materials science. *Angewandte Chemie, International Edition*, **40**, 4128–58.
43. Selvan, S.T., Patra, P.K., Ang, C.Y. and Ying, J.Y. (2007) Synthesis of silica-coated semiconductor and magnetic quantum dots and their use in the imaging of live cells. *Angewandte Chemie, International Edition*, **46**, 2448–52.
44. Yang, P., Ando, M. and Murase, N. (2009) Formation of two types of highly luminescent SiO_2 beads impregnated with multiple CdTe QDs. *New Journal of Chemistry*, **33**, 561–7.
45. Insin, N., Tracy, J.B., Lee, H., Zimmer, J.P., Westervelt, R.M. and Bawendi, M.G. (2008) Incorporation of iron oxide nanoparticles and quantum dots into silica microspheres. *ACS Nano*, **2**, 197–202.
46. Li, M.J., Chen, Z., Yam, V.W.-W. and Zu, Y. (2008) Multifunctional ruthenium(II) polypyridine complex-based core-shell magnetic silica nanocomposites: magnetism, luminescence, and electrochemiluminescence. *ACS Nano*, **2**, 905–12.
47. Yi, D.K., Selvan, S.T., Lee, S.S., Papaefthymiou, G.C., Kundaliya, D. and Ying, J.Y. (2005) Silica-coated nanocomposites of magnetic nanoparticles and quantum dots. *Journal of the American Chemical Society*, **127**, 4990–1.
48. Zhang, Y., Pan, S., Teng, X., Luo, Y. and Li, G. (2008) Bifunctional magnetic-luminescent nanocomposites Y_2O_3/Tb nanorods on the surface of iron oxide/silica core-shell nanostructures. *Journal of Physical Chemistry, C*, **112**, 9623–6.
49. Rossi, L.M., Shi, L., Quina, F.H. and Rosenzweig, Z. (2005) Stöber synthesis of monodispersed luminescent silica nanoparticles for bioanalytical assays. *Langmuir*, **21**, 4277–80.
50. Legrand, S., Catheline, A., Kind, L., Constable, E.C., Housecroft, C.E., Landmann, L., Banse, P., Pieles, U. and Wirth-Heller, A. (2008) Controlling silica nanoparticle properties for biomedical applications through surface modification. *New Journal of Chemistry*, **32**, 588–93.
51. Wang, D., He, J., Rosenzweig, N. and Rosenzweig, Z. (2004) Superparamagnetic Fe_2O_3 beads-CdSe/ZnS quantum dots core-shell nanocomposite particles for cell separation. *Nano Letters*, **4**, 409–13.
52. Roullier, V., Grasset, F., Boulmedais, F., Artzner, F., Cador, O. and Marchi-Artzner, V. (2008) Small bioactivated magnetic quantum dot micelles. *Chemistry of Materials*, **20**, 6657–65.
53. Yang, P., Ando, M. and Murase, N. (2009) Preparation of SiO_2 beads with highly luminescent and magnetic nanocrystals via a modified reverse micelle process. *New Journal of Chemistry*, **33**, 1457–61.
54. Murase, N. and Yang, P. (2009) Anomalous photoluminescence in silica-coated semiconductor nanocrystals after heat treatment. *Small*, **5**, 800–3.
55. Yu, W.W., Qu, L., Guo, W. and Peng, X. (2003) Experimental determination of the extinction coefficient of CdTe, CdSe and CdS nanocrystals. *Chemistry of Materials*, **15**, 2854–60.
56. Gaponik, N., Talapin, D.V., Rogach, A.L., Hoppe, K., Shevchenko, E.V., Kornowski, A., Eychmüller, A. and Weller, H. (2002) Thiol-capping of CdTe nanocrystals: an alternative to organometallic synthetic

routes. *Journal of Physical Chemistry, B*, **106**, 7177–85.

57 Rogach, A.L. (2000) Nanocrystalline CdTe and CdTe(S) particles: wet chemical preparation, size-dependent optical properties and perspectives of optoelectronic applications. *Materials Science and Engineering: B*, **69–70**, 435–40.

58 Schooss, D., Mews, A., Eychmüller, A. and Weller, H. (1994) Quantum-dot quantum well CdS/HgS/CdS: theory and experiment. *Physical Review. B, Condensed Matter*, **49**, 17072–8.

59 Battaglia, D., Blackman, B. and Peng, X. (2005) Coupled and decoupled dual quantum systems in one semiconductor nanocrystal. *Journal of the American Chemical Society*, **127**, 10889–97.

60 Jin, T., Fujii, F., Yamada, E., Nodasaka, Y. and Kinjo, M. (2006) Control of the optical properties of quantum dots by surface coating with calix[n]arene carboxylic acids. *Journal of the American Chemical Society*, **128**, 9288–9.

61 Yang, P., Ando, M. and Murase, N. (2009) Morphology- and color-tunable bright fibers with high concentration of CdTe nanocrystals assembled through sol-gel reaction. *Advanced Materials*, **21**, 4016–9.

62 Niu, H. and Gao, M. (2006) Diameter-tunable CdTe nanotubes templated by 1D nanowires of cadmium thiolate polymer. *Angewandte Chemie, International Edition*, **45**, 6462–6.

63 Mićić, O.I., Curtis, C.J., Jones, K.M., Sprague, J.R. and Nozik, A.J. (1994) Synthesis and characterization of InP quantum dots. *Journal of Physical Chemistry*, **98**, 4966–9.

64 Mićić, O.I., Sprague, J., Lu, Z.H. and Nozik, A.J. (1996) Highly efficient band-edge emission from InP quantum dots. *Applied Physics Letters*, **68**, 3150–2.

65 Mićić, O.I., Jones, K.M., Cahill, A. and Nozik, A.J. (1998) Optical, electronic, and structural properties of uncoupled and close-packed arrays of InP quantum dots. *Journal of Physical Chemistry, B*, **102**, 9791–6.

66 Mićić, O.I., Ahrenkiel, S.P. and Nozik, A.J. (2001) Synthesis of extremely small InP quantum dots and electronic coupling in their disordered solid films. *Applied Physics Letters*, **78**, 4022–4.

67 Guzelian, A.A., Katari, J.E.B., Kadavanich, A.V., Banin, U., Hamad, K., Juban, E., Alivisatos, A.P., Wolters, R.H., Arnold, C.C. and Heath, J.R. (1996) Synthesis of size-selected, surface-passivated InP nanocrystals. *Journal of Physical Chemistry*, **100**, 7212–19.

68 Langof, L., Fradkin, L., Ehrenfreund, E., Lifshitz, E., Mi i , O.I. and Nozik, A.J. (2004) Colloidal InP/ZnS core-shell nanocrystals studied by linearly and circularly polarized photoluminescence. *Chemistry and Physics*, **297**, 93–8.

69 Matsumoto, T., Maenosono, S. and Yamaguchi, Y. (2004) Organometallic synthesis of InP quantum dots using tris(dimethylamino)phosphine as a phosphorus source. *Chemistry Letters*, **33**, 1492–3.

70 Qian, Y.T. (1999) Solvothermal synthesis of nanocrystalline III-V semiconductors. *Advanced Materials*, **11**, 1101–2.

71 Yan, P., Xie, Y., Wang, W.Z., Liu, F.Y. and Qian, Y.T. (1999) A low-temperature route to InP nanocrystals. *Journal of Materials Chemistry*, **9**, 1831–3.

72 Haubold, S., Haase, M., Kornowski, A., Weller, H. and Luminescent, S. (2001) InP/ZnS core-shell nanoparticles. *ChemPhysChem*, **2**, 331–4.

73 Xie, R., Battaglia, D. and Peng, X.G. (2007) Colloidal InP nanocrystals as efficient emitters covering blue to near-infrared. *Journal of the American Chemical Society*, **129**, 15432–3.

74 Talapin, D.V., Gaponik, N., Borchert, H., Rogach, A.L., Haase, M. and Weller, H. (2002) Etching of colloidal InP nanocrystals with fluorides: photochemical nature of the process resulting in high photoluminescence efficiency. *Journal of Physical Chemistry, B*, **106**, 12659–63.

75 Bharali, D.J., Lucey, D.W., Jayakumar, H. and Pudavar, H.E. and Prasad, P.N. (2005) Folate-receptor-mediated delivery of InP quantum dots for bioimaging using confocal and two-photon microscopy. *Journal of the American Chemical Society*, **127**, 11364–71.

76 Zimmer, J.P., Kim, S.W., Ohnishi, S., Tanaka, E., Frangioni, J.V. and Bawendi,

M.G. (2006) Size series of small indium arsenide-zinc selenide core-shell nanocrystals and their application to *in vivo* imaging. *Journal of the American Chemical Society*, **128**, 2526–7.

77 Shavel, A., Gaponik, N. and Eychmüller, A. (2004) Efficient UV-blue photoluminescing thiol-stabilized water-soluble alloyed ZnSe(S) nanocrystals. *Journal of Physical Chemistry, B*, **108**, 5905–8.

78 Li, C.L., Nishikawa, K., Ando, M., Enomoto, H. and Murase, N. (2007) Highly luminescent water-soluble ZnSe nanocrystals and their incorporation in a glass matrix. *Colloids and Surfaces. A*, **294**, 33–9.

79 Li, C.L., Nishikawa, K., Ando, M., Enomoto, H. and Murase, N. (2008) Synthesis of Cd-free water-soluble $ZnSe_{1-x}Te_x$ nanocrystals with high luminescence in the blue region. *Journal of Colloid and Interface Science*, **321**, 468–76.

80 Li, C.L., Nishikawa, K., Ando, M., Enomoto, H. and Murase, N. (2007) Blue-emitting type-II semiconductor nanocrystals with high efficiency prepared by aqueous method. *Chemistry Letters*, **36**, 438–9.

81 Li, C.L., Ando, M., Enomoto, H., Murase, N. and Luminescent, H. (2008) Water-soluble InP/ZnS nanocrystals prepared via Reactive phase transfer and photochemical processing. *Journal of Physical Chemistry, C*, **112**, 20190–9.

82 Li, C.L., Ando, M., Enomoto, H. and Murase, N. (2008) Facile preparation of highly luminescent InP nanocrystals by a solvothermal route. *Chemistry Letters*, **37**, 856–7.

83 Talapin, D.V., Rogach, A.L., Mekis, I., Haubold, S., Kornowski, A., Haase, M. and Weller, H. (2002) Synthesis and surface modification of amino-stabilized CdSe, CdTe and InP nanocrystals. *Colloids and Surface A*, **202**, 145–54.

84 Pradhan, N., Reifsnyder, D., Xie, R., Aldana, J. and Peng, X. (2007) Surface ligand dynamics in growth of nanocrystals. *Journal of the American Chemical Society*, **129**, 9500.

85 Invitrogen. http://www.invitrogen.co.jp/qdot/.

12
Toxicology and Biosafety Evaluations of Quantum Dots
Pinpin Lin, Raymond H.S. Yang, Chung-Shi Yang, Chia-Hua Lin and Louis W. Chang

12.1
Introduction

There is no doubt that the development of nanotechnology is probably one of the most revolutionary industrial advancements of the past decade. New development of nanotechnology in the next decade will most likely focus on the applications of selected nanomaterials, including quantum dots (QDs), for multitasking purposes such as drug delivery devices, medical diagnostics, and sensors.

Quantum dots are fluorescent semiconductor nanocrystals (2–100 nm) with promising applications in both new electronic industry and in biomedical imaging usages. Several characteristics distinguish QDs from ordinary fluorophores, including in particular their long fluorescence lifetime which allows the separation of their signals from the background autofluorescence of some cells and tissues [1, 2]. Notably, QDs have clear advantages over common organic fluorophores that include a high luminescence, a resistance to photobleaching, and a broad range of fluorescent wavelengths. Because of their unique fluorescence spectrum, QDs can be conjugated with bioactive moieties such as receptor ligands to specific cells; this in turn allows such cells – including cancer cells – to be labeled [1, 3, 4]. Today, bioconjugated QDs are also being considered for site-specific drug deliveries [1, 5]. Because of these fluorescent properties, it was suggested that QDs might also be used for functional imaging with combinations of magnetic resonance imaging (MRI), positron emission tomography (PET), computed tomography (CT), and infrared (IR) fluorescence imaging [6]. As many applications and potential uses of QDs are presented and discussed in other chapters of this volume, they will not be detailed further here.

Whilst the many applications of nanomaterials (including QDs) are promising and exciting [6], potential problems relating to their toxic potential have been proposed by others [7, 8]. An excellent overview of QD toxicity was recently provided by Hardman [1] which summarizes the various toxicity studies conducted on QDs between 2002 and 2005 (Table 12.1). However, the point is made that most of these earlier studies (pre-2005) were conducted by nanoscientists rather

Table 12.1 Summary of major studies on QD toxicity between 2002–2005.

QD	Model	Exposure conditions/administration	QD concentration	Exposure duration	Toxicity	Reference
CdSe/ZnS-S+1 SA	EL-4 cells	1×10^6 cells/well	0.1–0.4 mg ml^{-1}	0–24 h	Cytotoxic: 0.1 mg ml^{-1} altered cell growth; most cells nonviable at 0.4 mg ml^{-1}	[9]
CdSe/ZnS-SSA	EL-4 cells	200 μl cell suspension injected (iv) into mice	0.1 mg/ml QDs per 5×10^7 cells	2 h to 7 days	No toxicity in mice in vivo	[9]
CdSe/ZnS conjugates: NH$_2$, OH, OH/COOH, H$_2$/OH, MUA, COOH	WTK1 cells	5×10^4 cells/ml	1–2 μM	12 h	2 μM QD-COOH induced DNA damage at 2 h DNA repair on prolonged incubation (12 h)	[10]
CdSe/ZnS-MUA	Vero, HeLa, and primary human hepatocytes	100 μl QDs/3×10^4 cells	0–0.4 mg ml^{-1}	24 h	Cytotoxic: 0.2 mg ml^{-1}, Vero; 0.1 mg ml^{-1}, HeLa; 0.1 mg ml^{-1}, hepatocytes;	[10]
CdTe	Rat pheochromocytoma cells, murine microglial cells	1×10^5 cells cm^{-2}	0.01–100 μg ml^{-1}	2–24 h	10 μg ml^{-1} cytotoxic	[11]
CdSe–MAA, TOPO QDs	Primary rat hepatocytes		62.5–1000 μg ml^{-1}	1–8 h	Cytotoxic: 62.5 μg ml^{-1} cytotoxic under oxidative/photolytic conditions No toxicity on addition of ZnS cap	[7]

QD type	Cell/organism	Delivery/amount	Concentration	Time	Observations	Ref.
QD micelles: CdSe/ZnS QDs in (PEG–PE) and phosphatidylcholine	*Xenopus* blastomeres	5×10^9 QDs per cell (~0.23 pmol per cell)	1.5–3 nl of 2.3 μM QDs injected; ~2.1×10^9 to 4.2×10^9 injected QDs per cell	Days	5×10^9 QDs per cell: cell abnormalities, altered viability and motility. No toxicity at 2×10^9 QDs per cell	[12]
CdSe/ZnS amp-QDs, and mPEG QDs	Mice	200-μl tail vein injection	Injections; ~180 nM QD, ~20 pmol QD g^{-1} animal weight	15 min cell incubations, 1–133 days *in vivo*	No signs of localized necrosis at the sites of deposition	[13]
CdSe/ZnS–DHLA	*Dictyostelium discoideum* and HeLa cells		400–600 nM	45–60 min	No effects on cell growth	[14]
Avidin-conjugated CdSe/ZnS QDs	HeLa cells		0.5–1.0 μM	15 min	No effect on cell growth, development	[14]
CdSe/ZnS– amphiphilic micelle	Mice	Tail vein injection	60 μM QD g^{-1} animal weight, 1 μM and 20 nM final QD concentration	Not given	Mice showed no noticeable ill effects after imaging	[15]
CdSe/ZnS–DHLA QDs	Mice, B16F10 cells	5×10^4 B16F10 cells with 10 μl QDs (~10 pmol), tail vein (iv) injection	100 μl of B16F10 cells used for tail vein injection, ~2×10^5 to 4×10^5 cells injected	4–6 h cell incubation, mice sacrificed at 1–6 h	No toxicity observed in cells or mice	[16]
CdSe/ZnS–MUA QDs; QD–SSA complexes	Vero cells	0.4 mg ml^{-1}	0.24 mg ml^{-1}	2 h	0.4 mg ml^{-1} MUA/SSA-QD complexes did not affect viability of Vero cells	[17]
CdSe/ZnS	HeLa cells	1×10^6 cells	10 pmol QDs per 1×10^5 cells (~10 nM)	10 days (cell culture)	10 nM QD had minimal impact on cell survival	[18]

From Ref. [1].

than by trained toxicologists or biomedical researchers. Hence, it is suggested that major uncertainties remain with regards to QD toxicity and biosafety, and new information is urgently needed in this area.

Nanotechnology research is a rapidly developing field, and since the publication of Hardman's review many excellent new studies have provided a wealth of updated information and concepts on QD toxicity. In this chapter, the aim is to provide not only a more current overview of QD toxicology, but also the most contemporary concepts on the biosafety evaluation of these important nanocrystals. Hence, the information provided will be in accordance with the following topics, deemed important to the toxic potential of QDs:

- The physico-chemical properties of QDs
  - surface chemistry and coatings
  - charge and size of QDs
  - core metals and biodegradation of QDs
- Toxicity and role of oxidative stress
- Absorption, distribution, metabolism, and elimination (ADME) and biosafety evaluations
- Mitochondria as the prime target of QD toxicity
- Environmental and ecological concerns
- Concluding comments and future perspectives

It is our hoped that this chapter will provide readers with essential information related to the toxicity of QDs, as well as a comprehensive idea of conducting biosafety evaluations on nanomaterials in general.

12.2
The Physico-Chemical Properties of Quantum Dots

The toxicity of chemicals is frequently related to, or influenced by, their physico-chemical properties, and this is particularly true in the case of nanomaterials, including QDs. The most important properties are probably the size, charge, and coating chemistry of the nanocrystals [1, 9, 19, 20]. Quantum dots are semiconductor nanocrystals with sizes ranging from 2 to 100 nm. Structurally, most QD nanocrystals consist of a metal or metalloid core-complex with an outer shell that increases their bioavailability. A variety of metal complexes have been used for the QD core, including noble metals and transitional metals such as zinc–selenium (ZnSe), cadmium–selenium (CdSe), and cadmium–tellurium (CdTe) [21]. Recently, newer metalloid complexes, such as CdTe/CdSe, CdSe/ZnTe, and PbSe, have also been introduced as core materials for QDs [22].

When synthesized, QDs are basically hydrophobic in nature and therefore are not biologically compatible or useful for biological or medical applications. In order to increase their biological compatibility, an external hydrophilic coating, such as polyethylene glycol (PEG), is added to render the QDs water-soluble and biologically compatible (Figure 12.1). Other specific bioactive moieties can be

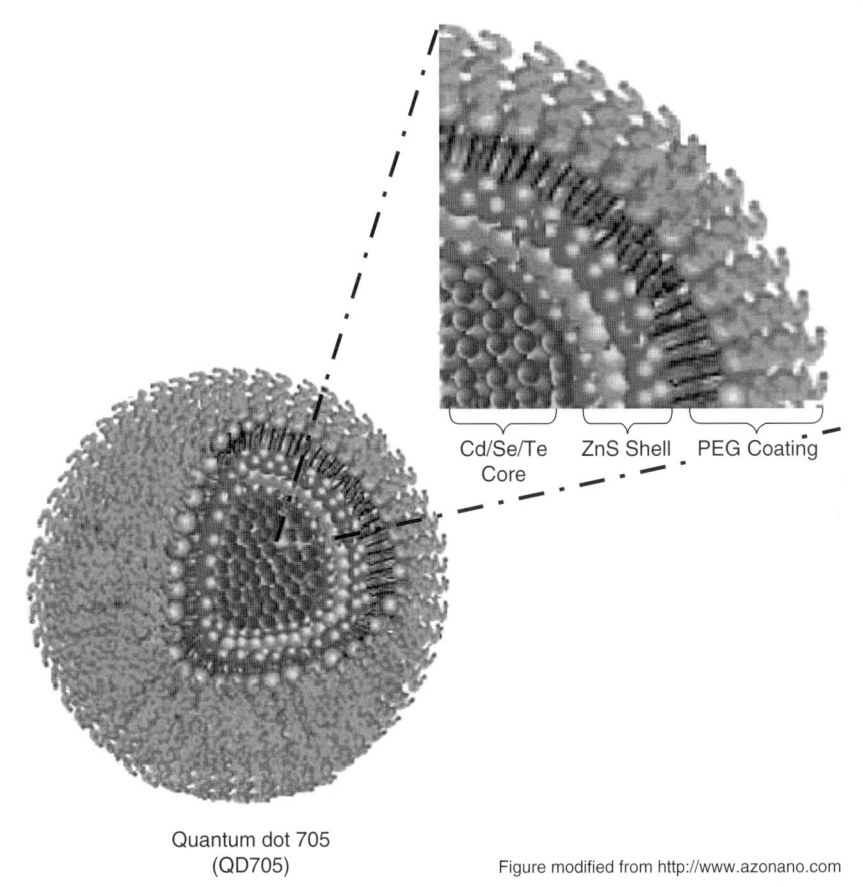

Figure 12.1 Schematic representation of a typical QD structure. Modified from: www.azonano.com.

further conjugated to this coating for specific diagnostic or therapeutic purposes. Such physico-chemical characteristics of QDs are sometimes referred to as "core–shell-conjugates," specific forms of which can be created for the specific functional uses of QDs. Unfortunately, the very physico-chemical characteristics of QDs, the hydrophilic coatings and the metalloid core complexes that render QDs functional may also be potentially toxic.

12.2.1
Surface Chemistry and Coatings

Because toxic metals such as Cd, Te, and Se constitute the inner cores of most QDs, naked (uncoated) QDs or QDs with defective coatings are thought to be cytotoxic [7, 23]. Various outer coatings have therefore been designed to provide a stable shielding for the inner cores of QDs from biological exposure, and to render the QD more biocompatible. However, it is now recognized that surface coatings which may render the QDs functional might also influence their toxicity [19].

Hoshino et al. [9] tested the toxicities of CdSe/ZnS QDs with five different surface modified coatings, namely MUA (QD-COOH), cysteamine (QD-NH$_2$), thioglycerol (QD-OH), and those with functional groups such as QD-OH/COOH and QD-NH$_2$/OH. The QD-COOH and QD-OH/COOH were most highly negatively charged, QD-OH was negatively charged, and the QD-NH$_2$ and QD-NH$_2$/OH were both positively charged. When a Comet test for DNA damage was applied, the toxicity of QDs was found to vary greatly with their coatings: MUA caused severe cytotoxicity, cysteamine was weakly genotoxic, and QD-OH was least toxic. It was concluded that some of these hydrophilic coatings were cytotoxic or genotoxic, and that the surface treatments of nanocrystals would influence their biological behavior. The study conducted by Hoshino and coworkers was important in that it demonstrated convincingly, for the first time, that surface coatings on QDs are important not only for functional purposes but also with regards to the toxic potential of the nanocrystals. Based on their results, Hoshino et al. claimed that the toxicity of QDs could be attributed primarily to their surface coatings, and not to their cores.

Surface coatings determining cytotoxicity of QDs were also reported by Ryman-Ramussen et al. [24], who indicated that QDs coated with carboxylic acid or PEG-amine were cytotoxic, whereas PEG-coated QDs displayed a low cytotoxicity. Wang et al. [25], however, noted that at low pH (gastric acidity) conditions, PEG-coated QDs could also become toxic, and cautioned that the route of exposure (ingestion) might alter the stability of the surface coating and influence the degree of toxicity of QDs. In a more recent study, Hoshino et al. [26] affirmed that the biological behavior and toxic potentials of QDs could be manipulated by modifications on their surface coatings. Hoshino's group concluded that some capping surfaces of the QDs were actually responsible for the biological effects of the nanocrystal QD–complex. Unless the core structure of the QD-complex was disintegrated, the surface coating materials might be even more toxic than the inner core itself. Thus, it became apparent that both the outer coatings and the inner core of QDs could contribute to QD toxicity, depending on the chemistry of the coatings and the integrity of the inner core complex. The importance of core integrity in QD toxicity will be discussed further later in the chapter.

The different "toxicities" of various surface coatings may occur because these surface molecules can significantly influence the amount of nanoparticles being taken into the cells by endocytosis [7, 27, 28]. During endocytosis, the nanoparticles are encapsulated by endosomes and carried to various compartments of the cell [29]. By using QDs with different surface molecules (PEG-substituted and non-PEG-substituted), Chang et al. [28] demonstrated that the cellular uptake of QDs with different surface chemistries were different, and that the amount of nanoparticles uptaken into cells (intracellular dose) correlated directly with the toxicity outcome. It was concluded that the cytotoxicity of QDs correlated well with intracellular levels of QDs rather than extracellular levels. Thus, modifications of the surface chemistry of QDs were seen to be important not only for functional aspects of the QDs but also in endocytosis of the QDs, and thus any eventual toxicity.

In an *in vivo* study in mice, Ballou *et al.*, [13] showed that the long-chain PEG coating increased the circulating half-time, and also reduced the nonspecific (macrophage) uptake of QDs; thus, the QD surfaces can influence the serum lifetime and the pattern of deposition of QDs. A similar finding was also observed by Fischer *et al.* [30], who showed that the pharmacokinetics and biodistribution of two QDs with different coatings – one with mercaptoundecanoic acid-lysine and the other with bovine serum albumin (BSA) – were different. The results of this study provided important *in vivo* evidence that the surface chemistry of QDs could influence both the bioapplications and biosafety of QDs.

12.2.2
Charge and Size of QDs in Relationship to Potential Toxicity

Surface charge and final size are two important properties of surface coatings. In consideration of the surface charge of QDs and their toxicity, both Shiohara *et al.* [10] and Hoshino *et al.* [9] showed less stability of positively charged CdTe/ZnS QDs with higher cytotoxicity than with negatively charged QDs. Lovoric *et al.* [11], on the other hand, claimed that charge, rather than the size, of the nanoparticle was the influencing factor in toxicity induction. It was suggested by these investigators that, because negatively charged QDs were smaller in size than positively charged QDs, they were able to enter cells (especially the nuclei) more readily than positively charged QDs, for toxicity induction. Smaller-sized nanoparticles were also shown to undergo a more rapid uptake by cells [31]. Male *et al.* [20] also indicated that the cytotoxicity of CdTe QDs was more prominent with QDs of a smaller size than with equally charged QDs of a larger size. As a consequence, these authors also believed that an increased toxicity would be more size-related than surface-charge related. Today, it is generally agreed that the size of nanoparticles is an important factor influencing cellular response or toxic outcome [11, 20, 31].

Indeed, the main characteristic of nanomaterials is their final *size*, which frequently induces different interactions, absorptions, and metabolism in biological systems, compared to their larger counterparts. It was reported that ultrafine particulates (<100 nm), especially in coal and silica, enhanced the induction of oxidative stress, inflammatory cytokines, and cytotoxic cellular responses among workers in occupational situations [32–35]. These observations indicated that, as the particle size shrank, there was a tendency for greater biological effects. This situation arose because, as the particle size decreased, the total surface area actually increased, thus allowing a greater proportion of molecules or reactive groups of the element to be exposed. Thus, in general, reductions in particle size constitute a greater toxicological concern. In a recent study conducted by Hoshino *et al.* [26], the cytotoxicity of QDs was seen to be both dose- and size-dependent, with the degree of cytotoxicity being proportional to the number of QDs that entered the cells. This correlation of QD cytotoxicity with intracellular levels of QDs was also demonstrated by Chang *et al.* [28].

Clearly, the *charges* of nanoparticles (including QDs) are not totally uninvolved with their potential toxicity. Negatively charged nanoparticles (e.g., COOH-coated QDs) are readily taken up by macrophages via endocytosis [9]. In contrast, NH_2 (PEG) QDs are not readily taken up by macrophages because the PEG-coatings provide a "steric-replusive barrier" against cell interactions [36–38], which reduces their uptake by phagocytic cells. It should be noted that whilst the increased phagocytic removal of nanoparticles by macrophages may protect cells against any toxic effects of the nanoparticles, it might also reduce the intended "therapeutic function." Whilst positively charged QDs are not stable within cells [10], negatively charged QDs (e.g., carboxylated QDs) appear to be more stable due to their confinement in endosomes after entering the cells [9]. The instability of positively charged QDs within cells is believed to be one of the reasons for the greater toxicity of these QDs [19].

In a recent study, Geys et al. [39] showed that both negatively charged (carboxyl-coated) and positively charged (amine-coated) CdSe/ZnS-QDs could induce pulmonary thrombosis *in vivo*, albeit to different severities and via different mechanisms. Negatively charged QDs had a more pronounced pulmonary thrombotic effect than positively charged QDs. It was also suggested that the thrombotic *processes* induced by negatively or positively charged nanoparticles were also different. The proposal was that negatively charged QDs would trigger coagulation induction, most likely via contact activation, leading to fibrin formation and simultaneous platelet activation by thrombin [39]. Positively charged nanoparticles, on the other hand, were shown to induce thrombosis via an enhancement of ADP-triggered platelet aggregation [40], although unfortunately, nonphysiological high doses of QDs were used in this study. The actual toxic effect of QDs (pulmonary thrombosis), as claimed by these investigators, was therefore difficult to assess, although differences in the surface charge of nanoparticles in relation to the induction of different biological responses were still clearly demonstrated.

In addition to the uptake by cells and toxicity of QDs being influenced by the surface coatings [9, 26], so too can their elimination. Yang et al. [41] and Lin et al. [42] each showed that, in mice, PEG-coated CdTeSe-QDs had an extremely long body retention (up to six months), without any urinary or fecal excretion. This long bioaccumulation and retention of QDs, especially in the kidneys, constituted an increased toxic risk of the nanocrystals. Chen et al. [43] reported that CdSeS-QDs coated with a hydroxyl-modified silica network could greatly enhance the hepatic and renal elimination of QDs via the feces and urine. It was claimed that the hydroxyl silica-coated QDs has strong fluorescence properties, *in vivo* stability, and an absence of free cadmium release from the core. Whilst such development of a "new breed" of QDs is encouraging, a full spectrum of ADME studies on hydroxyl silica-QDs should be conducted in order to affirm their usefulness and biosafety.

Within the workplace, the dermal exposure to nanomaterials is a practical concern, with skin lotions such as sun-screen or sun-block lotions containing nanomaterials. Thus, the dermal penetration of QDs to the systemic circulation must be evaluated. When Ryman-Rasmussen et al. [44] tested QDs with diverse physico-chemical properties (size, shape, charge, surface coating) for their dermal penetrating abilities, QD 565 (round-shaped) with a carboxylic acid coating were shown to penetrate the skin more efficiently than QD 655 (ellipsoid) with a similar coating, despite their similar hydrodynamic sizes. Furthermore, both PEG (neutral) and PEG-amine-coated (cationic) QDs showed better dermal penetration abilities than those with carboxylic acid coatings (anionic). These observations showed not only that the physico-chemical properties (size, shape, charge, chemistry) could influence the biological behaviors of QDs, but they also contradicted the conventional belief that the dermal barrier of intact skin would be impervious to nanomaterials unless the skin had been abraded or injured [35, 45–47]. In a subsequent study conducted by Mortensen et al. [48], it was shown that the dermal penetration of nanomaterials could be enhanced by UV exposure, again raising concern that nanomaterials with toxic potential might be used in sun-screen lotions. However, specific safety evaluations in this area must be undertaken before any firm conclusions can be drawn.

In summary, the surface coatings of QDs – in terms of their chemistry, charge, shape and size – can influence the properties of, and cellular responses to, QDs; this includes the endosomal intake of QDs into the cells, their phagocytic removal, and their elimination. Notably, all of these factors play contributory roles in the induction of toxicity by QDs in biological systems.

12.2.3
Core Metals and Biodegradation of QDs

In order to enhance their biocompatibility and to shield their potentially toxic core from exposure, most – if not all – QDs are capped with either amphiphilic compounds or hydrophilic thiols that insulate the inner core from its surroundings. One important characteristic of QDs is the use of a metal or metalloid as the core for the nanocrystal, for which metals such as cadmium (Cd), tellurium (Te), lead (Pb), arsenic (As), copper (Cu), zinc (Zn), and selenium (Se) are known to be used, either singly or in combination. Some of these metals, notably Cd, Te, As, Cu, and Pb, are toxic to biological systems. Deterioration of the outer coating, by processes such as photolysis, oxidation, or low pH, either outside or within the biological system, might expose the inner metallic core to the biological environment, and thus constitute a toxic risk to the cells and tissues [7, 49, 50].

Exposure of the inner core material, or leakage of the core metals, are generally the result of QD instability. Consequently, the stability of QDs during their synthesis, storage, and after their introduction to a biological system (in vivo environment), become critically important with regards to QD integrity. Both, photolysis and oxidation are known to induce the instability of QDs, with each process causing deterioration of the outer coating and the core–shell, such that the toxic shell ("capping" material) and inner core metalloid are exposed, and a toxic effect is induced. Derfus [7] has reported that a prolonged exposure to air, ultraviolet light and to hydrogen peroxide, enhanced the toxicity of MAA-TOPO-capped CdSe QDs towards cultured hepatocytes. It was concluded that the prolonged exposure to an oxidative or photolytic environment would promote deterioration of the nanocrystal, thereby releasing free cadmium from the deteriorated or degraded QDs. Although the addition of one or two layers of ZnS as a shell to the QDs may reduce ambient air oxidation damage, this would be ineffective for eliminating any enhanced cytotoxicity due to photo-oxidation. These findings confirmed the observations of Aldana et al. [49], that the thiol-coated CdSe QDs were photochemically unstable. The photochemical stability of QDs can be influenced by the thickness of the ligand monolayer coating, and also by the manner in which the nanocrystals are stored. Hence, care must be taken during the manufacture and storage of QD materials so as to avoid any degradation of the material before its biomedical application.

As QDs have shown great promise with regards to applications in biomedicine, their stability in biological systems is critical. In fact, the stability of QDs *in vivo* may be one of the most important factors for the "bio-safety" of QDs. Although, many studies have suggested intracellular degradation of the QD complex (notably the inner core complex) within biological systems, few have actually demonstrated such degradation. *Autofluorecence*, which is a unique property of intact QDs, can be used to reflect the stability or "intactness" of the QD particle. Deterioration of the outer coating usually leads to a reduction in fluorescence which, in biological systems, may be caused by a reduction or loss of QDs in the cells and tissues by exocytosis (the expulsion of particles from cells for bioelimination from the biological system). Alternatively, autofluorescence may be caused by an intracellular deterioration of the QD surface coating [19] via the digestive actions of lysosomes and peroxisomes [7, 29]. A recent study conducted by Mancini et al. [51] affirmed that the fluorescence quenching of QDs was indeed largely caused by localized surface defects via oxidation processes due to reactive oxygen species (ROS) such as hydrogen peroxide (H_2O_2) and hypochlorous acid (HOCl). Phagocytes such as monocytes and macrophages are most likely responsible for such actions. It has also been noted that HOCl and H_2O_2, being neutral in charge, can diffuse across the outer polymer coating layer and induce chemical oxidation of the inner QD surface. It is believed that this "etching" process not only causes quenching of the fluorescence but may also lead to damage of the inner core, allowing soluble metal ions (e.g., Cd, Te, Se, Zn) to leak from the QD complex.

Both, the outer organic coating component and the inner core, when damaged, can contribute to cytotoxicity [9, 19]. A reduction in fluorescent intensity was indeed reported when QDs were introduced *in vitro* to cell cultures for more than 10 days. The bio-oxidation of QDs, with deterioration of the outer shell or coating, was suggested to be responsible for this situation [9]. A time-correlated loss of fluorescence was also observed in tissues when QDs were introduced *in vivo* to live animals [3, 52]. Whilst reductions in the fluorescence of QDs, both *in vitro* and *in vivo*, have provided an intriguing indication for the biodeterioration of QD coatings within biological systems, evidence for biodegradation of the metal core complex (CdTe complex) in cells and tissues remains to be demonstrated.

Based on a "fixed ratio" of Cd:Te in CdTe-based QDs, Lin et al. [53] have demonstrated elegantly that the Cd:Te ratio changed (increased) with time in the renal tissues of mice when CdTe-QDs were introduced intravenously (Figure 12.2). Such

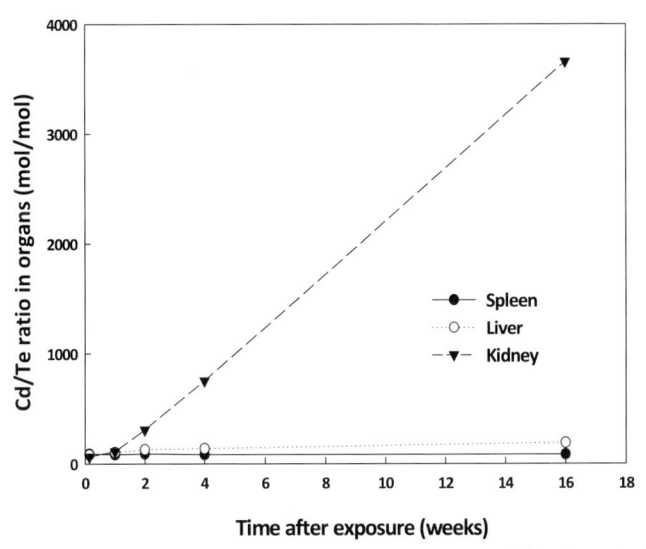

Figure 12.2 Time course study on changes in the Cd/Te molar ratio (Cd/Te ratio) in the spleen, liver, and kidneys of ICR mice treated with 40 pmol of QD705. No significant change in Cd/Te ratio was observed in the spleen and liver over 16 weeks, but the Cd/Te ratio increased sharply in the kidneys. This indicated a steady disintegration of the QD705 complex with release of Cd in the kidneys, but not in the spleen and liver. Reproduced with permission from Ref. [53].

changes in the Cd:Te ratio provide a good *chemical index* for degradation of the CdTe complex within a biological system. Free cadmium – but not bound cadmium (cadmium in a CdTe complex) – is a potent inducer of metallathionine (MT-1) expression in renal epithelial cells. When the CdTe–QD complex was introduced into cells, an elevation in MT-1 signified the release or leakage of free cadmium from the QD complex. By using MT-1 expression as a biological marker for free cadmium, Lin et al. [53] successfully demonstrated a significant elevation in MT-1 expression (increased levels of free cadmium) in the renal tubular epithelial cells of QD-treated mice (Figure 12.3). Thus, by means of both a *chemical index* (Cd:Te ratio) and a *biological marker* (MT-1 expression), it was possible to demonstrate, for the first time, the "chemical fate" of QDs *in vivo*, namely a gradual disintegration of the QD complex with the release of free cadmium within living cells *in vivo*. It was interesting to note that the change in Cd:Te ratio was found primarily in the kidney tissues, and significantly less in the liver or spleen, where high concentrations of QDs were also localized. These observations suggested that the biodegradation of QDs *in vivo* might be tissue-specific, and that the kidney appears to be the prime target organ for QD breakdown and toxicity.

Because cadmium is the most recognized toxic component in QDs, cadmium from the QD core has always been assumed to be the primary material responsible for QD toxicity when the complex disintegrates [7, 54]. In a comparative study of CdTeSe-based QDs and equimolar cadmium chloride ($CdCl_2$), Lin et al. [53]

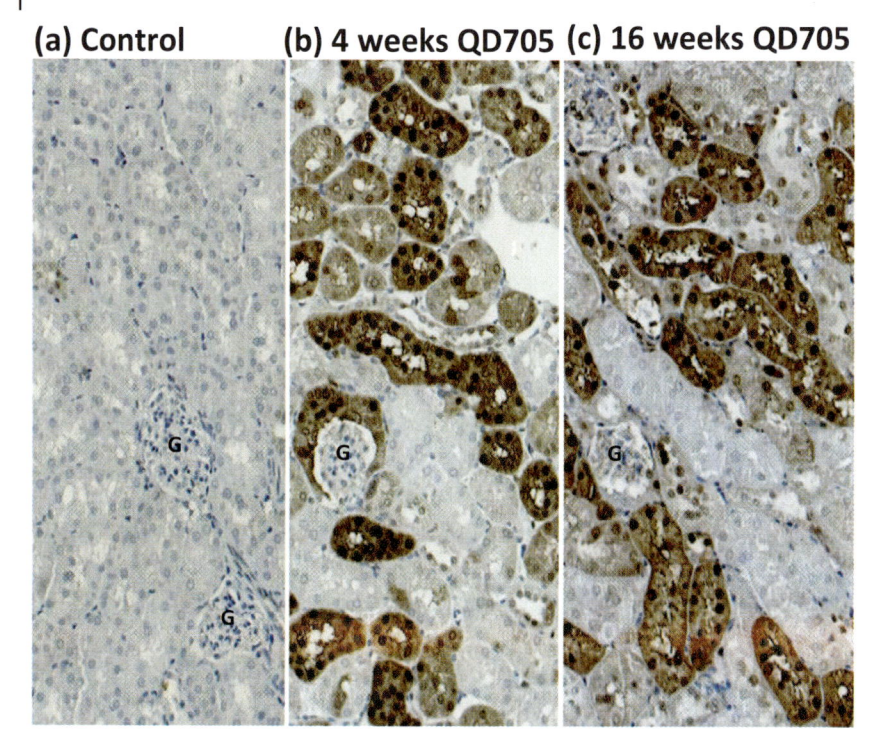

Figure 12.3 Immunostaining for metallathionine (MT) protein in kidney tissues from animals with or without QD705 exposures. (a) Control (no QD705 exposure); kidney tissue showed no MT-1 staining; (b) Kidney tissue from animals exposed to QD705, four weeks. Many renal epithelial cells and proximal convoluted tubules (PCTs) were positively stained (brownish color), indicating an induction of MT; (c) Strong MT stainings were still observed among PCT epithelial cells at 16 weeks after QD705 exposure. The glomerulus (G) was found to be free of MT staining. Original magnifications ×200. Scale bars = 50 μm. Reproduced with permission from Ref. [53].

demonstrated specific alterations and degenerations of mitochondria in the renal tubular epithelial cells of mice after exposure to both QDs and inorganic $CdCl_2$. In this case, the mitochondrial changes induced by QDs were similar, but not identical, to those induced by $CdCl_2$. This suggested that, whilst cadmium had clearly contributed to the QD toxicity, other factors or elements (such as tellurium and selenium in the core, or ROS and oxidative stress induced by other factors) might also have played contributory roles influencing the overall toxic outcome of the QDs.

As the QD is not a simple chemical, no single factor can be held responsible for its toxicity. Recently, Cho et al. [29] were unable to show any correlation between intracellular cadmium levels and the cytotoxicity induced by QD exposure. Likewise, after exposing porcine renal epithelial cells to CdSe-QDs for 8–24 h, Stern et al. [55] failed to show any induction in MT-1 production (i.e., free cadmium release) or oxidative stress in the cultures, despite clear cellular changes (marked cellular vacuolization and autophagic vacuoles with cellular debris) being observed. These autophagic vacuoles were interpreted as "lysosomal disorders," denoting the active removal and digestion of toxicants and degenerative cell components by lysosomes. The presence of QDs within the cells or autophagic vacuoles, however, was not demonstrated. Based on these findings, it was considered that QD-induced cell changes, after such a short exposure time, might depend on the properties of the particle as a whole, rather than on only the metal materials.

12.3
Toxicity and Role of Oxidative Stress

Although, as indicated previously [1], many studies have been reported relating to the toxicity of QDs, concern has been raised that many of these were conducted by nanoscientists rather than by toxicologists or biomedics with expertise in biosafety evaluation. It was also noted that most of the studies reporting negative findings [9, 12–16] had included only short-term exposures (minutes to hours), despite cytotoxicity generally being identified only after longer exposures (days to weeks) [9–11]. As most QDs were identified as cytotoxic (both *in vitro* and *in vivo*) only *after* deterioration of their surface coatings [3, 9, 18, 52], it was clear that, within a biological system, QDs would demonstrate toxic effects only after their biodegradation. This biodegradation process may be slow and time-dependent, with physiologically based pharmacokinetic (PBPK) studies having demonstrated extremely long body retentions of CdTeSe-QDs in mice (up to 16 weeks), without significant excretion, after a single intravenous injection [41] [42]. The biodegradation of CdTeSe-QDs was also shown to occur slowly and time dependently in renal tubular epithelial cells [53]. Clearly, QDs have a very long body-retention and slow degradation *in vivo*, in addition to a preferential organ distribution and tissue-specific degradation.

Although the importance of the physico-chemical properties of the outer coatings (surface chemistry, size, charge) and the composition of the inner core (Cd, Te, Se, Pb) on the toxicity of QDs has been well recognized, the "biological" factors, events and mechanisms leading to QD toxicity remain unclear. Previously, the cytotoxicity of QDs has been shown to be closely affected by their endocytosis, and to correlate directly with their intracellular levels [28]. During endocytosis, the nanoparticles are encapsulated within vesicles (endosomes) and carried to various cytoplasmic compartments within the cell. Chang *et al.* [28] showed that different surface coatings of QDs could induce different degrees of endocytotic uptake, with the extent of cellular uptake (i.e., intracellular exposure) being the primary criterion influencing the ultimate toxicity of the QDs.

Because the inner cores of QDs contain toxic metals (e.g., Cd, Te, Se, Pb), it has been widely proposed that such toxicity would be related to the exposure of the cellular environment to these materials. Although, based on its well-known toxicity, cadmium has received most attention [7, 54], there is increasing evidence that whilst cadmium may have a role in QD toxicity, such toxicity cannot be explained on the basis of cadmium release alone [29, 53, 55], and that other factors must be involved.

The generation of ROS and oxidative stress by QDs has attracted much attention in recent years as one of the prime bases for QD toxicity [11, 29, 56, 57]. Recently, Samia et al. [58] showed that singlet oxygen could be generated by bare CdSe QDs in toluene and in water when conjugated with a photosensitizer. Other ROS, such as $^{\cdot}OH$ and $^{\cdot}O_2$, have also been detected in solutions containing QDs [59], while DNA nicking and ROS generation from both CdSe and CdSe/ZnS QDs have also been reported [60]. Lovric et al. [11] demonstrated the production of ROS in live cells exposed to naked CdTe QDs, that correlated with damage of the plasma membrane, mitochondria, and cell nuclei, leading to apoptosis. These changes may be suppressed in the presence of antioxidants, however.

Localized areas of oxidative stress within cells, caused by internalized QDs, have been described by Funnel and Maysinger [61]. Oxidative stress has been shown to regulate autophagy via post-translational modification [62], while autophage induction by QDs (as discussed above) can be influenced by endocytotic activity [28] and is thus closely associated with the toxicity of QDs. The presence of QDs has also been demonstrated in the nuclei [11, 24], and this might possibly affect gene regulation, including those genes involved in autophagy. It has been proposed that QDs might be perceived by cells as an endosomal pathogen, and targeted towards an autophagic pathway for destruction [63].

While various studies have demonstrated the adverse effects (cell growth, apoptosis, etc.) of QDs, proposals regarding the "mechanistic" actions of QDs related to such toxicity have been few in number. Recently, several major mechanistic proposals for QD toxicity have been made [11, 29, 57], although these study models were all established with cultured cells *in vitro*. Lovric et al. [11] provided convincing evidence for the production of ROS in live cells incubated with "naked" CdTe QDs (CdTe QDs with degraded outer coats). Here, it was proposed that core–shell QDs could undergo several degradation processes in cells via oxidation with the generation of ROS, which would trigger a series of events leading to plasma membrane damage, mitochondrial changes, and ultimately to apoptosis. When Cho et al. [29] tested four QDs with different chemical compositions, charges and surface modifications, it was concluded that the core–shell CdSe/ZnS QDs actually presented little damaging effects to cells (MCF-7 cells). Moreover, it was suggested that the toxicity of CDTe QDs observed was probably (at least in part) a consequence of the contamination of QDs by free Cd ions prior to incubation. These free Cd ions could enter the cell, generate ROS, and attack the mitochondria, although intact CdTe QDs could also enter the cell via endosomal activity. The QDs would then be degraded in the lysosomes, with additional cadmium release and ROS generation (Figure 12.4). These investigators suggested that the cadmium (either from "contamination" or from intracellular release) and the photo-oxidative breakdown of QDs to form the basis of ROS generation and subsequent QD toxicity. However, no attempt was made to differentiate between ROS generated from cadmium and the photo-oxidation of QDs. Although cadmium (either from contamination or intracellular release) is considered to be important in QD cytotoxic-

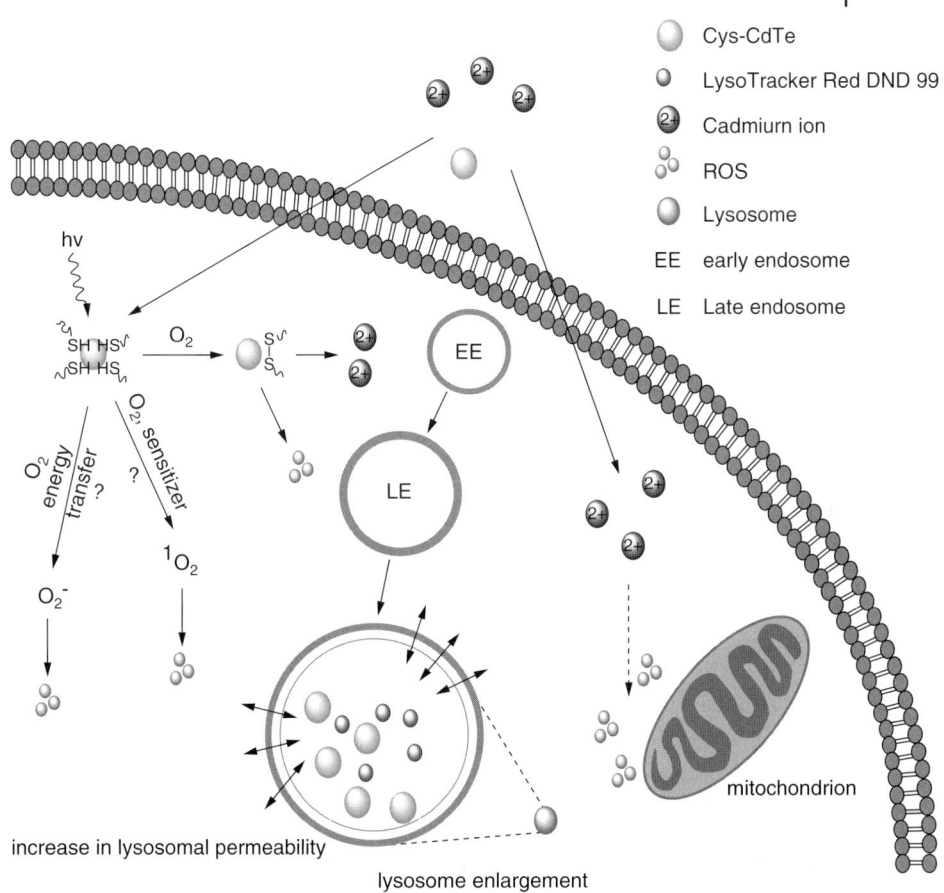

Figure 12.4 Schematic representation of the mechanistic pathways implicated in the cytotoxicity of CdTe QDs in live cells, highlighting the salient changes in cellular morphology, the chemical species involved, and the chemical reactions that can lead to ROS and free Cd^{2+} ion release. Reproduced with permission from Ref. [29].

ity, no correlation between the intracellular cadmium and cell viability was identified in this study. Furthermore, no explanation was offered as to how cadmium could generate ROS within the cells.

In another study, the same group demonstrated elegantly the importance of oxidative stress in CdTe QD toxicity, with or without antioxidant (N-acetylcysteine; NAC) conjugation or capping [29]. By exposing neuroblastoma cells to various modified CdTe QDs, the observed CdTe toxicity was attributed to two main mechanisms: (i) an upregulation of the Fas receptor, leading to apoptosis; and (ii) lipid peroxidation leading to cell membrane damage and an impairment of mitochondrial function. Both of these events can be triggered by ROS. The study also showed that both Fas expression and lipid peroxidation could be effectively

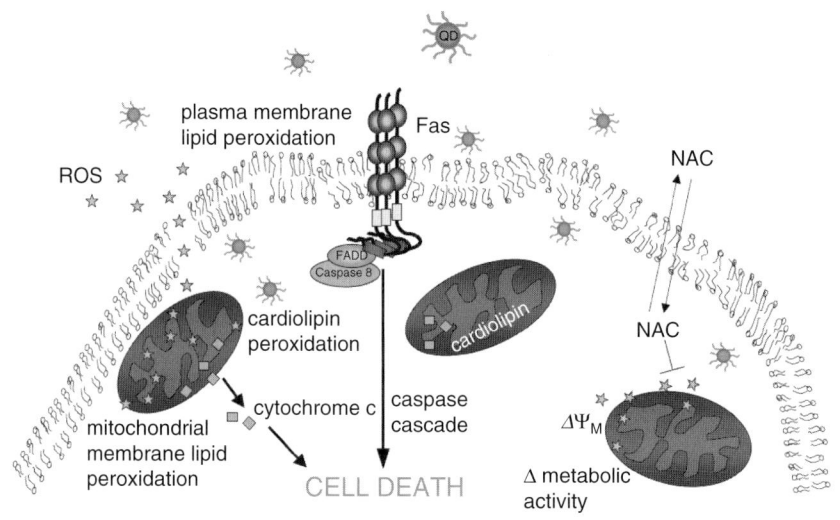

Figure 12.5 Proposed mechanism of QD induced cell death involving Fas, lipid peroxidation and mitochondrial impairment. Reproduced with permission from Ref. [57].

decreased by surface-modified CdTe QDs with NAC, with consequential reductions in apoptosis and mitochondrial damage. This indicated that oxidative stress indeed played an important role in QD toxicity. Ultimately, the investigators suggested that ROS could be generated extracellularly, thus upregulating Fas (a membrane receptor) for activation of the apoptotic cascade. The ROS generated, after entering the cell, would also induce plasma membrane damage via lipid peroxidation and mitochondrial changes (Figure 12.5). This well-conducted study provided a more intriguing mechanistic proposal than had been proposed previously [29] although, again, no suggestion was made as to how the QDs induced ROS production extracellularly. As this was an *in vitro* study, the extracellular ROS were generated in the culture media, possibly via "contaminated cadmium" in the media, or during the photo-oxidative breakdown of QDs in the culture media [29]. The actual generation of extracellular ROS *in vivo* remains to be demonstrated. Based on these results, the cytotoxic effects of QDs were reduced – but not totally eliminated – by NAC, which in turn suggested that oxidative stress contributed only partially to the induced QD cytotoxicity, but was not entirely responsible for it. This interesting *in vitro* cell model requires further validation with *in vivo* studies.

Although there is increasing evidence that ROS generation and oxidative stress are involved in QD toxicity, further studies are required to better define the source of ROS (either extracellularly or intracellularly), the nature of the ROS, the role of cadmium or other elements, and the role of the mitochondria (villain or victim of ROS generation?). These aspects will be further discussed later in the chapter.

12.4
ADME and Biosafety Evaluation

Whilst many excellent investigations have been carried out on the toxic effects and mechanisms of QDs in recent years, most were conducted under *in vitro* conditions [11, 29, 55, 57]. Major differences exist between *in vitro* and *in vivo* exposures, and it is well known that both cells and chemicals behave very differently under *in vivo* and *in vitro* conditions. This is particularly true for nanoparticles, many of which tend to aggregate in solutions or in culture media; such aggregation alters the physico-chemical properties of the nanomaterial, and thus its toxicity. Recently, serious concerns have been raised regarding the assessment of nanomaterials by *in vitro* exposure alone since, as noted above, some nanomaterials (e.g., QDs) tend to be organ- and tissue-specific in terms of their distribution and toxicity. As the final toxicity may be dose-, duration-, and metabolically related, animal studies capable of providing details of both the biological and chemical fates of nanomaterials within an intact biological system, are important. Those factors which require consideration between *in vitro* and *in vivo* studies include:

- The duration of exposure: *in vitro* exposures are usually much shorter than *in vivo* exposures.

- The dose of exposure: much higher or even unrealistic doses tend to be used in many *in vitro* studies.

- Cell specificity in response to toxic challenges: epithelial cells may respond differently from fibroblasts, or nerve cells *in vivo* to toxic chemicals. Different cell types will also show differences in vulnerability, resistance, or response to oxidative stress.

- Metabolism of the chemicals by cells: the metabolic activity and ability of a single-cell culture (e.g., neurons) will differ from that of cells in intact tissues; that is, with supporting cells such as the extracellular matrix (ECM).

In other words, whilst *in vitro* studies provide either good reference information or a cell model for understanding the potential toxic actions of QDs, the correct selection of cell type for the investigations is critical. With regards to tissue and cell specificity for the distribution and toxicity for QDs, the selection of a "wrong" cell type for an *in vitro* study may totally misdirect the results and conclusions. Thus, the validation of *in vitro* information with animal models (*in vivo* studies) is essential.

The pharmacokinetic study of a chemical forms a fundamental base for its safety assessment. Typically, such a study will address the absorption (the rate of uptake of the chemical by the biological system), distribution (deposition and accumulation of the chemical in the biological system), metabolism (degradation or "handling" of chemicals within the biological system), and elimination (excretion of the chemical or its metabolites from the biological system). Data from these absorption, distribution, metabolism, and elimination (ADME) studies help to define the biological and chemical fates of a chemical once it has entered the biological system; thus, ADME constitutes the most important base in biosafety evaluation. It is also clear that such ADME information can only be obtained via animal (*in vivo*) studies, and not *in vitro* studies. The importance of ADME studies in the biosafety evaluation of QDs has long been recognized and encouraged [1]. However, because they are laborious, time-consuming, and expensive (long-term animal studies are always expensive), properly conducted ADME studies on QDs are few in number.

Ballou *et al.* [13], by using whole-animal fluorescence imaging, provided the first qualitative kinetic assessment of QDs *in vivo*, and showed the kinetics of QDs to be surface chemistry-dependent. Similarly, Fischer *et al.* used Sprague-Dawley rats and tracked QDs (as cadmium) in the tissues via inductively coupled mass spectrometry (ICP-MS) analysis [30]; this was the first comprehensive attempt to provide quantitative information on the plasma clearance, tissue/organ distribution and excretion (urinary and fecal) of QDs after the intravenous administration of two QDs, namely QD-LM (QDs with a mercaptoundecanoic acid coating) and QD-BSA (QD-LM + additional BSA). The results showed the plasma half-life of QD-LM to be significantly longer than for QD-BSA (58.5 and 38.7 min, respectively). Moreover, whilst QD-LM was shown to accumulate to a lesser degree than QD-BSA in the liver and spleen, the pattern was reversed in the lungs and kidneys. Clearly, the surface chemistry of QDs can have a major influence on their consequential pharmacokinetics and tissue accumulation. In the liver and spleen, the QDs were localized at the edge of the sinusoids and in the red pulp area, which suggested an active macrophage uptake of QDs in these organs. Interestingly, QDs (measured as cadmium) appeared not to be excreted in either the urine or feces, even by day 10 after administration. Unfortunately, the study duration was very short (10 days), and no tissue redistribution kinetics and calculation on mass balance was performed that may have provided an overall metabolic profile of QDs in these animals.

QD-705 (QD with a CdTeSe core, ZnS shell, and PEG-5000 coating) represents one of the most recently synthesized and most commonly seen QDs in industry [64]. Both, Yang *et al.* [41] and Lin *et al.* [42] conducted detailed PBPK studies on QD-705, using mice. As in the study of Fischer *et al.*, cadmium (representing QDs) was measured and tracked with ICP-MS analysis for up to 180 days, during which time the plasma half-life of QD-705 was found to be 18.5 h, and the three major organs of accumulation to be the spleen, liver, and kidneys. These findings were consistent with those reported by Fischer *et al.* [30]. Whole-body mass balance studies performed at 1 day, 28 days, and six months post-dosing revealed a considerable time-dependent redistribution from the body mass to the liver and kidneys (Figure 12.6), with the accumulation of QDs in most tissues slowing or declining with time, but QD levels in the kidneys continuing to increase. The mass balance recoveries at all time points (1 day, 4 weeks, 16 weeks) were close to 100% without any significant loss of the injected QDs through fecal or urinary elimination, and the exceedingly long-term body retention of QDs being extraordinary. Both, the liver and spleen form part of the reticuloendothelial system (RES), which

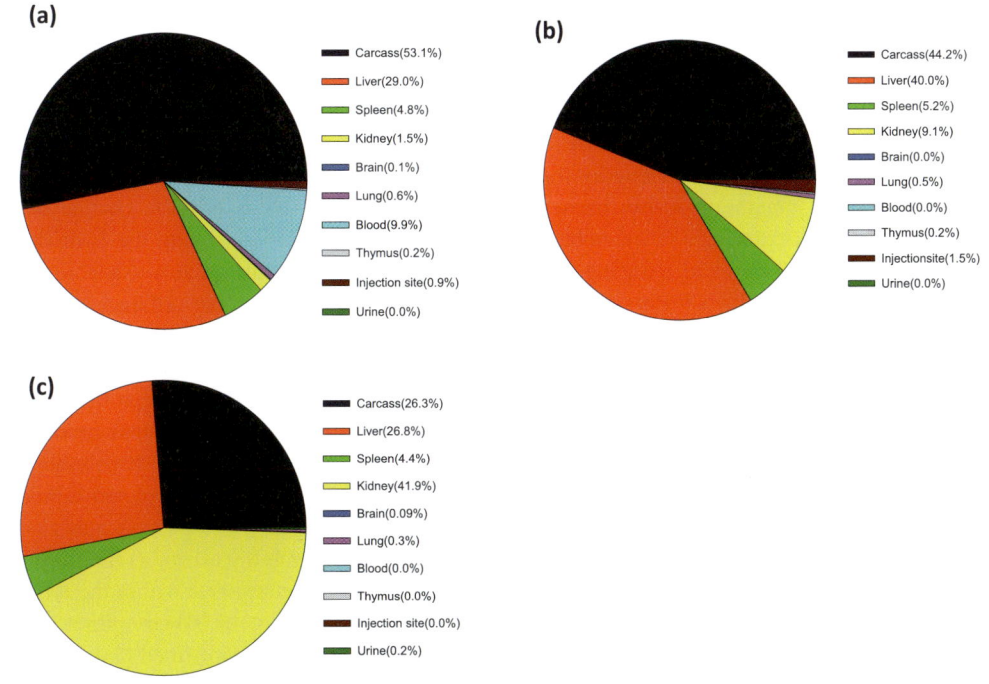

(a)
- Carcass (53.1%)
- Liver (29.0%)
- Spleen (4.8%)
- Kidney (1.5%)
- Brain (0.1%)
- Lung (0.6%)
- Blood (9.9%)
- Thymus (0.2%)
- Injection site (0.9%)
- Urine (0.0%)

(b)
- Carcass (44.2%)
- Liver (40.0%)
- Spleen (5.2%)
- Kidney (9.1%)
- Brain (0.0%)
- Lung (0.5%)
- Blood (0.0%)
- Thymus (0.2%)
- Injection site (1.5%)
- Urine (0.0%)

(c)
- Carcass (26.3%)
- Liver (26.8%)
- Spleen (4.4%)
- Kidney (41.9%)
- Brain (0.09%)
- Lung (0.3%)
- Blood (0.0%)
- Thymus (0.0%)
- Injection site (0.0%)
- Urine (0.2%)

Figure 12.6 Tissue distribution of an intravenous dose of QD705 in ICR mice. Studies were carried out at 1 day (a), 28 days (b) and 6 months (c) after dosing. Overall recoveries for 1 day, 28 days and 6 months were, respectively, 102.2%, 108.8%, and 40–60%. Panels (a) and (b) reproduced with permission from Ref. [41].

has strong macrophage activities; hence, the early accumulation of QDs in these two organs most likely reflect the active phagocytic activities of the macrophages rather than actual uptake by the parenchymal cells of these organs (hepatocytes of the kidneys, lymphocytes in the white pulp of the spleen). Indeed, studies by both Fischer *et al.* [30] and Yang *et al.* [41] demonstrated QD accumulations (seen as red fluorescent particles under fluorescence microscopy) in areas where macrophages were known to be most active in these organs, notably the edge of the hepatic sinusoids and the red pulps of the spleen. In contrast, although initially the kidneys showed a lower content of QDs than the liver or spleen, they accumulated QDs at a steady rate such that, after six months, the level of QDs in the kidney (per g tissue) exceeded those in either the spleen or liver. Although the kidney is known not to possess an active macrophage system, uptake by these organs most likely occurred in the renal tubular epithelial cells. A recent study by Chen *et al.* [43] indicated that the elimination (both fecal and urinary) of QDs can be greatly enhanced when the QDs are coated with an hydroxyl-silica network. The liver and kidney remain the organs of preferential distribution for these hydroxyl-silica coated QDs. With PBPK modeling and computational toxicology calculations, Lin *et al.* [42] predicted the kinetic distribution and redistribution of QDs *in vivo*. This modeling study affirmed the kidneys to be the *target organs* for both the distribution and toxicity of QDs.

Based on data generated from studies conducted by Yang and Lin [41, 42], Lee *et al.* [65] furthered the PBPK calculations and stated that the uptake of QDs could be influenced directly by their adsorption to plasma proteins, or by their self-agglomeration within tissues. Agglomeration would be expected to change the anticipated QD tissue-to-blood partition coefficients by increasing the hydrodynamic size and reducing the number of particles and surface area available for interactions. Clusters of QDs that adsorb to plasma proteins or self-agglomerate may take longer to undergo endocytosis, or may never be transported into cells. In addition to the blood, the lymphatic system plays an important role in the movement and transport of QDs within the body. It was also noted that the accumulation of QDs in the kidney is an intriguing and a time-dependent process, with late increases in kidney QDs content believed to occur via redistributions from other body tissues. The special ability of renal proximal convoluted tubule (PCT) epithelial cells to sequester free cadmium for prolonged periods may also explain the increased and prolonged elevation of QD levels (measured as cadmium) in the kidneys. The release of free cadmium from QD complexes within the renal cells has been predicted [65] and, indeed, the disintegration of QD-complexes and the release of cadmium in renal epithelial cells has been confirmed [53]. When the same group also reported a rapid elevation in MT-1 levels in renal epithelial cells, it was suggested that the cadmium, which had been slowly released from QDs, could be rapidly and completely "captured" by MT-1, thus permitting the long-term storage of cadmium in renal cells.

While the ADME and pharmacokinetic studies conducted by Yang *et al.* [41] and by Lin *et al.* [42] have revealed the "biological fates" (plasma half-life, organ distribution, tissue retention and accumulation, excretion and elimination) of QDs, their "chemical fates" (the metabolism of QDs or integrity of the QD particles in the biological system) remain to be defined. Whilst the importance of the outer coat (surface chemistry) of QDs, and its relationship with the toxic potential of QDs, has been well demonstrated [9, 26], it is believed that the integrity of the metallic core would also influence the final destiny of QD toxicity. Disintegration of the inner core within the biological system, thereby releasing toxic metals, has been suggested [7, 49, 50]. Because QDs contain metallic components within their inner cores, they have always been "tracked" during ADME studies by monitoring the cadmium content through metal analysis [30, 41, 42]. Yet, one serious concern remains, namely that although ICP-MS can identify total cadmium levels, it cannot provide information on the integrity of the QDs within the biological system. However, in mice exposed to a CdTeSe-based QD (QD-705), by taking advantage of the relatively constant Cd:Te ratio in CdTe-QDs and the different biological half-lives of free Cd and Te in tissues, the change in Cd:Te ratio with time could be calculated for various tissues [53]. Because free Cd has a much longer biological half-life than free Te, a temporal increase in the Cd:Te ratio would represent the occurrence of biodegradation of CdTe-QDs in a biological system. Although Stern

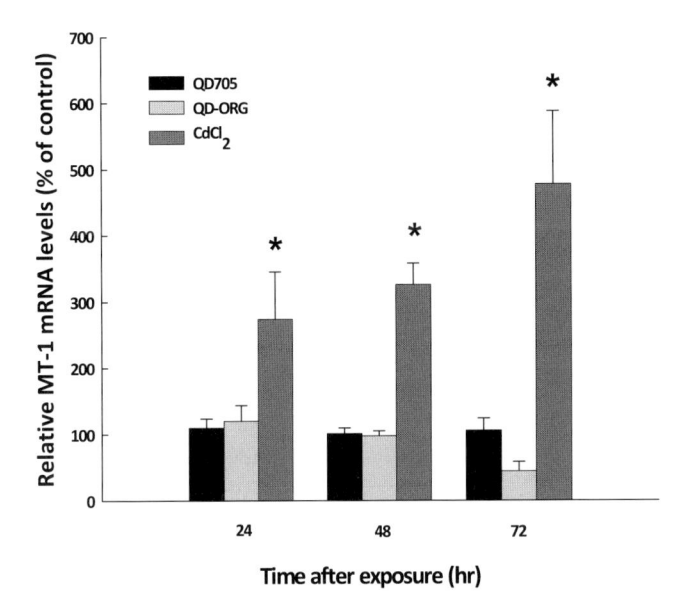

Figure 12.7 Comparative abilities in the induction of MT-1 expression by intact QD705 (QD705 with PEG coating), QD705 with exposed Cd/Se/Te complex (QD-ORG), and inorganic Cd salt (CdCl$_2$) in RAG cells. The results showed that only Cd from CdCl$_2$, but not the bounded-Cd in QD705 or QD-ORG, has the ability to induce MT-1 mRNA levels in mouse renal adenocarcinoma cells in a time-corresponding manner. Data represent the mean of eight replicates; * indicates $P < 0.05$ as compared with vehicle-treated control cells. Reproduced with permission from Ref. [53].

et al. [55] failed to detect MT-1 induction in porcine renal epithelial cell cultures after exposing the cells to intact CdTe-QDs for 8–24 h, a time-correlated enhancement of MT-1 expression (see Figure 12.3) was successfully demonstrated by 4–16 weeks in the renal PCT epithelial cells of mice treated with CdTeSe-based QDs [53]. As free cadmium, but not cadmium bound within the QD complex, can induce MT-1 expression (Figure 12.7), Lin et al. [52] affirmed that there was indeed a temporal degradation of the QD-complex in renal epithelial cells *in vivo*, and an increased release of free cadmium to the cells with time.

It is important to note that, although the liver, spleen and kidneys are target organs for QD accumulation, QD disintegration occurs primarily in the kidneys, with a much less significant or even no disintegration of QDs being detected in the liver or spleen, respectively, by 16 weeks and four months after exposure. It is likely that QDs in the liver and spleen are mostly "stored" in the macrophages, without degradation. Consequently, the macrophages in the liver and spleen may actually "shield" or protect the hepatocytes and splenic cells against QD toxicity *in vivo* by isolating the QD crystals in storage. (Of course, such a "protective" action of macrophages would not be noted *in vitro*, when QDs are added directly to cultured hepatocytes or splenic cells, in the absence of macrophages.) It appears, however, that the renal epithelial cells are able not only to take up QDs directly

[55] but also to break down the QD-complex after absorption [53]. Whilst this breakdown of the QD-complex may be intended as a detoxification action, such an action would in fact become more harmful than protective due to the release of toxic metals from the degraded QDs. Although the biomechanism for degradation of the QD-complex within renal epithelial cells remains unknown, it is recognized that both lysosomes and peroxisomes are abundant in these cells, and that the digestive disintegration of QDs by these organelles, via acid or oxidative actions, is possible [7].

12.5
The Mitochondrion as a Prime Target for QD Toxicity

Since 2005, many investigations have been devoted to exploring the "mechanisms" of QD toxicity. The involvement of ROS generation [11], c-Jun N-terminal kinase (JNK) activation [56], cadmium-related oxidative stress [29], ROS-induced Fas upregulation and lipid peroxidation [57] have each been proposed as the mechanistic bases for the "cytotoxicity" of CdTe-based QDs. All of these studies were performed *in vitro*, with different cell models using apoptosis as the criteria for cytotoxicity. All of the mechanisms proposed were related to ROS generation and oxidative stress in one way or another. However, when the results of these studies were examined more closely, aside from ROS generation there was another common finding, namely the changes that occur in the mitochondria of cells treated with QDs. These include the rounding of mitochondria, disintegration of the tubular mitochondrial network (oriented alignment of mitochondria), a loss of cytochrome c (a unique mitochondrial protein), a loss of mitochondrial membrane potential, and an impairment of mitochondrial function. Indeed, it would appear that the mitochondria are closely involved in QD toxicity.

Lin *et al.* examined kidneys from QD-treated mice by using both light and electron microscopy. Despite an absence of any remarkable histopathology, electron microscopy revealed specific alterations of mitochondria in the epithelial cells of the PCT. Mitochondrial swelling (Figure 12.8) and a loss of mitochondrial orientation, both of which have also been reported in cadmium nephropathy or in cadmium poisoning [66–70], were observed in the PCT cells of QD-treated mice. These changes usually reflect damage to the mitochondrial membrane, changes in mitochondrial permeability, and a loss of basal infoldings in the PCT. The ultrastructural changes observed using electron microscopy may be correlated with the "mitochondrial rounding" and "disintegration of tubular mitochondrial network" observed via MitoTracker Red staining with light microscopy [11, 29]. A significant depletion in cytochrome c oxidase (CCO) activity in many PCTs, as reported by Lin *et al.*, was also consistent with observations by others on the loss of cytochrome c and a reduction in mitochondrial function in cells treated with CdTe-QDs [11, 29, 56].

Whilst the swelling of mitochondria and loss of mitochondrial enzyme (cytochrome c) and function have been reported in cadmium toxicity, other mitochon-

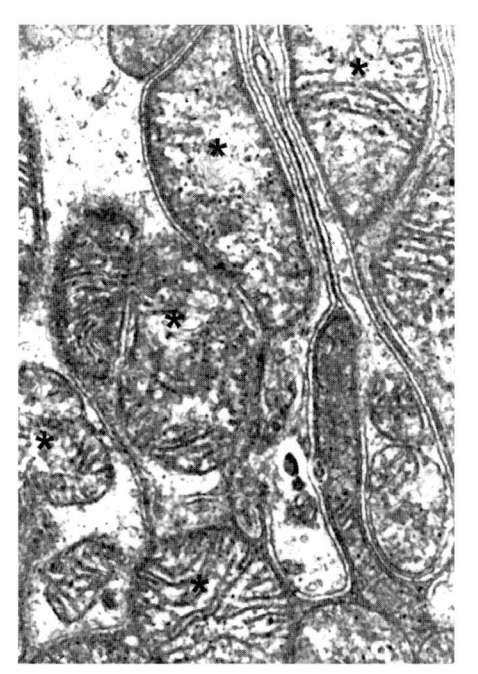

Figure 12.8 Mitochondrial swelling. Swollen (edematous) mitochondria (*) in a PCT epithelial cell. Note the watery matrix and broken inner cristae in many mitochondria. QD-treated mouse, four weeks. Original magnification, ×20000.

drial changes not seen in cadmium nephropathy have also been observed in QD toxicity. These include the formation of giant mitochondria and micro-mitochondria, an increased biogenesis of mitochondria (mitochondrial budding or division) leading to mitochondrial hyperplasia of (increased mitochondrial numbers and density in the cell), and membranous degeneration of the mitochondria. Giant mitochondria, unlike swollen mitochondria, not only maintain their matrical density and intact cristae but may also develop a tremendously proliferated inner membrane (cristae), giving the appearance of a "fingerprint" configuration (Figure 12.9a,b). Increased mitochondrial budding and the formation of micro-mitochondria (Figure 12.10a,b), leading to mitochondrial hyperplasia are also observed. These mitochondrial modifications have been reported in certain human diseases such as cardiomyopathy, congenital muscular dystrophy, and mitochondrial encephalomyopathies [71–73], where mitochondrial functions are deficient. It is believed that such mitochondrial alterations represent compensatory changes of the mitochondria due to significant losses of function or functional capability. The induction of cytochrome c loss [11, 29, 56] and a significant depletion of CCO induced by CdTeSe-QDs would clearly place such stress on the mitochondria.

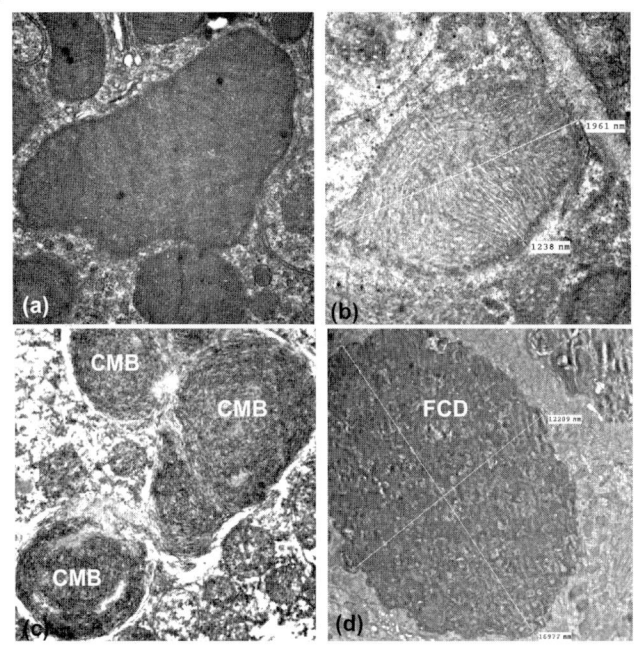

Figure 12.9 Compensatory giant mitochondria formation and decompensatory degeneration of mitochondria. (a) Giant mitochondria PCT epithelial cell of QD-treated mouse (16 weeks). Original magnification ×40 000. Note the intact nature of the inner cristae and density of the matrix; (b) The highly proliferative inner membrane cristae of a giant mitochondria which give this organelle a "finger-print" appearance (QD-treated, 16 weeks). Original magnification ×40 000; (c and d) Decompensatory degeneration of giant mitochondria leading to formation of cytoplasmic membrane bodies (CMB) and large focal cytoplasmic degradation (FCD) in the renal PCT epithelial cells (QD-treated, 24 weeks). Original magnifications ×15 000 and ×8000.

It should be pointed out that "decompensation" of the mitochondria (mitochondrial degeneration) also occurred when the compensatory action failed. Lin et al. showed that the giant mitochondria decompensated via membranous degeneration, giving rise to cytoplasmic membranous bodies and large cytoplasmic degradations (Figure 12.9c,d). It is also important to note that these mitochondrial alterations and degenerations occurred in the absence of, or prior to, cell death and apoptosis. Thus, mitochondrial damages represent early cell injury and can be used as early indicator of cell toxicity. As these changes occur at the organelle level, and are undetectable by means of traditional histopathological (hematoxylin and eosin) screening, other detection methods such as mitochondrial markers can be developed for more sensitive evaluations.

Although many studies have indicated ROS and oxidative stress as a basis of the "toxicity" of QDs [11, 29, 56], no specific indications were given as to the nature of the ROS and oxidative stress induced. Despite mitochondrial damages being referred to in all these studies, mitochondrial injury tended to be considered as a toxic consequence rather than as a specific target for QD toxicity. Lovric et al. [11]

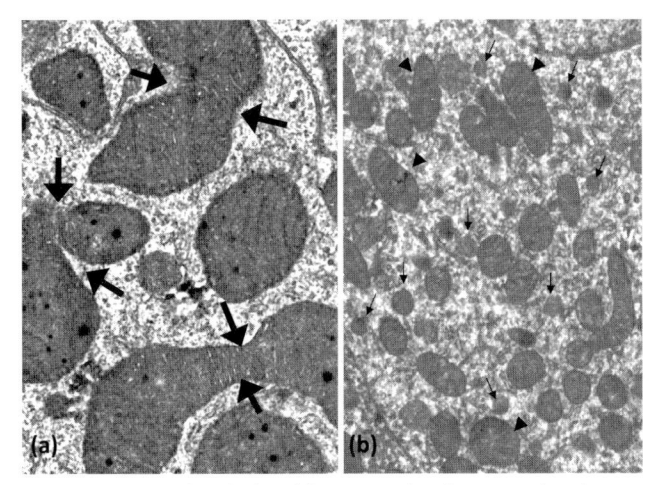

Figure 12.10 Mitochondrial proliferation and formations of micro-mitochondria. (a) Close examination revealed many mitochondria in budding (division) process within a PCT epithelial cell. Original magnification ×20 000; (b) PCT epithelial cell from QD-treated mouse (four weeks). A great variation in mitochondrial size was observed. Many of these mitochondria were extremely small in size (micro-mitochondria) (arrows) as compared to those of normal size (arrowheads). This representing newly generated mitochondria from active mitochondria biogenesis. Original magnification ×10 000.

first recognized the important role of mitochondria in QD toxicity, and have suggested that mitochondria are "... early targets of QD-induced stress and are severely damaged by QDs." A recent study by P. Lin et al. (unpublished results) further affirmed, using electron microscopy, that the mitochondria do indeed represent an early and specific target of QD toxicity.

The mitochondria can serve as either a "receptor" or a "donor" of ROS. Thus, for ROS production and oxidative stress-related toxicity, the mitochondria may be both "victim" and "villain." As a receptor, it can be influenced, triggered, or damaged by the cellular ROS status, whilst as a donor it can generate ROS which can then trigger other biological signal cascades, including that of apoptosis. The release of cadmium from Cd-based QDs has always been of concern to many investigators (see above) and, indeed, the disintegration of the CdTeSe complex with the release of free cadmium–at least within the renal epithelial cells *in vivo*–has been demonstrated convincingly [53]. Cadmium is known to induce alterations in antioxidant enzymes such as superoxide dismutase (SOD) [74] and catalase [75], with the generation of ROS [76–78] and lipid peroxidation [75]. The precise mechanism by which cadmium induces ROS formation and oxidative stress, however, is still not clear [79]. However, based on available information it appears that these disruptions of the antioxidant systems by cadmium [74] have a pivotal role in the enhancement of ROS accumulation via a reduction of ROS removal, rather than in ROS production *per se*. An enhanced ROS accumulation will eventually lead to oxidative stress.

The mitochondria are in fact the major source of ROS in cells, with 1–4% of the oxygen consumed by mitochondria being incompletely reduced and leading to the production of ROS, including superoxide radicals, hydrogen peroxide, and hydroxyl radicals [80, 81]. The interruption of correct mitochondrial function (including an ability to reduce the oxygen consumed) would certainly also enhance ROS production by mitochondria. In mitochondria, the most important sites of ROS formation are Complex I and Complex III [82], and any impairment of electron transfer through NADH, and through Complexes I and III, will induce enhanced superoxide radical formation [80, 81]. The treatment of cells with cadmium is known to result in specific mitochondrial changes, especially in hepatocytes and renal epithelial cells [80, 81, 83–87]. It is of interest to note that cadmium induces ROS generation only in Complex III (bc1 complex), a mitochondrial electron-transfer mechanism which is specifically related to a ubiquinone-CCO (oxidoreductase) system that oxidizes coenzyme Q by using cytochrome c as an electron acceptor to generate the membrane potential [88, 89]. Observations on the loss of cytochrome c in mitochondria [11, 18, 29] and the reduction in CCO in cells treated with QD, clearly suggest an impairment of Complex III of the mitochondrial electron-transfer chain by cadmium released from the QD complex [53]. This disruption of Complex III would lead to an enhanced production of ROS, and especially of harmful superoxide radicals. Cho et al. [29], however, failed to correlate intracellular cadmium levels in QD-treated cells with the degree of cytotoxicity induced. Lin et al. also indicated that mitochondrial changes in QD-treated mice were similar, but not identical, to those reported for cadmium toxicity. Thus, the influences of other factors, including the QD complex taken into the cells, cannot be excluded. The biodegradation of intracellular QDs via bio-oxidation or photooxidative processes with ROS generation has been proposed by Cho et al. [29]. The ROS generated may, in turn, interrupt the function of the mitochondria, thereby triggering further ROS production.

In summary, based on the available information, it appears that three types of ROS production in QD-treated cells should be considered: (i) a reduction of the antioxidant system in cells by the cadmium released inducing ROS accumulation [74, 75]; (ii) a direct disruption of the Complex III electron-transfer chain of mitochondria by the cadmium released by QDs [88, 89], thus enhancing the mitochondrial production of ROS; and (iii) an oxidative biodeterioration and biodegradation of the QDs in the cells, accompanied by ROS production [29]. All of the ROS produced, including those derived from the mitochondria themselves, can act upon the mitochondria so as to induce not only damage to the mitochondrial outer membrane but also changes in the confirmation of adenine nucleotide translocase, a protein involved in the mitochondrial permeability transition pore [11]. Changes in the mitochondrial outer membrane and permeability regulation may ultimately lead to mitochondrial swelling and a loss of cytochrome c, both of which are observed in the renal tubular cells of QD-treated mice.

To date, many reports have been made indicating that ROS can induce apoptotic cell death. Aside from ROS, cytochrome c released from the mitochondria also

12.5 The Mitochondrion as a Prime Target for QD Toxicity | 455

plays an important role in apoptotic signaling; indeed, cytochrome c has been referred to as the "machinery of death" [90]. The additional generation of ROS by the released cytochrome c has been suggested via the lipid peroxidation of cardiolipin, a phospholipid which is important for the association of cytochrome c with the inner mitochondrial membrane [91]. Consequently, a "vicious cycle" involving ROS action on the mitochondria, mitochondrial injury, loss of cytochrome c, reduction of mitochondrial Complex III function, further ROS generation and mitochondrial injury, may be established and lead to cumulative mitochondrial defects (outer membrane damage, loss of cytochrome c, mitochondrial swelling, etc.), a deficiency in mitochondrial function, mitochondrial compensatory alterations (giant/micro-mitochondria formation, mitochondrial biogenesis, mitochondrial hyperplasia) and eventual mitochondrial decompensation (degeneration), as demonstrated by several groups [11, 29, 56, 57]. All of these studies have provided compelling evidence for mitochondria being a prime target of QD toxicity.

An overall summary of the proposed events on QD toxicity leading to ROS production, mitochondrial changes, and cytotoxicity is presented in Figure 12.11.

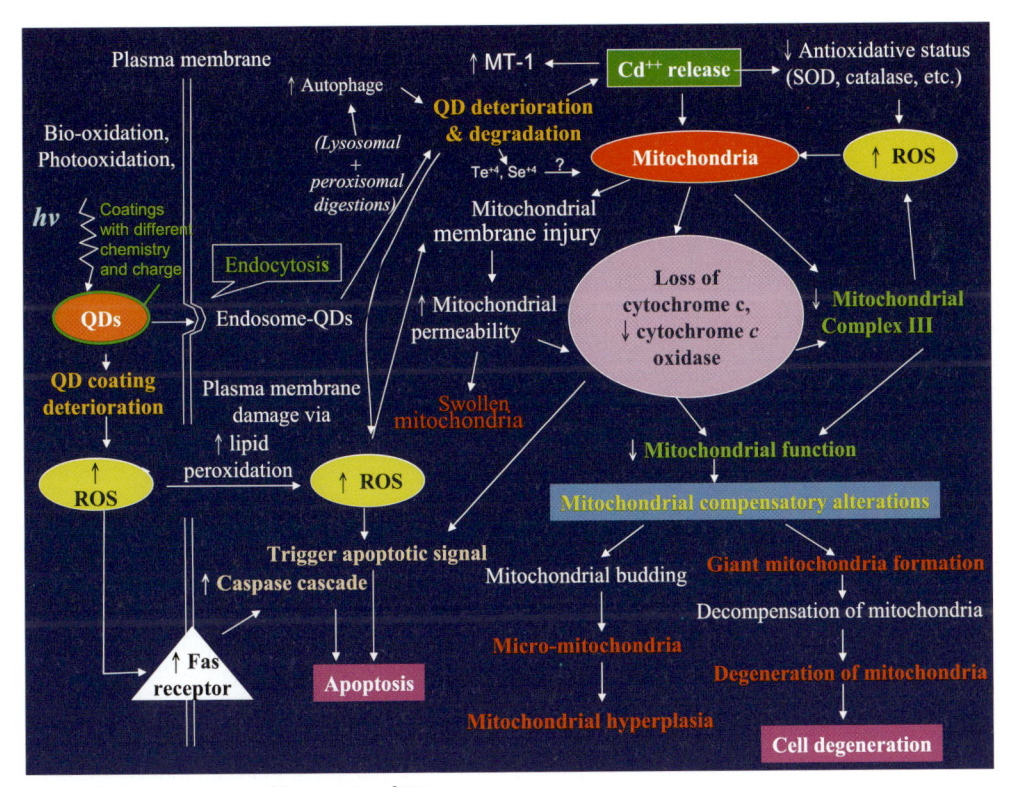

Figure 12.11 An overview of the toxicity of QDs.

12.6
Environmental and Ecological Concerns

With potential opportunities for exposure to QDs including workplace, environmental (including from products using QDs), and medical (diagnostic and therapeutic) usages, the main *routes* of exposure include inhalation, dermal contact, ingestion, and intravenous injection. Yet today, with very few ADME studies of QDs having been conducted, very little information is available relating QD toxicity to the route of exposure. Moreover, since most QDs include metallic components that are toxic (e.g., Cd, Te, Pb, Se), continued accumulation and long half-lives are to be expected when they enter environmental and ecological systems. Most importantly, these accumulations will, inevitably, pose both environmental and ecological risks.

As with most chemicals used in industry, QDs can be released into the environment via water or air, often during the manufacturing process. The storage or disposal of QD materials, as well as their transport within the factory or their release (whether by incineration or in waste water) into the environment, represent a major concern. Consequently, based on the current enthusiasm for nanotechnology and nanorelated products, it is likely that vast increases will emerge in the production and use of these materials, the safety aspects of which are as yet virtually unknown. Very little information is also currently available on where, and how, these nanomaterials would partition in the environment, whether in the air, water, soil, plants, and food chains. It is safe to say that not all QDs are alike, and with their tremendously diverse properties it is unlikely that the environmental partitioning of QDs will be established with any great ease. Yet, despite a lack of information on the environmental fate of QDs, their eventual degradation via photolytic or oxidative processes can be regarded as certain. Unfortunately, however, as toxic metals are released from the inner cores of the QDs, environmental contamination is likely to become a worry rather than an academic concern. Thus, aside from the bioevaluation of QDs in animal and human health, it is essential that attention be paid to the environmental and ecological impact of these materials.

12.7
Concluding Comments and Future Perspectives

Because of their unique properties, QDs may be considered ideal probes for *in vivo* imaging [13, 15, 92–95]. In this chapter, the various aspects of QDs toxicity, the physico-chemical properties, surface chemistry, core metals, *in vitro* and *in vivo* investigations (ADME), target tissues, mechanistic considerations, and effects on mitochondria, have been discussed, as have the most recent approaches to biosafety evaluation. The aim here has not been to prove that QDs have toxic potential and thus be eliminated from future use, but rather to provide information to create a new generation of QDs that would be more useful, safer,

12.7 Concluding Comments and Future Perspectives

and less expensive. The modifications of both surface coatings and inner cores to create safer products should certainly be possible using current state-of-the-art techniques.

Nanotechnology represents one of the most exciting technological developments of the past decade, and will hopefully continue to provide many benefits for years to come, without making any sacrifice to health and safety. Unfortunately, however, the technologies of safety evaluation are not advancing as rapidly as their industrial counterparts, with century-old techniques such as histopathology and cytotoxicity/genotoxicity screening still being used to assess these most contemporary of materials. Clearly, it is vital that more sensitive and specific methods for safety assessment are developed.

Whilst nanomaterials have preferential sites of distribution (target organ) and deposit (target tissue), metal-based QDs tend to target site the mitochondria, where they impart their toxic action. Therefore, as histopathological methods to detect cell degeneration and death may be insufficiently sensitive or specific to define the "cytotoxicity" of QDs, mitochondrial markers must be developed to assume this role. Most importantly, as different QDs will behave differently in the same biological system, the mitochondria might represent a favorite target for most metals [68–70, 96–100], with mitochondrial markers ultimately being used to assess the biosafety of metal-based QDs and other nanomaterials.

As noted by Hardman [1], toxicity should be assessed by toxicologists, and not simply by conducting studies based on "standard protocols." Poorly conducted studies and/or interpretations cause more harm than merit to science and to industry if misleading data and wrongful interpretations are quoted and requoted as if they were factual and correct. Such "literature pollution" must be carefully guarded against and, with nanoresearch having gained much popularity in recent years, it is essential that only valid data are identified and selected.

Although *in vitro* investigations are of value, they are not a substitute for *in vivo* studies. As noted above, nanoparticles tend to aggregate when incubated under *in vitro* conditions, and consequently their physico-chemical properties must be carefully monitored. Animal studies must also be conducted to establish the basic ADME data for these relatively unknown materials, and mechanistic studies conducted under *in vitro* conditions. Moreover, *in vivo* findings must be validated with *in vitro* investigations whenever possible.

Whilst animal studies are important, differences in the physiology and metabolism of animals and humans may have a major influence on the outcome of toxicity studies. When extrapolating from animals to humans, both PBPK [65] (also R.S.H. Yang *et al.*, unpublished results) and computational toxicology [42] modeling can be used, and these are valuable for assessing the effects of nanomaterial exposure in humans. However, blood-flow-limited models alone are inadequate to explain the complex pharmacokinetics of QDs in biological systems, and more detailed descriptions of nanoparticle flux into specific tissues, including any tissue-specific vascular and lymphatic effects and mechanisms of cellular uptake, may be required [65]. This area of nanomaterials research requires further development. Although this chapter is devoted to the toxicology and biosafety of QDs, the basic principles and concepts presented may apply equally well to the assessment of other nanomaterials.

Without doubt, human health is the most important issue in the safety evaluation of nanomaterials. However, with the tremendous development during recent years of new factories and industries using nanomaterials, attention must be focused on ensuring the safety of the environment and its ecology.

References

1 Hardman, R. (2006) A toxicologic review of quantum dots: toxicity depends on physicochemical and environmental factors. *Environmental Health Perspectives*, **114**, 165–72.
2 Alivisatos, P. (2004) The use of nanocrystals in biological detection. *Nature Biotechnology*, **22**, 47–52.
3 Gao, X., Cui, Y., Levenson, R.M., Chung, L.W. and Nie, S. (2004) *In vivo* cancer targeting and imaging with semiconductor quantum dots. *Nature Biotechnology*, **22**, 969–76.
4 Wu, X., Liu, H., Liu, J., Haley, K.N., Treadway, J.A., Larson, J.P., Ge, N., Peale, F. and Bruchez, M.P. (2003) Immunofluorescent labeling of cancer marker Her2 and other cellular targets with semiconductor quantum dots. *Nature Biotechnology*, **21**, 41–6.
5 Yu, S. and Chow, G.M. (2004) Carboxyl group (-CO_2H) functionalized ferrimagnetic iron oxide nanoparticles for potential bio-applications. *Journal of Materials Chemistry*, **14**, 2781–6.
6 Michalet, X., Pinaud, F.F., Bentolila, L.A., Tsay, J.M., Doose, S., Li, J.J., Sundaresan, G., Wu, A.M., Gambhir, S.S. and Weiss, S. (2005) Quantum dots for live cells, *in vivo* imaging, and diagnostics. *Science*, **307**, 538–44.
7 Derfus, A.M., Chan, W.C.W. and Bhatia, S.N. (2004) Probing the cytotoxicity of semiconductor quantum dots. *Nano Letters*, **4**, 11–18.
8 Nel, A., Xia, T., Madler, L. and Li, N. (2006) Toxic potential of materials at the nanolevel. *Science*, **311**, 622–7.
9 Hoshino, A., Fujioka, K., Oku, T., Suga, M., Sasaki, Y.F., Ohta, T., Yasuhara, M., Suzuki, K. and Yamamoto, K. (2004) Physicochemical properties and cellular toxicity of nanocrystal quantum dots depend on their surface modification. *Nano Letters*, **4**, 2163–9.
10 Shiohara, A., Hoshino, A., Hanaki, K., Suzuki, K. and Yamamoto, K. (2004) On the cyto-toxicity caused by quantum dots. *Microbiology and Immunology*, **48**, 669–75.
11 Lovric, J., Cho, S.J., Winnik, F.M. and Maysinger, D. (2005) Unmodified cadmium telluride quantum dots induce reactive oxygen species formation leading to multiple organelle damage and cell death. *Chemistry and Biology*, **12**, 1227–34.
12 Dubertret, B., Skourides, P., Norris, D.J., Noireaux, V., Brivanlou, A.H. and Libchaber, A. (2002) *In vivo* imaging of quantum dots encapsulated in phospholipid micelles. *Science*, **298**, 1759–62.
13 Ballou, B., Lagerholm, B.C., Ernst, L.A., Bruchez, M.P. and Waggoner, A.S. (2004) Noninvasive imaging of quantum dots in mice. *Bioconjugate Chemistry*, **15**, 79–86.
14 Jaiswal, J.K., Mattoussi, H., Mauro, J.M. and Simon, S.M. (2003) Long-term multiple color imaging of live cells using quantum dot bioconjugates. *Nature Biotechnology*, **21**, 47–51.
15 Larson, D.R., Zipfel, W.R., Williams, R.M., Clark, S.W., Bruchez, M.P., Wise, F.W. and Webb, W.W. (2003) Water-soluble quantum dots for multiphoton fluorescence imaging *in vivo*. *Science*, **300**, 1434–6.
16 Voura, E.B., Jaiswal, J.K., Mattoussi, H. and Simon, S.M. (2004) Tracking metastatic tumor cell extravasation with quantum dot nanocrystals and fluorescence emission-scanning

microscopy. *Nature Medicine*, **10**, 993–8.

17 Hanaki, K.-I., Momo, A., Oku, T., Komoto, A., Maenosono, S., Yamaguchi, Y., and Yamamoto, K. (2003) Semiconductor quantum dot/albumin complex is a long-life and highly photostable endosome marker. *Biochemical and Biophysical Research Communications*, **302** (3), 496–501.

18 Chen, F. and Gerion, D. (2004) Fluorescent CdSe/ZnS nanocrystal–peptide conjugates for long-term, nontoxic imaging and nuclear targeting in living cells. *Nano Letters*, **4**, 1827–32.

19 Clift, M.J., Rothen-Rutishauser, B., Brown, D.M., Duffin, R., Donaldson, K., Proudfoot, L., Guy, K. and Stone, V. (2008) The impact of different nanoparticle surface chemistry and size on uptake and toxicity in a murine macrophage cell line. *Toxicology and Applied Pharmacology*, **232**, 418–27.

20 Male, K.B., Lachance, B., Hrapovic, S., Sunahara, G. and Luong, J.H. (2008) Assessment of cytotoxicity of quantum dots and gold nanoparticles using cell-based impedance spectroscopy. *Analytical Chemistry*, **80**, 5487–93.

21 Dabbousi, B.O., Rodriguez-Viejo, J., Mikulec, F.V., Heine, J.R., Mattoussi, H., Ober, R., Jensen, K.F. and Bawendi, M.G. (1997) (CdSe)ZnS core–shell quantum dots: synthesis and characterization of a size series of highly luminescent nanocrystallites. *Journal of Physical Chemistry B*, **101**, 9463–75.

22 Kim, S., Fisher, B., Eisler, H.-J. and Bawendi, M. (2003) Type-II quantum dots: CdTe/CdSe(core/shell) and CdSe/ZnTe(core/shell) heterostructures. *Journal of the American Chemical Society*, **125**, 11466–7.

23 Chan, W.C. and Nie, S. (1998) Quantum dot bioconjugates for ultrasensitive nonisotopic detection. *Science*, **281**, 2016–18.

24 Ryman-Rasmussen, J.P., Riviere, J.E. and Monteiro-Riviere, N.A. (2007) Surface coatings determine cytotoxicity and irritation potential of quantum dot nanoparticles in epidermal keratinocytes. *Journal of Investigative Dermatology*, **127**, 143–53.

25 Wang, L., Nagesha, D.K., Selvarasah, S., Dokmeci, M.R. and Carrier, R.L. (2008) Toxicity of CdSe nanoparticles in Caco-2 cell cultures. *Journal of Nanobiotechnology*, **6**, 11.

26 Hoshino, A., Manabe, N., Fujioka, K., Suzuki, K., Yasuhara, M. and Yamamoto, K. (2007) Use of fluorescent quantum dot bioconjugates for cellular imaging of immune cells, cell organelle labeling, and nanomedicine: surface modification regulates biological function, including cytotoxicity. *Journal of Artificial Organs*, **10**, 149–57.

27 Goodman, C.M., McCusker, C.D., Yilmaz, T. and Rotello, V.M. (2004) Toxicity of gold nanoparticles functionalized with cationic and anionic side chains. *Bioconjugate Chemistry*, **15**, 897–900.

28 Chang, E., Thekkek, N., Yu, W.W., Colvin, V.L. and Drezek, R. (2006) Evaluation of quantum dot cytotoxicity based on intracellular uptake. *Small*, **2**, 1412–17.

29 Cho, S.J., Maysinger, D., Jain, M., Roder, B., Hackbarth, S. and Winnik, F.M. (2007) Long-term exposure to CdTe quantum dots causes functional impairments in live cells. *Langmuir*, **23**, 1974–80.

30 Fischer, H.C., Liu, J., Pang, K.S. and Chang, W.C.W. (2006) Pharmacokinetics of nanoscale quantum dots: in vivo distribution, sequestration, and clearance in the rat. *Advanced Functional Materials*, **16**, 1299–305.

31 Rejman, J., Oberle, V., Zuhorn, I.S. and Hoekstra, D. (2004) Size-dependent internalization of particles via the pathways of clathrin- and caveolae-mediated endocytosis. *Biochemical Journal*, **377**, 159–69.

32 Donaldson, K., Stone, V., Tran, C.L., Kreyling and W., Borm, P.J. (2004) Nanotoxicology. *Occupational and Environmental Medicine*, **61**, 727–8.

33 Donaldson, K. and Tran, C.L. (2002) Inflammation caused by particles and fibers. *Inhalation Toxicology*, **14**, 5–27.

34 Nel, A. (2005) Atmosphere. Air pollution-related illness: effects of particles. *Science*, **308**, 804–6.

35 Oberdorster, G., Oberdorster, E. and Oberdorster, J. (2005) Nanotoxicology: an emerging discipline evolving from studies of ultrafine particles. *Environmental Health Perspectives*, **113**, 823–39.

36 Gref, R., Minamitake, Y., Peracchia, M.T., Trubetskoy, V., Torchilin, V. and Langer, R. (1994) Biodegradable long-circulating polymeric nanospheres. *Science*, **263**, 1600–3.

37 Peracchia, M.T., Fattal, E., Desmaele, D., Besnard, M., Noel, J.P., Gomis, J.M., Appel, M., D'Angelo, J., and Couvreur, P. (1999) Stealth PEGylated polycyanoacrylate nanoparticles for intravenous administration and splenic targeting. *Journal of Controlled Release*, **60**, 121–8.

38 Porter, C.J., Moghimi, S.M., Illum, L. and Davis, S.S. (1992) The polyoxyethylene/polyoxypropylene block co-polymer poloxamer-407 selectively redirects intravenously injected microspheres to sinusoidal endothelial cells of rabbit bone marrow. *FEBS Letters*, **305**, 62–6.

39 Geys, J., Nemmar, A., Verbeken, E., Smolders, E., Ratoi, M., Hoylaerts, M.F., Nemery, B. and Hoet, P.H. (2008) Acute toxicity and prothrombotic effects of quantum dots: impact of surface charge. *Environmental Health Perspectives*, **116**, 1607–13.

40 Nemmar, A., Hoylaerts, M.F., Hoet, P.H., Dinsdale, D., Smith, T., Xu, H., Vermylen, J. and Nemery, B. (2002) Ultrafine particles affect experimental thrombosis in an *in vivo* hamster model. *American Journal of Respiratory and Critical Care Medicine*, **166**, 998–1004.

41 Yang, R.S., Chang, L.W., Wu, J.P., Tsai, M.H., Wang, H.J., Kuo, Y.C., Yeh, T.K., Yang, C.S. and Lin, P. (2007) Persistent tissue kinetics and redistribution of nanoparticles, quantum dot 705, in mice: ICP-MS quantitative assessment. *Environmental Health Perspectives*, **115**, 1339–43.

42 Lin, P., Chen, J.W., Chang, L.W., Wu, J.P., Redding, L., Chang, H., Yeh, T.K., Yang, C.S., Tsai, M.H., Wang, H.J. *et al.* (2008) Computational and ultrastructural toxicology of a nanoparticle, Quantum Dot 705, in mice. *Environmental Science & Technology*, **42**, 6264–70.

43 Chen, Z., Chen, H., Meng, H., Xing, G., Gao, X., Sun, B., Shi, X., Yuan, H., Zhang, C., Liu, R. *et al.* (2008) Bio-distribution and metabolic paths of silica coated CdSeS quantum dots. *Toxicology and Applied Pharmacology*, **230**, 364–71.

44 Ryman-Rasmussen, J.P., Riviere, J.E. and Monteiro-Riviere, N.A. (2006) Penetration of intact skin by quantum dots with diverse physicochemical properties. *Toxicological Sciences*, **91**, 159–65.

45 Tinkle, S.S., Antonini, J.M., Rich, B.A., Roberts, J.R., Salmen, R., DePree, K. and Adkins, E.J. (2003) Skin as a route of exposure and sensitization in chronic beryllium disease. *Environmental Health Perspectives*, **111**, 1202–8.

46 Zhang, L.W. and Monteiro-Riviere, N.A. (2008) Assessment of quantum dot penetration into intact, tape-stripped, abraded and flexed rat skin. *Skin Pharmacology and Physiology*, **21**, 166–80.

47 Zhang, L.W., Yu, W.W., Colvin, V.L. and Monteiro-Riviere, N.A. (2008) Biological interactions of quantum dot nanoparticles in skin and in human epidermal keratinocytes. *Toxicology and Applied Pharmacology*, **228**, 200–11.

48 Mortensen, L.J., Oberdorster, G., Pentland, A.P. and Delouise, L.A. (2008) *In vivo* skin penetration of quantum dot nanoparticles in the murine model: the effect of UVR. *Nano Letters*, **8**, 2779–87.

49 Aldana, J., Wang, Y.A. and Peng, X. (2001) Photochemical instability of CdSe nanocrystals coated by hydrophilic thiols. *Journal of the American Chemical Society*, **123**, 8844–50.

50 Aldana, J., Lavelle, N., Wang, Y. and Peng, X. (2005) Size-dependent dissociation pH of thiolate ligands from cadmium chalcogenide nanocrystals. *Journal of the American Chemical Society*, **127**, 2496–504.

51 Mancini, M.C., Kairdolf, B.A., Smith, A.M. and Nie, S. (2008) Oxidative quenching and degradation of polymer-encapsulated quantum dots: new insights into the long-term fate and toxicity of nanocrystals *in vivo*. *Journal of*

the American Chemical Society, **130**, 10836–7.
52 Akerman, M.E., Chan, W.C., Laakkonen, P., Bhatia, S.N. and Ruoslahti, E. (2002) Nanocrystal targeting in vivo. *Proceedings of the National Academy of Sciences of the United States of America*, **99**, 12617–21.
53 Lin, C.H., Chang, L.W., Chang, H., Yang, M.H., Yang, C.S., Lai, W.H., Chang, W.H. and Lin, P. (2009) The chemical fate of the Cd/Se/Te-based quantum dot 705 in the biological system: toxicity implications. *Nanotechnology*, **20**, 215101.
54 Kirchner, C., Liedl, T., Kudera, S., Pellegrino, T., Munoz Javier, A., Gaub, H.E., Stolzle, S., Fertig, N. and Parak, W.J. (2005) Cytotoxicity of colloidal CdSe and CdSe/ZnS nanoparticles. *Nano Letters*, **5**, 331–8.
55 Stern, S.T., Zolnik, B.S., McLeland, C.B., Clogston, J., Zheng, J. and McNeil, S.E. (2008) Induction of autophagy in porcine kidney cells by quantum dots: a common cellular response to nanomaterials? *Toxicological Sciences*, **106**, 140–52.
56 Chan, W.H., Shiao, N.H. and Lu, P.Z. (2006) CdSe quantum dots induce apoptosis in human neuroblastoma cells via mitochondrial-dependent pathways and inhibition of survival signals. *Toxicology Letters*, **167**, 191–200.
57 Choi, A.O., Cho, S.J., Desbarats, J., Lovric, J. and Maysinger, D. (2007) Quantum dot-induced cell death involves Fas upregulation and lipid peroxidation in human neuroblastoma cells. *Journal of Nanobiotechnology*, **5**, 1.
58 Samia, A.C., Chen, X. and Burda, C. (2003) Semiconductor quantum dots for photodynamic therapy. *Journal of the American Chemical Society*, **125**, 15736–7.
59 Ipe, B.I., Lehnig, M. and Niemeyer, C.M. (2005) On the generation of free radical species from quantum dots. *Small*, **1**, 706–9.
60 Green, M. and Howman, E. (2005) Semiconductor quantum dots and free radical induced DNA nicking. *Chemical Communications (Cambridge, England)*, 121–3.
61 Funnell, W.R. and Maysinger, D. (2006) Three-dimensional reconstruction of cell nuclei, internalized quantum dots and sites of lipid peroxidation. *Journal of Nanobiotechnology*, **4**, 10.
62 Scherz-Shouval, R., Shvets, E. and Elazar, Z. (2007) Oxidation as a post-translational modification that regulates autophagy. *Autophagy*, **3**, 371–3.
63 Huang, J. and Klionsky, D.J. (2007) Autophagy and human disease. *Cell Cycle*, **6**, 1837–49.
64 Lim, Y.T., Kim, S., Nakayama, A., Stott, N.E., Bawendi, M.G. and Frangioni, J.V. (2003) Selection of quantum dot wavelengths for biomedical assays and imaging. *Molecular Imaging*, **2**, 50–64.
65 Lee, H.A., Leavens, T.L., Mason, S.E., Monteiro-Riviere, N.A. and Riviere, J.E. (2009) Comparison of quantum dot biodistribution with a blood-flow-limited physiologically based pharmacokinetic model. *Nano Letters*, **9**, 794–9.
66 Bai, S. and Xu, Z. (2006) Effects of cadmium on mitochondrion structure and energy metabolism of *Palteobagrus fulvidraco* gill. *Ying Yong Sheng Tai Xue Bao*, **17**, 1213–17.
67 Chan, W.Y. and Rennert, O.M. (1981) Cadmium nephropathy. *Annals of Clinical and Laboratory Science*, **11**, 229–38.
68 Kukner, A., Colakoglu, N., Kara, H., Oner, H., Ozogul, C. and Ozan, E. (2007) Ultrastructural changes in the kidney of rats with acute exposure to cadmium and effects of exogenous metallothionein. *Biological Trace Element Research*, **119**, 137–46.
69 Matsuura, K., Takasugi, M., Kunifuji, Y., Horie, A. and Kuroiwa, A. (1991) Morphological effects of cadmium on proximal tubular cells in rats. *Biological Trace Element Research*, **31**, 171–82.
70 Thevenod, F. (2003) Nephrotoxicity and the proximal tubule. Insights from cadmium. *Nephron Physiology*, **93**, 87–93.p
71 Casademont, J. and Miro, O. (2002) Electron transport chain defects in heart failure. *Heart Failure Reviews*, **7**, 131–9.
72 Nishino, I., Kobayashi, O., Goto, Y., Kurihara, M., Kumagai, K., Fujita, T.,

Hashimoto, K., Horai, S. and Nonaka, I. (1998) A new congenital muscular dystrophy with mitochondrial structural abnormalities. *Muscle and Nerve*, **21**, 40–7.

73 Siciliano, G., Mancuso, M., Pasquali, L., Manca, M.L., Tessa, A. and Iudice, A. (2000) Abnormal levels of human mitochondrial transcription factor A in skeletal muscle in mitochondrial encephalomyopathies. *Neurological Sciences*, **21**, S985–987.

74 Hussain, T., Shukla, G.S. and Chandra, S.V. (1987) Effects of cadmium on superoxide dismutase and lipid peroxidation in liver and kidney of growing rats: *in vivo* and *in vitro* studies. *Pharmacology and Toxicology*, **60**, 355–8.

75 Shaikh, Z.A., Vu, T.T. and Zaman, K. (1999) Oxidative stress as a mechanism of chronic cadmium-induced hepatotoxicity and renal toxicity and protection by antioxidants. *Toxicology and Applied Pharmacology*, **154**, 256–63.

76 Hassoun, E.A. and Stohs, S.J. (1996) Cadmium-induced production of superoxide anion and nitric oxide, DNA single strand breaks and lactate dehydrogenase leakage in J774A.1 cell cultures. *Toxicology*, **112**, 219–26.

77 Bagchi, D., Joshi, S.S., Bagchi, M., Balmoori, J., Benner, E.J., Kuszynski, C.A. and Stohs, S.J. (2000) Cadmium- and chromium-induced oxidative stress, DNA damage, and apoptotic cell death in cultured human chronic myelogenous leukemic K562 cells, promyelocytic leukemic HL-60 cells, and normal human peripheral blood mononuclear cells. *Journal of Biochemical and Molecular Toxicology*, **14**, 33–41.

78 Szuster-Ciesielska, A., Stachura, A., Slotwinska, M., Kaminska, T., Sniezko, R., Paduch, R., Abramczyk, D., Filar, J. and Kandefer-Szerszen, M. (2000) The inhibitory effect of zinc on cadmium-induced cell apoptosis and reactive oxygen species (ROS) production in cell cultures. *Toxicology*, **145**, 159–71.

79 Ercal, N., Gurer-Orhan, H. and Aykin-Burns, N. (2001) Toxic metals and oxidative stress part I: mechanisms involved in metal-induced oxidative damage. *Current Topics in Medicinal Chemistry*, **1**, 529–39.

80 Raha, S. and Robinson, B.H. (2000) Mitochondria, oxygen free radicals, disease and ageing. *Trends in Biochemical Sciences*, **25**, 502–8.

81 Turrens, J.F. (1997) Superoxide production by the mitochondrial respiratory chain. *Bioscience Reports*, **17**, 3–8.

82 Tahara, E.B., Navarete, F.D. and Kowaltowski, A.J. (2009) Tissue-, substrate-, and site-specific characteristics of mitochondrial reactive oxygen species generation. *Free Radical Biology and Medicine*, **46**, 1283–97.

83 Dudley, R.E., Svoboda, D.J. and Klaassen, C.D. (1984) Time course of cadmium-induced ultrastructural changes in rat liver. *Toxicology and Applied Pharmacology*, **76**, 150–60.

84 Early, J.L. 2nd, Nonavinakere, V.K. and Weaver, A. (1992) Effect of cadmium and/or selenium on liver mitochondria and rough endoplasmic reticulum in the rat. *Toxicology Letters*, **62**, 73–83.

85 Koizumi, T., Yokota, T., Shirakura, H., Tatsumoto, H. and Suzuki, K.T. (1994) Potential mechanism of cadmium-induced cytotoxicity in rat hepatocytes: inhibitory action of cadmium on mitochondrial respiratory activity. *Toxicology*, **92**, 115–25.

86 Tang, W. and Shaikh, Z.A. (2001) Renal cortical mitochondrial dysfunction upon cadmium metallothionein administration to Sprague-Dawley rats. *Journal of Toxicology and Environmental Health. Part A*, **63**, 221–35.

87 Toury, R., Boissonneau, E., Stelly, N., Dupuis, Y., Berville, A. and Perasso, R. (1985) Mitochondria alterations in Cd^{2+}-treated rats: general regression of inner membrane cristae and electron transport impairment. *Biology of the Cell*, **55**, 71–85.

88 Adam-Vizi, V. and Chinopoulos, C. (2006) Bioenergetics and the formation of mitochondrial reactive oxygen species. *Trends in Pharmacological Sciences*, **27**, 639–45.

89 Wang, Y., Fang, J., Leonard, S.S. and Rao, K.M. (2004) Cadmium inhibits the electron transfer chain and induces

reactive oxygen species. *Free Radical Biology and Medicine*, **36**, 1434–43.
90 Newmeyer, D.D. and Ferguson-Miller, S. (2003) Mitochondria: releasing power for life and unleashing the machineries of death. *Cell*, **112**, 481–90.
91 Shidoji, Y., Hayashi, K., Komura, S., Ohishi, N. and Yagi, K. (1999) Loss of molecular interaction between cytochrome c and cardiolipin due to lipid peroxidation. *Biochemical and Biophysical Research Communications*, **264**, 343–7.
92 So, M.K., Xu, C., Loening, A.M., Gambhir, S.S. and Rao, J. (2006) Self-illuminating quantum dot conjugates for *in vivo* imaging. *Nature Biotechnology*, **24**, 339–43.
93 Stroh, M., Zimmer, J.P., Duda, D.G., Levchenko, T.S., Cohen, K.S., Brown, E.B., Scadden, D.T., Torchilin, V.P., Bawendi, M.G., Fukumura, D. and Jain, R.K. (2005) Quantum dots spectrally distinguish multiple species within the tumor milieu *in vivo*. *Nature Medicine*, **11**, 678–82.
94 Tada, H., Higuchi, H., Wanatabe, T.M. and Ohuchi, N. (2007) *In vivo* real-time tracking of single quantum dots conjugated with monoclonal anti-HER2 antibody in tumors of mice. *Cancer Research*, **67**, 1138–44.
95 Yu, X., Chen, L., Li, K., Li, Y., Xiao, S., Luo, X., Liu, J., Zhou, L., Deng, Y., Pang, D. and Wang, Q. (2007) Immunofluorescence detection with quantum dot bioconjugates for hepatoma *in vivo*. *Journal of Biomedical Optics*, **12**, 014008.
96 Belyaeva, E.A., Dymkowska, D., Wieckowski, M.R. and Wojtczak, L. (2008) Mitochondria as an important target in heavy metal toxicity in rat hepatoma AS-30D cells. *Toxicology and Applied Pharmacology*, **231**, 34–42.
97 Huang, X.P., O'Brien, P.J. and Templeton, D.M. (2006) Mitochondrial involvement in genetically determined transition metal toxicity I. Iron toxicity. *Chemico-Biological Interactions*, **163**, 68–76.
98 Kumar, V., Bal, A. and Gill, K.D. (2009) Susceptibility of mitochondrial superoxide dismutase to aluminium induced oxidative damage. *Toxicology*, **255**, 117–23.
99 Mehta, R., Templeton, D.M. and O'Brien, P.J. (2006) Mitochondrial involvement in genetically determined transition metal toxicity II. Copper toxicity. *Chemico-Biological Interactions*, **163**, 77–85.
100 Peraza, M.A., Cromey, D.W., Carolus, B., Carter, D.E. and Gandolfi, A.J. (2006) Morphological and functional alterations in human proximal tubular cell line induced by low level inorganic arsenic: evidence for targeting of mitochondria and initiated apoptosis. *Journal of Applied Toxicology*, **26**, 356–67.

Index

a

absorption
– coefficient 68, 183, 255f., 260f., 411
– maximum 183
– multiphoton 302
– nonspecific 197, 200
– organic dyes 70f.
– photon 8, 148, 294f., 301ff.
– profile 148
– rate 148
– simultaneous two-photon 263
– spectra 68, 100, 163, 183f., 299, 333, 339, 396, 402f.
– two-photon 294f., 301ff.
– UV/Visible 343
accumulation
– macromolecule 97f.
– nanoparticle 98
– nonselective 96
– QD 36, 38, 47, 91f., 321, 449
– selective 26
adatom 373, 375
ADME (absorption, distribution, metabolism, elimination) 432, 445ff.
adsorption
– nonspecific 85
– obsonizing proteins 93
afterglow 281
aggregation
– nonspecific 223
– QDs 12, 32, 83, 90, 193, 262, 305
American Cancer Society 63
amplified stimulated emission (ASE), see emission
annular dark-field scanning transmission elecron microscopy (ADF-TEM) 408, 410
α-particle emitters 274
atomic force microscopy (AFM) 189, 196, 208f.
– protein nanoarray 208ff.
– self-assembled QD system 371
attenuation coefficient, see extinction coefficient
Auger electrons 260
autofluorescence 11, 29
– background 95, 353
– blood 75
– cell 89, 429
– collagen 32
– QD 98f.
– tissue 22, 28, 33f., 116, 129, 133, 148, 295, 302, 429

b

band
– absorption 10f.
– alignment 72, 74, 81
– conduction 8f., 64, 72f., 75, 78, 80, 83, 336, 354
– -edge 73, 76, 81, 294, 300f.
– -edge transition 299
– emission 10
– offsets 72, 77f., 83, 333, 354
– structure 74, 79f.
– valence 8f., 64, 72f., 75, 78, 80, 83, 333, 336, 354
band gap 64f., 69, 72, 263
– bulk 417
– CdS 334
– CdSe 75
– CdTe 75
– energy 8f., 345
– shell material 72
– ZnS 334
– ZnSe core 81
bathochromic shift 75, 77
beads 131, 150
– glass beads 394, 397ff.

Nanomaterials for the Life Sciences Vol.6: Semiconductor Nanomaterials.
Edited by Challa S. S. R. Kumar
Copyright © 2010 WILEY-VCH Verlag GmbH & Co. KGaA, Weinheim
ISBN: 978-3-527-32166-7

– magnetic   154
– multicolor encoded   150
binding
– cancer cell-specific   92
– multivalent targeted   131
– nonspecific   85, 88f., 126, 129, 155, 196, 201ff.
– specific   211, 313
bioconjugates
– purification   91
– QD–antibody   32, 37, 91, 95, 98f., 102, 120f., 130, 193, 304ff.
– QD–antigen   95
– QD–DNA   151, 306
– QD–oligonucleotide   120, 124ff.
– QD–peptide   32f., 93, 95f., 304, 313, 315
– QD–protein   120, 124ff.
– QD–streptavidin   119ff.
– QDs   17, 19, 23, 36f., 84ff.
– stop-and-go movement   99
bioconjugation   11, 14f., 85, 88ff.
– covalent bonding   89f., 99
– direct attachment   89f., 120
– electrostatic interaction   14f., 88ff.
– hydrophobic attachment   90
biodegradation   293f., 439, 448
biodistribution
– gold nanoparticles   260
– in vivo   103
– QDs   37, 47, 103, 313, 435
biological activity   90, 189
bioluminescence
– activity   99, 157
– BRET-based protease   160ff.
– resonance energy transfer (BRET)   157
bioluminescence resonance energy transfer (BRET)   99f., 168ff.
biomarker   7, 439
– analysis   8
– cancer   9f., 15f., 19
– clinical biomarker measurements   131ff.
– panel   8
– screening   8, 15f.
– tumor   121
biopsies   7f., 315
– needle core   121
– tissue   10, 7
– tumor   8, 20
biosafety   429ff.
blinking, see fluorescence intermittency
blocking layer   202ff.
Bohr exciton radius   9, 65, 263, 294, 299, 369, 377

β-particle emitters   274
brachytherapy
– high-dose-rate   260
– low-dose-rate   260
– lower-energy   260
– -nano   252, 271ff.
BSA   201ff.
buffer   85, 88, 413
bulk semiconductors   11, 72f., 117, 263, 415, 417
– size   393

c

cancer   5ff.
– antigen (CA)   7
– bladder   292
– bronchial   292
– detection   6f., 15f.
– diagnosis   6f., 15, 17
– head-and-neck   278, 292
– human breast   20f., 23, 91, 95f., 99, 261, 311
– liver   25
– lung   22, 292
– mesothelioma   25
– pancreatic   25, 167, 311
– prostate   7, 19, 98, 259, 310, 312f.
– skin   292
– subtypes   8
capping   12, 90, 119, 182, 193, 437
carbon nanotubes (CNTs)   281
– single walled (SWNTs)   282
cell   5
– apoptosis   5
– death   252, 292, 444
– embryonic stem (ESC)   168
– fixed   19, 21f., 119, 148
– growth rate   5
– killing   257, 259, 269, 294
– live   16, 19, 22, 119, 225f.
– membrane   23f., 85, 88, 251, 293
– membrane fusion   227f.
– migration   27
– penetrating peptides (CPPs)   307
– proliferation   28, 64, 251
– –surface marker   98, 131
– T-cell differentiation   129
channel
– hole-trap-mediated nonradiative   80
chemical index   439
chemotaxis   23
chemotherapy   313
chromophores   29, 242
clinical diagnostics   147

coating
- defective   433
- hydrophilic   118, 434
- hydrophobic   88
- ligand   88f.
- ligand-polymer   13, 88f.
- nonspecific   259
- phospholipid micelles   354
- polymeric   32, 39, 85, 88, 98, 118
- proton-absorbing polymeric   26, 44
- QDs   84ff.
- silica   85ff.
- streptavidin   119ff.
- ZnS   333f.
coding
- electrochemical   155
- multicolor   166
colocalization   142, 235, 237
- assay   231
- multiplexed detection   151f.
- nanoprobes   152
color
- -based tissue classification   125
- channels   165
- contrast   148
- false-color composite image   141
- four color genotyping   130
- intensity   152
- multicolor methods   129, 133
- seven-color QD detection   129
composite nanodevices   277
Compton scattering   255, 270
computed tomography (CT)   6, 28, 63, 66, 251, 254
conductivity
- electrical   8
confocal microscope   181
- fixed-point   241
- high-speed   37, 99
- laser scanning   133, 135, 211
- three-dimensional (3-D)   26
- virus tracking   229
contour map   251
contrast agents   7, 66f.
- angiographic   32
- fluorescence   32
- nanoparticle-based   7
- nontargeted imaging   31, 37f.
control cell lines   22
conversion
- internal   292
coupling
- electron–phonon   358

crosslinkers
- bifunctional   192
- covalent   191f., 220f., 404
- electrostatic   404
- 1-ethyl-3-(3-dimethylaminopropyl) carboiimide (EDC)   90
- heterobifunctional   15
- PEG (polythylene glycol)   32
- water-soluble   191
crosslinking chemistry   304
- avidin–biotin   304
- carbodiimide   304
cross
- channel configuration   200
- -interferences   155
- -reactivity   155
- -talk   182, 184
cross-section   256f.
- photoelectric   251
- two-photon absorption   294f., 303, 315f.
crystallinity   344
CT, see computed tomography
cut-off filters   182
cytotoxicity, see toxicity

d

dangling bonds   193
decay
- kinetics   182
- photoluminescence   393, 396
deconvolution, see signal
degradation
- chemical   147
- metabolic   229
- photodegradation   147
dendrimers
- delivery   276
- radioactive gold   277
- synthesis   275
detection
- BRET-based protease   160ff.
- efficiency   206
- electrochemical   152f., 155
- enzymatic activity   149
- FRET-based protease   157ff.
- in situ   124
- infection   239ff.
- limit   188, 196
- microarray-based multi-allele   151
- mRNA   168
- multiplex   117, 147ff.
- multitarget electrical DNA protocol   155
- protein immuno-   153
- protein microarrays   186f., 196, 206

– real-time 241ff.
– sensitivity 10, 151
– threshold 129
– virus 239ff.
differential interference contrast (DIF) 166
diffusion
– coefficient 395
– time 207, 412
DNA
– double stranded (dsDNA) 160, 260
– single strands (ssDNA) 151, 181
– strand breakage 273, 317
dorsal skinfold chamber 99
dose enhancement factor, see radiation therapy
dose-response calibration curve 201
dosimetry, see radiation
downstream image analysis 124
drug delivery 15, 17
– targeted 41, 182, 292
drug
– discovery 151
– loading capacity 43
– screening 147
dynamic light scattering (DLS) 225f.

e

electrochemical
– coding 155
– stripping transduction 155
electron
– density 255
electron-hole pair, see exciton
electron microscopy (EM) 65, 134
– freeze-fracture (ff-EM) 44
– high-resolution (HR) 337
– multiplex 135
electron transfer 261, 292f.
electronic
– energy levels 72f.
– spin-inversion 292f.
– transition 334, 338
– X transition 335f.
– XX transition 335f.
electrophoresis 150
electroporation 165, 305f.
emission
– amplified stimulated (ASE) 335f.
– band-edge 294
– maxima 72, 162
– onset 77
– photon 9
– size-tunable 11, 148f., 170, 183
– spectra 68, 98, 117, 183f., 299, 333
– trap 343
– wavelength 10, 65, 98, 116, 118, 148, 152, 162, 335
emulsion 88
– micro- 402f.
– reverse microemulsion synthesis 399
– water-in-oil micro- 398f.
endocytosis
– caveolae-mediated 309
– clathrin-mediated 308
– dynamin-dependent 137
– nonspecific 93, 307, 435
– receptor-mediated 24, 304f., 308
– QDs 434f., 441
– spontaneous 25
– virus 226f., 231
energy-dispersive X-ray (EDX) 408, 410
energy
– absorbance 257
– kinetic 256
– level diagram 72f., 299, 345
– metastable state 263
– pump 335f.
– states, see excited states
– thermal 292
energy transfer 256f., 292f., 318, 320
– efficiency 262, 269
– linear energy transfer (LET) 260, 273
– pathway 257, 265
enhanced permeability and retention (EPR) effect 35f., 91f
enhancement factors 151
enzymatic
– activity 157ff.
– inhibition 159
epidermal growth factor (EGF) 23
epifluorescence microscopy 166, 229
– virus tracking 229
epitaxial growth 74, 298, 333, 351, 367, 371f.
etching process 76, 438
exchange
– electrostatic 220
– ligand 12f., 83f., 89, 119, 190f., 220f.
– mercapto (–SH) 220f.
excitation 8, 10
– efficiency 270
– one-photon 303
– photon flux 148
– power 369, 378
– resonant 151
– two-photon 295, 303
– wavelength 116, 148, 157, 183, 269, 300f.

excited state 72f., 263, 292f., 300, 302
– transition, see electronic transition
– metastable state 263
– singlet 292f.
– triplet 292, 316
exciton 8, 64, 264, 369, 377, 382f.
– bi- 369f., 377, 382f.
– lifetime 9f., 68
– transition 370, 377f., 382f.
– wavefunction 344
extinction coefficient 9, 148, 262, 301, 411
extraction
– efficiency 379
ex vivo histological analysis 96

f

Feinberg's microspot test 180
finite-difference time-domain analysis 261
FISH (fluorescence in situ hybridization) 16, 19, 22, 125f.
– double-labled 127
– multiplex QD-based 22, 125f., 168
– single-step 22
flow cytometry, see multiplex
fluorescence
– angiography 31
– background 10
– efficiency 12, 116
– ex vivo 313
– fine-tuning 78
– free track area 27
– intermittency 134, 367, 372, 381
– kinetics 233
fluorescence line-narrowing spectroscopy 339
fluorescence
– lymphangiography 33ff.
– in situ hybridization, see FISH
– in vitro 28
– in vivo 28, 38
– organic dyes 70f.
– QDs 70f.
– resonance energy transfer (FRET) 69, 157f., 262, 265, 267ff.
– stability 17
– targeted 38
fluorescein
– calibration beads 131
– isothiocyanate (FITC) 129
fluoroimmunoassays 120
fluorophores 21f., 38, 116f.
– intrinsic 10f.
– organic 22, 63, 129, 182, 184
– overlapping spectra 116

formalin-fixed, paraffin-embedded tissue (FFPET) 116f., 126, 133
Fourier transform infrared (FTIR) spectroscopy 222
FRET, see fluorescence
full width at half maximum (FWHM) 10, 69f., 221, 359, 409, 412
functional
– bilayer films 84
– monolayer 84, 88

g

gene
– amplification 22
– copy number 22
– expression profiling 116, 124, 149, 183, 251
– mismatch repair 125, 132
– reporter 99
– signatures 116, 124, 141
– virus-mediated gene delivery 226
genetic engineering 91
glass beads 394, 397ff.
– biofunctional 404ff.
– complex structured 414f.
glass-coated QDs 401, 405, 414
glass fibers
– belt-like 414, 417
– solid 414
– tubular 414
glass
– sol–gel-derived 393ff.
– magnetic luminescence 405
– matrix 393ff.
– melting 394
Goeppert–Mayer (GM) units 303
γ-ray 260, 272
– -induced polymerization 277
ground state 9, 72, 263, 292, 301f.
– depletion (GSD) 65
– exciton 370
– triplet molecular $^3O_2$ 261

h

Hanburry–Brown–Twiss arrangement 381, 383
Helical TomoTherapy 253f., 257
heterostructures, see nanoheterostructures
high-throughput
– analysis 153, 179
– cytoplasmic delivery of QDs
– genomic data 8
– processing 22
– proteomic data 8

- multiplexed reverse-phase protein microarray analysis 130
- screening of biomarkers 8, 15f., 183
- studies 116
- technologies 7f., 19
high Z
- materials 255f., 259f.
Human genome Project 115
hybridization
- DNA 149ff.
- duplex QD–ISH 139
- in situ (ISH) 115, 124ff.
- kinetics 150
- sandwich-type 152f.
- signals 22, 129
- single QD–ISH 129, 135
- triplex QD–ISH 137, 140
hydrodynamic diameter 34, 99, 102, 225, 264, 321, 448
hydrogen-atom abstraction 261
hydrolysis 399
- enzymatic 26
hydrophilic
- lipid bilayer structure 273
- QDs 13, 84, 118f., 309, 394, 399, 414
- shell 12
hydrophobic
- alkyl chains 12
- lipid bilayer structure 273
- micelle 87
- QDs 12, 88, 189ff.

i
IHC, see immunohistochemistry
imaging
- analysis 132ff.
- biological 8, 65, 67, 119ff.
- cancer 3ff.
- cell–surface receptors 23
- cellular 14, 26
- dual-functionality tumor-targeted 15, 103
- dual modality PET/NIRF 103f.
- fixed cell 22f.
- hyperspectral 133
- IHC imaging system 21
- in vitro cancer 294f., 301, 304
- in vivo animal 148, 168
- in vivo blood cell 32, 100
- in vivo cancer 10, 15, 28f., 294ff.
- in vivo five-color lymphatic drainage 169
- intrasurgical spectral fluorescence 169
- live cell 16, 22f., 165
- long-term 31
- lymphatic drainage 35

- medical 6, 17
- membrane receptors 23
- molecular 36, 41, 74, 85, 88, 99, 264
- multicolor 36, 116, 166, 301
- multimodality cancer 39
- multiplex cancer boimarker 7f., 98, 100
- multiplexed 29, 115ff.
- multiplexed cellular 165ff.
- near-infrared, see NIR
- noninvasive 6, 63
- nontargeted 31, 38
- optical 6, 18, 26, 28, 37f.
- real-time intraoperative 34
- spectroscopic hepatoma 98
- targeted molecular 36f.
- targeted tumor 63ff.
- techniques 6ff.
- time-lapse 231, 238
- time-domain 148
- tissue biopsies 10, 17
- tumor vasculature 16, 31f.
- two-photon excited fluorescence 314f.
- viruses 219ff.
- whole animal tumor 16, 35
immobilization
- antibodies 187, 196
- co- 155
- proteins, 185, 188, 200, 210f.
- surface 208
- virus 222ff.
immunoassays 91, 149, 153ff.
- fluoro- 153
- lab-on-a-chip system 155
- QD-based multiplexed 153, 156
immunocytochemistry 272
immunofluorescence
- conventional 20
- QDs 116, 120
- quintuplet 122
- signals 21
- staining 104
- triplex 122
immunogenicity 91
- polystyrene (PS)
immunohistochemistry (IHC) 19f., 120ff.
- QD-based ISH/ICH 126, 135, 137
- multiplex 20, 115, 125
- Q- 123
- QDs-based 21f., 135
- quadruplex QD- 125
- scoring 132
immunoprecipitation reaction 180
inductively coupled plasma (ICP) 402
- mass spectrometry (ICP-MS) 446, 448

infrared (IR) active materials 331
– quantum dot core-shell 352f.
– type I core-shell structures 353ff.
– type II core-shell structures 355ff.
interaction
– antibody–antigen 23, 197, 206
– biotin–avidin 191
– biotin–streptavidin 130
– coulombic 369
– dipole–dipole 265, 382
– electrostatic 14f., 88f., 162, 220, 307
– exciton–exciton 335f.
– one-on-one 183
– protein–protein 135f., 181, 196, 198f., 201ff.
– protein–mRNA 135f., 225
– protein–surface 208
– receptor–ligand 23
– thiol–ZnS 84
– van der Waals 85
– X-ray-matter 255f.
intercalation 190
interface
– core–shell 74, 81
– solid–liquid 373
inter-system crossing 292
intracellular delivery 304ff.
– biochemical techniques 307ff.
– cell-penetrating peptide-mediated 307ff.
– chitosan-mediated 309
– liposome-mediated 309
– nonspecific intracellular delivery 307
– physical methods 305f.
intracellular pressure 259
intravital microscopy 32, 38, 41, 67
– high-resolution 32
ISH (in situ hybridization), see hybridization

k

kinetic energy 256

l

labeling
– cell–surface receptors 23
– cellular 16, 19, 26
– direct 187
– double- 20, 135, 165, 167
– efficiency 101
– enveloped viruses 230ff.
– enzyme-labled antibody methods 20
– ex vivo 29, 31
– fixed cell 19
– fluorescence 9
– in vitro 19
– in vivo 31
– live cell 19, 23, 28
– multiplex 20f., 25
– nonenveloped 231
– QD- 30, 32, 98f.
– selective 312
– single-virus tracking 228f.
– site-specific 231
– triple 135
– virus 230ff.
lab-on-a-chip system 150, 155
laser scanning cytometry (LCS) 17
laser
– high-speed laser scanner 17
– multiphoton laser excitation 29, 67
lattice
– adapter 337, 344, 346
– mismatch 74, 76, 78, 81, 83, 331, 334, 344, 354
– parameter 345, 351
LCS, see laser scanning cytometry
lethal total body irradiation 272
lifetime
– core–shell system 80
– -limited emission rate 148
– fluorescence 9f., 68, 162, 184, 334, 429
– mean decay 80
– multiplexing 69
– nonradiative 80
– radiative 80, 82
ligands
– amine-terminated 83
– aliphatic coordinating 83
– biocompatible 85
– cap 12, 90, 119, 182, 193
– carboxyl-terminated 83, 126
– exchange 12f., 83f., 89, 119, 190f., 220f.
– bifunctional 83
– functionalization 83
– hydrophilic 84, 119
– labile 12
– surface charge 92
– target-specific 87
– water-soluble 83
linear energy transfer (LET), see energy transfer
lipids
– bilayer structure 273ff.
– immuno- 311, 313
– liposome-mediated intracellular delivery 309
– paramagnetic 85, 87
lipodots, see lipids
liposomes, see lipids

liquid crystal
– phase   405
– tunable filter (LCTF)   133
lithography
– dip-pen nanolithography (DPN)   188f., 196, 208, 210f.
– electron beam   188
– focused ion-beam   188
– nanoimprint   188, 209
– optical   383
– soft   188
luciferases   99
lymph nodes
– in vivo targeting   314f.

m
magnetic field
– alternating (AMF)   275
magnetic resonance imaging (MRI)   6, 17, 28, 39f., 251
mapping
– gene   22
– sentinel lymph note (SLN)   16, 33f.
melting point   373
metal-oxide semiconductor field-effect transistors (MOSFETs)   280ff.
metatstasis   63f.
microarrays
– antibody   180f.
– -based energy transfer system   159
– cell motility   27
– chip scanner   181
– cytometer   28
– detection   185ff.
– DNA   151, 181
– fabrication of protein   193ff.
– fiber-optic microarray platform   152
– functional protein   185f.
– gene expression profiling   116
– genomic   115
– in vitro diagnostic   17, 26
– limitations   201ff.
– low-to-high-density protein   180, 188, 194, 208
– low-to-medium-density protein   180, 188
– mega dense   188
– multiplexed   28
– optical read-out   182, 187
– protein   179ff.
– QDs-based protein   19, 115, 132
– reverse-phase protein (RPPM)   130, 185f.
– scanner   199
– technology   8, 15
– tissue (TMAs)   16f., 21

– well-on-a-chip ProteoChip base plate   198, 203, 205
microcontact printing   188
microemulsion, see emulsion
microfluidic network (µFN)   199ff.
microfluidics device   181, 207
microinjection   25, 165, 305ff.
microplate assay   28, 179
– lable-free   181
microtiter-well plates   156
microtubulus   237, 302, 307f.
mitochondria   432, 450ff.
– damage   452
– swelling   450f.
molecular beacons (MBs)   27
molecular beam epitaxy (MBE)   372f., 385
molecular
– markers   7
– supra-   307
molecules
– biotin-tagged   14
– capture   185
– cell adhesion (CAMs)   31
– macro-   97
monitoring
– progression of viral infection   239ff.
– real-time   10, 183, 234, 238, 261
MRI, see magnetic resonance imaging
multifunctional nanoparticle system   43f.
multiphoton laser excitation, see laser
multiple
– targets   11
multiplex analysis
– in vitro   149
multiplex
– biomarker   7f.
– codes   150
– flow cytometry   115, 120, 129f., 165, 241
– gene expression measurements   116, 124, 141
– genomic analysis   130
– liquid-phase detection platform   129
– protein detection   130
– solid-phase detection platform   129f.
– staining   20, 120, 122, 132ff.
– tagging   301
– target detection   19
multiplexing   11, 15, 17, 119ff.
– capability   19, 29, 34
– experiments   18
– four-plex staining   123
– lifetime   69
– quintuplet   122

n

nanoarray
– fabrication of protein 194ff.
– multicomponent system-based 189
– protein 179f., 188, 194f., 206ff.
nanoassembly
– sandwiched 152
– self-assembled 274
nanocubes 361f.
nanocrystal
– encapsulation 406f., 410
– Fe_3O_4 275, 405ff.
– magnetic 405
nanoheterostructures
– characterization 350ff.
– compositions 367
– GaN/AlGaN 385f.
– nanowire growth 371ff.
– spherical 33
– nonspherical 349f.
nanolithography, see lithography
nanomaterial
– energy mediator 255, 261ff.
nanoneedle 375ff.
– axis 377
– strain-relaxed 377
nano-on-micro (NOM) combinations 155f.
nanoparticles, see quantum dot (QD)
nanorods 349ff.
nanowire 367ff.
– applications 378ff.
– CdS 379
– CdSe/ZnS 368
– electronic devices 379f.
– field-effect transistor (FET) 379f.
– Ge/Si 380
– growth 367, 371ff.
– high-temperature single-photon emission 381fff.
– laser-active materials 379
– light-emitting devices (LEDs) 379f.
– polarization behavior 383
– sensors 379
– single-photon generation 380f.
– ZnO 379
– ZnSe 375, 382
National Cancer Institute (NCI) 3f.
National Institutes of Health (NIH) 3
– roadmap's new Nanomedicine Initiatives 3
National Science Foundation (NSF) 3
near-infrared, see NIR

Nierdre
– killing 270
– limit 269
NIR (near-infrared) 28f.
– -emitting QDs 29, 31f., 34
– in vivo dual modality NIR fluorescence (NIRF) imaging 39f., 97
– optical window 28
– penetration depth 75
– probes 29
– scanning optical microscopy (NSOM) 65

o

obsonization 93
ODMR measurements 343
optical biosensor
– self-referencing 181
optical coherence tomography (OCT) 63, 65
optical detection
– protein microarrays 186f.
optical imaging 6, 18
– dual-focus 26
optical read-out 182, 187, 194
optoelectronics 281
organic dyes 9f., 15
– photochemical stability 68
– thermal stability 68
Ostwald ripening 351
oxidative stress 432, 435, 440, 444, 452
oxygen
– singlet 269, 292, 304, 316

p

particle in the box model 9, 338
passive uptake 305
Peltier cooling 385
penetration
– cell 31, 165, 279, 305, 307f.
– dermal 436f.
– high-energy particles 253
– membrane 233
– NIR penetration depth 75
– shell 298
– tissue 98f., 261, 295f., 302
phase transfer
– TOPO-coated QDs 13
PET, see positron emission tomography
pH 22, 437
– -dependent FRET efficiency 160
– nanocrystal solution 405
– sensing 161
– sensitive QDs 219, 242f.
– sol–gel reaction 395

phagocytosis 91, 437
phagokinetic track 27
pharmacokinetic 320, 435, 445ff.
– physiologically based (PBPK) 441, 446ff.
– QD imaging probes 37
phase transfer
– QDs 83
phosphorescence 265, 293, 316
– nanometer-sized 281
photoactive localization microscopy (PALM) 65
photobleaching 9, 29, 148, 183, 206f., 229, 281
– transient 337
photocurrent
– short-circuit 281
photodynamic therapy (PDT) 101, 182
– nucleus-targeted 316
– QD alone 316f.
– radiation therapy 261ff.
photoelectric effect 256f., 259, 269
photoluminescence (PL) 75f., 159
– decy time 393, 396
– efficiency 394f., 403ff.
– InP 369f.
– intensity 79
– micro- 376, 378, 382
– nanoneedles 375f.
– peak width 76
– self-assembled QD system 371
– spectra 77, 81, 346, 375f., 396, 400
– time-resolved 337
photolysis 437
photon correlation spectroscopy 377
photon
– bosonic character 380
– bunches 380
– energy 255f., 260
– energy density 270
photo-oxidation 101, 344ff.
photosensitizer (PS)
– conventional 316
– first-generation 294
– –QDs conjugates 258, 263, 294f., 304, 318ff.
– second-generation 294
photothermally 261
Planck's constant 263
planning target volume, see target
plasma membrane barrier 24
Poisson
– photon number 380
– sub-Poissondistribution 381
polydispersity 334

polymer
– amphiphilic 12f., 85
– antibiofouling effect of PEG 92
– diblock copolymer 84, 191f.
– matrix 405
– PEG encapsulation 84
– PEGylation 85, 93
pore
– intracellular 14
positron emission tomography (PET) 6, 17, 28, 39, 63, 65f., 251
– in vivo dual modality PET fluorescence imaging 39f.
potential stepping 344
precipitation
– QDs 83f.
probes
– Au nanoparticle 196
– antibody/DNA 116
– clinical applications 197ff.
– dual-modality imaging 39f., 45
– fluorescence 17, 26, 36f.
– FRET-based 158, 160f.
– molecular imaging 74, 88
– multifunctional QDs-based 39, 102ff.
– multimodal imaging 85, 87f.
– nanoprobes 160
– QD-based optical/MR dual modality 85
– QD-labeld oligonucleotide 125ff.
– QD nanoprobes 152
profiling
– cancer molecular 137ff.
– gene 8
– gene expression 116, 124, 149
– in vitro biomolecular 15
– protein 8
prostate-specific antigen (PSA) 7
protein
– capture 200ff.
– denaturization 253
– engineering 91
– expression profiling 125, 203
– green fluorescent (GFP) 136
– yellow fluorescent (YFP) 136
– X-linked inhibitor of apoptosis (XIAP) 137, 140
proton
– relaxation time 251
– sponge effect 307
proximal convoluted tubules (PCTs) 440
plasmon-enhanced fluorescence
– microscopy 151
– spectroscopy 151
proton sponges 26, 44

q

quantitation
– clinical biomarker measurements   131ff.
– signal   133
quantitative analysis
– biodistribution   39
– biomarkers   21
– FISH signals   126
– fluorescence gene expression   134
– IHC imaging system   21
– NIR fluorescence imaging   40
– PET   103
– protease activity   159
– tumor cell behavior   30
quantum confinement effect   9, 65, 182, 264, 294, 369
quantum cryptography   380f.
quantum dot (QD)
– beads   130f.
– biocompatible   45, 119, 258f., 264, 301, 303f., 432
– brightness   11, 15, 21, 26, 63, 65, 68, 75f., 78, 81, 85, 87, 102, 117ff.
– cadmium-free   394, 414ff.
– carboxylated   92, 436
– calibration   131, 134
– cluster   393, 412f.
– colloidal   181, 183f., 295, 339, 394f.
– composition   9f., 72, 153, 10
– -conjugated immunoliposome system (QD–Ils)   43f.
quantum dot core   12, 69
– radius   81f., 117
– size   77, 80, 101f.
quantum dot core-shell   44, 69, 73f., 295ff.
– highly luminescent InP   418ff.
– infrared materials, see infrared
– –shell (CSS)   81f., 344ff.
– type I structures   332ff.
– type II structures   334ff.
– type II–VI structures   345
– type III–V structures   345, 418ff.
quantum dot delivery
– biochemical techniques   307ff.
– nontargeted intracellular   304ff.
– physical methods   305f.
– process   91ff.
quantum dot
– diameter   34
– disintegration   449f.
– distribution   95
– dual-functionality   41ff.
– encapsulation   31, 84, 191, 224f., 401, 403, 406f.
– endocytosis   91
– fluorescence emission   9ff.
– GaN/AlGaN   385f.
– gold   258f.
– gold-NP–QD   159, 164
– growth   371ff.
– hydrophobic-capped   295f., 304, 309, 432
– hydroxylated   92
– immunomagnetic nanoparticle (IMN)   275
– incorporation in glass beads   394ff.
– intracellular delivery   25f.
– lipid-enclosed   30
– –liposome hybrid nanoparticle   31
– magnetic   41f., 275, 405
– metal oxide   258
– movement   23f., 91ff.
– multiplexing   11, 15
– multifunctional   43f.
– phagocytosing   39
– –Photofrin   265ff.
– photoluminescence   264ff.
– photophysical properties   8ff.
– –photosensitizer (PS) pair   258, 263, 294f., 304, 318
– photostability   9, 14f., 21, 26, 68, 73, 78, 81, 183, 203, 207, 294f., 301
– physico-chemical properties   432ff.
– retention   95
– self-assembled   367, 371f., 382, 384f.
quantum dot shell
– barrier layer   82
– critical thickness   370
– layer   78
– mixed structures   295
– multiple structures   295, 337ff.
– silica   85f.
– silica–gold   261
– thickness   79f., 298, 333, 338, 341, 346ff.
quantum dot
– size   8f., 102, 117, 170, 382, 432, 435ff.
– size-dependent process   91, 93, 97
– size variation   72, 119, 148f.
– stability   118, 436f.
– storage stability   14
– Stranski–Krastanow (KS) materials   372
– structure   432
– surface area   41, 100, 203, 294
– surface charge   83, 85, 432, 435ff.
– surface modification   83, 119, 259
– surface-to-volume ratio   299, 331, 379, 393

quantum dot synthesis   11f., 119f., 295ff.
– Cd-based   295f., 298ff.
– core-only NIR   297
– core-shell   119f., 297ff.
– InP   296
– NIR   296f
– visible   295f.
quantum dot
– thermal stability   68
– tunable   117, 120, 148f., 170, 264, 294, 296, 299
– two-color   241f.
quantum dot types   64ff.
– group II–V core–shell   76f.
– group II–VI core–shell   74ff.
– group IV–VI core–shell   78
– reverse type I   78ff.
– reverse type II   78ff.
– type I core–shell   73ff.
quantum dot uptake   95
– antibody-mediated   97ff.
– nonspecific   435
– peptide-mediated tumor   95ff.
quantum dot
– water-dispersble   68, 83f., 189, 191, 304, 393f., 418
– water-insoluble   13
– water-soluble   8, 11f., 27, 102, 119, 295, 432
– water stability   84
– quantum well (QDQW)   9, 337ff.
– wires   9
– yield (QY), see brightness
quantum
– efficiency   101, 332, 345, 360
– information processing   380
quantum mechanical models   9, 412
– probability distribution function   412f., 422
quantum
– repeating   380
– rods   349f.
– teleportation   380
quenching
– collisional   316
– efficiency   157
– energy   267f.
– fluorescence   193, 438
– rapid   339
– self-quenching effect   229, 234

r
radiation
– damage   277
– dose   252f., 269
– dosimetry   252, 280ff.
– electromagnetic (EM)   252f.
– energy 252, 255
– ionizing   252f., 260, 278f.
– nonionizing   252, 260f., 292
– –photodynamic therapy   258
– photoluminescent nanopartivles   264ff.
– repetitive   254
– therapeutic ratio   253f., 259
radiation therapy   251ff.
– dose enhancement factor (DEF)   259f.
– enhanced   255ff.
– gold nanoparticles   257ff.
– photodynamic therapy (PDT)   261ff.
radioimmunology   272f.
radioimmunotherapy   273f.
radical
– centers   316
– free   252, 254, 261, 278f.
– hydroxyl   292f., 317
– oxidative   265
– scavenging ability   279f.
– -superoxide   261, 278f., 292f., 454
radionuclides   273, 275, 277
– nanoparticle-facilitated delivery   274f.
radioprotective
– nanoparticles   255, 258, 277ff.
– thiols   254
radioresistance
– hypoxia-assisted   254
radiosensitivity   254
radiosensitizers
– chemical   254f.
– physical   255ff.
Rayleigh scattering   29
reactive oxygen species (ROS)   46, 48, 279
recombination   73
– electron–hole   80, 369
– nonradiative   12, 384
– radiative   9
– surface-related   12
reflecting high-energy electron refraction (RHEED)   373f.
reflux   409, 412f., 416f.
– duration   414f.
refractive index   256, 382
relaxation
– nonradiative   292f.
– radiative   292f.
resolution 66f.
– cellular   116
– color   117, 132
– single-molecule   208

– spatial   28, 41, 65, 116, 396
– subcellular   97
– temporal   65, 263
– time-resolved   26
reverse micelle method   394, 398ff.
RNA
– interference (RNAi)   43
– small interfering (siRNA)   17, 26, 43f.
ROS (reactive oxygen species)   292f., 316, 438, 440, 442ff.

s

saturated structured-illumination microscopy (SSIM)   65
scanning electron microscopy (SEM)   180
– glass fibers   414, 416f.
– nanoneedles   375
– nanowire   373ff.
scanning probe microscopy (SPM)   180, 372
Schrödinger equation   263
screening
– cancer   17
– in vitro   15
– multiple therapeutic agents   30
– multiplexed   150
– rapid   150, 188
second-order correlation function   380f.
self-assembled monolayer (SAM)   196, 211
sensor, see detection
sentinel lymph node (SLN) mapping   16, 33f., 102, 358f.
– mapping   358
– NIR fluorescence   34
– QD-mediated   34
signal
– acquisition   10
– amplification   69
– deconvolution   116f., 153
– distribution   133
– intensity   133, 140
– multiple fluorescent   117
signal-to-noise-ration (SNR)   10, 23, 100, 116, 130, 133, 147, 168, 198, 203, 205
– NIR QD–RGD   313
signal transduction   23
single-molecule analysis   69
single nucleotide polymorphisms (SNPs)   149, 153
– genotyping   151
single-particle spectroscopy   349, 396
single-photon emission computed tomography (SPECT)   6, 28, 39, 63, 66
SLN, see sentinel lymph note

sol–gel synthesis   393ff.
solid-phase
– biomarker   203
– detection platform   129f.
solubility
– methods   14
– organic dye   68
– QDs   11f., 68
SPECT, see single photon emission computed tomography
staining
– antibody   236
– counter-   235
– immuno-   104, 235f., 312, 440
– multiplex   20, 120, 122, 132ff.
– MT (metallathionine)   440
stimulated emission depletion (STED)   65
stochastic optical reconstruction microscopy (STORM)   65
Stöber method   394, 397f., 401, 404
Stokes shift   10f., 63, 68, 148, 184, 281
stop-and-go movement, see bioconjugates
Stranski–Krastanow (KS) QD materials, see quantum dot (QD)
successive ion layer adsorption and reaction (SILAR)   74, 340, 359
superdex gel filtration   126
surface
– capping   101
– chemistry   69, 433ff.
– defects   11, 65, 393
– energy   394
– engineering   15
– functionalization   119, 183, 192f.
– kinetics   182
– passivation   11f., 69, 182
– receptors   379
– -related recombination   12
surface-to-volume-ratio   12
susceptibility
– third-order nonlinear   394
synaptic regulation   23
synchrotron   257, 352

t

target
– delineation   251
– planning target volume (PTV)   272
targeting
– active cellular   92, 95, 98
– antibody-mediated active   92
– efficiency   95f., 100
– in vitro   96, 294, 310ff.
– in vivo   96, 98, 102, 294, 313f.

- passive 91ff.
- tumor 95, 103
therapeutic ratio, see radiation
thermoluminescent dosimeters (TLDs) 280ff.
thermometry 261
third harmonic generation (THG) microscopy 30
- epi-detection 30
threshold voltage 282
time-correlated lifetime imaging spectroscopy 10
tissue
- damage 254
- radioresistant 253
TMAs, see microarrays
total internal reflection fluorescence (TIRF) 24, 65, 67
- virus tracking 229
toxicity
- cellular 265
- computational toxicology calculations 447
- cyto- 14, 32, 45, 100, 430f., 433f., 443
- inorganic QDs 100
- organic dye 68
- QDs 12, 14, 25, 46f., 68, 100f., 264, 320f., 429ff.
- studies 430f.
transconductance 380
transfection
- chemical-mediated 25, 165
- co- 232
- genetic materials 305
- post- 234
- transient 235
transformation
- carcinogenic 7
- cellular 64
- genetic materials 305
tracking
- cell 29
- in vitro 16
- in vivo 20
- long-term 31
- QDs 30, 99
- viruses 219ff.
transient hole burning spectroscopy 339
transmission elecron microscopy (TEM) 77, 86, 226
- CdS/HgS/CdS 339f.
- CdSe/CdS core-shell QDs 333, 345, 347, 370
- cerium oxide nanoparticles 278
- glass beads 402, 406ff.

- high-resolution (HR-TEM) 79, 339f., 351
- nanoneedles 375
- nanorods 351
- nanowire 373
- self-assembled QD system 371
trap emission 343
trap states 74, 76, 282, 299, 339
- surface 333
trapping 12, 65
trioctyl phosphine (TOP) 83, 119, 295ff.
trioctyl phosphine oxide (TOPO) 11f., 83, 119, 189, 191f., 295ff.
triplet state 261
tumor
- ablation 101
- angiogenic 31, 97
- biopsy 8, 20
- classification 7
- delineation 251
- targeted imaging 63ff.
- -to-background ratio 314
tumor vasculature 96ff.
- in vivo targeting 314f.
tunneling 182, 333, 382, 413
two-photon microscopy 32, 38, 68

u

ultrasound (US) imaging 6, 28, 63, 65f.
ultraviolet (UV)
- band 252
- excitation 101
- fluorescence 75
- illumination 321, 397, 412, 415
- irradiation 419ff.
- UVA light 262
- /Visible spectrum 75, 77, 331, 343, 350

v

vapor–liquid–solid (VLS) growth 372ff.
vesicles
- two-dimensional (2-D) trajectories 26
virus
- addition 223
- capsids 224f., 233, 236
- enveloped 230, 233
- imaging 219ff.
- immobilization 222ff.
- infection 239ff.
- internalization 234
- nonenveloped 228, 231, 233
- patterning 224
- polymerization 223
- –QD networks 222f., 236ff.
- respiratory syncytial (RSV) 239ff.

– singlevirus tracking   225ff.
– tracking   219ff.
– trafficking   226f., 231, 234f., 238
voltametry
– stripping method   152f., 155

w
Western blotting   180
wide-field microscopy (WF)   65

x
X-ray
– bremsstrahlung   256
– crystallography   231

X-ray diffraction (XRD)   350f.
– Fe_3O_4 408f.
– glass fibers   414
X-ray
– electrons   260
– energy   269f.
– imaging   6, 28
– monochromatic   257
X-ray photoelectron spectroscopy   340, 352

z
zeta potential   307, 401